Lecture Notes in Computer Science

Edited by G. Goos, J. Hartmanis and J. van Leeuwen

Advisory Board: W. Brauer   D. Gries   J. Stoer

Springer
Berlin
Heidelberg
New York
Barcelona
Budapest
Hong Kong
London
Milan
Paris
Santa Clara
Singapore
Tokyo

Fabrizio d'Amore   Paolo G. Franciosa
Alberto Marchetti-Spaccamela   (Eds.)

Graph-Theoretic Concepts in Computer Science

22nd International Workshop, WG '96
Cadenabbia, Italy, June 12-14, 1996
Proceedings

Springer

Series Editors

Gerhard Goos, Karlsruhe University, Germany

Juris Hartmanis, Cornell University, NY, USA

Jan van Leeuwen, Utrecht University, The Netherlands

Volume Editors

Fabrizio d'Amore
Paolo Giulio Franciosa
Alberto Marchetti-Spaccamela
Università degli Studi di Roma "La Sapienza"
Dipartimento di Informatica e Sistemistica
Via Salaria 113, I-00198 Roma, Italy
E-mail: damore@dis.uniroma1.it
        pgf@dis.uniroma1.it
        alberto@dis.uniroma1.it

Cataloging-in-Publication data applied for

Die Deutsche Bibliothek - CIP-Einheitsaufnahme

Graph theoretic concepts in computer science : 22th
international workshop ; proceedings / WG '96, Cadenabbia,
Italy, June 12 - 14, 1996 / Fabrizio D'Amore ... (ed.). - Berlin ;
Heidelberg ; New York ; Barcelona ; Budapest ; Hong Kong ;
London ; Milan ; Paris ; Santa Clara ; Singapore ; Tokyo :
Springer, 1997
  (Lecture notes in computer science ; Vol. 1197)
  ISBN 3-540-62559-3
NE: D'Amore, Fabrizio [Hrsg.]; WG <22, 1996, Cadenabbia>; GT

CR Subject Classification (1991): G.2.2, F.2, F.1.2-3, F.3-4, E.1, I.3.5

ISSN 0302-9743
ISBN 3-540-62559-3 Springer-Verlag Berlin Heidelberg New York

© Springer-Verlag Berlin Heidelberg 1997
Printed in Germany

Typesetting: Camera-ready by author
SPIN 10548979   06/3142 – 5 4 3 2 1 0   Printed on acid-free paper

Preface

The 22nd International Workshop on Graph-Theoretic Concepts in Computer Science (WG '96) was held in Cadenabbia, Italy, from June 12 to June 14, 1996, in the facilities offered by Villa "La Collina", the retreat of Konrad Adenauer, first chancellor of the Federal Republic of Germany. It was organized by the *Dipartimento di Informatica e Sistemistica* of the *Università di Roma "La Sapienza"*.

The workshop was attended by 57 participants, coming from several countries. In addition to the beautiful scenery of *Lago di Como*, they enjoyed the quiet and pleasant atmosphere of Villa "La Collina", which favored discussion and cooperation.

The Program Committee of WG '96 consisted of:

G. Ausiello, Rome (I), Co-Chair

H. Bodlaender, Utrecht (NL)

M. Habib, Montpellier (F)

L. Kirousis, Patras (GR)

L. Kučera, Prague (CZ)

A. Marchetti-Spaccamela, Rome (I),
   Co-Chair

E. Mayr, Munich (D)

R. Möhring, Berlin (D)

M. Nagl, Aachen (D)

H. Noltemeier, Würzburg (D)

O. Sýkora, Bratislava (SK)

G. Tinhofer, Munich (D)

P. Widmayer, Zurich (CH)

The Program Committee selected 30 among the 65 submitted papers, after a careful reviewing process. The present volume includes these papers, together with the abstract of the invited lecture by Andrew Yao (Princeton University).

We wish to thank the authors of submitted papers and the reviewers: their contributions made the workshop possible. A list of reviewers is given at the end of this volume.

Many thanks are also due to Paola Alimonti, Serafino Cicerone, Silvana Di Vincenzo, Daniele Frigioni, Roberto Giaccio, Giulio Pasqualone for their work before and during the workshop.

We gratefully acknowledge the financial support provided by CNR (National Research Council, Italy) and by the TMR Programme (Training and Mobility of Researchers – Euroconferences) of the European Union.

December 1996

F. d'Amore
P. G. Franciosa
A. Marchetti-Spaccamela

The 22 WGs and their Chairs

List of Reviewers

D. Aclioptas (Patras)
P. Alimonti (Rome)
G. Ausiello (Rome)
H. Bodlaender (Utrecht)
Y. C. Stamatiou (Patras)
C. Capelle (Montpellier)
A. Clementi (Rome)
K. Cremer (Aachen)
P. Crescenzi (Rome)
E. Dahlhaus (Berlin)
F. d'Amore (Rome)
N. D. Dendris (Patras)
G. Di Battista (Rome)
F. M. Donini (Rome)
P. Ezequel (Montpellier)
M. Flammini (Rome)
B. de Fluiter (Utrecht)
D. Frigioni (Rome)
P. G. Franciosa (Rome)
R. Giaccio (Rome)
E. Graedel (Aachen)
Y. Guo (Aachen)
J. Gustedt (Berlin)
M. Habib (Montpellier)
S. Hartmann (Berlin)
C. de la Higuera (Montpellier)
P. Janssen (Montpellier)
G. Kant (Utrecht)
O. Karch (Würzburg)
L. Kirousis (Patras)
T. Kloks (Utrecht)
E. Kranakis (Patras)
L. Kučera (Prague)
J. van Leeuwen (Utrecht)
R. Möhring (Berlin)
J. Magun (Zurich)
E. Malesinska (Berlin)
A. Marchetti-Spaccamela (Rome)
E. Mayr (Munich)
R. McConnell (Berlin)
M. Mueller-Hannemann (Berlin)
M. Nagl (Aachen)
H. Noltemeier (Würzburg)
L. Nourine (Montpellier)

W. Oberschelp (Aachen)
F. Parisi-Presicce (Rome)
H. la Poutré (Utrecht)
A. Radermacher (Aachen)
P. Scheffler (Berlin)
I. Schiermeyer (Aachen)
A. Schuerr (Aachen)
A. S. Schulz (Berlin)
M. Skutella (Berlin)
O. Sýkora (Bratislava)
L. Therese (Montpellier)
D. M. Thilikos (Utrecht)
G. Tinhofer (Munich)
M. Veldhorst (Utrecht)
L. Viennot (Montpellier)
P. Vismara (Montpellier)
P. Widmayer (Zurich)
A. Winter (Aachen)
G. M. Ziegler (Berlin)

Contents

Hypergraphs and Decision Trees[*]

Andrew C. Yao

Department of Computer Science
Princeton University
Princeton, New Jersey 08544, USA

Abstract

We survey some recent results on a space decomposition problem that has a close relationship to the size of algebraic decision trees. These results are obtained by establishing connections between the decomposition problem and some extremal questions in hypergraphs.

Let $d, n \geq 1$ be integers. A d-*elementary cell* is a set $D \subseteq R^n$ defined by a finite set of constraints $f_i(x) = 0$, $g_j(x) > 0$ where f_i, g_j are polynomials of degrees not exceeding d. For any set $S \subseteq R^n$, let $\kappa_d(S)$ be the smallest number of disjoint d-elementary cells that S can be decomposed into. It is easy to see that any degree-d (ternary) algebraic decision tree for solving the membership question of S must have size no less than $\kappa_d(S)$. Thus, any lower bound to $\kappa_d(S)$ yields also a lower bound the size complexity for the corresponding membership problem.

Let $A_n = \{(x_1, x_2, \cdots, x_n) \,|\, x_i \geq 0\}$. A well-known result of Rabin states that any algebraic decision tree for the membership question of A_n must have height at least n. In this talk we discuss a recent result by Grigoriev, Karpinski and Yao [GKY], which gives an exponential lower bound to $\kappa_d(A_n)$ for any fixed d, and hence to the size of any fixed degree (ternary) algebraic decision tree for solving this problem. The proof utilizes a new connection between $\kappa_d(A_n)$ and the maximum number of minimal cutsets for any rank-d hypergraph on n vertices. We also discuss an improved lower bound by Wigderson and Yao [WY]. Open questions are presented.

References

[GKY] D. Grigoriev, M. Karpinski and A. Yao, "An exponential lower bound on the size of algebraic decision trees for MAX," preprint, December 1995.

[WY] A. Wigderson and A. Yao, manuscript, March 1996.

---

[*] This work was supported in part by the National Science Foundation under grant CCR-9301430.

Improved Approximations of Independent Dominating Set in Bounded Degree Graphs [*]

Paola Alimonti[1] and Tiziana Calamoneri[2]

[1] Dipartimento di Informatica e Sistemistica, Università di Roma "La Sapienza", via Salaria 113, 00198 Roma, Italy - alimon@dis.uniroma1.it.
[2] Dipartimento di Scienze dell'Informazione, Università di Roma "La Sapienza", via Salaria 113, 00198 Roma, Italy - calamo@dsi.uniroma1.it.

Abstract. We consider the problem of finding an independent dominating set of minimum cardinality in bounded degree and regular graphs. We first give approximate heuristics for MIDS in cubic and at most cubic graphs, based on greedy and local search techniques.

Then, we consider graphs of bounded degree B and B-regular graphs, for $B \geq 4$. In particular, the greedy phase proposed for at most cubic graphs is extended to any B and iteratively repeated until the degree of the remaining graph is greater than 3. Finally, the algorithm for at most cubic graphs is executed.

Our algorithms achieve approximation ratios:
- 1.923 for cubic graphs;
- 2 for at most cubic and 4-regular graphs;
- $\frac{(B^2-2B+2)(B+1)}{B^2+1}$ for B-regular graphs, $B \geq 5$;
- $\frac{(B^2-B+1)(B+1)}{B^2+1}$ for graphs of bounded degree $B \geq 4$.

Keywords: Minimum Independent Dominating Set, Approximation Algorithms, Bounded Degree Graphs, Regular Graphs, Cubic Graphs, Greedy, Local Search.

1 Introduction

It is widely known that many NP-complete graph problems remain NP-complete even if restricted to bounded degree and regular graphs [3]. On the other hand, variation in which the degree of the graph is bounded by a constant often allows to achieve different results with respect to the approximation properties. Namely, problems that for general graphs cannot be approximated within any constant approximation ratio (e.g. maximum independent set, minimum dominating set, minimum independent dominating set) have been shown to be in

[*] Work supported by: the CEE project ALCOM-IT ESPRIT LTR, project no. 20244, "Algorithms and Complexity in Information Technology"; the Italian Project "Algoritmi, Modelli di Calcolo e Strutture Informative", Ministero dell'Università e della Ricerca Scientifica e Tecnologica.

APX for bounded degree graphs [2, 5, 6, 7, 8, 10]. Moreover, for some approximable NP optimization problems (e.g. min vertex cover), better approximation ratios have been achieved [1, 10].

An independent dominating set in a graph is a collection of vertices such that it is adjacent to all other vertices, and vertices in the collection are mutually nonadjacent.

The problem of finding an independent dominating set of minimum cardinality (MIDS) is NP-hard, even if we restrict to graphs of degree bounded by a constant $B \geq 3$. Furthermore, the problem for general graphs cannot be approximated within any constant approximation ratio, while it has been shown to be APX-complete and approximable within $B + 1$ for bounded degree graphs [5, 6, 7].

In this work we give approximate heuristics for MIDS in cubic and at most cubic graphs, based on greedy and local search techniques. Namely, both proposed algorithms are composed of two phases. In the former one, the algorithms greedily select vertices of degree 3 and remove them and all their adjacent vertices until the graph becomes of degree 2. In the latter one, a sort of local search phase is performed to complete the solution. When the graph is cubic a preprocessing phase is also executed in order to improve the value of the solution.

Then, we consider graphs of bounded degree B and B-regular graphs, for $B \geq 4$. In particular, the greedy phase proposed for at most cubic graphs is extended to any B and iteratively repeated until the degree of the remaining graph is greater than 3. Finally, the algorithm for at most cubic graphs is executed.

Our algorithms achieve approximation ratios:
- 1.923 for cubic graphs;
- 2 for at most cubic and 4-regular graphs;
- $\frac{(B^2-2B+2)(B+1)}{B^2+1}$ for B-regular graphs, $B \geq 5$;
- $\frac{(B^2-B+1)(B+1)}{B^2+1}$ for graphs of bounded degree $B \geq 4$;
improving the previously known ratios of $B + 1$ [6], as shown in table 1.

	$B = 3$	$B = 4$	$B = 5$	$B = 6$	$B = 7$	$B = 8$	$B = 9$	$B = 10$
previous results	4	5	6	7	8	9	10	11
bounded degree graphs	2	3.824	4.846	5.865	6.880	7.892	8.902	9.911
regular graphs	1.923	2	3.923	4.919	5.920	6.923	7.927	8.931

Table 1. Table of results

The remainder of the paper is organized as follows. In Section 2, we state basic definitions and notations, and we discuss preliminary results. In Section 3, approximate algorithms for MIDS in at most cubic and cubic graphs are presented and analyzed. In Section 4, we generalize the previous results to any bounded degree and regular graphs.

2   Definitions and Preliminaries

Throughout this paper we follow the standard graph-theoretic terminology of
[9], and we consider only finite, simple, loopless and possibly not connected but
without isolated vertices graphs. In fact, since isolated vertices belong to any
independent dominating set, the previous assumption is not restrictive.

Definition 2.1 *A graph $G = (V, E)$ has* bounded degree B *if each vertex $v \in V$
has degree up to B, and is B-regular if each vertex $v \in V$ has exactly degree B.
A* cubic *graph is a 3-regular graph. A graph is* at most cubic *if it has bounded
degree 3.*

Definition 2.2 *Given a graph $G = (V, E)$ the* minimum independent dominating set *problem (MIDS) is the problem of finding the smallest possible set $S^* \subseteq V$
of vertices such that for all $u \in V - S^*$ there is a $v \in S^*$ for which $(u, v) \in E$,
and such that no two vertices in S^* are joined by an edge in E. Variation in
which the degree of G is bounded by a constant B is denoted by MIDS-B.*

Given a graph $G = (V, E)$, we denote by S^* a MIDS and by S the solution
determined by our algorithms.

Lemma 2.1 *If G is a graph of bounded degree B, then $|S^*| \geq \frac{n}{B+1}$.*

Proof. The claim follows from the fact that S^* is a dominating set, and each
vertex $v \in S^*$ can dominate at most B vertices.

Remark 1 Given a connected graph $G = (V, E)$ of bounded degree 2, $|S^*| \leq
|V|/2$. Suppose now that k vertices are forbidden to be in the MIDS. The optimal
value of such a constrained solution (if it exists) remains no greater than $|V|/2$,
but the k vertices are the even vertices of an odd length chain (see fig. 1). From
now on we will denote by *peaks* such even vertices.

Fig. 1. An example of optimal constrained solution of cardinality greater than $|V|/2$

Let $G = (V, E)$ be a graph with bounded degree B. We denote by $adj(v) =
\{u \in V | \langle u, v \rangle \in E\}$ and by $adj^2(v) = \bigcup_{u \in adj(v)} adj(u) - \{v\} - adj(v)$.

The proposed algorithms use an auxiliary graph $G' = (V', E')$, at the beginning equal to G. Let V'_k be the set of vertices of degree $k \leq B$ in G'.

3 MIDS in at Most Cubic Graphs

Despite the apparent simplicity of cubic and at most cubic graphs, many graph problems are no easier to solve, when restricted to them (see [4] for a complete survey on cubic graphs). Nevertheless, (at most) cubicity often allows to achieve better approximation performance for many NP-hard graph problems.

In the following we first give an approximate algorithm for MIDS-3. We then prove that the proposed heuristic approximates MIDS for at most cubic graphs within 2. Finally we show that a slight variation of the algorithm achieves a guaranteed performance ratio 1.923 for cubic graphs.

In view to make clear the exposition, we first sketch the basic strategy and then analyze its performance.

3.1 Overview of Algorithm 1

The overall strategy we use to solve MIDS-3 is the following (see fig. 2):

- *while-loop* (lines 4-11)
  In this phase vertices of degree 3 in G' are sequentially considered. Given a vertex $v \in V_3'$, it is put in the independent dominating set S. Then v and its adjacent vertices are removed from G'. Indeed, every vertex $u \in adj(v)$ is dominated by v, and adjacent vertices cannot belong to an independent set. At the end of this phase, $G' = (V', E')$ has bounded degree 2. After the execution of the while-loop, an optimal MIDS could be found in G'. Since all isolated vertices of any graph must be put in the MIDS, it would be desirable to reduce their number as much as possible, even if, at the same time, a small disadvantage is introduced.

- *swap-step* (lines 12-32)
  The aim of this step is to benefit by a sort of local search to slow the number of isolated vertices of G' down, providing that the drawback is not too large. We define the "neighborhood structure" of any vertex $v \in S$ as the set of independent dominating sets of the subgraph of G induced by $adj(v)$.
  Then, for each neighbor of v, we compute a profit function p. We say N_v a neighbor corresponding to the maximum $p(N_v)$. Moreover we indicate by $adj(N_v) = \bigcup_{u \in N_v} adj(u) - \{v\}$ the set of vertices adjacent in G to vertices in N_v but v (see fig. 3).
  Each vertex v put in S in the while-loop is now processed. If $p(N_v) \leq 0$, v is left in S, otherwise the pair (v, N_v) is swapped, that is v is removed from S, all vertices in N_v are put in S, and all vertices in $adj(N_v)$ are removed from the current graph G'.
  In particular, among the vertices removed from the graph there will be some isolated vertices and there could be some non-isolated vertices. Function $p(N_v)$ takes into account both the benefit of the deletion of isolated vertices (stored in I_v) and of the disadvantage possibly introduced by removing non-isolated vertices (T-components stored in T_v).

```
        Algorithm 1
        Input:  G = (V, E)
        Output:  S
 1.   begin
 2.       G' = (V', E') ← G = (V, E)
 3.       S ← ∅
 4.       while (V'₃ ≠ ∅) do
 5.               choose v ∈ V'₃
 6.               S ← S ∪ {v}
 7.               V' ← V' - {v} - adj(v)
 8.               Iᵥ ← ∅
 9.               Tᵥ ← ∅
10.               for each k do update V'ₖ
11.       endwhile
12.       C ← {C s.t. C is a connected component of G'}
13.       for each v ∈ S do
14.               for each neighboor N̄ᵥ of v do
15.                       for each C ∈ C s.t. ∃w ∈ C ∩ adj²(v) do
16.                          case of C
17.                              T-component of v w.r.t. N̄ᵥ:  T̄ᵥ ← T̄ᵥ ∪ C
18.                              isolated vertex w:  Iᵥ ← Iᵥ ∪ {w}
19.                          C ← C - C
20.                       endfor
21.                       Nᵥ ← N̄ᵥ s.t. p(N̄ᵥ) is max
22.                       Tᵥ ← T̄ᵥ
23.               endfor
24.               if (p(Nᵥ) > 0) then
25.                   S ← S ∪ Nᵥ - {v}
26.                   V' ← V' - adj(Nᵥ)
27.                   for each k do
28.                           update V'ₖ
29.                   update C
30.               endif
31.       endfor
32.       S ← S ∪ optimal MIDS for G'
33. end.
```

Fig. 2. Algorithm for MIDS-3

- *final-step* (line 33)
  Finally, an optimal MIDS for the remaining graph G' is found and added to S.

3.2   Analysis of Algorithm 1

We first show that the proposed heuristic finds a feasible solution of MIDS-3, and then prove that it approximates the problem within 2. Finally we focus on

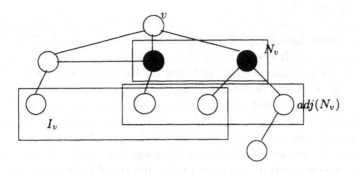

Fig. 3. An example of N_v, I_v and $adj(N_v)$

cubic graphs and show that a better guaranteed approximation ratio is achieved with a slight variation in the while-loop.

Theorem 3.1 *Given an at most cubic graph G, Algorithm 1 finds an independent dominating set.*

Proof. In order to show that S is a dominating set, observe that, during the execution of the while-loop, each vertex v put in S dominates vertices that are removed with it, i.e. it dominates all vertices in $\{v\} \cup adj(v)$. Then, in the swap-step, every time a pair (v, N_v) is swapped, the set N_v dominates all vertices in $\{v\} \cup adj(v)$, and all vertices in $adj(N_v)$ are removed from G'. Finally, an optimal MIDS is determined in the remaining graph.

To prove that S is an independent set, first consider that the while-loop finds an independent set: when a vertex v is put in S, its adjacent vertices are deleted from G'. Then, in the swap-step, every time a pair (v, N_v) is swapped, all vertices in $adj(N_v)$ are removed from G'. Then, at each iteration of the swap step vertices in the current G' cannot be adjacent to any vertex in S (line 26 of Algorithm 1). Finally, an optimal MIDS is determined in the remaining graph.

Before proving the performance ratio of the algorithm, we discuss the definition of the profit function leading the swap-step.

Let (v, N_v) be a pair where v is a vertex in S after the while-loop and N_v an independent dominating set in the subgraph induced in G by $adj(v)$ such that the corresponding $p(N_v)$ is maximum. In order to figure out how the profit function is defined, we will show in which way p evaluates whether (v, N_v) must be swapped or not. Notice that, as far as N_v is defined, if the pair (v, N_v) is not swapped, then no other pair (v, N_v') is.

Let $w \in adj(N_v) \cap V'$ be a vertex in a connected component in G', then one of the following holds:

a. $w \in V_0' \cap adj(N_v)$

a. $w \in V_0' \cap adj(N_v)$

Let v be the general vertex put in S in the while-loop. In the computation of the profit function $p(N_v)$, only isolated vertices in $adj(N_v)$ must be considered, say I_v. It is easy to see that, as far as Algorithm 1 is concerned, the sets I_v are mutually disjoint and that for each v there is up to one vertex in $V_0' \cap adj^2(v)$ which does not belong to I_v.

Suppose (v, N_v) is not swapped, then w would be put in S in the final-step. Otherwise, if (v, N_v) is swapped, w is deleted from G'. It follows that it is suitable to swap (v, N_v) if the number of isolated vertices in $adj(N_v)$ overcomes the increase of $|S|$ due to the swap (i.e. $|N_v| - 1$) and the possible negative contribution produced by the change of G'(see item c).

b. $w \in V_1' \cap adj(N_v)$

Consider the connected component C containing w in G'. Since G' has bounded degree 2 and w has degree 1, then C must be a chain, and w is one of its extremes. Hence the cardinality of the optimal independent dominating set in the chain cannot increase by swapping (v, N_v), that corresponds to the deletion of w. Then, no contribution is given by such a vertex in the computation of p.

c. $w \in V_2' \cap adj(N_v)$

Consider the connected component C containing w in G'. As G' has bounded degree 2, and w has degree 2, then C must be either a cycle or a chain and w is an internal vertex.

Notice that, since we are interested in computing an independent dominating set of cardinality not greater than $n/2$, from remark 1, we can restrict to consider only the case in which w is a peak of an odd chain. Indeed, only the deletion of all peaks in an odd chain leads the value of the solution to overcome the bound of half of the vertices. Therefore, we define T-*component* of v with respect to N_v either a chain of length 3 such that its peak is w, or a chain of length 5 such that both its peaks belong to $adj(N_v)$, or a chain of length 7 such that its three peaks belong to $adj(N_v)$. Because of the cubicity of the graph, no T-components of length greater than 7 must be considered. From now on, and when no confusion arises, we will simply call T_v the set of all T-components of v with respect to N_v.

It remains to analyze the case in which the peaks of an odd chain belong to different $adj(N_{v_j})$, where v_j is a vertex put in S during the while-loop. Such an odd chain becomes a T-component in T_{v_j} if and only if the pairs (v_i, N_{v_i}) swap, $i \neq j$ (see fig. 4). Therefore, the assignment of T-components to vertices put in S in the while-loop must be done when they are processed in the swap-step.

As far as Algorithm 1 is concerned, after the execution of the swap-step, graph G is implicitly partitioned into disjoint subgraphs, in the following way:

- for each vertex v put in S during the while-loop, subgraph G_v induced in G by $v \cup N_v \cup I_v \cup T_v$.
- the possibly not connected subgraph G_R induced in G by the remaining vertices.

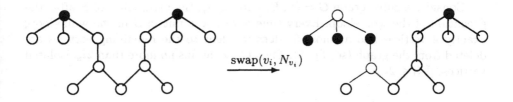

Fig. 4. "Shared" T-components

The previous remarks make us able to define the profit function p, that leads the execution of the swap-step, $p(N_v) = |I_v| - |N_v| + 1 - |T_v|$.

We will prove that $|S| \leq n/2$ by showing that, for each subgraph, up to half of their vertices belong to S found by the algorithm.

Lemma 3.2 *Let v be a vertex put in S in the while-loop and n_v be the number of vertices in G_v. Then, at most $n_v/2$ vertices belong to S.*

Proof. Notice that if (v, N_v) is not swapped, then v, all vertices in I_v and all peaks of T-components in T_v are put in S. Otherwise, if (v, N_v) is swapped, then all vertices in N_v and all non-peaks of T-components in T_v are put in S.

Since the previous sets are disjoint, at least one of them has cardinality no greater than $n_v/2$. Function $p(N_v) = |I_v| - |N_v| + 1 - |T_v|$ leads the choice.

Lemma 3.3 *Let n_R be the number of vertices in G_R. Then, at most $n_R/2$ vertices belong to S.*

Proof. After the swap-step, vertices in G_R are partitioned in the following way:
− n_d vertices adjacent to N_v for some swapped pair (v, N_v), and thus already dominated;
− n_i vertices made isolated by the swap-step;
− $n_R - n_i - n_d$ vertices of bounded degree 2.

In view of remark 1, up to $\frac{n_R - n_i - n_d}{2} + n_i = \frac{n_R + n_i - n_d}{2}$ vertices must be put in S.

The only case in which one dominated vertex leaves more than one isolated vertex is due to T-components. Since T-components have already been considered in subgraphs G_v, then $n_i \leq n_d$ holds. The thesis follows.

Theorem 3.4 *Algorithm 1 approximates MIDS-3 within 2.*

Proof. In view of lemmas 2.1, 3.2 and 3.3:

$$\frac{|S|}{|S^*|} \leq \frac{n/2}{n/4} = 2$$

Consider a cubic graph $G = (V, E)$ and say i_w the cardinality of S after the execution of the while-loop. Every time a vertex is put in S in the while-loop, exactly 4 vertices and at most 6 edges incident to the remaining vertices are deleted from the graph (see fig. 5). Then G' ccontains no more than $2i_w$ isolated vertices, $|V_0'| \leq 2i_w$.

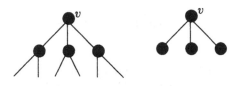

Fig. 5. Extremal cases in the while-loop

As a consequence of such a bound for $|V_0'|$, the swap-step is not crucial anymore to achieve guaranteed approximation ratio 2. Indeed, if we run the final-step right after the while-loop the cardinality of the solution is $|S| \leq i_w + |V_0'| + (n - 4i_w - |V_0'|)/2 \leq n/2$.

Finally, we will show that for cubic graphs the performance ratio achieved by Algorithm 1 can be improved by running a preprocessing phase. Informally speaking, it would be desirable to put in S several vertices adjacent to three vertices not dominated yet. The preprocessing phase exploits cubicity of the graph choosing vertices that can be put in the solution without leaving isolated vertices. Then Algorithm 1 is run on the remaining possibly not connected at most cubic graph.

Let $G = (V, E)$ be a cubic graph. In the preprocessing phase vertices of degree 3 are selected and put in the solution S, similarly to the while-loop of Algorithm 1, but a further condition must hold in order to guarantee that no isolated vertex is left in the remaining graph G'.

Namely, vertices put in S must have distance at least 5. Because of the cubicity of the graph no isolated vertex is produced but in the case drawn in fig. 6, and then no isolated vertex is left in the remaining graph by assuming that in this case the produced isolated vertex is also put in S.

Theorem 3.5 *Algorithm 1 with the preprocessing phase approximates MIDS for cubic graphs within 25/13.*

Proof. As far as the preprocessing phase is concerned, vertices are put in S in the following way:
1. either one vertex is put in S and up to 45 vertices are forbidden; (fig. 6.a);
2. or two vertices are put in S and up to 24 are forbidden (fig. 6.b).

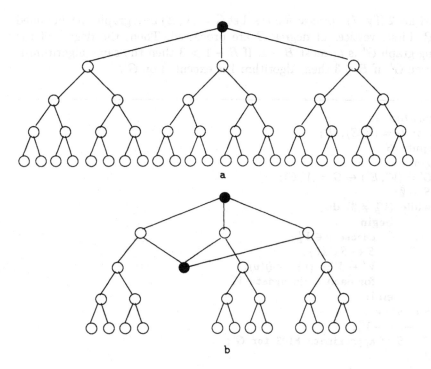

Fig. 6. Extremal cases in the preprocessing phase

Let n_1 be the number of vertices following 1. and n_2 be the number of pair of vertices following 2. Since $46n_1 + 26n_2 \leq n$ after the execution of Algorithm 1 we have:

$$|S| \leq n_1 + n_2 + \frac{n - 4n_1 - 5n_2}{2} \leq \frac{25}{52}$$

Therefore, in view of lemmas 2.1

$$\frac{|S|}{|S^*|} \leq 1.923$$

4 MIDS in Bounded Degree Graphs

In this section we generalize the above results to any bounded degree graphs $(B \geq 4)$, and give an heuristic that approximates MIDS-B within

- $K_B(B + 1)$ for bounded degree graphs, $B \geq 4$
- 2 for 4-regular graphs
- $(K_B - \frac{B-2}{B^2+1})(B + 1)$ for B-regular graphs, $B \geq 5$

  where $K_B = \frac{(B^2 - B + 1)}{B^2 + 1}$.

Algorithm 2 (fig. 7) works as follows. Let $G = (V, E)$ be a graph with bounded degree B. First, vertices of degree B are processed. Then, the degree of the remaining graph G' is at most $B - 1$. If $B - 1 > 3$ then the same algorithm is executed on G'; if $B \leq 3$ then Algorithm 1 is executed on G'.

```
    Algorithm 2;
    Input: G = (V, E), B;
    Output: S;
1.  begin
2.      G' = (V', E') ← G = (V, E);
3.      S ← ∅;
4.      while (V'_B ≠ ∅) do
5.              begin
6.                  choose v ∈ V'_B;
7.                  S ← S ∪ {v};
8.                  V' ← V' − {v} − adj(v);
9.                  for each k do update V'_k
10.             end;
11.     S ← S ∪ V'_0;
12.     V' ← V' − V'_0;
13.     S ← S ∪ approximate MIDS for G';
14. end.
```

Fig. 7. Algorithm for MIDS-B

Theorem 4.1 *Given a bounded degree graph, Algorithm 2 finds an independent dominating set.*

Proof. Similar to proof of theorem 3.1.

Consider just one iteration of Algorithm 2 (lines 1-12) on a graph of bounded degree B. Say G' the remaining graph after the iteration. Let i_w be the number of vertices put in S and $|V'_0|$ be the number of vertices made isolated during the while-loop of this iteration.

Lemma 4.2 *The following inequalities hold:*

a. $i_w \geq \frac{|V'_0|}{B(B-1)}$

b. $i_w \leq \frac{n - |V'_0|}{B+1}$

c. $|V'_0| \leq \frac{B(B-1)n}{(B^2+1)}$

Proof. During the execution of the while-loop (line 4-10), every time a vertex is put in S, exactly $B + 1$ vertices and at most $B(B - 1)$ edges incident to the remaining vertices are deleted from the graph. Therefore:

a. the graph contains no more than $B(B-1)i_w$ isolated vertices.
b. the graph has at least $|V_0'|$ vertices, and thus the number of vertices removed from the graph is $i_w(B+1) \leq n - |V_0'|$
c. From a. and b. the thesis follows.

Theorem 4.3 *Algorithm 2 approximates MIDS-B within $K_B(B+1)$, for $B \geq 4$, where $K_B = \frac{(B^2-B+1)}{B^2+1}$.*

Proof. By induction on the maximum degree B, we will show that, for each $B \geq 4$, Algorithm 2 determines S such that

$$|S| \leq K_B n = \left(1 - \frac{B}{B^2+1}\right) n$$

Then, from lemma 2.1,

$$\frac{|S|}{|S^*|} \leq \frac{(B^2-B+1)(B+1)}{B^2+1}$$

which concludes the proof.

Basis: $B = 4$. Algorithm 2 finds a MIDS S of cardinality

$$|S| \leq i_w + |V_0'| + \frac{1}{2}(n - 5i_w - |V_0'|)$$

Indeed, after the while-loop, the remaining at most cubic graph G' has $(n - 5i_w - |V_0'|)$ vertices. Then, Algorithm 1 is executed, and at most half of such vertices are put in S.

Therefore, since $B = 4$ and in view of lemma 4.2.a. and c. the basis of the induction is proved.

Inductive Step: Consider now $B > 4$; after each iteration, Algorithm 2 is executed on G', then the inductive hypothesis can be used:

$$|S| \leq i_w + |V_0'| + K_{B-1}(n - (B+1)i_w - |V_0'|)$$

Since the coefficient of i_w is negative, from lemma 4.2.a. we have:

$$|S| \leq nK_{B-1} + \frac{|V_0'|}{B(B-1)}(1 - BK_{B-1} - K_{B-1}) + |V_0'|(1 - K_{B-1}) =$$

$$= nK_{B-1} + |V_0'|\frac{B^2 - K_{B-1}B^2 - B + 1 - K_{B-1}}{B(B-1)}$$

From lemma 4.2.c.

$$|S| \leq \left(1 - \frac{B}{B^2+1}\right) n$$

In view of particular properties of regular graphs, a better performance ratio can be achieved if the input graph is B-regular.

Lemma 4.4 *For B-regular graphs, the following inequalities hold:*

a. $i_w \geq \frac{n}{B^2+1}$

b. $|V_0'| \leq (B-1)i_w$

Proof. During the execution of the while-loop, every time a vertex is put in S:

a. up to $B^2 + 1$ vertices of degree B become of lower degree (see fig. 5);

b. exactly $B + 1$ vertices and at most $B(B-1)$ edges incident the remaining vertices are deleted from the graph. Then G' ccontains no more than $(B-1)i_w$ isolated vertices.

Theorem 4.5 *Algorithm 2 approximates MIDS for 4-regular graphs within 2 and for B-regular graphs within $(K_B - \frac{B-2}{B^2+1})(B+1)$, for $B \geq 5$, where $K_B = \frac{(B^2-B+1)}{B^2+1}$.*

Proof. We first prove that the solution S found by Algorithm 2 for B-regular graphs $(B \geq 5)$ is:

$$|S| \leq \left(1 - \frac{2B-1}{B^2+1}\right)n$$

Since after the while-loop Algorithm 2 is run on the graph G' of bounded degree $B - 1$, we use the result of theorem 4.3.

$$|S| \leq i_w + |V_0'| + K_{B-1}\left(n - (B+1)i_w - |V_0'|\right)$$

From lemma 4.4.b.

$$|S| \leq nK_{B-1} + i_w(1 - BK_{B-1} - K_{B-1}) + (1 - K_{B-1})(B-1)i_w =$$

$$= nK_{B-1} + i_w(B - 2BK_{B-1})$$

Therefore, in view of lemma 4.4.a. and of the definition of K_B:

$$|S| \leq \frac{n}{B^2+1}\left(K_{B-1}(B-1)^2 + B\right) \leq \left(1 - \frac{2B-1}{B^2+1}\right)n$$

Similarly to the proof of theorem 4.3, the thesis follows.

The proof for 4-regular graphs is analogous to the previous one by assuming $K_3 = 1/2$, and therefore it is omitted for sake of brevity.

5 Conclusions

In this paper we have considered the minimum independent dominating set problem in bounded degree graphs. We have given approximate heuristics based on greedy and local search techniques that improve the previously known approximation ratios.

The algorithm proposed for the minimum independent dominating set problem in cubic and at most cubic graph (MIDS-3) is made up by two phases. In the former greedy phase a partial solution is constructed by sequentially selecting vertices of degree 3 and removing them and their adjacent vertices from the graph. Then a local search phase (swap-step) is performed in order to obtain a better complete solution.

For the minimum independent dominating set problem in any bounded degree and regular graphs (MIDS-B) the proposed heuristic iteratively works like the greedy phase in the algorithm for MIDS-3. Then, when the degree of the graph is bounded by 3, the algorithm for MIDS-3 is applied.

Unfortunately it is not easy to extend the local search phase to graphs of bounded degree higher than 3. Indeed, when the degree of the graph increases (even for $B = 4$) connected components giving a negative contribution to each swap can be more complex and shared among vertices put in the solution in the first step. This fact leads to the lower improvement in the quality of the solution achieved by our heuristics for graphs of degree higher than 3.

Besides, it is worth to notice that for cubic and at most cubic graphs we have found independent dominating sets of size at most $0.48n$ and $0.5n$, respectively. To find independent dominating sets of size less or equal than half of the vertices could be not possible for any bounded degree graph. Indeed, for degree higher than 3 there exist instances for which the cardinality of optimal solutions is greater than half of the vertices (e.g. see fig. 8).

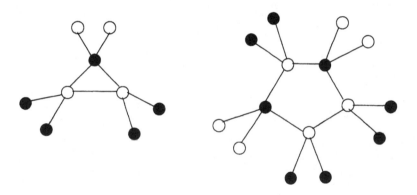

Fig. 8. Optimal MIDS for graphs of bounded degree 4

Acknowledgments: We thank Viggo Kann for a careful reading of earlier drafts and for encouraging us to work on this subject. Also many thanks to Giorgio Ausiello for his helpful suggestions on improving the paper. A special thank to Johan Hastad for suggesting the idea that led us to theorem 3.5.

References

1. P. Berman, M. Fürer, Approximating Maximum Independent Set in Bounded Degree Graphs, *Proc. 5-th Ann. ACM-SIAM Symp. on Discrete Algorithms*, ACM-SIAM, 365-371, 1994.

2. M. Bellare, O. Goldreich, M. Sudan, Free Bits, PCPs and non-approximability-towards tight results, *Proc. of the 36th Annual IEEE Conference on Foundations of Computer Science*, 1995.
3. P. Crescenzi and V. Kann, A Compendium of NP Optimization Problems, *Manuscript*, 1995.
4. R. Greenlaw and R. Petreschi, Cubic graphs, *ACM Computing Surveys*, 27, 471–495, 1995.
5. M.M. Halldorsson, Approximating the minimum maximal independence number, *Inform. Process. Lett.*, 46, 169-172, 1993.
6. V. Kann, On the Approximability of NP-Complete Optimization Problems, *PhD Thesis*, Department of Numerical Analysis and Computing Science, Royal Institute of Technology, Stockolm, 1992.
7. V.Kann, Polynomially bounded minimization problems that are hard to approximate, *Nordic J. Computing*, 1, 317-331, 1994.
8. C. Lund, and M. Yannakakis, On the Hardness of Approximating Minimization Problems, J. ACM 41, 960-981, 1994.
9. C. Papadimitriou, and K. Steiglitz, *Combinatorial Optimization Algorithms and Optimization*, Prentice-Hall, Englewood Cliffs, New Jersey, 1982.
10. C. Papadimitriou, and M. Yannakakis, Optimization, Approximation, and Complexity Classes, *Journal of Computer and System Sciences*, 43, 425-440, 1991.

A New Characterization of P_4-connected Graphs

Luitpold Babel[1]* and Stephan Olariu[2]**

[1] Technische Universität München, 80290 München, Germany
[2] Old Dominion University, Norfolk, VA 23529, U.S.A.

Abstract. A graph is said to be P_4-connected if for every partition of its vertices into two nonempty disjoint sets, some P_4 in the graph contains vertices from both sets in the partition. A P_4-chain is a sequence of vertices such that every four consecutive ones induce a P_4. The main result of this work states that a graph is P_4-connected if and only if each pair of vertices is connected by a P_4-chain. Our proof relies, in part, on a linear-time algorithm that, given two distinct vertices, exhibits a P_4-chain connecting them. In addition to shedding new light on the structure of P_4-connected graphs, our result extends a previously known theorem about the P_4-structure of unbreakable graphs.

1 Introduction

Very recently, B. Jamison and S. Olariu [11] introduced the notion of P_4-connectedness. Specifically, a graph $G = (V, E)$ is P_4-*connected* if for every partition of V into nonempty disjoint sets V_1 and V_2, some chordless path on four vertices and three edges (i.e. a P_4) contains vertices from both V_1 and V_2.

The concept of P_4-connectedness leads to a structure theorem for arbitrary graphs in terms of P_4-connected components and suggests, in a quite natural way, a tree representation unique up to isomorphism. The leaves of the resulting tree are the P_4-connected components and the weak vertices, that is, vertices belonging to no P_4-connected component.

This structure theorem and the corresponding tree representation, on the one hand, provide tools for the study of graphs with a simple P_4-structure, such as P_4-reducible [8], P_4-extendible [9], P_4-sparse [10] and, more generally, (q, t) graphs [1]. On the other hand, and more importantly, the structure theorem lays the foundation of the so-called *homogeneous decomposition* of graphs [11], a decomposition which can be seen as an extension of the well known *modular* decomposition (also called *substitution* decomposition [13]). In this context, it is of particular interest to investigate graphs which are prime with respect to the homogeneous decomposition. It is an immediate consequence of the structure theorem that those graphs are P_4-connected.

---

* This author was supported by the Deutsche Forschungsgemeinschaft (DFG)
** This author was supported in part by NSF grant CCR-94-07180 and by ONR grant N00014-95-1-0779.

Further strong motivation for the study of P_4-connected graphs is provided by their relationship to unbreakable graphs. These graphs are known to play a central role in attempts to reveal the structure of minimal counterexamples to the famous, and yet unresolved, Strong Perfect Graph Conjecture.

The concept of P_4-connectedness obviously generalizes the usual connectedness of graphs, since a graph $G = (V, E)$ is connected, in the usual sense, if for every partition of V into nonempty disjoint sets V_1 and V_2 there exists an edge with one endpoint in V_1 and the other in V_2. A more common characterization states that each pair of different vertices is connected by a path. The main contribution of this paper is to point out a similar characterization of P_4-connected graphs in terms of so-called P_4-chains.

The remainder of this work is organized as follows. Section 2 establishes terminology and summarizes previous art. Section 3 reviews a number of results about unbreakable graphs and shows that unbreakable graphs are P_4-connected. In Section 4 we introduce the concept of a P_4-chain and present our main result, a characterization of P_4-connected graphs in terms of P_4-chains. As a corollary, we obtain an extension of a result of Chvátal [5] on unbreakable graphs. In Section 5, we present a linear-time algorithm that, given a pair of vertices in a P_4-connected graph, constructs a P_4-chain connecting them.

2 Basics and Terminology

All graphs in this paper are finite, with no loops nor multiple edges. In addition to standard graph-theoretical terminology, compatible with [2], we need new terms that we are about to define.

Let $G = (V, E)$ be a graph with vertex-set V and edge-set E. For a vertex v of G, $N(v)$ denotes the set of all vertices adjacent to v. If $U \subseteq V$ then $G(U)$ stands for the graph induced by U. Occasionally, to simplify the exposition, we shall blur the distinction between sets of vertices and the subgraphs they induce, using the same notation for both. The complement of G is denoted by $\overline{G}$. A *clique* is a set of pairwise adjacent vertices, a *stable set* is a set of pairwise nonadjacent vertices. G is termed a *split* graph if its vertices can be partitioned into a clique and a stable set.

We say that two sets X and Y of vertices of G are *nonadjacent* if no edge has one endpoint in X and the other in Y. X and Y are *totally adjacent* if every vertex in X is adjacent to all vertices in Y. Finally, sets X and Y that are neither nonadjacent nor totally adjacent are termed *partially adjacent*.

A vertex v is said to *distinguish* a set U of vertices if v is partially adjacent to U. A subset Z of V with $1 < |Z| < |V|$ is termed a *homogeneous set* if no vertex outside Z distinguishes Z, i.e. each vertex outside Z is nonadjacent or totally adjacent to Z. The graph obtained from G by shrinking every maximal homogeneous set to one single vertex is called the *characteristic graph* of G.

As usual, we let P_k stand for the chordless path with k vertices and $k - 1$ edges. In a P_4 with vertices u, v, w, x and edges uv, vw, wx, vertices v and w are referred to as *midpoints* whereas u and x are called *endpoints*.

Adapting the terminology of [11], a graph $G = (V, E)$ is P_4-*connected*, or *p-connected* for short, if for every partition of V into nonempty disjoint sets V_1 and V_2 there exists a *crossing* P_4, that is, a P_4 containing vertices from both V_1 and V_2. The *p-connected components* of a graph are the maximal induced subgraphs which are *p*-connected. Note that the *p*-connected components are closed under complementation and are connected subgraphs of G and $\overline{G}$. Vertices which are not properly contained in a *p*-connected component are called *weak vertices*. It is easy to see that each graph has a unique partition into *p*-connected components and weak vertices.

A *p*-connected graph is termed *separable* if its vertex-set V can be partitioned into two nonempty disjoint sets V_1 and V_2 in such a way that each crossing P_4 has its midpoints in V_1 and its endpoints in V_2. This partition is commonly written as (V_1, V_2). It is obvious that the complement of a separable *p*-connected graph is also separable. The partition (V_1, V_2) of G becomes (V_2, V_1) in $\overline{G}$. The next statement, presented in [11], gives detailed information about the structure of separable *p*-connected graphs.

Theorem 2.1. *Let G be separable p-connected with separation (V_1, V_2). The subgraph of G (respectively $\overline{G}$) induced by V_2 (respectively V_1) is disconnected. Furthermore, every component of the subgraph of G (respectively $\overline{G}$) induced by V_2 (respectively V_1) with at least two vertices is a homogeneous set in G.*

This result immediately implies a very useful characterization of separable *p*-connected graphs.

Corollary 2.2. *A p-connected graph is separable if and only if its characteristic graph is a split graph.*

The following structure theorem of B. Jamison and S. Olariu [11] shows that separable *p*-connected graphs play a crucial role in the decomposition of arbitrary graphs.

Theorem 2.3. (Structure Theorem). *For an arbitrary graph G exactly one of the following conditions is satisfied:*

(1) G is disconnected;
(2) $\overline{G}$ is disconnected;
(3) There is a unique proper separable p-connected component H of G with a partition (H_1, H_2) such that every vertex outside H is adjacent to all vertices in H_1 and to no vertex in H_2;
(4) G is p-connected.

As shown in [11], this theorem allows one to represent an arbitrary graph by a labeled rooted tree constructed in the obvious recursive way. The labels of the interior nodes of the tree correspond to the first three conditions in the structure theorem, while the leaves are the *p*-connected components and the weak vertices.

3 Unbreakable Graphs

A nonempty subset C of vertices in a graph G is termed a *star-cutset* if $G - C$ is disconnected and some vertex in C is adjacent to all the remaining vertices in C. Chvátal [4] proposed to call a graph *unbreakable* if neither the graph nor its complement contains a star-cutset. The study of unbreakable graphs is motivated by attempts at resolving the famous Strong Perfect Graph Conjecture (SPGC, for short) which is still open [7]. A graph is called *perfect* if, for the graph and all of its induced subgraphs, the chromatic number equals the size of a maximum clique. The SPGC states that a graph is perfect if and only if it contains no induced odd chordless cycle of length at least five and no complement of such a cycle. In [4] Chvátal proved that every minimal imperfect graph is unbreakable. Thus, if the SPGC is false, every minimal counterexample must be unbreakable.

The following lemma points out the relationship between unbreakable and p-connected graphs.

Lemma 3.1. *The class of p-connected graphs strictly contains the unbreakable graphs.*

Proof. Let G be an unbreakable graph. It is well known and straightforward to verify that G does not contain a homogeneous set. Therefore, both G and $\overline{G}$ are connected. If G is not p-connected, then the Structure Theorem implies that there is a unique proper separable p-connected component H of G with a partition (H_1, H_2) such that every vertex outside H is adjacent to all vertices in H_1 and to no vertex in H_2. By virtue of Theorem 2.1 and Corollary 2.2, H must be a split graph. Moreover, there is only one vertex outside H, say w. Clearly, w is adjacent to all vertices of the clique and to no one in the stable set of H. This means that $N(w)$ is a star-cutset which is a contradiction.

Since the P_4 is p-connected but not unbreakable, the containment is strict. □

The P_4-*structure* of a graph G is a hypergraph defined on the same vertex-set as G, with edges consisting of the vertex-sets of the induced P_4s of G. It has been conjectured by Chvátal [3] and proved by Reed [15] that the perfection of a graph depends only on its P_4-structure: if two graphs have the same P_4-structure then either both are perfect or none of them is. This result, known as the Semi-Strong Perfect Graph Theorem, in particular motivated the study of the P_4-structure of unbreakable graphs.

In this context, the central question is how the P_4s of a graph relate to each other. Let $X_1, X_2, \ldots, X_s$, $s \geq 1$, denote sets of vertices of a graph G such that

- for all i, $(1 \leq i \leq s)$, X_i induces a P_4, and
- for all i, $(2 \leq i \leq s)$, X_{i-1} and X_i share exactly three vertices.

Sets X_1 and X_s satisfying the two conditions above are said to be 3-*chained*. The following theorem was first proved by Chvátal [5] and, later, in a more general setting, by Olariu [14].

Theorem 3.2. *In an unbreakable graph every two P_4s are 3-chained.*

In the next section we shall offer an extension of this result to the class of p-connected graphs.

4 p-connected Graphs

Consider a graph $G = (V, E)$ and two vertices x and y in V. A P_4-chain, or p-chain for short, *of length $k - 1$ connecting x and y* is a sequence of distinct vertices $(v_1, v_2, \ldots, v_{k-1}, v_k)$ such that

- $x = v_1$, $y = v_k$, and
- for all i, $(1 \le i \le k - 3)$, $X_i := \{v_i, v_{i+1}, v_{i+2}, v_{i+3}\}$ induces a P_4.

Two vertices x and y are said to be *p-connected* if $x = y$ or there exists a p-chain connecting both vertices.

Simplest examples of p-chains are the chordless paths P_k and their complements $\overline{P_k}$ for $k \ge 4$. Several further examples are depicted in Fig. 1.

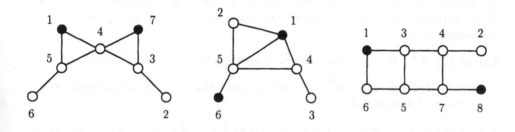

Fig. 1. Some examples of p-chains

It is an important and useful property that p-chains are invariant under complementation. In other words:

Observation 4.1. *A p-chain in G is also a p-chain in $\overline{G}$. Thus, two vertices are p-connected in G if and only if they are p-connected in $\overline{G}$.*

The p-connectedness generalizes the usual connectedness of graphs since a graph is connected if for every partition of the vertex-set there is some edge having one endpoint in each set of the partition. The common definition states that a graph is connected if and only if each pair of vertices is connected by a path, i.e. a sequence of vertices such that any two consecutive vertices induce an edge. A p-chain, where any four consecutive vertices induce a P_4, can be seen as an analogue to a path in the context of p-connectedness. Here is the main result of this paper, a characterization of p-connected graphs by means of p-chains.

Theorem 4.2. *A graph is p-connected if and only if every pair of vertices in the graph is p-connected.*

Proof. Let $G = (V, E)$ be a graph such that every pair of vertices in G is p-connected. Let further V_1 and V_2 be an arbitrary partition of the vertex-set of G. Choose two arbitrary vertices x, y from V_1 and V_2, respectively. By assumption there exists a sequence of vertices $(x = v_1, v_2, \ldots, v_k = y)$ such that, for $i = 1, 2, \ldots, k - 3$, the set $X_i = \{v_i, v_{i+1}, v_{i+2}, v_{i+3}\}$ induces a P_4. If X_1 contains vertices from both V_1 and V_2, then X_1 induces a crossing P_4. Likewise, if X_{k-3} contains vertices from both V_1 and V_2, then we are done. Otherwise, since $X_1 \subseteq V_1$, $X_{k-3} \subseteq V_2$ and $X_{i-1} \cap X_i \neq \emptyset$, there must exist a set X_j with vertices from both V_1 and V_2: this is the desired crossing P_4.

The proof of the converse implication will be given in Section 5 in the form of an algorithm. This algorithm, given an arbitrary pair of distinct vertices in a p-connected graph, returns a p-chain connecting them. $\square$

This novel characterization of p-connected graphs extends, in a certain sense, Theorem 3.2. We shall say that two vertices x and y are 3-*chained* if there are P_4s X_1 and X_s, $s \geq 1$, containing x respectively y, which are 3-chained. Obviously, if two vertices are p-connected then they are also 3-chained. Thus, we have the following result.

Corollary 4.3. *In a p-connected graph every two vertices are 3-chained.*

For further reference, we now state a result that follows immediately from the proof of Theorem 4.2.

Corollary 4.4. *A graph is p-connected if and only if some vertex of the graph is p-connected with all other vertices.*

It is easy to show that p-connectedness of vertices in a graph G is an *equivalence relation* on the vertex-set V of G. Reflexivity and symmetry follow immediately from the definition, transitivity is implied by the corollary to the following statement. With this knowledge, it is obvious that the *equivalence classes* of V are precisely the p-connected components of G.

Lemma 4.5. *Let U and W, $U \cap W \neq \emptyset$, be sets of vertices of a graph G. If both U and W induce p-connected subgraphs of G, then the subgraph induced by $U \cup W$ is p-connected.*

Proof. Let z be an arbitrary vertex in $U \cap W$. Since $G(U)$ is p-connected, Theorem 4.2 implies that z is connected with every vertex $x \in U$ by a p-chain. Similarly, z is connected with every vertex $y \in W$ by a p-chain. Now the conclusion follows by Corollary 4.4. $\square$

Corollary 4.6. *Let x, y, z be distinct vertices of a graph. If x and z, respectively z and y are p-connected, then x and y are p-connected.*

Proof. Let U and W be the sets of vertices belonging to the p-chains connecting x and z respectively z and y. Then $G(U)$ and $G(W)$ are p-connected and $z \in U \cap W$. By Lemma 4.5, the graph $G(U \cup W)$ is p-connected. Now, Theorem 4.2 implies that there exists a p-chain connecting x and y in $G(U \cup W)$. □

5 Finding a p-chain in Linear Time

As a preamble to our algorithm we want to point out that p-connectedness of a graph can be checked very efficiently.

Lemma 5.1. *The task of testing whether a graph is p-connected can be performed in time linear in the size of the graph.*

Proof. It is well known that the modular decomposition of a graph G can be obtained in time linear in the size of G using one of the algorithms presented in [6] or [12]. These algorithms, in particular, compute the connected components of G respectively $\overline{G}$ and, if both graphs are connected, provide all the maximal homogeneous sets of G. This immediately allows us to verify conditions (1) and (2) of the Structure Theorem. Condition (3) can be checked by shrinking every maximal homogeneous set of G to one single vertex. If condition (3) holds then, by virtue of Theorem 2.1 and Corollary 2.2, the characteristic graph of G consists of a split graph together with one further vertex adjacent to all vertices of the clique and to none in the stable set. This, again, can be tested in linear time using standard techniques (see e.g. [7]). □

We shall now spell out the details of an algorithm that constructs a p-chain between any pair of vertices in a p-connected graph. The algorithm is based primarily on a reduction technique which transforms the problem to a smaller graph. To be more precisely, each maximal homogeneous set of the graph is replaced by one single vertex. The following statement implies that this operation preserves p-connectedness of the graph.

Lemma 5.2. *A graph is p-connected if and only if its characteristic graph is p-connected.*

Proof. Let G be p-connected. We show that shrinking a homogeneous set to a single vertex or replacing a vertex by a homogeneous set preserves p-connectedness.

Let G^* be obtained from G by shrinking a homogeneous set Z to a single vertex z. In order to prove that G^* is p-connected, let X be an arbitrary P_4 in G with some vertices outside Z. If X has no vertices in Z then X must be present in G^*, too. If X has vertices in Z then, obviously, it has exactly one vertex in Z. Since Z is homogeneous, the P_4 obtained from X by removing the unique vertex in Z and replacing it with z is still a P_4 in G^*. Therefore, if G^* is not p-connected, then we can find a partition V_1, V_2 in G^* without a crossing P_4. But now, in G, Z is completely included in one of V_1 or V_2, and by the previous argument there can be no crossing P_4 in G between the corresponding sets augmented by Z.

Conversely, let G^{**} be obtained from G by replacing a vertex z by a homogeneous set Z. Consider a partition V_1, V_2 of the vertex-set of G^{**}. If Z is completely included in one of these sets, say V_1, then let X be a crossing P_4 for the partition $V_1 - Z \cup \{z\}, V_2$ in G. If X does not contain z then X is also present in G^{**}. If X contains z then the P_4 obtained from X by removing z and replacing it with an arbitrary vertex from Z is a P_4 in G^{**}. Obviously, X is crossing between V_1 and V_2. If Z is not completely included in one of the sets V_1 or V_2 then let X be an arbitrary P_4 in G containing z. One of the P_4s which are obtained by replacing z by the vertices from Z is crossing between V_1 and V_2. This shows that G^{**} is p-connected. $\qquad\square$

Hence, it suffices to study the characteristic graph. The remainder of the algorithm dissects this graph and tries to find a p-chain by successively incorporating the parts of the graph. The p-connectedness and the fact that there are no homogeneous sets provide information about the structure of the graph which finally allows to exhibit a p-chain. Here is a formal description of the algorithm. It is formulated for the case of adjacent vertices. The case of nonadjacent vertices involves a similar argument in $\overline{G}$.

Algorithm CHAIN$(G;x,y)$

Input: A p-connected graph $G = (V, E)$ and two adjacent vertices $x, y \in V$.
Output: A p-chain connecting x and y.

1.1. Find all maximal homogeneous sets of G.
1.2. If x and y belong to a maximal homogeneous set Z **then**
　　　　Find a crossing P_4 between Z and $V - Z$.
　　　　Let (z, u, v, w) denote this P_4 with $z \in Z$ and $u, v, w \notin Z$.
　　　　STOP with the p-chain (x, u, v, w, y).
1.3. Shrink every maximal homogeneous set to one single vertex.

2.1. Let $V' := V - \{x, y\}$. **Compute** the sets
　$A := \{v \in V' : vx \in E, vy \notin E\}$, $B := \{v \in V' : vx \notin E, vy \in E\}$,
　$U := \{v \in V' : vx \in E, vy \in E\}$, $W := \{v \in V' : vx \notin E, vy \notin E\}$.

3.1. If A is not totally adjacent to B **then**
　　　　Find nonadjacent vertices $a \in A$ and $b \in B$.
　　　　STOP with the p-chain (x, a, b, y).

3.2. If A (or B) is adjacent to W **then**

> **Find** adjacent vertices $a \in A$ (or B) and $w \in W$.
> **STOP** with the p-chain (x, a, w, y).

3.3. If $|W| = 0$ **then**

> **Find** a shortest path $x, u_1, u_2, \ldots, u_k, y$ between x and y in $\overline{G}$.
> **STOP** with the p-chain $(x, u_1, u_2, \ldots, u_k, y)$.

4.1. Find the connected components $W_1, W_2, \ldots, W_r$ of W in G.

4.2. If there is a set W_i with $|W_i| \geq 2$ **then**

> **Find** adjacent vertices $w, w' \in W_i$ and a vertex $u \in U$
> such that $uw \in E$ and $uw' \notin E$.
> **STOP** with the p-chain (x, u, w, w', y).

5.1. Find the connected components $U_1, U_2, \ldots, U_s$ of U in $\overline{G}$.

5.2. Let $U_1, U_2, \ldots, U_l$ denote those sets which are nonadjacent to W and **let** $U^W := U_{l+1} \cup \ldots \cup U_s$.

5.3. If there is a set U_i, $l + 1 \leq i \leq s$, with $|U_i| \geq 2$ **then**

> **If** there are nonadjacent vertices $u, u' \in U_i$ and a vertex $w \in W$
> such that $uw \in E$ and $u'w \notin E$ **then**
> > **STOP** with the p-chain (x, u, u', w, y).
>
> **If** there are nonadjacent vertices $u, u' \in U_i$ and a vertex $a \in A$
> such that $ua \in E$ and $u'a \notin E$ **then**
> > **Find** a common neighbor w of u and u' in W.
> > **STOP** with the p-chain (x, w, a, u', u, y).
>
> **If** there are nonadjacent vertices $u, u' \in U_i$ and a vertex $b \in B$
> such that $ub \in E$ and $u'b \notin E$ **then**
> > **Find** a common neighbor w of u and u' in W.
> > **STOP** with the p-chain (x, u, u', b, w, y).

6.1. If $A = \emptyset$ or $B = \emptyset$ **then goto** 7.1.

6.2. If there is a set U_i, $1 \leq i \leq s$, which is not totally adjacent to A or B **then**

> **Find** a shortest path $x, u_1, u_2, \ldots, u_k, y$ between x and y
> in the complement of $G(\{x, y\} \cup A \cup B \cup U_i)$.
> **STOP** with the p-chain $(x, u_1, u_2, \ldots, u_k, y)$.

6.3. Find vertices $u, u' \in U^W, a \in A, b \in B$
such that $ua \in E$, $ub \notin E$ and $u'a \notin E$, $u'b \in E$.

> **If** u and u' have a common neighbor $w \in W$ **then**
> > **STOP** with the p-chain (x, u', a, w, b, u, y).
>
> **else**
> > **Find** vertices $w, w' \in W$ with $wu \in E$ and $w'u' \in E$.
> > **STOP** with the p-chain $(x, a, w', u', u, w, b, y)$.

7.1. If $B = \emptyset$ **then let** $B := A$ and **exchange** the roles of x and y.

7.2. Find the connected components $B_1, B_2, \ldots, B_q$ of B in G.

7.3. If there is a set B_i with $|B_i| \geq 2$ **then**

> **If** there are adjacent vertices $b, b' \in B_i$ and a vertex $u \in U^W$
> such that $bu \in E$, $b'u \notin E$ **then**

Find a neighbor w of u in W.

STOP with the p-chain (x, b, b', u, w, y).

7.4. Find the vertex-sets U^1 and B^1 of all components of $U - U^W$ resp. B which are partially adjacent to a component of B resp. $U - U^W$.

For $j \geq 2$ find

the set U^j of all vertices of $U - U^W$ which are totally adjacent to $B^1, \ldots, B^{j-2}$ and not totally adjacent to B^{j-1} and the set B^j of all vertices of B which are nonadjacent to $U^1, \ldots, U^{j-2}$ and adjacent to U^{j-1}.

7.5. Determine the smallest j such that U^W is not totally adjacent to B^j (for convenience of notation, let j be odd; j even is settled analogously).

7.6. Determine a sequence $u_{j+1}, b_j, u_{j-1}, \ldots, b_5, u_4, b_3, u_2$ of vertices with $u_{j+1} \in U^W$, $b_i \in B^i$, $u_i \in U^i$ and partially adjacent components of B^1, U^1, say B_1, U_1, such that u_{i+1}, $i > 2$, is nonadjacent to b_i and totally adjacent to $b_{i-2}, b_{i-4}, \ldots, b_3$ and B_1, u_2 is nonadjacent to B_1, b_i is adjacent to u_{i-1} and nonadjacent to $u_{i-3}, u_{i-5}, \ldots, u_2$ and U_1.

Find a neighbor w of u_{j+1} in W.

7.7. If there are adjacent vertices $b_1, b_1' \in B_1$ and a vertex $u_1 \in U_1$ such that $b_1 u_1 \in E$, $b_1' u_1 \notin E$ **then**

STOP with the p-chain $(x, b_1', u_1, b_1, u_2, b_3, u_4, \ldots, u_{j-1}, b_j, u_{j+1}, w, y)$.

7.8. If there are nonadjacent vertices $u_1, u_1' \in U_1$ and a vertex $b_1 \in B_1$ such that $b_1 u_1 \in E$, $b_1 u_1' \notin E$ **then**

STOP with the p-chain $(x, u_1', u_1, b_1, u_2, b_3, u_4, \ldots, u_{j-1}, b_j, u_{j+1}, w, y)$.

Lemma 5.3. *Algorithm CHAIN correctly computes a p-chain between any two adjacent vertices x and y in a p-connected graph G.*

Proof. We shall verify the correctness of the algorithm by stating a number of observations and remarks. Note that, due to Observation 4.1 and the fact that a graph is p-connected if and only if its complement is p-connected, we can arbitrarily switch between G and $\overline{G}$.

Assume that x and y belong to a homogeneous set Z. Since G is p-connected, there is a crossing P_4 between Z and $V - Z$. As pointed out before, each such P_4 contains exactly one vertex from Z. If $\{z, u, v, w\}$ induces such a P_4 with $z \in Z$ then, since Z is homogeneous, also $\{x, u, v, w\}$ and $\{y, u, v, w\}$ induce P_4s. This shows that (x, u, v, w, y) is a p-chain connecting x and y.

If x and y do not belong to a common homogeneous set then we proceed with the characteristic graph of G which, by Lemma 5.2, is p-connected. Obviously, a p-chain between x and y in the characteristic graph of G is also a p-chain between x and y in G.

In the status of the algorithm after passing Step 3 we are dealing with a p-connected graph (for convenience we again denote this graph by G) which contains no homogeneous sets and which has the following properties:

- A is totally adjacent to B;
- A and B are nonadjacent to W;
- $W \neq \emptyset$.

In other words, x and y do not belong to a common P_4 and are connected by a path consisting of three vertices in $\overline{G}$. This implies the following facts:

- $U \neq \emptyset$ and U is adjacent to W [otherwise G is disconnected, thus not p-connected];
- $A \cup B \neq \emptyset$ [otherwise $\{x, y\}$ is a homogeneous set].

After passing Step 4 we conclude that:

- W is a stable set [if W is not stable then there is a connected component W_i of W which contains at least two vertices. This component is distinguished by a vertex u from outside W, otherwise W_i is a homogeneous set. Obviously, u must belong to U and, since W_i is connected, u distinguishes two adjacent vertices w, w' from W_i. The algorithm finds such vertices which, together with x and y, induce the desired p-chain].

After passing Step 5 we observe:

- $U^W \neq \emptyset$ and U^W is a clique [there are vertices which are adjacent to W, hence U^W is nonempty. Assume that U^W is not a clique. Then there exists a nontrivial connected component U_i of U^W in $\overline{G}$. This component is distinguished by a vertex from one of the sets W, A or B, otherwise U_i is a homogeneous set. Since U_i is connected in $\overline{G}$, such a vertex in particular distinguishes two nonadjacent vertices from U_i. If no vertex from W distinguishes U_i then, clearly, any two vertices from U_i have a common neighbor in W. The algorithm finds such vertices. It is easy to verify that these vertices induce a p-chain between x and y].

Assume that both A and B are nonempty. If there is a set U_i, $1 \leq i \leq s$, which is not totally adjacent to A or to B then, due to the connectivity of U_i in $\overline{G}$, x and y are connected by a chordless path consisting of at least five vertices in the complement of the graph which is induced by $\{x, y\} \cup A \cup B \cup U_i$. This path is a p-chain. If there is no such set then:

- U^W is neither totally adjacent to A nor to B [assume without loss of generality that U^W is totally adjacent to A. Renumber the sets $U_1, \ldots, U_l$ in such a way that $U_1, \ldots, U_k$ are not totally adjacent to B but totally adjacent to A, and $U_{k+1}, \ldots, U_l$ are totally adjacent to B. It is easy to see that the set $\{y\} \cup A \cup U_{k+1} \ldots \cup U_l$ is homogeneous which is a contradiction].

The last fact implies the existence of vertices u and u' from U^W as desired in Step 6.3. Note further that each vertex from U^W has a neighbor in W. Using this information the algorithm constructs a p-chain connecting x and y.

After passing Step 6 we know that one of the sets A or B is empty and the other one is nonempty. We can assume w.l.o.g. that A is empty. After passing Step 7.3 it is not hard to realize that:

- $U^1, B^1 \neq \emptyset$ [if there is a connected component of B with at least two vertices which is distinguished by a vertex from U^W then Step 7.2 of the algorithm finds a p-chain between x and y. Otherwise, this component can only be distinguished by a vertex from $U - U^W$. Similarly, a nontrivial connected component of $U - U^W$ in $\overline{G}$ can only be distinguished by a vertex from B. Clearly, all components of B resp. $U - U^W$ which do not belong to B^1 and U^1 are trivial. If U^1 and B^1 are empty then U is a clique and B is a stable set. This, however, implies that there is no P_4 containing x in contradiction to the p-connectedness of G. Hence U^1 and B^1 are nonempty];
- U^W is not totally adjacent to $\bigcup_{j \geq 1} B^j$ [otherwise, it is straightforward to realize that $\{x\} \cup \bigcup_{j \geq 1} U^j \cup \bigcup_{j \geq 1} B^j$ is a homogeneous set].

These observations imply the existence of a sequence of vertices as pointed out in Step 7.6. Finally, since B_1 and U_1 are partially adjacent, we can either find a vertex from U_1 which distinguishes two adjacent vertices from B_1 or a vertex from B_1 which distinguishes two nonadjacent vertices from U_1. In each case we obtain a p-chain as described in the last steps of the algorithm. □

Theorem 5.4. *Algorithm CHAIN can be implemented to run in time linear in the size of the graph G.*

Proof. We shall sketch the main features of the implementation without stating all the details. The crucial point in a linear implementation is to avoid the computation of $\overline{G}$. All steps of the algorithm must be performable in time linear in the size of G regardless whether we consider G or $\overline{G}$.

As pointed out before the maximal homogeneous sets can be obtained in time linear in the size of the graph. In order to find a crossing P_4 between a homogeneous set Z and $V - Z$ it suffices to consider the graph where Z is replaced by one single vertex z (recall that each crossing P_4 has precisely one vertex in Z). A P_4 containing z is a crossing P_4 in the original graph. The following procedure finds such a P_4 if one exists. If the first step is successful then we obtain a P_4 having z as an endpoint, otherwise z is a midpoint.

Procedure Find_$P_4(G;z)$

Input: A graph $G = (V, E)$ and a vertex $z \in V$.
Output: The vertex-set of a P_4 containing z.

1. Find the connected components $X_1, X_2, \ldots, X_r$ of $V - \{z\} - N(z)$ in G.

  If there is a vertex $u \in N(z)$ which is partially adjacent to a set X_i **then**

    Find adjacent vertices $v, w \in X_i$ with $uv \in E$ and $uw \notin E$.

    STOP with the set $\{z, u, v, w\}$.

2. Find the connected components $Y_1, Y_2, \ldots, Y_s$ of $N(z)$ in $\overline{G}$.

If there is a vertex $u \notin N(z)$ which is partially adjacent to a set Y_i **then**

Find nonadjacent vertices $v, w \in Y_i$ with $uv \in E$ and $uw \notin E$.

STOP with the set $\{z, u, v, w\}$.

The connected components and a shortest path between a pair of vertices in a graph $G = (V, E)$ are usually computed in time $O(|V| + |E|)$ using Breadth First Search. The following procedure performs Breadth First Search in $\overline{G}$.

Procedure BFS($\overline{G}$;v)

Input: Adjacency lists of a graph $G = (V, E)$ and a start vertex $v \in V$.
Output: A partition of V into layers.

1. Let $V_0 := \{v\}$, $R := V - V_0$ and $i := 0$.

2. While $R \neq \emptyset$ **do**

Compute $V_{i+1} := \{u \in R \,|\, u \text{ not totally adjacent to } V_i \}$.

Let $R := R - V_{i+1}$, $i := i + 1$.

Using the adjacency lists of the vertices from V_i, we can find the vertices from R which are adjacent to V_i and compute the number of their neighbors in V_i. In order to check whether a vertex u is totally adjacent to V_i we only have to compare this number to the cardinality of V_i. This shows that V_{i+1} can be computed in time $O(\sum_{w \in V_i} |N(w)|)$. As a consequence, the procedure is of complexity $O(|V| + |E|)$.

It is now straightforward how to find the connected components of $\overline{G}$ in time linear in the size of G. If $\overline{G}$ is connected, then a shortest path between a pair of vertices x, y in $\overline{G}$ is constructed as follows. Start the procedure with one of the vertices, say x. Let $y = v_k \in V_k$ for some $k \geq 1$. Now, for $i = k, k-1, \ldots, 1$, find a nonneighbor v_{i-1} of v_i in V_{i-1}. Obviously, this again can be done in time $O(|V| + |E|)$.

With these ingredients it is standard routine to realize an implementation of algorithm CHAIN which requires time linear in the size of the input. $\square$

Acknowledgment. This work was done, in part, while the first author was visiting Old Dominion University.

30

References

1. L. Babel, S. Olariu, On the isomorphism of graphs with few P_4s, in: M. Nagl, ed., *Graph-Theoretic Concepts in Computer Science, 21th International Workshop, WG'95*, Lecture Notes in Computer Science 1017, 24–36, Springer-Verlag, Berlin 1995.

2. J.A. Bondy, U.S.R. Murty, *Graph Theory with Applications,* North-Holland, Amsterdam, 1976.

3. V. Chvátal, A semi-strong perfect graph conjecture, *Annals of Discrete Mathematics* 21 (1984) 279–280.

4. V. Chvátal, Star-cutsets and perfect graphs, *Journal of Combinatorial Theory (B)* 39 (1985) 189–199.

5. V. Chvátal, On the P_4-structure of perfect graphs III. Partner decompositions, *Journal of Combinatorial Theory (B)* 43 (1987) 349–353.

6. A. Cournier, M. Habib, A new linear time algorithm for modular decomposition, *Trees in Algebra and Programming*, Lecture Notes in Computer Science 787, 68–84, Springer-Verlag 1994.

7. M.C. Golumbic, *Algorithmic Graph Theory and Perfect Graphs,* Academic Press, New York, 1980.

8. B. Jamison, S. Olariu, P_4-reducible graphs, a class of uniquely tree representable graphs, *Studies in Applied Mathematics* 81 (1989) 79–87.

9. B. Jamison, S. Olariu, On a unique tree representation for P_4-extendible graphs, *Discrete Applied Mathematics* 34 (1991) 151–164.

10. B. Jamison, S. Olariu, A unique tree representation for P_4-sparse graphs, *Discrete Applied Mathematics* 35 (1992) 115–129.

11. B. Jamison, S. Olariu, p-components and the homogeneous decomposition of graphs, *SIAM Journal on Discrete Mathematics* 8 (1995) 448–463.

12. R. McConnell, J. Spinrad, Linear-time modular decomposition and efficient transitive orientation of comparability graphs, *Fifth Annual ACM-SIAM Symposium of Discrete Algorithms* 536–545, 1994.

13. R. H. Möhring, Algorithmic aspects of comparability graphs and interval graphs, in: I. Rival, ed., *Graphs and Orders,* Dordrecht, Holland, 1985.

14. S. Olariu, On the structure of unbreakable graphs, *Journal of Graph Theory* 15 (1991) 349–373.

15. B.A. Reed, A semi-strong perfect graph theorem, *Journal of Combinatorial Theory (B)* 43 (1987) 223–240.

Node Rewriting in Hypergraphs *

Michel Bauderon, Hélène Jacquet

Laboratoire Bordelais de Recherche en Informatique
Université Bordeaux I
33405 Talence Cedex
email : (bauderon,jacquet)@labri.u-bordeaux.fr

Abstract. Pullback rewriting has recently been introduced as a new and
unifying paradigm for vertex rewriting in graphs. In this paper we show
how to extend it to describe in a uniform way more rewriting mechanisms
such as node and handle rewriting in hypergraphs.

1 Introduction

After more than twenty years of (hyper)graph rewriting, a large number of
rewriting mechanisms have been introduced which generate various classes of
(hyper)graphs with different properties. A huge number of papers has been
devoted to the classification and comparison of all those rewriting techniques,
proposing various kind of encoding - sometimes fairly complicated - to help compare different classes of languages. Quoting them would significantly increase the
size of this paper but we can refer the reader to at least [7, 10, 22, 18].

Still, since most of those works propose their own mechanism, their own formalism - in general an *ad hoc* set-theoretic one - this comparison is quite difficult
and the necessity of a unifying framework is getting clearer and clearer. Of the
proposals for such a framework, the double pushout approach has probably been
the most successful, although it can deal only with a limited number of cases.

In a series of recent papers by the first author ([1, 2, 3], it has been shown how
a new categorical paradigm - *pullback rewriting* - can be used to provide a unifying and - we think - elegant description of various graph rewriting mechanisms
such as :

- node label controlled rewriting [16], neighbourhood controlled embedding
  [17] (for node rewriting in graphs),
- double pushout [13] or hyperedge rewriting [4, 15] (for edge or hyperedge
  rewriting in hypergraphs).

While filling a gap in the general pattern of graph rewriting [2] since no categorical rewriting mechanism (similar to double pushout for edge rewriting) was
available for node rewriting, it actually stands as a good candidate for a universal
generating mechanism.

---

* This work has been supported by the Esprit BRA "Computing with graph
transformations"

This paper is a new step towards the elaboration of a uniform theory of graph rewriting, following the lines described above. We shall be interested here in node rewriting in hypergraphs which is - to our knowledge - an almost untouched issue (see [19]) and shall briefly indicate how it can be used to describe as well (hyper)handle rewriting (replacement of a hyperedge together with all its incident vertices in an NCE-like way [9, 18]).

Before going into the details, we would like to emphasise one of the peculiarities of our framework : our notion of a labelling. Indeed, in all classical approaches to rewriting (words, trees, graphs or hypergraphs), a fundamental role is played by a notion of labelling and by the alphabet in which the labels are taken. Labels are names which are put on the basic items which constitute the object to be rewritten, in order to characterise the behaviour of each item through each rewriting step - in a uniform way since two items with the same label normally behave exactly in the same way. The labelling (which may change or not through the rewriting) is then given by a simple mapping from the object into an alphabet set A which associates a unique letter to each item of the object we want to label. The alphabet is generally completely unstructured, at most it may be ranked (see for instance [4]).

Our notion of a labelling will be completely different, since we need a more precise way to name the items in such a way that we shall be able to distinguish not only the item itself but also its neighbourhood and the rest of the graph. This is why we shall use as an alphabet not a mere set but a very structured object - namely a graph -, as a label not a letter but a graph morphism and as a labelling, not a mapping but a (coherent) family of such morphisms. By the way, the use of such a complicated labelling will make totally useless that of classical labels - yet they will sometimes appear in drawings, simply to facilitate the understanding.

Although this approach is quite new, it must be noted that it has recently been used independently by other authors in some - totally unrelated - works on graph colourings respecting orientation [21]. In the same way, it seems that products and pullbacks are becoming more and more popular after years of omnipresence of sums and pushouts [6].

The paper is organised as follows. Section 2 sets the basic definitions we choose for hypergraphs and hypergraph morphisms and describes the basic rewriting mechanism we shall use : *pullback rewriting*. Node rewriting is then developed in sections 3 and 4. We are quite aware that our approach may look rather unfamiliar to many readers. However, it is quite impossible to be anything like self-contained in such a short space and we have only recalled the basic definitions we need, referring the reader to other papers for more details. As a consequence, this paper contains mainly definitions and examples and the only important result we give is stated in a rather informal way. A full version of this work is in preparation [5].

2   A Category for Hypergraphs

Let $\mathbb{N}$ denote the set of non negative integers and $\mathbb{N}_+$ the set of positive ones. In the field of hypergraph rewriting, several definitions have been given and used of what actually is a hypergraph (see for instance [4, 15]). As we shall see in the sequel, for us a hypergraph will merely be a bipartite graph (up to some details) and we shall use graph morphisms between hypergraphs, *not* hypergraphs morphisms in the usual sense. Let us then choose a fairly simple notion of hypergraph (which could easily be enriched) and show how we treat it as a graph.

To make understanding slightly easier for readers unfamiliar with products and pullbacks, simple examples may be found in [2]. For more details on category theory we refer the reader to standard textbooks such as [20].

Definition 1. A *hypergraph* is a pair $G = \langle S_G \cup H_G, E_G \rangle$ where $S_G \cup H_G$ such as $S_G \cap H_G = \emptyset$ is the set of vertices and $E_G \subseteq S_G \times H_G$ is the set of edges. An edge between vertices u and v will be denoted by $[u, v]$.

We shall use the words of *node* (or *vertex*) for the elements of S_G, *hyperedge* for those of H_G and call *edges* the elements of E_G (which are often called *tentacles* in the literature).

For each hyperedge $e \in H_G$, we denote by $vert_G(e)$ the set of all its adjacent nodes. We shall say that the hyperedge $e \in H_G$ *links* the vertices of $vert_G(e)$ and that any element of $vert_G(e)$ *belongs* to e.

A vertex (or a node) is *isolated* if it belongs to no hyperedge ; a hyperedge $e \in H_G$ with an empty sequence of $vert_G(e)$ is an isolated hyperedege. The number of neighbours of a hyperedege e is the arity of e.

By $\dot{u}$ we denote an item (a vertex or a hyperedge) and all its neighbours (resp. incident hyperedges and incident vertices).

It is clear that this definition of hypergraphs corresponds to the standard coding of a hypergraph as a bipartite graph. Figure 1 represents [2] a hypergraph with

Fig. 1. An example of hypergraph.

3 vertices and 4 hyperedges : arities of the hyperedges are $3, 2, 1$ and 0 (there is one isolated hyperedge).

---

[2] In drawings, dots represent vertices and squares stand for hyperedges.

So far we have kept to the classical definition for hypergraphs (or bipartite graphs), but we shall diverge when choosing morphisms since they will not respect the arities of the hyperedges.

Definition 2. A hypergraph morphism $h : G \to G'$ is a pair $h = \langle h_V, h_E \rangle$ with

$$h_V : S \to S' \quad \text{and} \quad h_E : E \to E'$$
$$H \to H'$$

such that $h_E([u, v]) = [h_V(u), h_V(v)]$.

Clearly, hypergraph morphisms defined in definition 2 are graphs morphisms in the usual sense, but graph morphisms which respects the bipartition. It is easy to verify that the good properties of hypergraph morphisms (from definition 2) turn the set of hypergraphs (according to the definition 1) into a category that we shall denote by $\mathcal{HG}$.

Proposition 3. *The category $\mathcal{HG}$ has arbitrary limits. In particular, $\mathcal{HG}$ has products and pullbacks. The hypergraph with one node, one hyperedge and one edge is a terminal object simply denoted by $\odot$: it is a neutral element for the product.* □

Let us simply recall that if G_1 and G_2 are two graphs, their *categorical product* $G_1 \times G_2$ is defined by its sets of vertices $S \cup H$ and edges E, in the following way :

- $S = S_1 \times S_2$ and $H = H_1 \times H_2$
- $E = \{[u_1 u_2, v_1 v_2] | [u_1, v_1] \in E_1 \wedge [u_2, v_2] \in E_2\}$.

The pullback of two graph morphisms $f_i : G_i \longrightarrow F, i = 1, 2$ is a pair of arrows $h_i : G \longrightarrow G_i, i = 1, 2$ where G is the subgraph of the product consisting of exactly those items (nodes and edges) on which $f_i \circ \pi_i$ coincide.

It is easily checked that $\odot$ is a unit. Note that the unique vertex of $\odot$ creates all the vertices of the product, while the unique hyperedge creates all the hyperedges and the unique edge creates all the edges. This explains the importance of what we call "vertices adjacent to one hyperedge of degree one" in all what follows. Of course their role is mainly technical. Note also that if F is $\odot$, the pullback is equal to the product.

Given those definitions and results, the basic framework that we want to use for hypergraph rewriting is the following :

Definition 4. A *production rule* is a hypergraph morphism $p : R \longrightarrow L$, where L is called the left-hand side of the rule, while R is the right-hand side. An *occurrence* of the left-hand side of p in the hypergraph G is a morphism $x : G \longrightarrow L$. The rewriting of G by p at occurrence x is the pullback of x and p.

Clearly, this definition is a little bit too general to be easily studied and will need specialising in particular through an appropriate choice of the hypergraph L which relates the two morphisms. This is the hypergraph that we shall call the *alphabet* .

3 Node rewriting in hypergraphs

3.1 The alphabet

Let us now fix the alphabet hypergraph $\mathcal{A}$, which plays a central role in the rewriting mechanism. It structure is guided by a very simple idea : build a hypergraph whose property will enable us to define morphisms which "label" the hypergraphs in such a way that we can distinguish the items to be rewritten, the context of the rewriting and the interface between them. It will therefore always contain three main "areas" which shall correspond to those three objectives.

Definition 5. The *alphabet hypergraph* $\mathcal{A}$ is the infinite hypergraph with

- vertices : $S_{\mathcal{A}} \cup H_{\mathcal{A}}$ where,
  - $S_{\mathcal{A}} = \{(i, s) | i \in \mathbb{N}\} \cup \{(-i, s) | i \in \mathbb{N}_+\}$
  - $H_{\mathcal{A}} = \{(i, h) | i \in \mathbb{N}\} \cup \{(-i, h) | i \in \mathbb{N}_+\}$
- edges : $E_{\mathcal{A}}$ is the union of the following sets of edges :
  - $\{[(j, s), (1, h)] \mid j \in \mathbb{N}_+\}$
  - $\{[(j, s), (i, h)] \mid i, j \in \mathbb{N}_+\}$
  - $\{[(-j, s), (i, h)] \mid i, j \in \mathbb{N}_+\}$
  - $\{[(-i, s), (-i, h)] \mid i \in \mathbb{N}_+\}$

In other words, $\mathcal{A}$ will contains the following "areas" :

- items composed by nodes $(-i, s), (-i, h)$ (for fixed $i \in \mathbb{N}_+$) (which will model the unknowns) ; we will called this sub-hypergraph the *unknown part* of $\mathcal{A}$,
- nodes (i, s) (for $i \in \mathbb{N}_+$) which are the neighbours,
- nodes (i, h) (for $i \in \mathbb{N}_+$) which represent the hyperedges linking the unknowns to their neighbours,
- item composed by nodes $(0, s), (0, h)$ which will model the context ; we will called this item the *context part* of $\mathcal{A}$.

With regard to edges, the context is linked to all the neighbours, neighbours and hyperedges $\{(i, h) | i \in \mathbb{N}_+\}$ form a complete bipartite graph and it is also the case for nodes $\{(i, h) | i \in \mathbb{N}_+\}$ and the nodes of the unknown part of $\mathcal{A}$. Such an

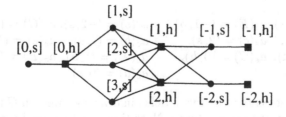

Fig. 2. The alphabet hypergraph $\mathcal{A}_{3,2,2}$

alphabet allows us to take into account an arbitrary number of distincts letters and variables. If we only need a finite number m of neighbours, n of hyperedges and t of variables we can restrict to its subgraph $\mathcal{A}_{m,n,t}$. The hypergraph $\mathcal{A}_{3,2,2}$ is represented on figure 2.

We now define the notion of an unknown on a hypergraph G as a certain kind of morphism from G to $\mathcal{A}$ which will "colour" the vertices of G by vertices of $\mathcal{A}$ in such a way as to distinguish the node to be replaced, its neighbours and the context.

Definition 6. Let $G = \langle S_G \cup H_G, E_G \rangle$ be a hypergraph and u be a node of S_G which is adjacent to one hyperedge $e \in H_G$ of degree one. A hypergraph morphism $a_{(G,u)} : G \longrightarrow \mathcal{A}$ is an *unknown* on u in G if there exists a unique integer $x \in \mathbb{N}_+$ such that :

- $a_{(G,u)}^{-1}(-x,s) = \{u\}$ and $a_{(G,u)}^{-1}(-x,h) = \{e\}$ and for all $i \neq x$, $i \in \mathbb{N}_+$, $a_{(G,u)}^{-1}(-i,s), a_{(G,u)}^{-1}(-i,h)$ are empty,
- either $a_{(G,u)}^{-1}(i,s)$ for $i \in \mathbb{N}_+$ is empty or it consists only of immediate neighbours of u,

The vertex u is called an unknown of type x.

Intuitively, an x-label $a_{(G,u)}^{-1}$ on u distinguishes between the unknown u, its immediate neighbours - vertices which are mapped to the vertices $\{(i,s) \mid i \in \mathbb{N}_+\}$ -, the adjacent hyperedges which are mapped to the vertices $\{(i,h) \mid i \in \mathbb{N}_+\}$ and the rest of the hypergraph mapped onto the context part of $\mathcal{A}$: in addition it associates to an unknown a type that we denote by $\tau(a_{(G,u)})$.

Definition 7. Let $G = \langle S_G \cup H_G, E_G \rangle$ be a hypergraph and lab_G a subset of S_G such as each element of lab_G is adjacent to one hyperedge of degree one, the pair $\langle G, lab_G \rangle$ is a *labelled hypergraph* if there is a unique x-label on each of its vertices which appear in lab_G.

Example 1. As an exemple, let us consider the labelled hypergraph G with $S_G = \{b, c, d, e\}$, $H_G = \{X, C, E\}$, $E_G = \{[e, X], [b, X], [c, X], [e, E], [e, C]\}$ and the set $lab_G = \{e, c\}$. We may for instance define on G a 1-label on e and a 2-label on c (for short a_e and a_c) such that :

$$
\begin{array}{ll}
a_e(e) = (-1, s), \ a_e(E) = (-1, h), & a_c(c) = (-2, s), \ a_c(C) = (-2, h), \\
a_e(b) = (1, s), \quad a_e(c) = (2, s), & a_c(b) = (3, s), \quad a_c(e) = (1, s), \\
a_e(X) = (1, h), \ a_e(C) = (0, h), & a_c(X) = (1, h), \ a_c(E) = (0, h), \\
a_e(d) = (0, s), & a_c(d) = (0, s).
\end{array}
$$

Figures 3 and 4 represents those two labels : in drawings, item in G projects onto one item which is in the same column. Note that graph G has been represented in two different ways in the upper row of Figures 3 and 4.

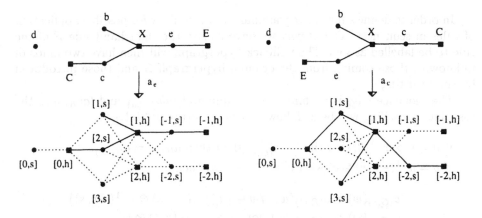

Fig. 3. The 1-label on e.   **Fig. 4.** The 2-label on c.

3.2 The rules

Definition 8. A *VRinH-rule* (for Vertex Replacement in Hypergraphs) is a morphism $r : R \longrightarrow \mathcal{A}$ if there exists a unique integer $x \in \mathbb{N}_+$ such that :

- $[-x, s]$ has an inverse image by r,
- each item u of the set $\{(i, s) | i \in \mathbb{N}\} \cup \{(0, h)\}$ is such as $r^{-1}(u)$ has a unique element,
- either $r^{-1}(i, h)$ for $i \in \mathbb{N}_+$ is empty or it consists of a set of hyperedges,
- if R is a labelled hypergraph, all the elements of lab_R project on $(-x, s)$.

We say that x is the *type* of the rule r and that R is its *right hand side* (although it is on the left).

Let $\underline{G}$ be a hypergraph such that $G \xleftarrow{a'} \underline{G} \xrightarrow{r'} R$ is the pullback of $R \xrightarrow{r} \mathcal{A} \xleftarrow{a} G$. We denote $\{a^{-1}(x) \otimes r^{-1}(x)\}$ the set of vertices v of $\underline{G}$ such that $r(r'(v)) = a(a'(v)) = x$ and $x \otimes y$ denotes the vertex $v \in \underline{G}$ such that $a'(v) = x$, $r'(v) = y$ and $a(x) = r(y)$. The rewriting mechanism is now defined by the following rule.

Definition 9. Let $a_{(G,u)}$ be a x-label on u in G and r be a VRinH-rule of type x : the application of r to G at $a_{(G,u)}$ is the pullback of the pair $(a_{(G,u)}, r)$. We let $\underline{G}$ denote the hypergraph built as a pullback.

Intuitively, the inverse image of the item described by nodes $(-x, s), (-x, h)$ and the edge $[(-x, s), (-x, h)]$ is the hypergraph to be substituted to the rewritten node, the connection relation is described by the items which project on $\mathbb{N}_+ \times \{s, h\}$ and the corresponding edges, the inverse image of the context part of $\mathcal{A}$ is the sub-graph of G which will not be affected by the rewriting and $a^{-1}(i, s)$ for $i \in N^+$ are those neighbours which shall be connected to "new hyperedges" after the rewriting.

In order to define a notion of grammar, we must allow for possible application of rules in sequences. This is why we have chosen the right hand side R of the rule to be labelled as well. The pullback hypergraph will then have two kinds of unknowns, those coming from the original hypergraph G and those introduced by the rewriting.

The sequence $lab_{\underline{G}}$ is nothing but the union of $lab_{G-\{\dot{u}\}}$ and lab_R, and the associated labelling will be as follow. For each vertex $v \in lab_{\underline{G}}$,

- if $v = v' \otimes r^{-1}(0,s)$ with $v' \in a_{(G,u)}^{-1}(0,s)$ then $a_{(\underline{G},v)} = a_{(G,v')}$.
- if $v = a^{-1}(-x,s) \otimes v'$ with $v' \in r^{-1}(-x,s)$ then

$$a_{(\underline{G},v)}(w) = a_{(R,v')}(w) \quad \forall w \in \{a_{(G,u)}^{-1}(-x,s) \otimes r^{-1}(-x,s)\}$$
$$a_{(\underline{G},v)}(w) = a_{(R,v')}(w') \quad \text{for} \ \ w \in a^{-1}(r(w')) \otimes w'$$
$$\text{and} \ \ r(w') = (i,s) \ \text{for} \ \ i \in \mathbb{N}_+$$

4 Applications

4.1 VRinH grammars.

The previous definitions can now be extended to that of a grammar, that we shall call VRinH.

Definition 10. A VRinH-grammar is a system $VR\!g = \langle A, Z, P \rangle$ where :

1. A is the alphabet hypergraph,
2. $Z \in \mathcal{HG}$ is the axiom : Z is a labelled hypergraph,
3. P is a finite set of pair (R, r) such that $R \xrightarrow{r} A$ is a VrinH-rule.

Letting $\to$ denote a step of direct derivation - the pullback of an unknown and a rule (section 3.2) -, the transitive and reflexive closure of $\to$ is denoted by $\xrightarrow{*}$ and a sequence of i direct derivation steps by $\xrightarrow{i}$.
A hypergraph $\tilde{K}$ such that $Z \xrightarrow{*} \tilde{K}$ is called a sentential form of $VR\!g$: a terminal hypergraph K is a sentential form of $VR\!g$ such that $|lab_K| = 0$. The language generated by $VR\!g$, $\mathcal{L}(VR\!g)$, is the set $\{K \in \mathcal{HG} / Z \xrightarrow{*} K \wedge |lab_K| = 0\}$. To give a simple example, let us consider the special case where each sentential form K has a unique unknown.

Definition 11. A VRinH grammar $VR\!g = \langle A, Z, P \rangle$ is a linear grammar if both $|lab_Z| = 1$ and for all pair $(R, r) \in VR\!g$, we have $|lab_R| = 1$.

Example 2. The following grammar [3] $VR\!g$ generates the set of all rank- and degree-unbounded labelled hypergraphs of the form shown in figure 5. It only requires two distinct unknowns of types 1 and 2. For clarity, the first is coloured by a, the second by b and its "adjacent hyperedge of degree one" are denoted

---

[3] This grammar has been borrowed from [18].

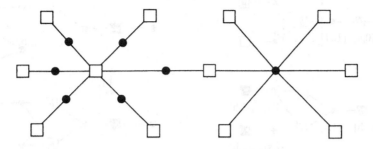

Fig. 5. A generated labelled hypergraph.

by □; in the same way, # and ⋆ denote two distinct kind of hyperedges [4].

One must keep in mind that VRg is a linear grammar, which implies that for each hypergraph K (a sentential form, or a hypergraph such that $(K, r) \in P$), there is a unique vertex in lab_K. Its label - the graph morphism named a or b - is built up so that the unknown is mapped to $(-1, s)$ or $(-2, s)$ and its respectively "adjacent hyperedge of degree one" to $(-1, h)$ and $(-2, h)$, the other adjacent hyperedges are mapped to the hyperedges of $\mathcal{A}_{1,2,2}$ (denoted by # and ⋆), the neighbours are mapped onto $[1, s]$ and the rest of the graph onto items $(0, s), (0, h)$(wheather it is a node or a hyperedge).

The axiom is a 1-labelled hypergraph $Z = \;$ ⬛#━━━●━━━⬛ , and P consists of the four VRinH-rules given in figure 6.
Our drawing of rewriting rules respects some very strict conventions which should make their interpretation straight forward.
First of all, item in the upper part of a rule projects onto an item which is in the same column. Second, a node with a certain label (# or ⋆) always projects on a node with the same label (we only use labels to show which node projects where).

With those conventions, in a labelled hypergraph R_i such that $r_i : R_i \rightarrow \mathcal{A}_{1,2,2}$ we distinguish three parts (as shown by the vertical dotted line on the drawing of r_1 in figure 6.**a** :

- the rightmost sub-hypergraph of R_i which is the graph which will be substituted to the rewritten node. Its image in $\mathcal{A}$ is drawn with solid lines,
- the central part which depicts the connection relation. Hyperedges of this sub-hypergraph are mapped onto nodes # or ⋆,
- and at least the sub-hypergraph composed by nodes $(0, s), (0, h)$ and $(1, s)$ which are mapped to the corresponding nodes $(0, s), (0, h)$ and $(1, s)$ of the alphabet $\mathcal{A}$.

---

[4] Remember that we do not really need those "labels" since the labelling of node are provided by morphisms into $\mathcal{A}_{1,2,2}$, but we shall use then to make the drawings slightly more intuitive.

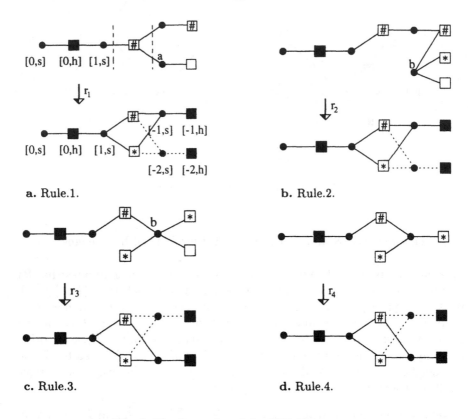

a. Rule.1. **b.** Rule.2.

c. Rule.3. **d.** Rule.4.

Fig. 6. The four rules of the VRinH-grammar.

As an example figure 7 describes $Z \xrightarrow{r_1} \underline{Z}$ as the pullback of $(a_{(Z,a)}, r_1)$.

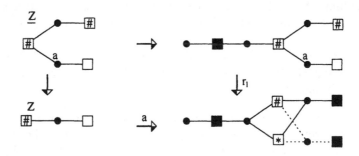

Fig. 7. A rewriting step.

4.2 Single and double pullback rewriting

It was shown in [2] how two steps of pullback rewriting could be combined into what was called *double pullback rewriting* because of its similarity with double pushout rewriting (actually a duality in the category theoretic sense) and that, within the framework described in that paper, double pullback rewriting was a category theoretic model for NCE-rewriting in graphs in the sense of [17], while single pullback rewriting modelled NLC-rewritings [16].

A similar construction can be done here, but for hypergraphs. It must by now be clear that single pullback rewriting in $\mathcal{HG}$ gives a reasonable definition of what node rewriting in a hypergraph should be. Then of course, double pullback rewriting can be defined along the lines of [2]. Definitions and examples given so far should make it clear that it will describe the rewriting of a sub-hypergraph into another one with a connection defined in an NCE way.

Rewriting (hyper)handles - namely hyperedges together with their adjacent vertices - has been proposed in several papers (e.g. [9, 18]) as a paradigm to rewrite hyperedges in hypergraphs with an NLC- or rather NCE-like way of connecting the rewritten part to the context in the original graph. This is clearly a special case of double pullback rewriting.

The basic result of this paper could be the following :

Theorem 12. *Let $\mathcal{HG}$ be the category of hypergraphs and $\mathcal{A}$ be the alphabet described in Section 3.1. Single pullback rewriting models node rewriting in hypergraphs in the sense of [19] while double pullback rewriting models NCE-subhypergraph rewriting in general and hyperhandle rewriting in the sense of [9, 18] in particular.*

Of course, within the size limits of this paper this result is more a claim than a theorem : we can not even recall the definitions found in the quoted papers. Still, although the complete proof is quite long it is very easy to convince oneself that it is true by simply trying to encode some of the examples found there.

5 Conclusion

This paper is a new step in the development of a global theory of rewriting items in graphs and hypergraphs using a single mechanism, pullback rewriting. We have shown that it could be used to describe a new kind of rewriting - nodes in hypergraphs - and have indicated how its double pullback variation would generate much larger classes of languages including those generated by (hyper)handle rewriting.

A lot of work remains to be done to reach a precise picture of the situation and we shall simply conclude with figure 8 which summarises the relationship between the various kind of rewriting studied so far.

Type of rewriting	Graph	Hypergraph
Single Pullback	NLC-rewriting	Node rewriting
Double Pullback	NCE-rewriting	Sub hypergraph, Handle rewriting

Fig. 8. Vertex replacement in graphs and hypergraphs

References

1. M. Bauderon, A categorical approach to vertex replacement : the generation of infinite graphs, *5th International Conference on Graph Grammars and their applications to Computer Science, Williamsburg, November 1994, Lect. Notes in Comp. Sci. N 1073*, 1996, 27-37.

2. M. Bauderon, A uniform approach to graph rewriting : the pullback approach, in Proceedings WG'95, *Lect. Notes in Comp. Sci n 1017*.

3. M. Bauderon, Parallel rewriting through the pullback approach, in SEGRA-GRA'95, to appear, *Elect. Notes in Theor. Comp. Sci.*

4. M. Bauderon, B. Courcelle, Graph expressions and graph rewriting, *Math. Systems Theory* 20 (1987), 83-127

5. M. Bauderon, H. Jacquet, Node rewriting in hypergraphs (full version), *Submitted.*

6. A. Bottreau, Y. Métivier, Kronecker product and local computations in graphs, *CAAP'96, Lect. Notes in Comp. Sci. N 105*, 1996, 02-16.

7. B. Courcelle, An axiomatic definition of context-free rewriting and its applications to NLC graph grammars, *Theor. Comput. Sci* 55 (1987), 141-181.

8. B. Courcelle, Graph rewriting : an algebraic and logic approach, in *J. Van Leeuwen, ed, Handbook of Theoretical Computer Science*, Vol. B (Elsevier, Amsterdam, 1990) 193-242

9. B. Courcelle, J. Engelfriet, G. Rozenberg, Handle-Rewriting Hypergraph Grammars, Jour. Comp. Sys. Sci. 46, (1993), 218-270

10. J. Engelfriet, G. Rozenberg, A comparison of boundary graph grammars and context-free hypergraph grammars, *Inf. Comp.* 84, 1990, 163-206

11. J. Engelfriet, G. Rozenberg, Graph grammars based on node rewriting : an introduction to NLC grammars, *Lect. Notes in Comp Sci* N 532, 1991, 12-23.

12. J. Engelfriet, L. Heyker, G. Leih, Context free graph languages of bounded degree are generated by apex graph grammars, *Acta Informatica* 31, 341-378 (1994)

13. H. Ehrig, Introduction of the algebraic theory of graph grammars, in Graph Grammars and their applications to Computer Science, *Lect. Notes in Comp Sci* N 73, 1979, 1-69

14. M. Farzan, D.A. Waller, Kronecker products and local joins of graphs, *Can. J. Math.* Vol. XXIX, No 2 1977, 255-269

15. A. Habel, Hyperedge Replacement : Grammars and Languages, *Lect. Notes in Comp Sci* N 643, 1992.

16. D. Janssens, G. Rozenberg, On the structure of node-label-controlled graph languages, *Inform. Sci.* 20 (1980), 191-216.

17. D. Janssens, G. Rozenberg, Graph grammars with neighboorhood-controlled embedding, *Theor. Comp. Sci.* 21 (1982), 55-74.

18. C. Kim, T.E. Jeong, HRNCE Grammars, -A hypergraphs generating system with an eNCE way of rewriting, *5th International Conference on Graph Grammars and*

their applications to Computer Science, Williamsburg, November 1994, Lect. Notes in Comp. Sci. 1073, 383-396.

19. R. Klempien-Hinrichs, Node Replacement in Hypergraphs : Simulation of hyperedge replacement and decidability of confluence, *5th International Conference on Graph Grammars and their applications to Computer Science, Williamsburg, November 1994, Lect. Notes in Comp. Sci.* 1073, 397-411.

20. S. McLane, *Categories for the working mathematician*, Springer, Berlin, 1971.

21. E. Sopena, The chromatic number of oriented graphs, Research Report 1083-95, LaBRI, Bordeaux

22. W. Vogler, On hyperedge replacement and BNLC graph grammars, *Discrete Applied Mathematics* 46 (1993) 253-273.

On κ-partitioning the n-cube *

Sergej L. Bezrukov

Department of Math. and Comp. Sci.
University of Paderborn
D-33095 Paderborn, Germany

Abstract. Let an edge cut partition the vertex set of the n-cube into k subsets $A_1, ..., A_k$ with $||A_i| - |A_j|| \leq 1$. We consider the problem to determine minimal size of such a cut and present its asymptotic as $n, k \to \infty$ and also as $n \to \infty$ and k is a constant of the form $k = 2^a \pm 2^b$ with $a \geq b \geq 0$.

1 Introduction

Graph partition problems often arise in solving large problems on multiprocessor computing systems. The solution of differential equations, using the finite elements method requires partitioning the area into some simple figures (e.g. triangles or rectangles) and then assigning the nodes of the obtained network to processors of a computing system. The assignment should be made in such a way that the load of each processor is possibly equal (i.e. almost the same number of network nodes should be assigned to each processor). Moreover, the data exchange between the processors should be minimized in order not to affect the speed of solution. Thus, one comes to a problem of partitioning the underlying network into equal parts by cutting possibly small number of edges.

A modification of this problem is in minimization of the maximum of the number of edges connecting a part of the network with another parts of the partition. Such a problem arises in electronics to satisfy the pin requirements of VLSI design [3].

Most of the graphs appearing in applications are complicated enough and highly irregular, which makes evaluating partition quality difficult. Many theoretical papers deal with the design of partition algorithms (cf. [4, 5, 6]) and the analysis of their optimality [1], as well as with the complexity of partition problems. For the evaluation of the quality of partition algorithms it is helpful to know exact results concerning partitioning of some graph classes, or at least good lower bounds for parameters of a partition. Results of such types exist for partitioning graphs into two parts [8, 9] and for some special graphs [2, 3].

In our paper, we study partitioning the n-cube into $k > 2$ parts, which after all is an interesting combinatorial problem. It has been known for a long time

---

* This work was partially supported by the German Research Association (DFG) within the project SFB 376 "Massive Parallelität: Algorithmen, Entwurfsmethoden, Anwendungen".

how to partition this graph into two equal parts with respect to the number of vertices, cutting as few edges as possible (bisection width). It is also easy to partition it into 2^a parts. For arbitrary k it is known that the minimal number of cutting edges is $\Theta(2^n \log k)$ [3, 10]. In our paper, we strengthen this result presenting asymptotic of the minimal number of cutting edges.

Consider the set $Q^n = \{(x_1, ..., x_n) \mid x_i \in \{0,1\}, i = 1, ..., n\}$. For $u, v \in Q^n$ the Hamming distance $\rho(u, v)$ is defined as the number of entries where u and v differ. We use the standard definition of the n-cube as a graph on the vertex set Q^n, where two vertices u, v are adjacent iff $\rho(u, v) = 1$. Throughout the text we also use the standard definition of a subcube of Q^n of dimension t, $0 \le t \le n$, and call its vertex set the face of dimension t.

For fixed n and k, consider a partition of Q^n into k parts A_i, $i = 1, ..., k$, satisfying $||A_i| - |A_j|| \le 1$, or in other words

$$\left\lfloor \frac{2^n}{k} \right\rfloor \le |A_i| \le \left\lceil \frac{2^n}{k} \right\rceil. \tag{1}$$

For such a partition $\mathcal{A} = \{A_1, ..., A_k\}$ denote $\nabla \mathcal{A} = \{(u, v) \mid u \in A_i, v \in A_j, i \ne j, \rho(u, v) = 1\}$. We say that a partition $\mathcal{A}$ is *minimal* (with respect to given n, k) if $|\nabla \mathcal{A}|$ is minimal possible and denote $\nabla(n, k) = \min_{\mathcal{A}} |\nabla \mathcal{A}|$.

We deal with the problem of finding the asymptotic of $\nabla(n, k)$ with k fixed and $n \to \infty$, however our methods also provide exact results at least for small k. The paper is organized as follows. In the next section we mention some known facts and present auxiliary results used throughout the paper. In Section 3 we present bounds for $\nabla(n, k)$, which differ in a multiplicative constant. Sections 4 and 5 are devoted to the asymptotic of $\nabla(n, k)$ in the cases when k is represented by the sum or the difference of two powers of 2 respectively. Some remarks conclude the paper in Section 5.

2 Optimal and quasioptimal subsets

Let $A \subseteq Q^n$. Denote $\partial A = \{(u, v) \mid u \in A, v \in Q^n \setminus A, \rho(u, v) = 1\}$. We call a set A optimal if $|\partial A| \le |\partial B|$ for any $B \subseteq Q^n$, $|B| = |A|$. Introduce the *lexicographic number* $\ell(u)$ of a vertex $u = (x_1, ..., x_n) \in Q^n$ defined by $\ell(u) = \sum_{i=1}^n x_i 2^{n-i}$. Denote $L_m^n = \{u \in Q^n \mid 0 \le \ell(u) < m\}$. We say that subsets $A, B \subseteq Q^n$ are *congruent* (denotation $A \cong B$) if B is the image of A in some automorphism of Q^n.

Theorem 1 (Harper [7]). *L_m^n is an optimal set for any $m = 1, ..., 2^n$. Moreover, if A is an optimal subset of Q^n with $|A| = m$, then $A \cong L_m^n$.*

Thus, optimal subsets are unique up to isomorphism.

Corollary 2. $\nabla(n, 2) = \frac{1}{2} 2^n.$ □

From now on we assume that $k > 2$. Denote $g(n, m) = |\partial L_m^n|$.

Lemma 3. *For any n and k,*

$$\nabla(n,k) \geq \frac{k}{2} \min\left\{ g\left(n, \left\lfloor \frac{2^n}{k} \right\rfloor\right), g\left(n, \left\lceil \frac{2^n}{k} \right\rceil\right) \right\}. \tag{2}$$

Proof. For $i \neq j$ denote $c_{i,j} = |\{(u,v) \mid u \in A_i, v \in A_j, \rho(u,v) = 1\}|$ and put $c_{i,i} = 0$, $i = 1, ..., k$. Considering ∂A_i one has

$$\sum_{j=1}^{k} c_{i,j} = |\partial A_i| \geq g(n, |A_i|). \tag{3}$$

Now summarize (3) for $i = 1, ..., k$. Since $c_{i,j} = c_{j,i}$, one has

$$\sum_{i=1}^{k}\sum_{j=1}^{k} c_{i,j} = 2|\nabla \mathcal{A}| \geq \sum_{i=1}^{k} g(n, |A_i|).$$

The lemma follows by taking into account (1) and that $g(n, |A_i|)$ is not less than the minimum in (2). □

Corollary 4. *Let $\mathcal{A} = \{A_1, ..., A_k\}$ be a partition of Q^n, satisfying (1). If each subset A_i is optimal, then the partition $\mathcal{A}$ is minimal.*

Indeed, one has equality in (3) for $i = 1, ..., k$.

Corollary 5. $\nabla(n, 2^a) = \dfrac{a}{2} 2^n$.

To show this, just partition Q^n into 2^a faces of dimension $n - a$ and notice that each face is an optimal subset.

Let us mention some simple properties of optimal subsets.

Proposition 6. *Let Q^n be partitioned into 2 faces Q', Q'' of dimension $n - 1$ and $A' \subseteq Q'$, $A'' \subseteq Q''$ with $A' \cong L_m^{n-1}$ (in Q'), $A'' \cong L_m^{n-1}$ (in Q'') for some m, $1 \leq m \leq 2^{n-1}$, and let the subsets A' and A'' be isomorphic. Then, in Q^n, it holds:*

a. $A' \cong A'' \cong L_m^n$;
b. $A' \cup A'' \cong L_{2m}^n$;
c. $Q' \cup A'' \cong Q'' \cup A' \cong L_{2^{n-1}+m}^n$. □

As soon as we are interested in the asymptotic of $\nabla(n, k)$ only, it is convenient to operate with partitions, each part of which may not be an optimal subset, but is in a sense close to one of them. To be more exact, let $A \subseteq Q^n$, $|A| = m$ and c be some fixed constant. We say that A is a *quasioptimal* set (with respect to the constant c), if there exists $B \subseteq Q^n$ with $B \cong L_m^n$, such that $|A \Delta B| \leq cn/m$, where Δ denotes the symmetric difference. Quasioptimal subsets obey similar properties as in Proposition 6.

Proposition 7. *Let Q^n be partitioned into 2 faces Q', Q'' of dimension $n-1$ and $A' \subseteq Q'$, $A'' \subseteq Q''$ be quasioptimal subsets (in corresponding subcubes) with $|A'| = |A''| = m$. Then, in Q^n, it holds:*

a'. A' and A'' are quasioptimal;
b'. $A' \cup A''$ is quasioptimal;
c'. $Q' \cup A''$ and $Q'' \cup A'$ are quasioptimal. □

In the sequel, we construct partitions $\mathcal{A} = \{A_1, ..., A_k\}$ of Q^n with growing n into k quasioptimal subsets A_i of cardinality (1). In this case for each subset A_i it holds $|A_i \Delta L^n_{|A_i|}| \leq c'$, where the constant c' depends on c and k only, but not on n. We will not specify the constants c and c' exactly in our constructions, we will just make sure that such constants exist. Clearly, for each quasioptimal set $A_i \subseteq Q^n$ of cardinality (1) it holds $|\partial A_i| - g(n, |A_i|) \leq c'n$. According, we call the partition $\mathcal{A}$ quasiminimal, if $|\nabla \mathcal{A}| - \frac{k}{2} g(n, \lfloor 2^n/k \rfloor) \leq c''n$, where the constant c'' depends on k and c' only.

Corollary 8. *Let k be fixed and $\mathcal{A} = \{A_1, ..., A_k\}$ be a partition of Q^n, satisfying (1). If each subset A_i is quasioptimal, then the partition $\mathcal{A}$ is quasiminimal.*

Indeed, similarly as in the proof of Lemma 3 one has

$$2|\nabla \mathcal{A}| = \sum_{i=1}^{k} |\partial A_i| \leq \sum_{i=1}^{k} (g(n, |A_i|) + c'n) \leq k\, g(n, \lfloor 2^n/k \rfloor) + k(c'n + 1).$$

Clearly, $g(n, \lfloor 2^n/k \rfloor)$ is exponential on n if k is fixed (cf. Theorem 9 in the next Section). Therefore, the asymptotic formula $\nabla(n, k) \sim \frac{k}{2} g(n, \lfloor 2^n/k \rfloor)$ as $n \to \infty$ holds if one can partition Q^n into k quasioptimal subsets.

3 Bounds for $\nabla(n, k)$

Theorem 9. *Let $2^{p-1} < k < 2^p$ and $n \geq 2(p-1)$. Then*

$$\frac{p-1}{2} \leq \frac{\nabla(n, k)}{2^n} \leq p + 1 \tag{4}$$

Proof. To get the lower bound we apply Lemma 3 and estimate the minimum in (2). Let $m = \lfloor 2^n/k \rfloor$ and partition Q^n into faces of dimension $n - p + 1$. Now L^n_m and L^n_{m+1} are proper subsets of one such face, say Q. Therefore,

$$\min\{g(n, m), g(n, m+1)\} \geq m(p-1) + \min\{|\partial L^n_m \cap Q|, |\partial L^n_{m+1} \cap Q|\}$$
$$\geq m(p-1) + n - p + 1.$$

Since $m \geq 2^n/k - 1$, one has

$$\nabla(n, k) \geq \frac{k}{2}(m(p-1) + n - (p-1)) \geq \frac{k}{2}\frac{2^n}{k}(p-1) + \frac{k}{2}(n - 2(p-1)) \geq 2^n \frac{p-1}{2}.$$

To get an upper bound we first partition Q^n into two faces Q', Q'' of dimension $n - 1$ and then partition isomorphically Q' and Q'' into k parts $\{A'_1, ..., A'_k\}$ and $\{A''_1, ..., A''_k\}$ respectively. Setting $A_i = A'_i \cup A''_i$ for $i = 1, ..., k$ we get a partition $\mathcal{A} = \{A_1, ..., A_k\}$ of Q^n. Since k is not a power of two, $|A'_i|, |A''_i| \in \{m, m+1\}$ with $m = \lfloor \frac{2^{n-1}}{k} \rfloor$, and so $A_i \in \{2m, 2m+2\}$. In order to get a partition satisfying (1), we take a vertex from each part of cardinality $2m + 2$ and add it to some part of cardinality $2m$. This leads to an increase of $|\nabla \mathcal{A}|$ at most on $\lfloor \frac{k}{2} \rfloor n$, so one has

$$\nabla(n, k) \leq 2\nabla(n - 1, k) + kn/2, \qquad (5)$$

which gives

$$\nabla(n, k) \leq 2^{n-p} \nabla(p, k) + \frac{k}{2} \left(n + 2(n - 1) + \cdots + (p + 1) 2^{n-p-1}\right) \qquad (6)$$

To compute $\nabla(p, k)$ notice that in any minimal partition of Q^p exactly $2k - 2^p$ parts have cardinality 1, and the remaining $2^p - k$ parts have cardinality 2. This gives

$$\nabla(p, k) = p \, 2^{p-1} - (2^p - k) = \frac{p - 2}{2} \, 2^p + k. \qquad (7)$$

Substituting (7) into (6) and taking into account

$$\sum_{i=1}^{t} i \, 2^i = (t - 1) \, 2^{t+1} + 2$$

and $k < 2^p$, one gets

$$\nabla(n, k) \leq \left\lfloor \frac{p - 2}{2} + \frac{k}{2^p} + \frac{k(p + 2)}{2^{p+1}} \right\rfloor 2^n \leq (p + 1) \, 2^n. \qquad \square$$

Corollary 10. *If k is fixed, then the limit $\lim\limits_{n \to \infty} \nabla(n, k)/2^n$ exists.*

Proof. By Theorem 9 the sequence $\beta_n = \frac{\nabla(n,k)}{2^n}$ is bounded and using (5) one get $\beta_n - \beta_{n-1} = 2^{-n}(\nabla(n, k) - 2\nabla(n - 1, k)) \leq kn \, 2^{-(n+1)}$. Thus, $\beta_n - \beta_{n-1} \to 0$ as $n \to \infty$. This implies $\liminf\limits_{n \to \infty} \beta_n = \limsup\limits_{n \to \infty} \beta_n$ and hence $\lim\limits_{n \to \infty} \beta_n$ exists. $\square$

Therefore, we represent $\nabla(n, k)$ in the form $\nabla(n, k) \sim c(k) \cdot 2^n$ as $n \to \infty$ and concentrate our attention on the function $c(k)$.

Theorem 11. *Let Q^n be partitioned into k quasioptimal subsets. Then, for any constant r and $n \to \infty$*

$$\nabla(n + r, 2^r \cdot k) \sim r \, 2^{n+r-1} + 2^r \, \nabla(n, k). \qquad (8)$$

Proof. Given a partition of Q^n into k quasioptimal subsets, we construct a partition of Q^{n+r} into $2^r \cdot k$ quasioptimal subsets. To do this, we first partition Q^{n+r} into 2^r faces of dimension n and then partition each Q^n into k quasioptimal subsets, assuming that the partitions of all the faces are isomorphic. It is remained to notice that if A is quasioptimal in Q^n, then it is also quasioptimal in Q^{n+r} by Proposition 7. □

Therefore, in order to determine $c(k)$ it is sufficient to partition Q^n into k quasioptimal subsets just for odd values of k. Under assumptions of Theorem 11, it follows that $c(2^r \cdot k) = r/2 + c(k)$.

Now let us turn to the asymptotic of $\nabla(n, k)$. In the next proposition we, for the only time in our paper, assume that k is not a constant.

Corollary 12. *Let $n \to \infty$ and $k = k(n) = 2^{\alpha_n} \beta_n \le 2^n$ for some integer α_n, β_n such that $\alpha_n \to \infty$ and $\log_2 \beta_n / \alpha_n \to 0$. Then,*

$$\nabla(n, k(n)) \sim \frac{\log_2 k(n)}{2} 2^n.$$

Indeed, the proof of Theorem 11 implies

$$\nabla(n, 2^r \cdot s) \le r \, 2^{n-1} + 2^r \, \nabla(n - r, s). \tag{9}$$

Applying (9) with $r = \alpha_n$, $s = \beta_n$ and taking into account (4), one has

$$\frac{\alpha_n}{2} + \frac{\log_2 \beta_n}{2} - 1 \le \frac{\nabla(n, k)}{2^n} \le \frac{\alpha_n}{2} + \frac{\nabla(n - \alpha_n, \beta_n)}{2^{n-\alpha_n}} \le \frac{\alpha_n}{2} + \log_2 \beta_n + 2,$$

and the assertion follows.

4   The case $k = 2^a + 2^b$

Proposition 13. *There exists a partition of Q^n into 3 optimal subsets satisfying (1).*

Proof. Partition Q^n into 2 faces Q', Q'' of dimension $n-1$. Let $m = \lfloor 2^{n-1}/3 \rfloor$ and avoiding trivial cases assume $n \ge 3$. Denote $A' = L_m^{n-1}$ in Q' and $A_1 = Q' \setminus A'$. Similarly denote $A'' = L_m^{n-1}$ in Q'' and $A_2 = Q'' \setminus A''$ and let $A_3 = A' \cup A''$.

By Proposition 6 each part A_i in the constructed partition $\{A_1, A_2, A_3\}$ is an optimal subset of Q^n. However, $|A_1| = |A_2| = \lceil 2^n/3 \rceil$, but $\lfloor 2^n/3 \rfloor - 1 \le |A_3| \le \lfloor 2^n/3 \rfloor$. If $|A_3| = \lfloor 2^n/3 \rfloor$, the desired partition is obtained. Otherwise, replace the set A' in the construction with L_{m+1}^{n-1}. The obtained partition is minimal by Corollary 4. □

The construction in Proposition 13 leads to the following generalization:

Theorem 14. *For any integer constant $a \ge 1$ and $k = 2^a + 1$, it holds*

$$c(k) = \frac{a(k-1)}{2(k-2)}.$$

Proof. Let $n \geq 2a + 1$ and consider first a partition of Q^n into $k - 1 = 2^a$ faces $Q_1, ..., Q_{k-1}$ of dimension 2^{n-a}. Now let $A'_i = L^{n-a}_{m_1} \subseteq Q_i$ with $m_1 = \left\lfloor \frac{2^{n-a}}{k} \right\rfloor$, and denote $A_i = Q_i \setminus A'_i$ for $i = 1, ..., k - 1$ and $A_k = A'_1 \cup \cdots \cup A'_{k-1} \subseteq Q^n$.

By Proposition 6 all the subsets A_i are optimal, but this partition may not satisfy (1), because $|A_k|$ may be too small. Since a is a constant, one can reassign a constant number (depending on a only) of vertices between the parts A_k and A_i with $k < i$ similarly as is was done in the proof of Proposition 13 so, that each obtained subset $\tilde{A}_i$ will be quasioptimal. The obtained partition is quasiminimal by Corollary 8.

To compute $c(k)$ we first compute $g_n = g(n, m_2)$ with $m_2 = \lfloor 2^n/k \rfloor$. Let us partition Q^n into the faces Q_i again. Clearly the set $A = L^n_{m_2}$ has a nonempty intersection just with one of these face, say $A \subseteq Q_1$. Moreover, $|A| = \lfloor \frac{k-1}{k} 2^{n-a} \rfloor$, thus for the set $\overline{A} = Q_1 \setminus A$ one has $|\overline{A}| = \left\lceil \frac{2^{n-a}}{k} \right\rceil = g_{n-a} + O(n)$. Therefore, $g_n = g_{n-a} + \frac{a}{k} 2^n + O(n)$, which gives $g_n \sim \frac{a(k-1)}{k(k-2)} 2^n$ and the formula for $\nabla(n, 2^a + 1)$ follows from Lemma 3. $\qquad\square$

Notice, that making a careful exchange of vertices between the parts A_i in the proof, one can show that Q^n can be partitioned into $2^a + 1$ optimal subsets satisfying (1).

Corollary 15. *For any integer constants a, b with $a > b \geq 0$, it holds*

$$c\left(2^a + 2^b\right) = \frac{a\, 2^{a-1} - b\, 2^{b-1}}{2^a - 2^b}.$$

Indeed, taking into account Corollary 5, we assume without loss of generality that $a > b$. Then $k = 2^b(2^{a-b} + 1)$ and the assertion follows from Theorems 11 and 14.

5   The case $k = 2^a - 2^b$

Proposition 16. *There exists a partition of Q^n into 7 quasioptimal subsets satisfying (1).*

Proof. Let $n \geq 8$. We construct a partition of Q^n into 7 optimal subsets $A_1, ..., A_7$ the cardinalities of which differ in a constant, which does not depend on n. From this partition one can similarly to above construct a partition of Q^n into 7 quasioptimal subsets satisfying (1).

The construction is done in 3 steps. In the first step, we consider a special partition of Q^5 into 8 faces $S_1, ..., S_8$ of dimension 2.

This partition is shown in Fig. 5. The subcubes corresponding to $S_1, ..., S_8$ are depicted in bold lines. The graph H shown in Fig. 5 corresponds to the neighborhood structure between the vertices $v_j \in S_8$ and the faces $S_1, ..., S_7$. The edge (v_j, S_i) of H means that $\rho(x, S_i) = 1$. Note that each vertex v_j of H in Fig. 5 is incident with one vertex of degree 1, one vertex of degree 2, and one vertex of degree 4.

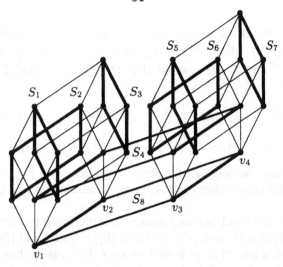

Fig. 1. A special partition of Q^5

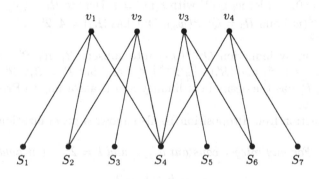

Fig. 2. The neighborhood structure of the partition

In the second step, we use the partition above to construct a partition $A_1, ..., A_7$. For that we represent Q^n as $Q^5 \times Q^{n-5}$ and for $u \in Q^5$ denote by $Q(u)$ the face of dimension $n - 5$ in this representation, which contains the vertex u. Now, for $i = 1, ..., 8$ put $Q_i = \bigcup_{u \in S_i} Q(u)$. One has $|Q_i| = 2^{n-3}$ and each Q_i is a face of Q^n of dimension $n - 3$ and thus an optimal set in Q^n. We are now going to partition the vertices of Q_8 into 7 almost equal parts and add one such a part to each set $Q_1, ..., Q_7$. Thus we will get a partition $A_1, ..., A_7$, with $||A_i| - \frac{1}{7}2^n| \le 10$, $i = 1, ..., 7$.

Thus, consider the face $Q^j = Q(v_j)$ and assume that we can partition each Q^j into 3 optimal (in Q^j) subsets P_1^j, P_2^j, P_3^j with $|P_1^j| = \lfloor \frac{4}{7}2^{n-5} \rfloor$, $|P_2^j| = \lfloor \frac{2}{7}2^{n-5} \rfloor$ and $|P_3^j| - \lceil \frac{1}{7}2^{n-5} \rceil \leq 1$ (we assume that the partitionings of $Q^1, ..., Q^4$ are isomorphic). Put:

$$A_1 = Q_1 \cup P_1^1 \qquad A_3 = Q_3 \cup P_1^2 \qquad A_5 = Q_5 \cup P_1^3 \qquad A_7 = Q_7 \cup P_1^4$$

$$A_2 = Q_2 \cup P_2^1 \cup P_2^2 \qquad A_4 = Q_4 \cup P_3^1 \cup P_3^2 \cup P_3^3 \cup P_3^4 \qquad A_6 = Q_6 \cup P_2^3 \cup P_2^4$$

Taking into account the neighboring properties above (cf. Fig. 1b) and Proposition 6, it is easily shown that each A_i is an optimal subset in Q^n and $||A_i| - \frac{1}{7}2^n| \leq 10$.

Therefore, the third and last step of our construction is to partition Q^t into 3 optimal subsets P_1, P_2, P_3 with $|P_1| = \lfloor \frac{4}{7}2^t \rfloor$, $|P_2| = \lfloor \frac{2}{7}2^t \rfloor$ and $|P_3| - \lceil \frac{1}{7}2^t \rceil \leq 1$. We construct such a partition by induction on t. It is easily shown that Q^2 and Q^3 can be partitioned into 3 optimal subsets of cardinalities $2, 1, 2$ and $5, 2, 2$ respectively. In such a partition of Q^1 one of the parts is an empty set. Now let $t \geq 4$.

Represent Q^t as $Q^3 \times Q^{t-3}$ and similarly to above for $u \in Q^3$ denote by $Q^{t-3}(u)$ the face of dimension $t - 3$ in this representation, which contains the vertex u. For $i = 0, ..., 7$ let $w_i \in Q^3$ with $\ell(w_i) = i$. Denote $R_1 = \bigcup_{i=0}^3 Q^{t-3}(w_i)$, $R_2 = \bigcup_{i=4}^5 Q^{t-3}(w_i)$ and $R_3 = Q^{t-3}(w_6)$. One has $|R_1| = 4 \cdot 2^{t-3}$, $|R_2| = 2 \cdot 2^{t-3}$ and $|R_3| = 2^{t-3}$.

Now partition by induction $Q^{t-3}(w_7)$ into subsets R_1', R_2', R_3' with $|R_1'| = \lfloor \frac{4}{7}2^{t-3} \rfloor$, $|R_2'| = \lfloor \frac{2}{7}2^{t-3} \rfloor$ and $|R_3'| - \lceil \frac{1}{7}2^{t-3} \rceil \leq 1$ and put $P_i = R_i \cup R_i'$, $i = 1, 2, 3$. Now each part P_i has the desired cardinality and is according to Proposition 6, an optimal subset. $\qquad \square$

The construction from Proposition 16 also allows a generalization:

Theorem 17. *For any integer constant $a \geq 3$ and $k = 2^a - 1$ it holds,*

$$c(k) = \frac{a(k+1) - 2}{2k}.$$

Proof. Our goal is to show that Q^n may be partitioned into quasioptimal subsets $A_1, ..., A_k$ with $||A_i| - 2^n/k| \leq const$, where the constant depends on k only. We follow the steps of the proof of Proposition 16, assuming $n \geq 2a$.

On the first step we construct a special partition of Q^{2a-1} into 2^a faces of dimension $a - 1$. Denote

$$Q^{\sigma\tau} = \left\{ (x_1, ..., x_{2a-1}) \in Q^{2a-1} \mid x_{2a-2} = \sigma, \ x_{2a-1} = \tau \text{ with } \sigma, \tau \in \{0, 1\} \right\}.$$

Clearly, each $Q^{\sigma\tau}$ is isomorphic to Q^{2a-3}. We proceed by induction on a, starting with partition of Q^5 as in Proposition 16. Consider isomorphic partitions of each $Q^{\sigma\tau}$ into faces $S_i^{\sigma\tau}$, $i = 1, ..., 2^{a-1}$. Now construct a partition S_i ($i = 1, ..., 2^a$) of Q^{2a-1} by setting $S_i = S_i^{00} \cup S_i^{01}$ for $i = 1, ..., 2^{a-1} - 1$ and $S_{2^{a-1}+i-1} = S_i^{10} \cup S_i^{11}$ for $i = 1, ..., 2^{a-1} - 1$. Finally, put $S_{2^a-1} = S_{2^{a-1}}^{01} \cup S_{2^{a-1}}^{11}$ and $S_{2^a} = S_{2^{a-1}}^{00} \cup S_{2^{a-1}}^{10}$.

Consider the bipartite graph H formed by the vertex sets $X = \{x \in S_{2^a}\}$ and $Y = \{S_1, ..., S_{2^a-1}\}$. A vertex $x \in X$ is incident to $y \in Y$ in H iff $\rho(x, S_i) = 1$ in Q^{2a-1} (here S_i is the face corresponding to y). Using induction on a it is easy to show the following properties of the graph H:

(i) $\deg(x) = a$ for any $x \in X$ and $\deg(y)$ is a power of 2 for any $y \in Y$;
(ii) Each vertex $x \in X$ is incident to exactly one vertex $y \in Y$ of degree 2^{j-1}, $i = 1, ..., a$.
(iii) Let $y \in Y$ with $\deg(y) = 2^j$ and $N(y) \subseteq X$ be the set of vertices incident to y in H. Then $N(y)$ is a face of S_{2^a} of dimension j.

In the second step represent Q^n as $Q^{2a-1} \times Q^{n-2a+1}$ and for $u \in Q^{2a-1}$ denote by $Q(u)$ the face of dimension $n - 2a + 1$ in this representation, which contains the vertex u. Now for $i = 1, ..., 2^a$ put $Q_i = \bigcup_{u \in S_i} Q(u)$. Each Q_i is a face of dimension $n - a$, so an optimal set in Q^n. Assume that we can partition each face $Q(u)$ with $u \in S_{2^a}$ into quasioptimal subsets $P_j(u)$, $j = 1, ..., a$, with $||P_j(x)| - \frac{2^{a-j}}{k} 2^{n-2a+1}| \le c$ and some constant c depending on a only.

Constructing the isomorphic partitions $\{P_j(u)\}$ of the faces $Q(u)$ ($u \in S_{2^a}$), we get the partition $A_1, ..., A_{2^a-1}$ of Q^n as follows. Let $y \in Y$ corresponds to the face S_i with $i < 2^a$, $\deg(y) = 2^{j-1}$ for some j, $1 \le j \le a$. Put $A_i = Q_i \bigcup_{u \in N(y)} P_j(u)$. Using the properties of the graph H above and Proposition 7 it is easy to show that each subset A_i is quasioptimal and $||A_i| - 2^n/k| \le const$ for some constant depending on a only.

Therefore, to complete the whole construction, we have to show that Q^t may be partitioned into a quasioptimal subsets P_j with $||P_j| - \frac{2^{a-j}}{2^a-1} 2^t| \le c$ for $j = 1, ..., a$, where the constant c does not depend on t. We carry out such a partition inductively. For $t = 0, ..., a - 1$ to construct such a partition is easy, some subsets are simply empty sets. Let $t \ge a$.

Represent Q^t as $Q^a \times Q^{t-a}$ and for $u \in Q^a$ denote as usual by $Q^{t-a}(u)$ the face of dimension $t - a$ in this representation containing the vertex u. Partition by induction the face $Q^{t-a}(u)$ with $\ell(u) = 2^a - 1$ into a quasioptimal subsets R'_j, $j = 1, ..., a$, with $||R'_j| - \frac{2^{a-j}}{2^a-1} 2^{t-a}| \le c$. Now partition the set $\{0, ..., 2^a - 1\}$ into a blocks B_j of 2^{a-j} consecutive numbers (i.e. $B_1 = \{0, ..., 2^{a-1} - 1\}$, $B_2 = \{2^{a-1}, ..., 2^{a-1} + 2^{a-2} - 1\}$ and so on) for $j = 1, ..., a$ and put $R_j = \bigcup_{\substack{u \in Q^a \\ \ell(u) \in B_j}} Q^{t-a}(u)$ and $P_j = R_j \cup R'_j$. Then, each P_j has the required cardinality and by Proposition 7 is a quasioptimal subset in Q^t.

In order to compute $c(k)$ we first compute $g_n = g(n, m)$ with $m = \lfloor 2^n/k \rfloor$. Let us partition Q^n into faces of dimension 2^{n-a+1}. Clearly, the set $A = L^n_m$ has a nonempty intersection with just one of these faces, say with face Q. Moreover, $|A| = 2^{n-a} + \lfloor \frac{1}{k} 2^{n-a} \rfloor$. Thus, one can partition Q into 2 faces Q' and Q'' of dimension $n - a$ in such a way that the set A is partitioned into 2 parts $A' \subseteq Q'$, $A'' \subseteq Q''$ with $|A'| = |Q'| = 2^{n-a}$ and for $\overline{A''} = Q'' \setminus A''$ it holds $|\overline{A''}| = \lceil \frac{2^{n-a}}{k} \rceil$.

Therefore, $g_n = g_{n-a} + \frac{a-1}{k} 2^n + \frac{k-1}{k} 2^{n-a} + O(n)$, which, taking into account that a is a constant, gives $g_n \sim \frac{a(k+1)-2}{k^2} 2^n$, and the formula for $\nabla(n, 2^a - 1)$ follows from Lemma 3. $\quad\square$

Corollary 18. *For any integer constants $a, b \geq 0$ with $a - b \geq 2$, it holds*

$$c\left(2^a - 2^b\right) = \frac{a\,2^{a-1} - b\,2^{b-1} - 2^b}{2^a - 2^b}.$$

The proof is similar to the proof of Corollary 15.

6 Concluding remarks

The above results lead to the following table, where the sign "?" indicates a presently unknown value.

Table 1. Some asymptotic results

k	2	3	4	5	6	7	8	9	10	11	12	13	14	15	16	17	18	19	20
$c(k)$	$\frac{1}{2}$	1	1	$\frac{4}{3}$	$\frac{3}{2}$	$\frac{11}{7}$	$\frac{3}{2}$	$\frac{12}{7}$	$\frac{11}{6}$	?	2	?	$\frac{29}{14}$	$\frac{31}{15}$	2	$\frac{32}{15}$	$\frac{31}{14}$	?	$\frac{7}{3}$

The entries for $k = 2, 4, 8, 16$ follow from Corollary 5, $c(3), c(7)$ and $c(15)$ are given by Propositions 13,16 and Theorem 17 respectively. Constructions for $k = 5, 9, 17$ are provided by Theorem 14, and the values of $c(6), c(12), c(14), c(18)$ are implied by Theorem 11. It is interesting to note that the function $c(k)$ is not monotone.

It should be mentioned that our constructions allow us to obtain not only quasiminimal but also minimal partitions and exact formulas for $\nabla(n, k)$ at least for small k. We leave this aspect without further development here.

Finally, our results are applicable to the pin limitations problem [3], which requires to construct a partition $\mathcal{A} = \{A_1, ..., A_k\}$ minimizing $\max_i |\partial A_i|$. Since each part A_i in our constructions is a quasioptimal set, all the values $|\partial A_i|$ for considered k are asymptotically equal and can be easily obtained from Corollaries 15 and 18. Similarities in the structure of the parts A_i in the constructed partitions allow to apply the obtained here results also for designing a single VLSI "building block" chip (cf. [2]), which could be used for constructing the whole n-cube by wiring together its multiple copies in an appropriate way.

7 Acknowledgments

The author thanks anonymous referees for their helpful comments and for calling his attention to the papers [2] and [3].

References

1. Chung F.R.K., Yao S.-T.: A Near Optimal Algorithm for Edge Separators. In: *Proc. 26th Symp. Theory of Comp. ACM*, 1994.

2. Schwabe E.J.: Optimality of a VLSI Decomposition Scheme for the De-Bruijn Graph. *Parallel Process. Lett.*, **3** (1993) 261-265.
3. Cypher R.: Theoretical Aspects of VLSI Pin Limitations. *SIAM J. Comput.*, **22** (1993) 356-378.
4. Diekmann R., Lüling R., Monien B., Spräner C.: Combining Helpful Sets and Parallel Simulated Annealing for the Graph-Partitioning Problem. *Int. J. Parallel Algorithms and Applications*, (to appear).
5. Diekmann R., Meyer D., Monien B.: Parallel Decomposition of Unstructured FEM-Meshes. In: *Proc. of IRREGULAR'95*, Springer LNCS **980** (1995) 199-215.
6. Gilbert J.R., Miller G.L., Teng S.-H.: Geometric Mesh Partitioning: Implementation and Experiments. In: *Proc. of IPPS'95*, 1995.
7. Harper L.H.: Optimal assignment of numbers to vertices. *J. Soc. Ind. Appl. Math.*, **12** (1964) 131-135.
8. Lipton R.J., Tarjan R.E.: A Separator Theorem for Planar Graphs. *SIAM J. Appl. Math.*, **36** (1979) 177-189.
9. Rolim J., Sýkora O., Vrt'o I.: Optimal Cutwidth and Bisection Width of 2- and 3-Dimensional Meshes. In: *Proc. of WG'95*, 1995.
10. Seitz C.: The Cosmic Cube. *Comm ACM*, **28** (1985) 22-33.

Embedding Complete Binary Trees in Product Graphs

Adrienne Broadwater[1], Kemal Efe[1], and Antonio Fernández[2]

[1] Center for Advanced Computer Studies, University of Southwestern Louisiana,
Lafayette, LA 70504
[2] MIT Laboratory for Computer Science, 545 Technology Square,
Cambridge, MA 02139

Abstract. This paper shows how to embed complete binary trees in products of complete binary trees, products of shuffle-exchange graphs, and products of de Bruijn graphs. The main emphasis of the embedding methods presented here is how to emulate arbitrarily large complete binary trees in these product graphs with low slowdown. For the embedding methods presented here the size of the host graph can be fixed to an arbitrary size, while we define no bound on the size of the guest graph. This is motivated by the fact that the host architecture has a fixed number of processors due to its physical design, while the guest graph can grow arbitrarily large depending on the application. The results of this paper widen the class of computations that can be performed on these product graphs which are often cited as being low-cost alternatives for hypercubes.

1 Introduction

Let $G^r(N)$ denote the r-dimensional product graph obtained from the N-node graph $G(N)$. Note that $G^r(N)$ contains N^r nodes. (As a special case, every graph $G(N)$ is a one-dimensional product of itself, and we omit r when $r = 1$.) Let $T(N)$ be the N-node complete binary tree, where $N = 2^h - 1$. We prove the following results:

1. $T(2^{rh - \lceil \frac{r}{2} \rceil + l} - 1)$, where $l > 1$, can be embedded in the r-dimensional product of complete binary trees, $T^r(2^h - 1)$, with dilation 2, congestion 2, and load $2^l - 1$.

2. Given the r-dimensional product of shuffle-exchange graphs, $S^r(N)$,
   (a) $T(N^r 2^{l-1} - 1)$ can be embedded in it with dilation 3, congestion 2, and load $2^l - 1$.
   (b) $T((N2^l)^r - 1)$ can be embedded in it with dilation 4, congestion 4, and load 2^{rl}.

3. $T((N2^l)^r - 1)$ can be embedded in the r-dimensional product of de Bruijn graphs, $D^r(N)$, with dilation 2, congestion 2, and load 2^{rl}.

The first problem above, for unit load, was originally addressed in [3], where it was shown that $T(2^{r(h-1)+1} - 1)$ is a subgraph of $T^r(2^h - 1)$. When $r = 2$ this method embeds the largest possible tree for the number of nodes in $T^r(2^h - 1)$, but when $r > 2$ the size of the tree shrinks by a factor of 2^{r-1}. Thus, as r grows the method of [3] becomes less and less interesting. To utilize more nodes of the host, a unit-load embedding was presented in [2] with dilation 3 and congestion 3. Our emphasis here is how to embed arbitrarily-large complete binary trees in the fixed size host graph. It turns out that the dilation and congestion values can be reduced from 3 to 2 when the load is increased.

The second and third problems above were addressed in [8] for unit load, but the methods presented there only apply for two dimensions and use only about half of the nodes of the product graph. The method in the current paper utilizes all (but one) of the nodes of the product graph and it is applicable for any number of dimensions. Also, our methods yield perfectly-balanced loads for the nodes of the host graphs.

Since a parallel architecture has a fixed size by its physical design, these results have significant practical importance as they show a way for solving arbitrarily-large tree computations on fixed-size parallel computers. These important practical concerns appear to have been omitted in most of the papers in the literature except by a few researchers [1, 6, 7].

2 Definitions and Notation

The nodes of the N-node *complete binary tree* are assigned the labels $1, \ldots, N$. Each node u, $u < N/2$, is connected to nodes $2u$ and $2u + 1$. This labeling will be referred to as the *level-order* labeling of $T(N)$ (see Figure 1). The graph $T(2^h - 1)$ will often be also called the h-level complete binary tree.

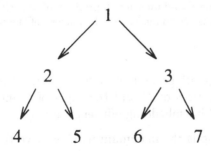

Fig. 1. Level-order labeling of the complete binary tree.

The N-node *shuffle-exchange graph*, denoted $S(N)$, contains $N = 2^n$ nodes, labeled $0, \ldots, N - 1$, and $3 \times 2^{n-1}$ edges connected as follows:

(a) (u, v) is an "exchange" edge if $v = u + 1$ where u is even or $v = u - 1$ where u is odd, or

(b) (u, v) is a "shuffle" edge if $v = 2u$ where $u < N/2$ or $v = (2u \bmod N) + 1$ where $u \geq N/2$.

The N-node *de Bruijn graph*, denoted $D(N)$, contains $N = 2^n$ nodes, labeled $0, \ldots, N - 1$, and 2^{n+1} edges connected as follows: (u, v) is an edge of $D(N)$ if $v = 2u \bmod N$ or $v = (2u \bmod N) + 1$.

Let $G = (V_G, E_G)$ and $H = (V_H, E_H)$ be two arbitrary graphs. Their *cartesian product* is the graph $P = G \otimes H$ whose vertex set is $V_G \times V_H$ and whose edge set contains all edges of the form $(x_1 x_0, y_1 y_0)$ such that either $x_1 = y_1$ and $(x_0, y_0) \in E_G$, or $x_0 = y_0$ and $(x_1, y_1) \in E_H$.

The *r-dimensional homogeneous product of an N-node graph* $G(N)$, denoted $G^r(N)$ is:

1. a single vertex with no labels and no edges if $r = 0$
2. $G(N) \otimes G^{r-1}(N)$ when $r > 0$.

Figure 2 illustrates this definition by presenting the construction of the two-dimensional product $S^2(8)$.

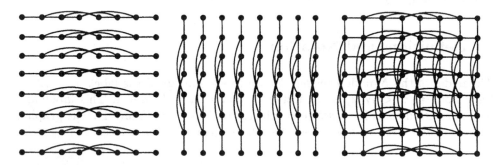

Fig. 2. Construction of the two-dimensional product of the shuffle-exchange graph $S(8)$. Both rows and columns are connected in the pattern of the basic shuffle-exchange graph.

An *embedding* of a "guest" graph G in a "host" graph H is a mapping of the vertices of G into the vertices of H and the edges of G into paths in H. The main cost measures used in embedding efficiency are [3]:

- *Load* of an embedding is the maximum number of vertices of G mapped to any vertex of H.
- *Dilation* of an embedding is the maximum path length in H representing an edge of G.
- *Congestion* of an embedding is the maximum number of paths (that correspond to the edges of G) that share any edge of H.

The level-order labeling of a complete binary tree as in Figure 1 defines an embedding of $T(N - 1)$ in $S(N)$ with dilation 2, congestion 2, and load 1 [5]. This labeling also shows that $T(N - 1)$ is a subgraph of $D(N)$ [8].

3 Embedding in the Product of Complete Binary Trees

In this paper we use the embedding method of [3] as part of the improved embedding method presented here. For easy reference this result is included here.

Theorem 1. $T(2^{r(h-1)+1} - 1)$ *is a subgraph of* $T^r(2^h - 1)$.

As an example, Figure 3 shows the embedding for $r = 2$.

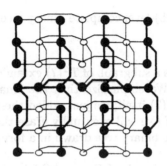

Fig. 3. Embedding the complete binary tree $T(31)$ in $T^2(7)$ by Theorem 1. The complete binary tree subgraph is highlighted by heavy dark lines.

The main result of this section is the following:

Theorem 2. $T(2^{rh-\lceil \frac{r}{2} \rceil + l} - 1)$, *where* $l > 1$, *can be embedded in* $T^r(2^h - 1)$ *with dilation 2, congestion 2, and load* $2^l - 1$.

Before proving the theorem, we will first distinguish a particular node in the $T^r(N)$ graph as follows:

- *Root of* $T^r(N)$: The node $v = v_{r-1}...v_1 v_0$ is the root of $T^r(N)$ if and only if $v_i = 1$ (that is, v_i is the root of $T(N)$), for all $0 \leq i \leq r - 1$.

First we show that a 63-node complete binary tree can be embedded in $T^2(7)$ with dilation 2, congestion 2, and load 3. A simple modification of this gives an embedding for $T(2^{l+5} - 1)$ in $T^2(7)$ with the same dilation and congestion, but the load is increased to $2^l - 1$, where $l > 1$. Next, we use induction on r to show that $T(2^{\lfloor \frac{5r}{2} \rfloor + 1} - 1)$ can be embedded in $T^r(7)$ with dilation 2, congestion 2, and load 3. Finally, by combining these results and Theorem 1 the claim of the theorem is obtained.

Lemma 3. $T(63)$ *can be embedded in* $T^2(7)$ *with dilation 2 and congestion 2, such that 10 nodes have load 3 and 33 nodes have load 1. The remaining 6 nodes of* $T^2(7)$ *are unused. In this embedding the root of the embedded tree coincides with the root of* $T^2(7)$.

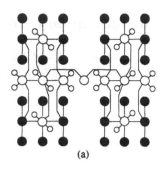 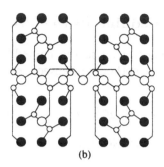

(a)   (b)

Fig. 4. Embedding the $(l+5)$-level complete binary tree in a subgraph of $T^2(7)$.

Proof. Figure 4.(a) presents a subgraph of $T^2(7)$ extended with some new nodes (the small empty nodes). We emphasize that the small empty nodes in Figure 4.(a) do not exist in $T^2(7)$ itself; we just added these nodes for convenience in the presentation of proof (we will eventually erase these nodes). Figure 4.(b) presents a 63-node complete binary tree drawn in a form suitable for the following discussion.

Consider embedding the graph of Figure 4.(b) in the graph of Figure 4.(a) by super-imposing the nodes of the two graphs on top of each other. It can be easily checked that any edge in Figure 4.(b) corresponds to a path of length no more than 3 in Figure 4.(a). Dilation-3 edges are those that connect the large dark nodes to small empty nodes in Figure 4.(b). It can be also easily seen that the maximum congestion of 3 is found in some of the edges connecting large empty nodes with small empty nodes in Figure 4.(a). (The reader can trace the connections sharing the edge from the large empty node to the small empty node at the rightmost column of Figure 4.(a).)

Finally, by contracting the edges between the large empty nodes and small empty nodes in Figure 4.(a) we obtain a real subgraph of $T^2(7)$, while we increase the load in the large empty nodes to 3. This process also reduces both the dilation and congestion values to 2. Since the tree of Figure 4.(b) has 6 levels we have obtained an embedding of $T(63)$ in $T^2(7)$ with dilation and congestion values of 2, and load 3. From the figure it is easily verified that the root of the embedded tree coincides with the root of $T^2(7)$.

Corollary 4. $T(2^{l+5} - 1)$, *where* $l > 1$, *can be embedded in* $T^2(7)$, *such that 32 nodes of* $T^2(7)$ *have load* $2^l - 1$, *10 nodes have load 3, and the root has load 1.*

This is obtained by simply replacing the dark nodes of Figure 4.(b) (the leaves of the embedded tree) by l-level complete binary trees, and then using the embedding method above.

The properties of the embedding highlighted in the statement of Lemma 3 are needed in Lemma 5 below. This lemma uses induction on r to increase the number of dimensions.

Lemma 5. $T(2^{\lfloor \frac{5r}{2} \rfloor +1} - 1)$ *can be embedded in* $T^r(7)$ *with dilation 2, congestion 2, and load 3. In this embedding the root of the embedded tree is the root of* $T^r(7)$ *and the leaves are in unit-load nodes.*

Proof. We prove the claim by induction on the number of dimensions, r. We will have two initial base cases (cases of $r = 1$ and $r = 2$) and an induction step that increases the number of dimensions by two. This allows to prove the claim for any number of dimensions, since depending on whether r is odd or even, we can use either $r = 1$ or $r = 2$ as the basis case, respectively.

The base cases are trivially verified. For $r = 1$, $T^1(7)$ is isomorphic to $T(2^{\lfloor \frac{5}{2} \rfloor +1} - 1)$. For $r = 2$, Lemma 3 above shows the embedding.

In the induction step, given an embedding of $T(2^{\lfloor \frac{5k}{2} \rfloor +1} - 1)$ in $T^k(7)$ with dilation 2, congestion 2, and load 3, we show that it is possible to embed $T(2^{\lfloor \frac{5(k+2)}{2} \rfloor +1} - 1)$ in $T^{k+2}(7)$ with the same dilation, congestion, and load. In this embedding the root of the embedded tree is the root of $T^{k+2}(7)$.

By removing all the edges along dimensions k and $k + 1$ from $T^{k+2}(7)$ we obtain 49 disjoint copies of $T^k(7)$. From the induction hypothesis, we can embed a disjoint copy of $T(2^{\lfloor \frac{5k}{2} \rfloor +1} - 1)$ in each of these copies.

Now consider only the roots of the embedded trees and reconnect them along dimensions k and $k + 1$. Considering only the dimensions k and $k + 1$, we have a graph isomorphic to $T^2(7)$. From Lemma 3, we know that a 6-level complete binary tree can be embedded in this graph. The leaves of this tree (the dark nodes of Figure Figure 4.(a)) correspond to the roots of embedded $T(2^{\lfloor \frac{5k}{2} \rfloor +1} - 1)$ graphs. (The trees whose roots fall in the large empty nodes are not considered.)

By this procedure, we have obtained an embedding of the $(2^{\lfloor \frac{5k}{2} \rfloor +1+5} - 1) = (2^{\lfloor \frac{5(k+2)}{2} \rfloor +1} - 1)$-node complete binary tree in $T^{k+2}(7)$ with dilation 2, congestion 2, and load 3, as claimed.

Proof of Theorem 2: If we remove the 2 lowest levels from every tree along each dimension in $T^r(2^h - 1)$ we obtain a graph isomorphic to $T^r(2^{h-2} - 1)$. From Theorem 1 we can embed a $(r(h-3)+1)$-level tree in this subgraph of $T^r(2^h - 1)$ such that the leaves of the tree are mapped to the leaves of $T^r(2^{h-2} - 1)$.

Similarly, if we remove the $h - 3$ top levels from every tree along each dimension we obtain a disconnected graph formed by $2^{r(h-3)}$ disjoint copies of $T^r(7)$. Then, by using Lemma 2, we embed a $(\lfloor \frac{5r}{2} \rfloor + 1)$-level tree in each copy of $T^r(7)$, where the roots of the embedded trees coincide with the roots of $T^r(7)$ graphs. The combination of both embeddings in $T^r(2^h - 1)$ yields an embedding of the $(\lfloor \frac{5r}{2} \rfloor + 1 + r(h - 3)) = (rh - \lceil \frac{r}{2} \rceil + 1)$-level complete binary tree in $T^r(2^h - 1)$ with dilation 2, congestion 2, and load 3. Note that in this tree the leaves are embedded with unit load.

Finally, by replacing the leaves of embedded tree with l-level trees (as in Corollary 1) we obtain a dilation 2 and congestion 2 embedding where the load is $2^l - 1$. ∎

This proves the first result claimed in the introduction and completes this section.

4 Embedding in the Product of Shuffle-Exchange Graphs

In this section we focus our attention on embeddings of complete binary trees of arbitrary size in $S^r(N)$. We start by presenting a method to embed $T(N^r - 1)$ in $S^r(N)$ with dilation 3, congestion 2, and unit load. We continue by showing how to extend this method for arbitrarily large trees with the same dilation and congestion values, thus proving the result 2.(a) claimed in the introduction.

However, in this embedding half of the nodes (minus one) of $S^r(N)$ have unit load, while the other half are collectively mapped most of the nodes of the embedded tree. In the next section we comment on a method to embed arbitrarily large trees with perfectly-balanced load distribution (result 2.(b)).

Theorem 6. $T(N^r - 1)$ can be embedded in $S^r(N)$ with dilation 3, congestion 2, and unit load.

Proof. We prove the theorem by induction on the number of dimensions. We already mentioned that $T(N - 1)$ can be embedded in $S(N)$ with dilation 2 and congestion 2, which proves the base case $r = 1$. We now illustrate the induction step by presenting the construction of the embedding of $T(N^2 - 1)$ in $S^2(N)$. The generalization of this process for arbitrary number of dimensions is similar and will be briefly described.

We begin by embedding $T(N - 1)$ in each of the subgraphs isomorphic to $S(N)$ that form the dimension-1 connections in $S^2(N)$. Since each node has a label of the form $v_1 v_0$, we can do this by using the level-order embedding of $T(N - 1)$ in $S(N)$ using the v_0 part of the label. Note that the roots of these N trees all have the form $v_1 1$ and that the nodes $v_1 0$ are all unused. See Figure 5 (looking at row connections only). We can now embed another $N - 1$ node complete binary tree using the level-order labeling in the nodes of the form $v_1 0$ using dimension-2 connections. This tree forms the "top" of the $N^2 - 1$ node complete binary tree. The root of this tree is at 10. The leaves of this tree are found in the nodes $k0$ where $N/2 \le k \le N - 1$. Each of these leaves now becomes the root of two subtrees as described next.

Let $k^l = 2k - N$ and $k^r = 2k - N + 1$. The left child of $k0$ is $k^l 1$ and the right child of $k0$ is $k^r 1$ (see Figure 5). The connection between $k0$ and $k^r 1$ is realized by a path of length 2 in $S^2(N)$. The path from $k0$ to $k^r 1$ is formed by the following edges:

1. $k0$ is connected to $k1$ by an exchange edge in dimension-1.
2. $k1$ is connected to $k^r 1$ by a shuffle edge in dimension-2. Since the binary form of k has a '1' in the most significant position, the shuffle of k results in the label value $2k - N + 1$.

The connection between $k0$ and $k^l 1$ is realized by a path of length 3. That path is formed by the following edges:

1. Traverse the two edges as described above, $k0$ to $k1$ to $k^r 1$.
2. $k^r 1$ is connected to $k^l 1$ by an exchange edge in dimension-2.

The dilation of this embedding is clearly 3. The congestion is 2 because the paths to the left and right child of $k0$ coincide with each other but do not coincide with any other path between adjacent nodes in the tree. This completes the case for $r = 2$.

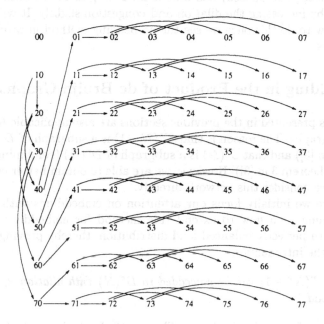

Fig. 5. Embedding of 63-node complete binary tree in the two-dimensional product of shuffle-exchange graphs.

Given that there is an embedding of an $(N^{r-1}-1)$-node complete binary tree in $S^{r-1}(N)$, with the root at node $10\ldots0$ and with congestion 2, and dilation 3, we can construct an embedding of the $N^r - 1$ node complete binary tree in $S^r(N)$ with these same properties. We do this by first embedding the $(N^{r-1}-1)$-node complete binary tree in the N subgraphs isomorphic to $S^{r-1}(N)$ formed if the highest dimension connections are not considered. All nodes within each subgraph have the same value v_{r-1} in their labels. We now embed an $(N-1)$-node complete binary tree in the new dimension in the subgraph isomorphic to $S(N)$ formed by the nodes of the form $v_{r-1}0\cdots0$. The root of this tree is at $10\cdots0$. We form the connections between the $N/2$ leaves of this tree and the roots of the N subtrees in the same manner as in the 2-dimensional case. This time only v_{r-1} and v_{r-2} will be considered when connecting $k0\cdots0$ to its descendents.

Corollary 7. $T(N^r 2^{l-1} - 1)$ can be embedded in $S^r(N)$ with dilation 3, congestion 2, and load $2^l - 1$.

This embedding is obtained by simply replacing the leaves of the embedded tree by an l-level complete binary tree, as in Corollary 4. This proves the result 2.(a) claimed in the introduction.

Note that if $l > 1$, the load of the embedding described in the above corollary is not fully balanced. Half the nodes of $S^r(N)$ will have load $2^l - 1$, while the other half (except one unused) has unit load. It is possible to obtain a better load balance by increasing the dilation and congestion slightly. It will be easier to explain how to do this once we see the embedding method in products of de Bruijn Graphs.

5 Embedding in the Product of de Bruijn Graphs

All the results presented in the previous sections are also applicable to products of de Bruijn graphs. The reason is that $T^r(N - 1)$ is a subgraph of $D^r(N)$ (from Theorem 13 in [3]) and that $S^r(N)$ is a subgraph of $D^r(N)$ (combining Theorem 2 in [4] and Theorem 3 in [3]). However, we are able to obtain better embeddings in $D^r(N)$ if we consider this network directly.

Again here we initially focus our attention on embeddings with unit load. Then we comment on how to extend this method for embedding arbitrarily large trees with perfectly balanced load distribution, thereby proving the result 3 claimed in the introduction.

Theorem 8. $T(N^r - 1)$ can be embedded in $D^r(N)$ with dilation 2, congestion 2, and unit load.

Proof. This proof is similar to that of Theorem 6. In the interest of brevity, we only sketch the basic idea pointing out the differences from the above case.

It was shown in [8] that $D(N)$ contains the $(N - 1)$-node tree as a subgraph. This result can be used for the first dimension connections of Figure 5. The connections in the second dimension require congestion 2, just as for $S^r(N)$, but a dilation of 2 instead of 3. This is because the connection between $k0$ and k^r1 is realized by a path of length 2 in $D^r(N)$. This path is formed by the following edges:

1. $k0$ is connected to $k1$ by an edge in dimension-1.
2. $k1$ is connected to k^r1 by an edge in dimension-2. Since the binary form of k has a '1' in the most significant position, the shuffle of k results in the value $2k - N + 1$.

The connection between $k0$ and k^l1 is realized by a path also of length 2. That path is formed by the following edges:

1. $k0$ is connected to $k1$ by an edge in dimension-1.
2. $k1$ is connected to k^l1 by the edge connecting k to label value $2k - N$ in dimension 2.

This completes the proof for the case of $r = 2$. For $r > 2$, similar arguments as in Theorem 6 apply.

We could use now this result to embed larger trees using the same technique used in Corollaries 4 and 7. Like in these results, the embedding obtained would not fully balance the load among the nodes of the host graph.

However, it is possible to map arbitrarily large complete binary trees to a fixed-size product $D^r(N)$ with perfectly-uniform load distribution. That is, if the product graph contains N^r nodes, we can embed $T((N2^l)^r - 1)$ in it with uniform load of 2^{rl} for all nodes of the product graph, with the exception of one node that will be mapped $2^{rl} - 1$ nodes.

The new embedding can be done in two steps. In the first step, we embed $T((N2^l)^r - 1)$ in $D^r(N2^l)$ with dilation 2, congestion 2, and load 1 by the method of Theorem 8. In the second step, we embed $D^r(N2^l)$ in $D^r(N)$ with dilation 1, congestion 1, and load 2^{rl} by the method given in Corollary 8 of [3]. This induces an embedding for $T((N2^l)^r - 1)$ in $D^r(N)$ with dilation 2, congestion 2, and load 2^{rl}, as claimed in the introduction (result 3).

This result can also be used to obtain an embedding of $T((N2^l)^r - 1)$ in $S^r(N)$ with perfectly-balanced load of 2^{rl} (result 2.(b)). To do so, we simply combine it with an embedding of $D^r(N)$ in $S^r(N)$ with dilation 2, congestion 2, and unit load [3, 5]. This leads to the dilation and congestion values of 4.

6 Remarks

The embedding methods in this paper can also be extended to product graphs made from graphs containing different numbers of nodes for different dimensions.

Theorem 2 implies that for any graph G, if G contains the complete binary tree as a subgraph, then its r-dimensional product can embed the complete binary tree with dilation 2 and congestion 2. Basically, the $G^r(N)$ contains the r-dimensional product of complete binary trees as a subgraph, so the embedding method of Theorem 2 can be applied to this subgraph.

Acknowledgments

The authors wish to thank Darren Broussard and Nancy Eleser for discussions on some of the ideas in this paper. K. Efe's research has been supported by a grant from the Louisiana Board of Regent, contract no: LEQSF(1995-97)-RD-A-33. A. Fernández is on leave from the Departamento de Arquitectura y Tecnología de Computadores, U. Politécnica de Madrid. His research has been partialy supported by the Spanish Ministry of Education under grant PF94 04166960.

References

1. K. Efe, "Embedding Large Complete Binary Trees in Hypercubes with Load Balancing," *Journal of Parallel and Distributed Computing*, vol. 35, no. 1, May 1996, pp. 104-109.
2. K. Efe and A. Fernández, "Mesh Connected Trees: A Bridge between Grids and Meshes of Trees," *IEEE Transactions on Parallel and Distributed Systems.* To appear in 1996.

3. K. Efe and A. Fernández, "Products of Networks with Logarithmic Diameter and Fixed Degree," *IEEE Transactions on Parallel and Distributed Systems*, vol. 6, pp. 963–975, Sept. 1995.
4. R. Feldmann and W. Unger, "The Cube-Connected Cycles Network is a Subgraph of the Butterfly Network," *Parallel Processing Letters*, vol. 2, no. 1, pp. 13–19, 1992.
5. A. Fernández, *Homogeneous Product Networks for Processor Interconnection*. PhD thesis, U. of Southwestern Louisiana, Lafayette, LA, Oct. 1994.
6. J. P. Fishburn and R. A. Finkel, "Quotient Networks," *IEEE Transactions on Computers*, vol. 31, pp. 288–295, Apr. 1982.
7. R. Koch, T. Leighton, B. Maggs, S. Rao, and A. L. Rosenberg, "Work-Preserving Emulations of Fixed-Connection Networks," in *Proceedings of the 21st Annual ACM Symposium on Theory of Computing*, (Seattle), pp. 227–240, May 1989.
8. A. L. Rosenberg, "Product-Shuffle Networks: Toward Reconciling Shuffles and Butterflies," *Discrete Applied Mathematics*, vol. 37/38, pp. 465–488, July 1992.

Clique and Anticlique Partitions of Graphs

Krzysztof Bryś and Zbigniew Lonc

Institute of Mathematics
Warsaw University of Technology
Pl. Politechniki 1
00-661 Warsaw, Poland

Abstract. In the paper we prove that, for a fixed k, the problem of deciding whether a graph admits a partition of its vertex set into k-element cliques or anticliques (i.e. independent sets) is polynomial.

1 Introduction

By a *clique* (respectively *anticlique*) we mean a subset of the vertex set of a graph consisting of pairwise adjacent (resp. nonadjacent) vertices.

The problem of deciding whether, for a fixed k, a graph can be partitioned into k-element cliques is NP-complete (see Garey and Johnson [3]) for $k \geq 3$. There are many results establishing the computional complexity of this problems in important subclasses of the class of all graphs. For example the problem is known to be NP-complete (for $k \geq 3$) in the class of line graphs (c.f. Cohen, Tarsi [2]) or comparability graphs (Lonc [6]). On the other hand its polynomiality was shown (for every k) in such classes as: complements of line graphs (c.f. Alon [1]), cographs and split graphs (c.f. Lonc [5]). In the class of complements of comparability graphs the problem is open (when $k \geq 3$) (see Möhring [8]).

In this paper we consider, for a fixed k, the computational complexity of the problem of existence of a partition of a graph into k-element cliques or anticliques. The following result by Lonc and Truszczyński [7] was the direct motivation of our research.

There exists a positive integer $n_0 = n_0(k)$ such that if a graph G has at least n_0 vertices and $|G| \equiv 0 \pmod{k}$ then the vertex set of G can be partitioned into k-vertex subgraphs $K_k, \overline{K_k}, K_{k-1} \cup K_1$ and $\overline{K_{k-1} \cup K_1}$ (see Figure 1).

We use the standard notation where K_m stands for the complete graph on m vertices, $\overline{G}$ for a complement of G and $G \cup H$ for a disjoint union of graphs G and H.

The key point of the proof of the above statement is an application of the famous Ramsey theorem. Two of the 4 graphs occurring in the above theorem by Lonc and Truszczynski are the graphs induced by a k-element clique and a k-element anticlique. The other two have a bit less regular structure. Therefore it seems to be interesting to ask which vertex sets of graphs can be partitioned into k-element cliques and anticliques alone. This question has been answered by Favaron *et al.* [4] for line graphs and $k = 3$. The paper [4] contains implicitly a complete list of 17 families of graphs for which the condition $|G| \equiv 0 \pmod 3$

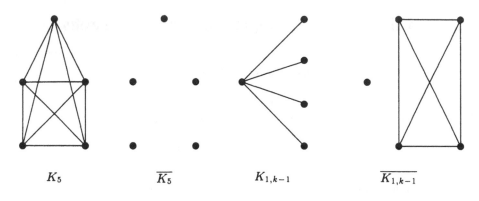

K_5 $\overline{K_5}$ $K_{1,k-1}$ $\overline{K_{1,k-1}}$

Fig. 1. The graphs $K_k, \overline{K_k}, K_{k-1} \cup K_1$ and $\overline{K_{k-1} \cup K_1}$, for $k = 5$.

is not sufficient for a partition of the vertex set of G into 3-element cliques or anticliques to exist. The recognition of the above mentioned families is linear.

The result of [4] was in some sense generalized by Lonc [6] who has shown that the problem of partition of the vertex set of a line graph into k-element cliques or anticliques is polynomial for an arbitrary but fixed k.

The main result of this paper is a theorem saying, for a fixed k (not a part of the instance), that we can check in polynomially many steps if the vertex set of an instance graph admits a partition into k-element cliques or anticliques. The number of steps of the algorithm we describe is enormously large, nevertheless polynomial with respect to the number of vertices in the instance graph. It was not our goal to push down the complexity of the algorithm. It could be done, but the proofs of our results would become much longer and more complicated.

We believe it is hopeless to characterize (for arbitrary k) all graphs G for which the condition $|G| \equiv 0 \pmod{k}$ is not sufficient for a partition of the vertex set of G into k-element cliques or anticliques to exist. However, for $k = 3$ such a characterization seems to be tractable. We show a few examples, when $k = 3$, of infinite families of such graphs in Figure 2. These examples can be generalized easily for larger values of k.

In this paper we mean by CA_k-*decomposition* of a graph G, a partition of the vertex set $V(G)$ of G into k-element cliques or anticliques.

2   Results

Define the following sequences g_n and f_n. Let $g_0 = 5^{5^k}$, $g_{n+1} = 5^{g_n}$, for $n = 0, 1, \ldots, 2k$, and $f_n = g_{2k+1-n}$, for $n = 1, 2, \ldots, 2k + 1$.

For any graph G we define a partition of the vertex set of G into at most $2k + 1$ subsets. Let $C_0 = D_0 = \emptyset$. For $i = 1, 2, \ldots, 2k + 1$ define C_i to be any clique of size f_i in $V(G) - \bigcup_{j=0}^{i-1} D_j$, if it exists. Let D_i be the set of vertices in

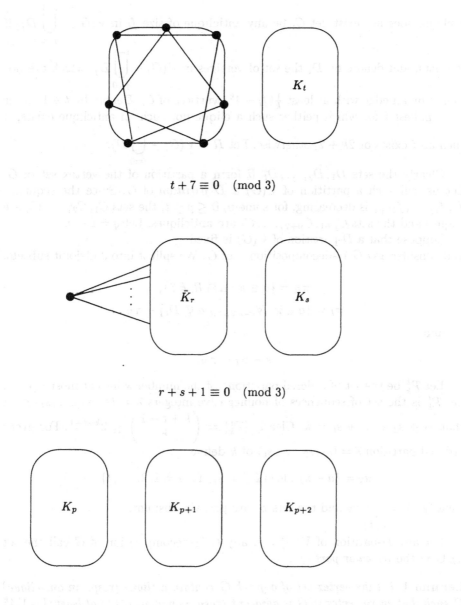

$$t + 7 \equiv 0 \pmod 3$$

$$r + s + 1 \equiv 0 \pmod 3$$

Fig. 2. Three infinite families of "exceptions" for $k = 3$.

$V(G) - \bigcup\limits_{j=0}^{i-1} D_j$ joined by an edge with at least $\frac{1}{k}(f_i - 4^k)$ vertices of C_i. If such

a clique does not exist, let C_i be any anticlique of size f_i in $V(G) - \bigcup\limits_{j=0}^{i-1} D_j$, if

it exists, and denote by D_i the set of vertices in $V(G) - \bigcup\limits_{j=0}^{i-1} D_j$, which are not

joined by an edge with at least $\frac{1}{k}(f_i - 4^k)$ vertices of C_i. Denote by $t + 1$ either the smallest i for which neither such a clique nor such an anticlique exists, if

such an i exists or $2k + 2$, otherwise. Let $R = V(G) - \bigcup\limits_{j=0}^{t} D_j$.

Clearly the sets $D_1, D_2, \ldots, D_t, R$ form a partition of the vertex set of G. Let us call such a partition of $V(G)$ a D-partition of G. Since the sequence $f_1, f_2, \ldots, f_{2k+1}$ is decreasing, for some p, $0 \leq p \leq t$, the sets $C_1, C_2, \ldots, C_p$ are cliques and the sets $C_{p+1}, C_{p+2}, \ldots, C_t$ are anticliques. Let $q = t - p$.

Suppose that a D-partition of $V(G)$ is fixed and consider any CA_k-decomposition π of G. We split π into 3 disjoint subsets: Let

$$\pi_R = \{a \in \pi : a \cap R \neq \emptyset\},$$
$$\pi_I = \{a \in \pi : \forall_{i=1,2,\ldots,t} \; a \nsubseteq D_i\} - \pi_R$$

and

$$\pi - \pi_I - \pi_R.$$

Let $\mathcal{P}_k^t$ be the set of ordered partitions of the number k into at most t parts, i.e. $\mathcal{P}_k^t$ is the set of sequences of nonnegative integers $\bar{s} = (s_1, s_2, \ldots, s_t)$ such that $s_1 + s_2 + \ldots + s_t = k$. Clearly $|\mathcal{P}_k^t| = \binom{k+t-1}{k} \leq 2^{k+t-1}$. For every ordered partition $\bar{s} = (s_1, s_2, \ldots, s_t)$ of k define

$$\pi_{\bar{s}} = \{a \in \pi_I : |a \cap D_i| = s_i, \text{ for } i = 1, 2, \ldots, t\}.$$

Clearly $\pi_I = \bigcup\limits_{\bar{s} \in \mathcal{P}_k^t} \pi_{\bar{s}}$ and the sets $\pi_{\bar{s}}$ are pairwise disjoint.

For any D-patition of $V(G)$ and any CA_k-decomposition of G call the set $\pi_R \cup \pi_I$ the *irregular part* of π.

Lemma 1 *Let the vertex set of a graph G contain a clique (resp. an anticlique) C such that every vertex in G is adjacent (resp. nonadjacent) to at least $(k-1)4^k$ vertices in C. If $|G| \equiv 0 \pmod{k}$ then G has a CA_k-decomposition.*

Proof. Let us prove the version of the lemma for cliques. The proof of the other version is dual.

Delete from $V(G) - C$ k-cliques and k-anticliques as many times as possible. By the Ramsey theorem the set of vertices R we obtain this way has less than

$R(k,k) \leq \binom{2k-2}{k-1} < 4^k$ vertices, where $R(k,k)$ stands for the appropriate Ramsey number.

Since every vertex x in R has at least $(k-1)4^k$ neighbours in C, we can choose $k-1$ of them adjacent to x. They form a k-clique which we delete. We can repeat this procedure for every vertex of R (the number $(k-1)4^k$ is sufficiently large). We end up with a clique C' of the size divisible by k. The k-cliques and k-anticliques form the required CA_k-decomposition of G. $\square$

Theorem 2 Let $D_1, D_2, \ldots, D_t, R$ be a D-partition of G and let $C_1, C_2, \ldots, C_t$ be the coresponding cliques and/or anticliques. If there are at least $k+1$ cliques or at least $k+1$ anticliques in the sequence $C_1, C_2, \ldots, C_t$ and $|G| \equiv 0 \pmod{k}$ then G admits a CA_k-decomposition.

Theorem 3 If G has a CA_k-decomposition then for every D-partition of $V(G)$, G has a CA_k-decomposition with the irregular part of size at most $4^{f_{t+1}}$.

We omit complicated proofs of these two theorems.

Corollary 4 The following conditions are equivalent
  (a) G has a CA_k-decomposition,
  (b) for every D-partition of $V(G)$ there is a subset U, $R \subseteq U \subseteq V(G)$, $|U| \leq k \cdot 4^{f_{t+1}}$, such that the graph induced by U has a CA_k-decomposition and

$$|D_i - U| \equiv 0 \pmod{k}, \text{ for } i = 1, 2, \ldots, t.$$

  (c) there exists D-partition of $V(G)$ such that there is a subset U, $R \subseteq U \subseteq V(G)$, $|U| \leq k \cdot 4^{f_{t+1}}$, such that the graph induced by U has a CA_k-decomposition and

$$|D_i - U| \equiv 0 \pmod{k}, \text{ for } i = 1, 2, \ldots, t. \tag{1}$$

Proof. $(a) \Rightarrow (b)$. Let U be the set of vertices of the members of the irregular part of the CA_k-decomposition of G appearing in the condition (b) of Theorem 3.
$(c) \Rightarrow (a)$. It suffices to show that $D_i - U$, for $i = 1, 2, \ldots, t$, has a CA_k-decomposition.
Suppose C_i is a clique (the other case is analogous), $i = 1, 2, \ldots, t$. Clearly

$$|C_i - U| \geq f_i - k \cdot 4^{f_{t+1}} \geq f_t - k \cdot 4^{f_{t+1}} = 5^{f_{t+1}} - k \cdot 4^{f_{t+1}} \geq$$
$$5^{f_{2k+1}} - k \cdot 4^{f_{2k+1}} = 5^{5^{5^k}} - k \cdot 4^{5^{5^k}} \geq (k-1) \cdot 4^k.$$

Thus every vertex of $D_i - U$ is joined to at least $(k-1)4^k$ vertices of the clique $C_i - U$. We are done by Lemma 1 and (1).

The trivial implication $(b) \Rightarrow (c)$ completes the proof. $\square$

We are now ready to give a polynomial algorithm deciding whether a given graph G admits a CA_k-decomposition.

Algorithm

1. Construct a D-partition $D_1, D_2, ..., D_t, R$ of $V(G)$
   and the cliques and/or anticliques $C_1, C_2, ..., C_t$.
2. Is the number of cliques or anticliques in the set $\{C_1, C_2, ..., C_t\}$ greater
   than k?
   If the answer is YES then STOP (G admits a CA_k-decomposition), otherwise
   go to 3.
3. For every subset $U \subseteq V(G)$ such that $|U| \le k \cdot 4^{f_{t+1}}$ and $|D_i - U| \equiv 0$
   (mod k), $i = 1, 2, ..., t$, check if the graph induced by U admits a CA_k-
   decomposition.
   If the answer is YES then STOP (G admits a CA_k-decomposition), otherwise
   go to 4.
4. STOP (G does not admits a CA_k-decomposition).

The correctness of the above algorithm follows easily from Theorem 2 and Corollary 4. Let us check the polynomiality. The construction of a D-partition of $V(G)$ and the sets $C_1, C_2, ..., C_t$ takes polynomially many steps with respect to the order n of the instance graph G which follows from the facts that the sizes of C_i's are bounded by a function of k not depending on n and that $t \le 2k + 1$. The number of sets U to be considered in step 3 is also bounded by a constant with respect to n because $|U| \le k \cdot 4^{f_{t+1}} \le k \cdot 4^{f_1}$. Moreover, for the same reason, it can be checked in polynomially many steps whether the graph induced by U admits a CA_k-decomposition (by exploring all possibilities). The polynomiality of the remaining steps of the Algorithm is obvious. Therefore we have shown the following theorem.

Theorem 5 *For every fixed positive integer, the problem of deciding whether a given graph G has a CA_k-decomposition is polynomial.*

References

1. N. Alon: A note on the decomposition of graphs into isomorphic matchings, *Acta Math. Acad. Sci. Hung.* **42** (1983), 221-223.
2. E. Cohen, M. Tarsi: NP-completeness of graph decomposition problem, *Journal of Complexity* **7** (1991), 200- 212.
3. M.R. Garey, D.S. Johnson, Computers and Intractability, A guide to the Theory of NP-Completeness, Freeman, San Francisco 1979.
4. O. Favaron, Z. Lonc, M. Truszczyński: Decomposition of graphs into graphs with three edges, *Ars Combinatoria* **20** (1985), 125-146.
5. Z. Lonc, Delta-system decompositions of graphs, to appear in *Discrete Applied Mathematics*.
6. Z. Lonc, On complexity of some chain and antichain partition problem, in Graph-Theoretic Concepts in Computer Science (ed. G. Schmidt and R. Berghammer), Lecture Notes in Computer Science **570**, Springer-Verlag 1992, 97-104.
7. Z. Lonc and M. Truszczyński, Decomposition of large uniform hypergraphs, *Order* **1** (1985) 345-350.
8. R.H. Möhring, Problem 9.10, in Graphs and Order (ed. I. Rival), Reidel Publishing Co., Dordrecht 1985, 583.

Optimal Parallel Routing in Star Graphs[*]
(Extended Abstract)

CHI-CHANG CHEN[1] and JIANER CHEN[2]

[1]Department of Computer Science and Engineering, Tatung Institute of Technology,
Taipei 10451, Taiwan, ROC
[2]Department of Computer Science, Texas A&M University
College Station TX 77843-3112, USA

Abstract. Star graphs have been recently proposed as attractive alternatives to the popular hypercube for interconnecting processors on a parallel computer. In this paper, we present an efficient algorithm that constructs an optimal parallel routing in star graphs. Our result improves previous results for the problem.

1   Introduction

The star graph [2] has received considerable attention recently by researchers as a graph model for interconnection network. It has been shown that it is an attractive alternative to the widely used hypercube model. Like the hypercube, the star graph is vertex- and edge-symmetric, strongly hierarchical, and maximally fault tolerant. Moreover, it has a smaller diameter and degree while comparing with hypercube of comparable number of vertices.

The rich structural properties of the star graphs have been studied by many researchers. The n-star graph S_n is a degree $n-1$, $(n-1)$-connected, and vertex symmetric Cayley graph [1, 2]. Jwo, Lakshmivarahan, and Dhall [15] showed that the star graphs are hamiltonian. Qiu, Akl, and Meijer [22] showed that the n-star graph can be decomposed into $(n-1)!$ node-disjoint paths of length $n-1$, and can be decomposed into $(n-2)!$ node-disjoint cycles of length $(n-1)n$. Results in embedding hypercubes into star graphs have been obtained by Nigam, Sahni, and Krishnamurthy [20] and by Miller, Pritikin, and Sudborough [18]. Broadcasting on star graphs have also been studied recently [3, 4, 14, 23].

Routing on star graphs was first studied by Akers and Krishnamurthy [2] who derived a formula for the length of the shortest path between any two nodes in a star graph and developed an efficient algorithm for constructing such a path. Recently, parallel routing, i.e., constructing node-disjoint paths, on star graphs has received much attention. Sur and Srimani [24] demonstrated that $n-1$ node-disjoint paths can be constructed between any two nodes in S_n in polynomial time. Dietzfelbinger, Madhavapeddy, and Sudborough [11] derived an improved algorithm that constructs $n-1$ node-disjoint paths of length bounded

---

[*] The research was supported in part by Tatung Institute of Technology under the grant B84017 and by the United States National Science Foundation grant CCR-9110824. The corresponding author is Jianer Chen, email: chen@cs.tamu.edu.

by 4 plus the diameter of S_n. The algorithm was further improved by Day and Tripathi [10] who developed an $O(n^2)$ time algorithm that constructs $n -$ 1 node-disjoint paths of length bounded by 4 plus the distance from u to v in S_n. The problem was also investigated by Jwo, Lakshmivarahan, and Dhall [16]. Misic and Jovanovic [19] derived a general algebraic expression for all (not necessarily node-disjoint) shortest paths between any two nodes in S_n. Palis and Rajasekaran [21], Dietzfelbinger, Madhavapeddy, and Sudborough [11], Qiu, Akl, and Meijer [22], and Chen and Chen [9] have considered the problem of node-disjoint paths between two sets of nodes in a star graph.

In this paper, we will improve the previous results on node-to-node routing in star graphs by developing an efficient algorithm that constructs optimal parallel routing between any two nodes in a star graph. More specifically, let u and v be any two nodes in the n-star graph S_n and let $dist(u,v)$ be the distance from u to v in S_n. The *bulk length* of a group of $n-1$ node-disjoint paths connecting u and v in S_n is defined to be the length of the longest path in the group. Define the *bulk distance $Bdist(u,v)$* between u and v to be the minimum bulk length over all groups of $n-1$ node-disjoint paths connecting u and v in S_n. We develop an $O(n^2 \log n)$ time algorithm that, given two nodes u and v in S_n, constructs a group of $n-1$ node-disjoint paths of bulk length $Bdist(u,v)$ that connect the nodes u and v in S_n.

Our algorithm involves careful analysis on the lower bound on the bulk distance $Bdist(u,v)$ between two nodes u and v in S_n, a non-trivial reduction from the parallel routing problem on star graphs to a combinatorial problem called *Partition Matching*, a subtle solution to the Partition Matching problem, and a number of routing algorithms on different kinds of pairs of nodes in a star graph. The basic idea of the algorithm can be roughly described as follows. Let u and v be two nodes in the n-star graph S_n. According to Day and Tripathi [10], the bulk distance $Bdist(u,v)$ is equal to $dist(u,v) + b$, where $b = 0$, 2, or 4. We first derive a necessary and sufficient condition for a pair of nodes u and v to have bulk distance $dist(u,v) + 4$. For a pair u and v whose bulk distance is less than $dist(u,v) + 4$, we develop an efficient algorithm that constructs a group of $n-1$ node-disjoint paths of bulk length $dist(u,v) + 2$ between u and v. Finally, an efficient algorithm, which is obtained from a reduction of an efficient algorithm solving the Partition Matching problem, is developed that constructs the maximum number of node-disjoint shortest paths (of length $dist(u,v)$) between u and v. Combining all these analysis and algorithms gives us an efficient algorithm that constructs an optimal parallel routing on star graphs. We should also point out that the running time of our algorithm is almost optimal (differs at most by a $\log n$ factor) since a lower bound $\Omega(n^2)$ on running time of parallel routing algorithms on star graphs can be easily derived.

2   Preliminary

A permutation $u = a_1 a_2 \cdots a_n$ of the elements in the set $\{1, 2, \ldots, n\}$ can be given by a product of disjoint cycles [5], which is called the *cycle structure* of the permutation. The cycle structure of a permutation u on $\{1, 2, \ldots, n\}$ can be easily constructed from u in time $O(n)$. A cycle is *nontrivial* if it contains more than one symbol. Otherwise the cycle is *trivial*. A $\pi[1, i]$ *transposition* on u is to exchange the positions of the first symbol and the ith symbol in u: $\pi[1, i](u) = a_i a_2 a_3 \cdots a_{i-1} a_1 a_{i+1} \cdots a_n$. It is sometimes more convenient to write the transposition $\pi[1, i](u)$ as $\pi[a_i](u)$ to indicate that the transposition exchanges the positions of the first symbol and the symbol a_i in u. Let us consider how a transposition changes the cycle structure of a permutation. Write u in its cycle structure

$$u = (a_{11} \cdots a_{1n_1} \, 1)(a_{21} \cdots a_{2n_2}) \cdots (a_{k1} \cdots a_{kn_k})$$

If a_i is not in the cycle containing symbol 1, then $\pi[1, i]$ "merges" the cycle containing the symbol 1 with the cycle containing a_i. More precisely, suppose that $a_i = a_{21}$ (note that each cycle can be cyclically permuted and the order of the cycles is not important), then the permutation $\pi[1, i](u)$ will have the following cycle structure:

$$\pi[1, i](u) = (a_{21} \cdots a_{2n_2} a_{11} \cdots a_{1n_1} \, 1)(a_{31} \cdots a_{3n_3}) \cdots (a_{k1} \cdots a_{kn_k})$$

If a_i is in the cycle containing the symbol 1, then $\pi[1, i]$ "splits" the cycle. More precisely, suppose that $a_i = a_{1j}, j > 1$ (note that $a_{11} = a_1$ and we assume $i > 1$), then $\pi[1, i](u)$ will have the following cycle structure:

$$\pi[1, i](u) = (a_{11} \cdots a_{1j-1})(a_{1j} \cdots a_{1n_1} 1)(a_{21} \cdots a_{2n_2}) \cdots (a_{k1} \cdots a_{kn_k})$$

Note that if a symbol a_i is in a trivial cycle in the cycle structure of a permutation $u = a_1 a_2 \cdots a_n$, then the symbol is in its "correct" position, i.e., $a_i = i$, and that if a symbol is in a nontrivial cycle, then the symbol is not in its correct position. Denote by ε the identity permutation $\varepsilon = 12 \cdots n$.

The n-*star graph* S_n is an undirected graph consisting of $n!$ nodes labeled with the $n!$ permutations on symbols $1, 2, \ldots, n$. There is an edge between two nodes u and v in S_n if and only if there is a transposition $\pi[1, i]$, $2 \leq i \leq n$, such that $\pi[1, i](u) = v$. The n-star graph is an $(n - 1)$-connected vertex-symmetric Cayley graph [2] (with generating set $\{\pi[1, 2], \pi[1, 3], \ldots, \pi[1, n]\}$ for the symmetric group of order n). Therefore, a parallel routing problem between two arbitrary nodes in S_n can be easily reduced in time $O(n^2)$ to the parallel routing problem between a node and the identity node ε in S_n. In the rest of this paper, we will only concentrate on the latter problem. A path from a node u in S_n to the identity node ε corresponds to a sequence of applications of the transpositions $\pi[1, i]$, $2 \leq i \leq n$, starting from the permutation u and ending at the permutation ε. We will write $dist(u)$ and $Bdist(u)$ for $dist(u, \varepsilon)$ and $Bdist(u, \varepsilon)$, respectively.

Let u be a node in the n-star graph with cycle structure $u = c_1 \cdots c_k e_1 \cdots e_m$, where c_i are nontrivial cycles and e_j are trivial cycles. If we further let $l = \sum_{i=1}^{k} |c_i|$, where $|c_i|$ denotes the number of symbols in the cycle c_i, then the distance from the node u to the identity node ε is given by the following formula [2].

$$dist(u) = \begin{cases} l + k & \text{if symbol 1 is in a trivial cycle} \\ l + k - 2 & \text{if symbol 1 is in a nontrivial cycle} \end{cases}$$

Combining this formula with the above discussion on the effect of applying a transposition on a permutation, we derive the following necessary rules for tracing a shortest path from the node u to the identity node ε in S_n.

Shortest Path Rules

Rule 1. If symbol 1 is in a trivial cycle in u, then in the next node on any shortest path from u to ε, the cycle $\{1\}$ is merged into a nontrivial cycle c_i in u. This corresponds to applying transposition $\pi[a]$ to u with $a \in c_i$;

Rule 2. If symbol 1 is in a nontrivial cycle $c_1 = (a_{11} \cdots a_{1d})$ in u, where $a_{1d} = 1$, then in the next node on any shortest path from u to ε, either the cycle c_1 is merged with a nontrivial cycle $c_i \neq c_1$ (this corresponds to applying transposition $\pi[a]$ to u, where $a \in c_i$), or the symbol a_{11} is deleted from the cycle c_1 (this corresponds to applying transposition $\pi[a_{12}]$ to u).

Fact 1. A shortest path from u to ε in S_n is obtained by a sequence of applications of the Shortest Path Rules, starting from the permutation u.

Fact 2. If a symbol $a \neq 1$ is in a trivial cycle in u, then a will stay in a trivial cycle in any node on a shortest path from u to ε.

Fact 3. If an edge $[u, v]$ in S_n does not lead to a shortest path from u to ε, then $dist(v) = dist(u) + 1$.

Parallel routing from a node u to ε in S_n is particularly simple when the symbol 1 is in a trivial cycle in u [10, 16]. For completeness, we roughly describe the routing process for this case. Let u be such a node in S_n. For each symbol i, $2 \leq i \leq n$, first apply the transposition $\pi[i]$ to u, then apply any sequence of Shortest Path Rules to construct a shortest path from the node $\pi[i](u)$ to ε. It is not hard to verify that this produces $n - 1$ node-disjoint paths from u to ε. Moreover, if i is in a nontrivial cycle in u, then the path constructed via $\pi[i](u)$ is a shortest path from u to ε, while if i is in a trivial cycle in u, then the path constructed via $\pi[i](u)$ has length $dist(u) + 2$. Therefore, if $\{1\}$ is the only trivial cycle in u then the algorithm constructs $n - 1$ node-disjoint paths of length $dist(u) = Bdist(u)$, and if there are other trivial cycles in u then the algorithm constructs $n - 1$ node-disjoint paths of length at most $dist(u) + 2$, which equals the bulk distance $Bdist(u)$ in this case.

Therefore, throughout the rest of this paper, we discuss the parallel routing problem in star graphs based on the following assumption:

Assumption A.
The node u in the n-star graph S_n has cycle structure $u = c_1 \cdots c_k e_1 \cdots e_m$, where c_i are nontrivial cycles and e_j are trivial cycles, and cycle c_1 is of form $(a_{11} a_{12} \cdots a_{1d})$, where $d \geq 2$ and $a_{1d} = 1$.

3 Nodes with bulk distance $dist(u) + 4$

According to Day and Tripathi [10], for any node u in a star graph, the bulk distance $Bdist(u)$ is equal to $dist(u) + b$, where $b = 0, 2$, or 4. In this section, we derive a necessary and sufficient condition for a node u to have bulk distance $dist(u) + 4$.

Let P be a path in S_n from u to ε. We say that the path P *leaves u with symbol a* if the second node on P is $\pi[a](u)$, and we say that the path P *enters ε with symbol a'* if the only nontrivial cycle in the node before ε on P is $(a'1)$.

Lemma 1. *Let u be a node in the n-star graph S_n as described in* **Assumption A.** *If $m > \min\{2^{k-2}, 1 + \sum_{i=2}^{k} |c_i|\}$, then any group of $n - 1$ node-disjoint paths between u and ε has bulk length at least $dist(u) + 4$.*

Proof. (Sketch)   Assume that P_1, ..., P_{n-1} are $n - 1$ node-disjoint paths between u and ε of bulk length bounded by $dist(u) + 2$. We will show that in this case we must have $m \leq \min\{2^{k-2}, 1 + \sum_{i=2}^{k} |c_i|\}$.

Suppose that a path P_i leaves u with a symbol e in a trivial cycle in u. By the Shortest Path Rules, the node $\pi[e](u)$ does not lead to a shortest path from u to ε. By Fact 3, $dist(\pi[e](u)) = dist(u) + 1$. On the other hand, the length of P_i is bounded by $dist(u) + 2$. Thus starting from the node $\pi[e](u)$, the path P_i must strictly follow the Shortest Path Rules. In particular, no node on the path P_i (including the node $\pi[e](u)$) can contain a cycle of form $(\cdots b1)$, where $b \neq a_{1d-1}$ and $b \notin \bigcup_{i=2}^{k} c_i$. This implies that the path P_i must enter ε with a symbol in the set $\{a_{1d-1}\} \cup (\bigcup_{i=2}^{k} c_i)$. Since there are exactly m of the paths P_1, ..., P_{n-1} leaving with symbols in trivial cycles in u, and since all these paths are node-disjoint, we conclude that there are at least m different symbols in the set $\{a_{1d-1}\} \cup (\bigcup_{i=2}^{k} c_i)$. That is, $m \leq 1 + \sum_{i=2}^{k} |c_i|$.

Now again consider the path P_i leaving u with a symbol e in a trivial cycle in u. Let u_i be the first node on P_i in which the symbol 1 is in a trivial cycle. Note that $u_i \neq u$ and $u_i \neq \pi[e](u)$. Moreover, since the node $\pi[e](u)$ has cycle structure $c_1' c_2 \cdots c_k$ (we ignore the trivial cycles here), where c_1' is obtained by merging the cycle c_1 with the cycle $\{e\}$ in u, and the path P_i strictly follows the Shortest Path Rules after the node $\pi[e](u)$, every nontrivial cycles in the node u_i is a cycle in the set $\{c_2, \ldots, c_k\}$. Therefore, the node u_i, hence the path P_i, corresponds to a subset of the set $\{c_2, \ldots, c_k\}$.

Similarly, if a path P_h in P_1, ..., P_{n-1} enters ε with a symbol e in a trivial cycle in u, then we can prove that there is a node u_j on P_h in which every nontrivial cycle is in the set $\{c_2, \ldots, c_k\}$.

As we have proved previously, no path in P_1, ..., P_{n-1} can both leave u and enter ε with symbols in trivial cycles in u. Thus, there are exactly $2m$ of these paths that either leave u or enter ε with a symbol in a trivial cycle in u. Now since each of these $2m$ paths corresponds to a subset of the set $\{c_2, \ldots, c_k\}$ and all these paths are node-disjoint, we conclude that the set $\{c_2, \ldots, c_k\}$ has at least $2m$ different subsets. That is, $2m \leq 2^{k-1}$, or equivalently, $m \leq 2^{k-2}$. $\square$

Now we show that when $m \leq \min\{2^{k-2}, 1 + \sum_{i=2}^{k} |c_i|\}$, we can always construct $n - 1$ node-disjoint paths between u and ε of bulk length $dist(u) + 2$.

For each symbol e in a trivial cycle in u, we construct a path that leaves u with e and enters ε with a symbol a in $\{a_{1d-1}\} \cup (\bigcup_{i=2}^{k} c_i)$ and another path that leaves u with a symbol a' in $\{a_{12}\} \cup (\bigcup_{i=2}^{k} c_i)$ and enters ε with e. Note that the symbol e always makes the paths not follow the Shortest Path Rules but the symbols a and a' so chosen to pair the symbol e are "nice" symbols so that they do not further increase the length of the paths, thus keeping the paths to have length $dist(u) + 2$. To make all these paths node-disjoint, we associate each such paths with a different subset of $\{c_2, \ldots, c_k\}$. Since $m \leq \min\{2^{k-2}, 1 + \sum_{i=2}^{k} |c_i|\}$, we have enough symbols in $\{a_{12}, a_{1d-1}\} \cup (\bigcup_{i=2}^{k} c_i)$ and enough subsets in $\{c_2, \ldots, c_k\}$ for this process.

For each a of the rest unused symbols in $\bigcup_{i=2}^{k} c_i$, we construct a path leaving u with a and entering ε with a as follows. We first apply the transposition $\pi[a]$ to merge the cycle containing a with the cycle c_1 (this follows the Shortest Path Rules). Then we apply the transposition $\pi[1]$ to make the symbol 1 in a trivial cycle (this does not follow the Shortest Path Rules). Finally we merge the cycle $\{1\}$ with the cycle containing a to form a cycle of form $(\cdots a1)$ (this again follows the Shortest Path Rules), and follow a shortest path from the resulting node to ε on which each node has a cycle of form $(\cdots a1)$. Since there is only one edge in the constructed path that does not follow the Shortest Path Rules, the path has length $dist(u) + 2$. Moreover, the fact that all nodes on the path (except the first three) have a distinguished cycle of form $(\cdots a1)$ makes this path node-disjoint from other constructed paths.

Finally, for each symbol a_{1h}, $h = 3, 4, \ldots, d$ in the cycle c_1 in u, we construct a path leaving u with a_{1h} and entering ε with a_{1h-1} as follows. We first apply the transposition $\pi[a_{1h}]$ on u to split the cycle c_1 into two cycles $(a_{11} \cdots a_{1h-1})$ and $(a_{1h} \cdots a_{1d})$ (this does not follow the Shortest Path Rules), then we use Rule 2 in the Shortest Path Rules to delete each symbol in the cycle $(a_{1h} \cdots a_{1d})$ until symbol 1 is in a trivial cycle. Now we merge the trivial cycle $\{1\}$ with the cycle $(a_{11} \cdots a_{1h-1})$ to get a cycle $(a_{11} \cdots a_{1h-1}1)$ (this follows the Shortest Path Rules). Now we follow a shortest path from the resulting node to ε and keep the cycle pattern $(\cdots a_{h-1}1)$ in each node along the path. Again the length of the constructed path is $dist(u) + 2$. Since the first part of the path keeps a distinguished cycle $(a_{11} \cdots a_{1h-1})$ and the second part of the path keeps a distinguished cycle of form $(\cdots a_{1h-1}1)$, the path is node-disjoint with the other constructed paths.

Lemma 2. *Let u be a node in the n-star graph S_n as described in* **Assumption A.** *If $m \leq \min\{2^{k-2}, 1 + \sum_{i=2}^{k} |c_i|\}$, then a group of $n - 1$ node-disjoint paths of bulk length $dist(u) + 2$ between u and ε can be constructed in time $O(n^2 \log n)$.*

Proof. (Sketch) According to the space limit here, we omit the formal description and analysis of the algorithm. The basic idea of the algorithm has been illustrated as above. For more details, the reader is referred to our complete paper [8]. □

Theorem 3. *Let u be a node in the n-star graph S_n as described in* **Assumption A**. *The bulk distance $Bdist(u)$ from the node u to the identity ε is $dist(u)+4$ if and only if $m > \min\{2^{k-2}, 1 + \sum_{i=2}^{k} |c_i|\}$.*

4 The maximum number of node-disjoint shortest paths

In this section we diverge to a slightly different problem. Let u be a node in the n-star graph S_n as described in **Assumption A**. How many node-disjoint shortest paths (of length $dist(u)$) can we find from u to ε?

This problem is closely related to a combinatorial problem, *Maximum Partition Matching*, formulated as follows:

> Let $S = \{C_1, C_2, \ldots, C_k\}$ be a collection of subsets of the universal set $U = \{1, 2, \ldots, n\}$ such that $\bigcup_{i=1}^{k} C_i = U$, and $C_i \cap C_j = \phi$ for all $i \neq j$. A *partition matching* (of order m) of S consists of two ordered subsets $L = \{a_1, a_2, \ldots, a_m\}$ and $R = \{b_1, b_2, \ldots, b_m\}$ of m elements of U (the subsets L and R may not be disjoint), together with a sequence of m distinct partitions of S: $(\mathcal{A}_1, \mathcal{B}_1), (\mathcal{A}_2, \mathcal{B}_2), \ldots, (\mathcal{A}_m, \mathcal{B}_m)$ such that for all $i = 1, \ldots, m$, a_i is contained in a subset in the collection $\mathcal{A}_i$ and b_i is contained in a subset in the collection $\mathcal{B}_i$. The *Maximum Partition Matching* problem is to construct a partition matching of order m for a given collection S with m maximized.

Theorem 4. *Maximum Partition Matching problem is solvable in time $O(n^2 \log n)$.*

Proof. (Sketch)   A highly nontrivial algorithm solving the Maximum Partition Matching problem is presented in [7]. The algorithm runs in time $O(n^2 \log n)$. Here we can only give a (very) rough description of the algorithm. The interested readers are referred to our complete paper [7].

Let $S = \{C_1, C_2, \ldots, C_k\}$ be the given collection. If k is large enough to satisfy the condition $2^k \geq 2n$, then we can use a technique similar to the one described for Lemma 2 to find a partition matching that pairs the maximum number of pairs of form (a, b), where a and b belong to different sets in the collection S.

When the number of sets in the collection S is very small (i.e., $2^k < 2n$), then we use a greedy strategy to find a partition matching. More specifically, we check each partition of S and see if it can be used to expand the current partition matching. A technique called *chain justification* should be used to find out those "implicitly useful" partitions of S. The process stops when there is neither explicitly useful nor implicitly useful partitions of S left. Now based on a technical condition, we can conclude either the constructed partition matching is the maximum or applying a process called *partition flipping* on the constructed partition matching gives a maximum partition matching. $\square$

Now we show how Theorem 4 can be used to find the maximum number of node-disjoint shortest paths from a node u to the identity node ε in star graphs.

Lemma 5. *Let u be a node as described in* **Assumption A.** *Then the number of node-disjoint shortest paths from u to ε cannot be larger than $1 + \sum_{i=2}^{k} |c_i|$.*

Proof. According to the Shortest Path Rules, only the path leaving u with a symbol in $\{a_{12}\} \cup (\bigcup_{i=2}^{k} c_i)$ can be a shortest path from u to ε. $\quad\square$

Lemma 6. *Let u be a node as described in* **Assumption A.** *Then the number of node-disjoint shortest paths from u to ε cannot be larger than 2 plus the number of partitions in a maximum partition matching in the collection $S = \{c_2, \ldots, c_k\}$, where c_i are regarded as sets of symbols in $\{1, \ldots, n\}$.*

Proof. Let P_1, ..., P_s be s node-disjoint shortest paths from u to ε. For each path P_i, let u_i be the first node on P_i in which the symbol 1 is in a trivial cycle. The node u_i is obtained by repeatedly applying *Rule 2* in the Shortest Path Rules, starting from the node u. It is easy to prove, by induction, that for any node v from u to u_i on the path P_i, the only possible nontrivial cycle in v that is not in the set $\{c_2, \ldots, c_k\}$ is the one that contains the symbol 1. In particular, all nontrivial cycles in the node u_i are in the set $\{c_2, \ldots, c_k\}$. Therefore, the node u_i, thus the path P_i, corresponds to a subcollection $\mathcal{B}_i$ of the collection $S = \{c_2, \ldots, c_k\}$.

Assume that the path P_i leaves the node u with a symbol b_i. Since P_i is a shortest path from u to ε, the symbol b_i is in the set $\{a_{12}\} \cup (\bigcup_{i=2}^{k} c_i)$ and b_i is contained in the cycle containing symbol 1 in the node $\pi[b_i](u)$ on the path P_i. According to *Rule 2* of the Shortest Path Rules, once b_i is contained in the cycle containing symbol 1, it will stay in the cycle containing symbol 1 until it is put into a trivial cycle. In particular, the symbol b_i is not in the set $\bigcup_{j \in \mathcal{B}_i} c_j$.

Now consider the last edge on the path P_i. Suppose that the path P_i enters ε with a symbol d_i. Then d_i is contained in a nontrivial cycle in the node before ε on the path P_i. By Fact 2, d_i is also in a nontrivial cycle in the node u_i, that is $d_i \in \bigcup_{j \in \mathcal{B}_i} c_j$. The only exception is $d_i = a_{1d-1}$ (in this case $u_i = \varepsilon$).

Now we let $\mathcal{A}_i = S - \mathcal{B}_i$. Then we can conclude that except at most two paths P_1 and P_2, each P_i of the other paths P_3, ..., P_s must leave u with a symbol $b_i \in \mathcal{A}_i$, and enter ε with a symbol $d_i \in \mathcal{B}_i$. (The two exceptional paths P_1 and P_2 may leave u with the symbol a_{12} or enter ε with the symbol a_{1d-1}.)

Since the s paths P_1, ..., P_s are node-disjoint, the symbols b_3, ..., b_s are all pairwise distinct, and the symbols d_3, ..., d_s are all pairwise distinct. Moreover, since all nodes u_3, ..., u_s are pairwise distinct, the collections $\mathcal{B}_3$, ..., $\mathcal{B}_s$ of cycles in $\{c_2, \ldots, c_k\}$ are also all pairwise distinct. Consequently, the partitions $(\mathcal{A}_3, \mathcal{B}_3)$, ..., $(\mathcal{A}_s, \mathcal{B}_s)$ form a partition matching of the collection $S = \{c_2, \ldots, c_k\}$.

This concludes that s cannot be larger than 2 plus the number of partitions in a maximum partition matching in the collection $S = \{c_2, \ldots, c_k\}$. $\quad\square$

Now we show how we construct the maximum number of node-disjoint shortest paths from the node u described in **Assumption A** to the identity node ε in the n-star graph. We first show how to route a single shortest path from u to ε, given a partition $(\mathcal{A}, \mathcal{B})$ of the collection $S = \{c_2, \ldots, c_k\}$, and a pair of symbols b and d, where b is in $\mathcal{A}$ and d is in $\mathcal{B}$. We allow b to be a_{12} — in this case $d \in \cup_{i=2}^{k} c_i$, $\mathcal{A} = \phi$, and $\mathcal{B} = \mathcal{S}$, and allow d to be a_{1d-1} — in this case $b \in \cup_{i=2}^{k} c_i$, $\mathcal{A} = \mathcal{S}$, and $\mathcal{B} = \phi$.

First apply transposition $\pi[b]$ to u to make b be contained in the cycle containing symbol 1. Then arbitrarily merge all other cycles in $\mathcal{A}$ to the cycle containing symbol 1. After this, we repeatedly delete symbols from the cycle containing symbol 1 until the symbol 1 is in a trivial cycle. If $\mathcal{A} = \mathcal{S}$, then we have reached the node ε. Otherwise, we merge the trivial cycle $\{1\}$ into the cycle containing symbol d to form a cycle of form $(\cdots d1)$. Now we arbitrarily merge all other cycles in $\mathcal{B}$ into the cycle containing symbol 1, followed by repeatedly deleting symbols from the cycle containing symbol 1. This completes the construction of a shortest path from u to ε. This process will be called the SINGLE ROUTING with partitions $(\mathcal{A}, \mathcal{B})$ and symbol pair (b, d).

Now we are ready for describing the main algorithm in this section. Consider the algorithm SHORTEST ROUTING given in Fig. 1.

Algorithm. SHORTEST ROUTING

INPUT: A node u in the n-star graph, as described in **Assumption A**.

OUTPUT: The maximum number of node-disjoint shortest paths from u to ε.

1. Construct a maximum partition matching $M[(b_1, d_1), \ldots, (b_s, d_s)]$ in the collection $\mathcal{S} = \{c_2, \ldots, c_k\}$ with the partitions $(\mathcal{A}_1, \mathcal{B}_1), \ldots, (\mathcal{A}_s, \mathcal{B}_s)$ of $\mathcal{S}$;
2. **if** $s = \sum_{i=2}^{k} |c_i|$ **then** construct $s + 1$ node-disjoint shortest paths as follows.
   2.1. Call the algorithm SINGLE ROUTING with the partition $(\phi, \mathcal{S})$ of $\mathcal{S}$ and the symbol pair (a_{12}, d_1);
   2.2. Call the algorithm SINGLE ROUTING with the partition $(\mathcal{S}, \phi)$ of $\mathcal{S}$ and the symbol pair (b_1, a_{1d-1});
   2.3. For $i = 2$ to s, call the algorithm SINGLE ROUTING with the partition $(\mathcal{A}_i, \mathcal{B}_i)$ of $\mathcal{S}$ and the symbol pair (b_i, d_i);
3. **if** $s < \sum_{i=2}^{k} |c_i|$ **then** construct $s + 2$ node-disjoint shortest paths as follows.
   3.1. Let b_0 be a symbol not in $\{b_1, \ldots, b_s\}$, and d_0 be a symbol not in $\{d_1, \ldots, d_s\}$;
   3.2. Call the algorithm SINGLE ROUTING with the partition $(\phi, \mathcal{S})$ of $\mathcal{S}$ and the symbol pair (a_{12}, d_0);
   3.3. Call the algorithm SINGLE ROUTING with the partition $(\mathcal{S}, \phi)$ of $\mathcal{S}$ and the symbol pair (b_0, a_{1d-1});
   3.4. For $i = 1$ to s, call the algorithm SINGLE ROUTING with the partition $(\mathcal{A}_i, \mathcal{B}_i)$ of $\mathcal{S}$ and the symbol pair (b_i, d_i);

Fig. 1. The algorithm SHORTEST ROUTING

Theorem 7. *The Algorithm* SHORTEST ROUTING *constructs the maximum number of node-disjoint shortest paths from node u to node ε in time* $O(n^2 \log n)$.

Proof. (Sketch)   From Lemma 5 and Lemma 6, we know that the number of shortest paths constructed by the algorithm SHORTEST ROUTING matches the maximum number of node-disjoint shortest paths from u to ε. What remains is to show that all these paths are node-disjoint.

Each shortest path P_i from u to ε is constructed by calling the algorithm SINGLE ROUTING on a distinct partition $(\mathcal{A}_i, \mathcal{B}_i)$ of $\mathcal{S}$ and a distinct symbol pair (b_i, d_i). The interior nodes of the path P_i can be split into three segments such that the nodes in each segment have a special pattern in their cycle structure. The nodes in the first segment have a cycle of form $(\cdots b_i \cdots b'_i a_{11} \cdots a_{1d-1} 1)$, the nodes in the second segment have cycle structure of form $(\cdots 1) c'_1 \cdots c'_h \cdots$, where $\mathcal{B}_i = \{c'_1, \ldots, c'_h\}$, and the nodes in the third segment contain a cycle of the form $(\cdots d_i 1)$.

It can be verified, by comparing the format of the cycle structure of each segment, that any two shortest paths constructed by the algorithm SHORTEST ROUTING are node-disjoint.

The running time of the algorithm SHORTEST ROUTING is dominated by step 1 of the algorithm, which takes time $O(n^2 \log n)$ according to Theorem 4.   □

Construction of the maximum number of node-disjoint shortest paths between two nodes in star graphs was previously studied in [16], which presents an algorithm that runs in exponential time in the worst case. More seriously, the algorithm seems based on an incorrect observation, which claims that when there are more than one nontrivial cycles in a node u, the maximum number of node-disjoint shortest paths between u and ε is always an even number. Therefore, the algorithm in [16] always produces an even number of node-disjoint shortest paths between u and ε. A counterexample to this observation has been constructed in [6].

5   Conclusion: an optimal parallel routing

Combining all the previous discussion in the present paper gives us an $O(n^2 \log n)$ time algorithm (see Fig. 2) that constructs $n - 1$ node-disjoint paths of bulk length
$Bdist(u)$ between any node u and the identity node ε in the n-star graph.

The correctness of the algorithm OPTIMAL PARALLEL ROUTING has been proved by Lemma 1, Lemma 2, Theorem 7, and the results in [10]. The fact that the running time of the algorithm is bounded by $O(n^2 \log n)$ also follows from these results.

We would like to make a few remarks on the complexity of our algorithm. The bulk distance problem on general graphs is NP-hard [13]. Thus, it is very unlikely that the bulk distance problem can be solved in time polynomial in the size of the input graph. On the other hand, our algorithm solves the bulk distance problem in time $O(n^2 \log n)$ on the n-star graph. Note that the n-star graph has $n!$ nodes. Therefore, the running time of our algorithm is actually a polynomial of the logarithm of the size of the input star graph. Moreover, our

Algorithm. OPTIMAL PARALLEL ROUTING

INPUT: A node u in the n-star graph.

OUTPUT: $n-1$ node-disjoint paths of length $Bdist(u)$ from u to ε.

1. **if** the symbol 1 is in a trivial cycle in u
   then apply the algorithm described in Section 2 to construct $n-1$ node-disjoint paths of bulk length $Bdist(u)$; STOP

2. { At this point, we assume that u satisfies the conditions in **Assumption A**.}

   case 1 $m > \min\{2^{k-2}, 1 + \sum_{i=2}^{k} |c_i|\}$
   use Day and Tripathi's algorithm [10] to construct $n-1$ node-disjoint paths of bulk length $dist(u) + 4 = Bdist(u)$;

   case 2 $m \leq \min\{2^{k-2}, 1 + \sum_{i=2}^{k} |c_i|\}$
   apply the algorithm SHORTEST ROUTING to construct the maximum number of node-disjoint shortest paths from u to ε;
   if the algorithm SHORTEST ROUTING returns $n-1$ paths
   then these are $n-1$ node-disjoint paths of bulk length $dist(u) = Bdist(u)$
   else apply the algorithm in Lemma 2 to construct $n-1$ node-disjoint paths of bulk length $dist(u) + 2 = Bdist(u)$

Fig. 2. The algorithm OPTIMAL PARALLEL ROUTING

algorithm is almost optimal (differs at most by a $\log n$ factor) since the following lower bound can be easily observed — the distance $dist(u)$ from u to ε can be as large as $\Theta(n)$. Thus, constructing $n-1$ node-disjoint paths from u to ε takes time at least $\Theta(n^2)$ in the worst case.

References

1. S. B. AKERS, D. HAREL, AND B. KRISHNAMURTHY, The star graph: an attractive alternative to the n-cube, *Proc. Intl. Conf. of Parallel Processing*, (1987), pp. 393-400.

2. S. B. AKERS AND B. KRISHNAMURTHY, A group-theoretic model for symmetric interconnection networks, *IEEE Trans. on Computers 38*, (1989), pp. 555-565.

3. N. BAGHERZADEH, M. DOWD, AND S. LATIFI, A well-behaved enumeration of star graphs, *IEEE Trans. on Parallel and Distributed Systems 6*, (1995), pp. 531-535.

4. N. BAGHERZADEH, N. NASSIF, AND S. LATIFI, A routing and broadcasting scheme on faulty star graphs, *IEEE Trans. on Computers 42*, (1993), pp. 1398-1403.

5. G. BIRKHOFF AND S. MACLANE, *A Survey of Modern Algebra*, The Macmillan Company, New York, 1965.

6. C. C. CHEN, *Combinatorial and algebraic methods in star and de Bruijn networks*, Ph.D. dissertation, Dept. Computer Science, Texas A&M University, 1995.

7. C. C. CHEN AND J. CHEN, The maximum partition matching problem with applications, *Tech. Report 96-001*, Dept. Computer Science, Texas A&M University, (1996).

8. C. C. CHEN AND J. CHEN, Optimal parallel routing in star networks, *Tech. Report 96-002*, Dept. Computer Science, Texas A&M University, (1996).

9. C. C. CHEN AND J. CHEN, Nearly optimal one-to-many parallel routing in star networks, *Tech. Report 96-003*, Dept. Computer Science, Texas A&M University, (1996).

10. K. DAY AND A. TRIPATHI, A comparative study of topological properties of hypercubes and star graphs, *IEEE Trans. Parallel, Distrib. Syst. 5*, (1994), pp. 31-38.

11. M. DIETZFELBINGER, S. MADHAVAPEDDY, AND I. H. SUDBOROUGH, Three disjoint path paradigms in star networks, *Proc. 3nd IEEE Symposium on Parallel and Distributed Processing*, (1991), pp. 400-406.

12. Z. GALIL AND X. YU, Short length versions of Menger's theorem, *Proceedings of 27th Annual ACM Symp. on Theory of Computing*, (1995), pp. 499-508.

13. M. R. GAREY AND D. S. JOHNSON, *Computers and Intractability: A Guide to the Theory of NP-Completeness*, Freeman, San Francisco, CA, 1979.

14. S. W. GRAHAM AND S. R. SEIDEL, The cost of broadcasting on star graphs and k-ary hypercubes, *IEEE Trans. on Computers 42*, (1993), pp. 756-759.

15. J. JWO, S. LAKSHMIVARAHAN, AND S. K. DHALL, Embedding of cycles and grides in star graphs, *Proc. 2nd IEEE Symp. Parallel and Distrib. Processing*, (1990), pp. 540-547.

16. J. JWO, S. LAKSHMIVARAHAN, AND S. K. DHALL, Characterization of node disjoint (parallel) path in star graphs, *Proc. 5th Intl. Parallel Processing Symp.*, (1991), pp. 404-409.

17. K. MENGER, Zur allgemeinen kurventheorie, *Fund. Math. 10*, (1927), pp. 96-115.

18. Z. MILLER, D. PRITIKIN, AND I. H. SUDBOROUGH, Near embeddings of hypercubes into Cayley graphs on the symmetric group, *IEEE Trans. on Computers 43*, (1994), pp. 13-22.

19. J. MISIC AND Z. JOVANOVIC, Routing function and deadlock avoidance in a star graph interconnection network, *J. Parallel and Distrib. Computing 22*, (1994), pp. 216-228.

20. M. NIGAM, S. SAHNI, AND B. KRISHNAMURTHY, Embedding hamiltonians and hypercubes in star interconnection graphs, *Proc. Intl. Conf. of Parallel Processing*, (1990), pp. 340-343.

21. M. A. PALIS AND S. RAJASEKARAN, Packet routing and PRAM emulation on star graphs and leveled networks, *J. Parallel and Distrib. Comput. 20*, (1994), pp. 145-157.

22. K. QIU, S. G. AKL, AND H. MEIJER, On some properties and algorithms for the star and pancake interconnection networks, *J. Parallel and Distrib. Comput. 22*, (1994), pp. 16-25.

23. J SHEU, C. WU, AND T. CHEN, An optimal broadcasting algorithm without message redundancy in star graphs, *IEEE Trans. Parallel, Distrib. Syst. 6*, (1995), pp. 653-658.

24. S. SUR AND P. K. SRIMANI, Topological properties of star graphs, *Computers Math. Applic. 25*, (1993), pp. 87-98.

Counting Edges in a Dag [*]

Serafino Cicerone[1] Daniele Frigioni[2][3] Umberto Nanni[3] Francesco Pugliese[4]

[1] Dipartimento di Ingegneria Elettrica, Università di L'Aquila, Monteluco di Roio,
I-67040 L'Aquila, Italy. cicerone@iinf02.ing.univaq.it
[2] Dipartimento di Matematica Pura ed Applicata, Università di L'Aquila, via Vetoio,
Coppito I-67010 L'Aquila, Italy. frigioni@smaq20.univaq.it
[3] Dipartimento di Informatica e Sistemistica, Università di Roma "La Sapienza", via
Salaria 113, I-00198 Roma, Italy. nanni@dis.uniroma1.it
[4] Telecom Italia DG, via Flaminia 189 - 00196 - Roma, Italy.

Abstract. In this paper we generalize a technique used by La Poutré
and van Leeuwen in [14] for updating the *transitive closure* and the *transitive reduction* of a dag, and propose a uniform approach to deal with
semi-dynamic problems on dags. We define a *propagation property* on
a binary relationship as a simple sufficient condition to apply this approach. The proposed technique is suitable for a very simple implementation which does not depend on the particular problem; in other words,
the same procedures can be used to deal with different problems on dags
by simply setting appropriate border conditions.
In particular we provide semi-dynamic algorithms and data structures for
maintaining a binary relationship defined over the vertices of a dag with n
vertices and m edges, requiring $O(n(q+m))$ total time, for any sequence
of q edge insertions (deletions). This gives $O(n)$ amortized time per operation over a sequence of $\Omega(m)$ edge insertions (deletions). Queries can
be answered in constant time. The space required is $O(n^2)$.
We apply the proposed technique to various problems about dominance,
providing the first known incremental and decremental solutions for
maintaining the *dominance relationship*, the *dominator tree*, and the
nearest common dominator of a dag.

1 Introduction

Many data structures have to be used in a dynamic environment, where both
updates and queries are performed. A naive approach to this requirement consists
in making fast updates and computing the answer to any query from scratch after
each update. Indeed, nothing better is known for important graph problems,
as in the case of network flow problems. For other problems, in order to give
fast answers to queries, *dynamic* algorithms have been devised, that update the
solution any time that a modification on the structure of the problem is required.

Restricting to graph problems we have that in some cases a *fully dynamic*
solution is known, i.e., supporting both insertions and deletions of edges (and

[*] Partially supported by EU ESPRIT Long Term Research Project ALCOM-IT under
contract no.20244, and by *Progetto Finalizzato Trasporti 2 (PFT 2)* of the Italian
National Research Council (*CNR*).

vertices): this is the case, for example, of connectivity, planarity testing, and minimum spanning tree. In some cases a fully dynamic solution has been proposed by trading off between query and update operations, i.e., the solution of the problem is kept in an implicit form, and one piece of the output is built only when it is explicitly required. This kind of solution has been provided for planar graphs, for the reachability problem [18], and for the shortest paths problem [13]. For some problems, when the explicit maintenance of the output data is necessary, only *semi-dynamic* solutions have been proposed, where input modifications are restricted to insertions (deletions) of edges. This is the case of the transitive closure problem [11, 12, 14] and the all pairs shortest path problem [2].

In this paper we propose a uniform approach to deal with semi-dynamic problems on directed acyclic graphs (*dags*), referred to as the *counting technique*, generalizing an approach used by La Poutré and van Leeuwen in [14] for updating the transitive closure and the transitive reduction of a dag. We introduce the *propagation property* on binary relationships as a simple sufficient condition to apply this approach. Furthermore, we generalize the proposed technique to also deal with any problem on dags that can be formulated in terms of binary relationships on the vertices of a dag, each satisfying the propagation property.

In particular, general semi-dynamic algorithms and data structures are provided for maintaining a binary relationship over the vertices of a dag with n vertices and m edges, requiring $O(n(q + m))$ total time for any sequence of q insertions (deletions) of edges. This gives $O(n)$ amortized time per operation over a sequence of $\Omega(m)$ edge insertions (deletions). Queries can be answered in constant time. The space required is $O(n^2)$, and the proposed solution has a very simple implementation (the main data structure is a $n \times n$ matrix of integers) which does not depend on the particular problem, whose peculiarities are confined in the initial border conditions.

We apply this approach to different problems about dominance. The *dominance* relationship in a digraph with source r can be defined as follows: a vertex x dominates a vertex y if any path from r to y contains vertex x. The *dominator tree* is a concise representation of the dominance relationship, where a vertex x is an ancestor of a vertex y if and only if x dominates y. Dominance relationship is an important tool used in several contexts, ranging from program structure analysis, optimization and verification [1], to fault tolerant communication networks [6]. Several algorithms have been developed in order to compute dominators [9, 15, 19]. An optimal off-line algorithm for computing the dominator tree of a flowgraph (a digraph with a source) is given in [9], and it requires $O(m + n)$ time and space. A parallel algorithm has been also proposed in [16].

The necessity of updating the dominator tree of a flowgraph arises in various dynamic contexts, such as incremental data flow analysis and incremental compilation [4, 17]. Further applications are known in the field of batch compilation for constructing the static single assignment representation of programs [7].

In [17] a fully-dynamic solution for maintaining the single source reachability tree and the dominator tree in a *reducible* flowgraph is given, working in

$O(m \log n)$ worst case time per update. A flowgraph is *reducible* if all edges $\langle u, v \rangle$ such that v does not dominate u induce a dag. The solution proposed in [17] is based on the observation that updating the dominator tree of such a dag is sufficient to update the dominator tree of the original reducible flowgraph.

Another problem about dominance is finding the *nearest common dominator* d of a given set of vertices U of a flowgraph G. A vertex $d \in V$ is the *nearest common dominator* of U if d dominates all vertices of U and there exists no vertex $d' \neq d$ that dominates all vertices of U and is dominated by d. If G represents a communication network, or a computer program where vertices are events, edges are computations, and U is the set of relays or events that are vulnerable for failure (see [6]), d is the best choice for a single recovery point. In [5] an optimal off-line algorithm for computing the nearest common dominator of a set of vertices in a rooted dag is given. Such algorithm takes $O(m')$ time, where m' is the number of edges that lie on paths from d to vertices of U, and hence it requires $O(m)$ worst case time.

In this paper we first show how the proposed technique can be used for maintaining the transitive closure and the transitive reduction of a dag, obtaining the same bounds of the best known algorithms both for the incremental and the decremental problem [11, 12, 14]. Then we show how the same technique is used in order to achieve the following improvements with respect to previous results about dominance. We give the first decremental solution for maintaining two different representations of the dominance relationship on dags: a $n \times n$ binary matrix and the dominator tree. The achieved bounds are $O(n)$ amortized time per edge deletion and $O(1)$ worst case time per query. This improves the best off-line solution [9], that requires $O(m + n)$ time to compute dominators and $O(n)$ time for each query. In the incremental case we obtain the same time bounds of the best known solution [3]. Furthermore, we give the first incremental and decremental solution for the nearest common dominator problem on dags, working in $O(n)$ amortized time per operation and $O(1)$ worst case time per query. This improves on the optimal off-line solution proposed in [5] that takes $O(m)$ worst case time.

The paper is organized as follows. The propagation property is introduced in Section 2 and the counting technique is described in Section 3. Then we first apply the proposed technique to transitive closure and transitive reduction in Section 4 and finally, in Section 5, we deal with dominance problems on dags.

2   Basic concepts and Propagation Property

We assume the standard graph theoretical terminology (e.g., see [8]). A *directed graph (digraph)* $G = (V, E)$ consists of a finite set V of *vertices*, with $|V| = n$, and a finite set $E \subseteq V \times V$ of *edges*, with $|E| = m$. We consider an edge $\langle x, y \rangle$ as directed from x to y, and say that it is *outgoing* from x and *incoming* to y. A *path* $\pi_{x,y}$ from x to y is a sequence of vertices $(x = v_0, v_1, v_2, \ldots, v_{k-1}, v_k = y)$ such that $\langle v_i, v_{i+1} \rangle \in E$ for each $i = 0, 1, \ldots, k - 1$, and whose *length*, denoted as $length(\pi_{x,y})$, is equal to k. A *cycle* is a nonempty path from a vertex to itself.

A digraph with no cycles is called a *directed acyclic graph (dag)*. A vertex y is *reachable* from a vertex x if there exists a path $\pi_{x,y}$. A *source* is a vertex with no incoming edges. A *rooted digraph* $G = (V, E; r)$ is a digraph with one source r referred to as the *root*. A rooted digraph is a *tree* if every vertex but the root has exactly one incoming edge. In a tree, if vertex y is reachable from vertex x, then x is an *ancestor* of y; on the other hand, if there exists edge $\langle x, y \rangle$ then x is the *parent* of y in the tree.

Now we introduce the *propagation property* as a generalization of the transitive property for a binary relationship. We consider the case in which a binary relationship R defined over the vertices of a digraph *propagates* along the edges.

Definition 1. Let $G = (V, E)$ be a digraph, R be a binary relationship defined over $V \times V$, and $R_0 \subseteq R$. *Relationship R satisfies the Propagation Property (PP) over G with boundary condition R_0 if, for any pair $(x, y) \in V \times V$, $(x, y) \in R$ if and only if either $(x, y) \in R_0$, or $x \neq y$ and there exists a vertex $z \neq y$ such that $(x, z) \in R$ and $\langle z, y \rangle \in E$.*

Due to the above definition relationship R is reflexive if and only if its boundary condition R_0 is reflexive. Furthermore the following lemma trivially holds.

Lemma 2. *Let $G = (V, E)$ be a digraph and R be a binary relationship satisfying PP over G. Then $(x, y) \in R$ if and only if either $(x, y) \in R_0$ or $x \neq y$ and there exists a vertex z in G such that $z \neq y$, $(x, z) \in R_0$ and y is reachable from z.*

Definition 3. Let $G = (V, E)$ be a digraph, R be a binary relationship satisfying PP over G, and x, y be two distinguished vertices in V. An edge $\langle z, y \rangle \in E$ is *useful* to the pair (x, y) if $(x, z) \in R$.

In the following we denote as $C(x, y) = \{\langle z, y \rangle \in E \mid (x, z) \in R\}$, the set of edges useful to pair (x, y), and as $C[x, y] = |C(x, y)|$ its cardinality. For any pair $(x, y) \in V \times V$ the following basic property of the *counter* $C[x, y]$ holds.

Lemma 4. *Let $G = (V, E)$ be a digraph, and R be a binary relationship satisfying PP over G with boundary condition R_0. Then $(x, y) \in R$ if and only if either $(x, y) \in R_0$ or $C[x, y] > 0$.*

3   The Counting Technique

In this section we describe a uniform approach, referred to as the *counting technique*, to deal with semi-dynamic problems on *dags* based on the propagation property. In particular, we first provide basic algorithms for maintaining a binary relationship that satisfies the propagation property over a dag in an incremental and a decremental fashion. Then, we show how these basic algorithms can also be used, together with some additional arguments, to deal with particular binary relationships that do not satisfy the propagation property.

A straightforward solution to the problem concerning the dynamic maintenance of a binary relationship R that satisfies the propagation property over a digraph G is suggested by Lemma 2. In fact, given matrices R_0 and TC, representing the boundary condition of R and the transitive closure of G, respectively, then Lemma 2 states that relationship R can be implicitly maintained by simply updating matrices R_0 and TC after each input modification. As far as matrix TC is concerned, it can be maintained in $O(n)$ amortized time per operation during sequences of edge insertions in digraphs and during sequences of edge deletions in dags, using solutions proposed in [11] and in [12], respectively. On the other hand, it is not possible to devise a general procedure to update matrix R_0 after an input modification. In fact, any boundary condition R_0 is strongly related to the corresponding relationship R, and, in general, it does not share any feature with the boundary conditions of other binary relationships. Hence, for each relationship R it is necessary to devise an ad-hoc procedure for updating the corresponding R_0. In the following, we show that, for each binary relationship considered in this paper, updating the corresponding boundary condition is quite simple, and it requires $O(n)$ worst case time after each input modification. Using this approach queries of kind "does pair (x, y) belong to relationship R ?" require $O(n)$ worst case time in order to verify the existence of a vertex z such that $(x, z) \in R_0$ and $(z, y) \in TC$.

In the following we propose another approach, based on Lemma 4, that allows us to explicitly maintain relationship R in an incremental and a decremental fashion, and hence to answer queries in constant time. Figures 1 and 2 show procedures Insert and Delete that update R after insertion and deletion of edges, respectively. We use the following data structures. Set OUT$[x]$ contains, for each vertex x, all outgoing edges from x; a $n \times n$ integer matrix that contains, for any pair (x, y) of vertices, the value of counter $C[x, y]$; a binary matrix that represents the boundary condition R_0. Furthermore, a queue Q_k, for any vertex k, is used to handle edges $\langle h, y \rangle$ useful to pair (k, y).

Any time an edge $\langle i, j \rangle$ is inserted in E the number of edges useful to any pair (k, y) can only increase. Procedure Insert finds such useful edges using queue Q_k, and properly updates counter $C[k, y]$. The behaviour of procedure Delete is analogous.

The correctness and the complexity of procedure Insert are stated in the following.

Theorem 5. *After the execution of* Insert $(\langle i, j \rangle)$, *for any pair of vertices* (k, h), *counter* $C[k, h]$ *equals the number of edges useful to pair* (k, h).

Proof. Let us suppose that $C[k, h]$ correctly counts the number of edges useful to pair (k, h) before inserting edge $\langle i, j \rangle$. Counter $C[k, h]$ can increase if and only if some edge $\langle t, h \rangle$ becomes useful to pair (k, h) due to the insertion of edge $\langle i, j \rangle$.

We prove that an edge $\langle t, h \rangle$ becomes useful to pair (k, h), due to the insertion of edge $\langle i, j \rangle$, if and only if it is inserted in queue Q_k during the execution of procedure Insert $(\langle i, j \rangle)$. On the other side, each edge $\langle t, h \rangle$ inserted in Q_k determines an increment of counter $C[k, h]$ by one (see line 12).

The proof is performed by induction on the number of edges inserted in Q_k.

(*basic step*). The new edge $\langle i, j \rangle$ is inserted into the empty queue Q_k (line 6) if and only if $(k, i) \in R$ (line 4) and hence it is useful to pair (k, j).

Procedure Insert $(\langle i, j \rangle : edge)$

1.   **begin**
2. $E' := E \cup \{\langle i, j \rangle\}$
3.   **for each** $k \in V$ **do**
4.   **if** $(k, i) \in R$
5.   **then begin**
6.   set-queue$(Q_k, \langle i, j \rangle)$
7.   **while** Q_k is not empty **do**
8.   **begin**
9.   dequeue$(Q_k, \langle l, h \rangle)$
10.   **if** $h \neq k$ **then**
11.   **begin**
12. $C[k, h] = C[k, h] + 1$
13.   **if** $C[k, h] = 1$ and $(k, h) \notin R_0$
14.   **then for each** $\langle h, y \rangle \in$ OUT$[h]$ **do**
15.   enqueue$(Q_k, \langle h, y \rangle)$
16.   **end**
17.   **end**
18.   **end**
19.   **end**

Fig. 1. Insertion of edge $\langle i, j \rangle$

(*inductive step*). By induction, let us suppose that an edge $\langle t, h \rangle$ has been inserted into queue Q_k if and only if it is useful to pair (k, h) due to the insertion of edge $\langle i, j \rangle$. When edge $\langle t, h \rangle$ is on the top of queue Q_k it is deleted (line 9) and vertices k and h are compared. According to Definition 3, if they are equal nothing has to be done, otherwise edge $\langle t, h \rangle$ is useful to pair (k, h), and hence $C[k, h]$ is incremented by one (line 12), and the following two cases may arise:

1. If $C[k, h] > 1$ or $(k, h) \in R_0$ then pair (k, h) was in R before inserting edge $\langle i, j \rangle$, and hence all edges $\langle h, y \rangle$ in OUT$[h]$ have already been inserted in $C(k, y)$, due to a previous edge insertion.
2. If $C[k, h]$ becomes equal to 1 then, by Lemma 4, the inserted edge $\langle i, j \rangle$ causes pair (k, h) to be inserted in R. Hence, any edge $\langle h, y \rangle$ in OUT$[h]$ becomes useful to pair (k, y) and it is correctly inserted in Q_k (line 15).

                                                                $\square$

Theorem 5 and Lemma 4 imply the correctness of Procedure Insert.

Theorem 6. *Let G be a dag with n vertices and m edges. The total time required by procedure* Insert *to perform q consecutive edge insertions in G is $O(n(q + m))$.*

Proof. Any basic operation performed by procedure Insert requires constant time. Therefore, to prove the theorem, it is sufficient to bound the number of queue's operations performed during an incremental sequence of edge insertions.

```
        Procedure Delete (⟨i, j⟩ : edge)
 1.    begin
 2.        E' := E − {⟨i, j⟩}
 3.        for each k ∈ V do
 4.            if (k, i) ∈ R
 5.            then begin
 6.                    set-queue(Q_k, ⟨i, j⟩)
 7.                    while Q_k is not empty do
 8.                    begin
 9.                            dequeue(Q_k, ⟨l, h⟩)
10.                            if k ≠ h then
11.                            begin
12.                                    C[k, h] = C[k, h] − 1
13.                                    if C[k, h] = 0 and (k, h) ∉ R_0
14.                                    then for each ⟨h, y⟩ ∈ OUT[h] do
15.                                            enqueue(Q_k, ⟨h, y⟩)
16.                            end
17.                    end
18.            end
19.    end
```

Fig. 2. Deletion of edge $\langle i, j \rangle$

An edge $\langle h, y \rangle$ is inserted in Q_k if and only if pair (k, h) has been added for the first time to relationship R (due to test in line 13) as a consequence of an edge insertion. This implies that, during a sequence of edge insertions, edge $\langle h, y \rangle$ can be inserted in Q_k at most once. In fact pair (k, h) can never leave relationship R due to a subsequent edge insertion. Since there are n queues, one for each vertex, and at most $q + m$ edges in the graph, it follows that the total time necessary to handle the whole sequence of q edge insertion is $O(n(q + m))$.

□

The correctness and the complexity of procedure Delete are stated by the following theorems. The corresponding proofs are omitted since they can be obtained by analogous techniques of the incremental case.

Theorem 7. *After the execution of* Delete $(\langle i, j \rangle)$, *for any pair of vertices* (k, h), *counter* $C[k, h]$ *equals the number of edges useful to pair* (k, h).

Theorem 8. *Let G be a dag with n vertices and m edges. The total time required by procedure* Delete *to perform q consecutive edge deletions in G is $O(n(q+m))$.*

The above algorithms allow us to maintain binary relationships that satisfy the propagation property in an incremental and a decremental fashion. On the other hand, there is a large class of binary relationships that, even though they do not satisfy the propagation property, can be maintained generalizing the approach of the counting technique. The following theorem shows how a relationship R in this class can be maintained, if some binary relationships satisfying the propagation property can be used as "building blocks" for yielding R.

Theorem 9. *Let $G = (V, E)$ be a dag with n vertices and m edges, $R \subseteq V \times V$ be a binary relationship, and $R_1, R_2, \ldots, R_k$ be binary relationships that satisfy PP over G.*

1. *If R can be formulated as a boolean (or arithmetic) expression on the binary relationships $R_1, R_2, \ldots, R_k$, and $R(i, j)$ can be computed in $O(t)$ time from $R_1, R_2, \ldots, R_k$, then R can be maintained incrementally and decrementally in $O(k \cdot n(m + q))$ total time over each sequence of q updates, and $O(t)$ time for each query.*
2. *If R can be updated using $R_1, R_2, \ldots, R_k$ after each update in $O(n)$ amortized time, then R can be maintained incrementally and decrementally in $O(n(m + q))$ total time over each sequence of q updates, and $O(1)$ time for each query.*

The proof of Theorem 9 is a straightforward consequence of Theorems 6 and 8. In the sequel we say that, if case 1 or case 2 of Theorem 9 occurs, relationship R can be maintained by the counting technique in an *implicit way* (since the information on R is distributed over $R_1, R_2, \ldots, R_k$), or in an *explicit way*, respectively.

4 Transitive closure and transitive reduction

In this section we give preliminary examples that show how the counting technique can be easily used in order to obtain the same results achieved on dags in [11, 12] for the Transitive Closure (TC) relationship, and in [14] for the Transitive Reduction (TR) relationship, respectively.

Given a digraph $G = (V, E)$, the digraph $G^+ = (V, E^+)$ such that an edge $\langle x, y \rangle$ belongs to E^+ if and only if there exists a path $\pi_{x,y}$ in G, is called the *transitive closure* of G. The following lemma shows how the transitive closure of a dag can be defined as a binary relationship over the vertices of the dag that satisfies the propagation property.

Lemma 10. *Let $G = (V, E)$ be a dag and $G^+ = (V, E^+)$ be its transitive closure. Relationship $TC = E^+$ satisfies PP over G with boundary condition $TC_0 = \{(x, x) \mid x \in V\}$.*

Proof. For any pair $(x, y) \in V \times V$, $(x, y) \in TC$ if and only if either $x = y$ (that is, $(x, y) \in TC_0$) or $x \neq y$ and there exists a vertex z such that $z \neq y, (x, z) \in TC$, and $\langle z, y \rangle \in E$. Hence, Definition 1 is verified. $\square$

Notice that, any time an input modification is performed on a dag G, the boundary condition TC_0 has not to be updated since it is independent from such modification. Hence, relationship TC can be maintained incrementally and decrementally over a dag G with the time bounds given in Theorems 6 and 8, respectively.

A transitive reduction TR of a digraph $G = (V, E)$ is a graph $G^- = (V, E^-)$ having the minimum number of edges and the same transitive closure of G. It is known that, for any dag G, the transitive reduction G^- is unique and is a subgraph of G (e.g., see [14]). In fact, if we call *non-trivial* any path whose length is greater or equal than 2, then an edge $\langle x, y \rangle \in E$ belongs to E^- if and only if there exists no non-trivial path from x to y. If we denote as NTP the binary relationship such that $(x, y) \in NTP$ if and only if there exists a non-trivial path from x to y, the following property for TR can be stated.

Proposition 11. *Let $G = (V, E)$ be a dag. The transitive reduction of G is the subgraph $G^- = (V, E^-)$ of G such that $E^- = \{\langle x, y \rangle \in E \mid \wedge (x, y) \notin NTP\}$.*

It is easy to verify that the relationship $TR = E^-$ does not satisfy the propagation property. Nevertheless, we are able to show that TR can be maintained in an implicit way by counting technique, as described in case 1 of Theorem 9. In fact, by Proposition 11 it follows that the transitive reduction of a dag $G = (V, E)$ can be formulated as: $TR = E \wedge \neg NTP$. Furthermore, if we denote as NTP_0 the set of all pairs of vertices (x, y) such that there exists a path of length 2 from x to y in G, the following lemma holds.

Lemma 12. *Let $G = (V, E)$ be a dag. Relationship NTP satisfies PP over G with boundary condition $NTP_0 = \{(x, y) \in V \times V \mid \exists z \in V, \ (x, z), (z, y) \in E\}$.*

Proof. For any pair $(x, y) \in V \times V$, $(x, y) \in NTP$ if and only if there exists a vertex $z \neq y$ such that either $(x, z), (z, y) \in E$ (that is, there is a non-trivial path having length 2 between vertices x and y), or $(x, z) \in NTP$ and $(z, y) \in E$ (that is, there is a non-trivial path having length greater than 2 between vertices x and y). Hence Definition 1 is verified. $\qquad\square$

The boundary condition NTP_0 can be easily updated after each input modification. Let us suppose to represent NTP_0 as a $n \times n$ integer matrix M that stores, for each pair of vertices (x, y), the number of paths having length 2 between x and y. Obviously, if we denote as $M[i, j]$ the generic element of matrix M in position (i, j), it follows that $(x, y) \in NTP_0$ if and only if $M[x, y] > 0$. Matrix M can be updated in $O(n)$ worst case time after each input modification as follows. Any time edge $\langle i, j \rangle$ is inserted (deleted), if there exists edge $\langle x, i \rangle$ or edge $\langle j, y \rangle$, increment (decrement) by one $M[x, j]$ or $M[i, y]$, respectively.

The following theorem summarizes the results of this section.

Theorem 13. *Let $G = (V, E)$ be a dag with n vertices and m edges. The transitive closure TC and the transitive reduction TR of G can be maintained incrementally and decrementally in $O(n(q + m))$ total time over each sequence of q updates, and $O(1)$ time for each query.*

5 Maintaining dominators in dags

In this section we apply the counting technique to get semi-dynamic solutions to various problems about dominance. First, let us briefly review some definitions.

Given a rooted digraph $G = (V, E; r)$ and two distinguished vertices x and y in G, vertex x is a *dominator* of vertex y (x *dominates* y) if any path $\pi_{r,y}$ contains x; if $x \neq y$ then x is a *proper dominator* of y. We denote as DOM(x) the set of all the dominators of vertex x. Vertex x is the *immediate dominator* of y, denoted as IDOM(y), if x is a proper dominator of y and no vertex $z \notin \{x, y\}$ exists such that $x \in$ DOM(z) and $z \in$ DOM(y). The *dominator tree* T_D of dag G is a concise representation of the dominance relationship, where, for each vertex x in G, the parent of x in T_D corresponds to IDOM(x). Hence, vertex x dominates vertex y in G if and only if x is an ancestor of y in T_D. Given a set $U \subseteq V$, a vertex $d \in V$ is the *nearest common dominator* of U if the following two conditions hold: i) d dominates all vertices of U; ii) there exists no vertex $d' \neq d$ that dominates all vertices of U and is dominated by d. If we denote the *dominance* relationship as the binary relationship $D = \{(x, y) \in V \times V \mid x \in$ DOM(y)$\}$, then it is easy to show that D does not satisfy the propagation property for any choice of the boundary condition D_0. Vice versa, if we call *non-dominance* the complementary relationship $\overline{D} = \{(x, y) \in V \times V \mid x \notin$ DOM(y)$\}$, the following theorem holds.

Theorem 14. *Let* $G = (V, E; r)$ *be a rooted dag. Non-dominance relationship* $\overline{D}$ *satisfies PP over* G *with boundary condition* $\overline{D}_0 = \{(x, y) \in V \times V \mid y = r \wedge x \neq r\}$.

Proof. We show that, according to Definition 1, pair $(x, y) \in V \times V$ is in $\overline{D}$ if and only if one of the following cases arises:
1. $(x, y) \in \overline{D}_0$;
2. $x \neq y$ and there exists a vertex z such that $z \neq y, (x, z) \in \overline{D}$ and $\langle z, y \rangle \in E$.
(if case). Case (1) is trivial, since no vertex dominates the root. In case (2) $(x, z) \in \overline{D}$, and then not all paths from r to z pass through x. Hence, since $\langle z, y \rangle \in E$, we have that not all paths from the root to y pass through x, and then $(x, y) \in \overline{D}$.
(only if case). Let us suppose that $(x, y) \in \overline{D}$, and hence $x \neq y$. If $y = r$ then case (1) occurs. If $y \neq r$, it follows that there exists a non-empty path $\pi_{r,y}$ from r to y not containing x, and hence a vertex $z \in \pi_{r,y}$, with $z \neq y$ and $\langle z, y \rangle \in E$. This implies that there exists a (possibly empty) path from r to z not containing x, and therefore case (2) above occurs. □

5.1 Maintenance of the dominators tree

We have already observed that the dominator tree T_D of a rooted dag G is only a concise representation of the dominance relationship D. A relevant difference arises when a query about dominance is performed. In fact, linear time in the number of vertices is needed to test if x dominates y using T_D and constant time using D; on the other hand, testing for the immediate dominator requires opposite bounds.

In this section we show how the dominator tree T_D of a dag G can be maintained in an explicit way by counting technique, as described in case 2 of Theorem 9. In particular, this is accomplished using relationships TC and D as building blocks, and procedures T_D-Insert and T_D-Delete reported in Figures 3 and 4, respectively.

We denote as $P_{x,y}$ the unique path between vertices x and y (if it exists) in T_D, and as $lca(x, y)$ the least common ancestor of x and y in T_D. Furthermore, let $G = (V, E; r)$ be a rooted dag and $\langle i, j \rangle$ the edge to be inserted in (deleted from) G to give the new dag $G' = (V, E'; r)$ with $E' = E \cup \{\langle i, j \rangle\}$ ($E' = E - \{\langle i, j \rangle\}$). The immediate dominator of a vertex x, the dominance relationship and the dominator tree in the new dag G', are denoted as $\text{IDOM}'(x)$, D' and T'_D, respectively.

Before analizing the incremental and the decremental maintenance of the dominator tree of a rooted dag in detail, we note that there are simple argumentation to deduce that some vertices do not change their own immediate dominator after an edge modification. In fact, the insertion (deletion) of edge $\langle i, j \rangle$ in a rooted dag G modifies only paths from root r to any vertex x reachable from j. Hence, only the set of dominators of such vertices can be modified by insertion (deletion) of edge $\langle i, j \rangle$. Moreover, any vertex dominated by j maintains its immediate dominator after that operation. This is formalized by the following observation.

Observation 15 *Let $G = (V, E; r)$ be a rooted dag, $\langle i, j \rangle$ the edge inserted in (deleted from) G and* $\text{CHANGE} = \{x \in V \mid (j, x) \in TC \wedge (j, x) \notin D\}$. *After the insertion (deletion) of edge $\langle i, j \rangle$, $\text{IDOM}'(x) = \text{IDOM}(x)$ for each $x \notin \text{CHANGE}$.*

In the sequel we denote as CHANGE_i and CHANGE_d the subsets of CHANGE containing the vertices that change the immediate dominator after an edge insertion and an edge deletion, respectively.

The incremental problem. Now we briefly explain how T_D is updated after each edge insertion as suggested by case 2 of Theorem 9, using relationship TC as a building block. Let $\langle i, j \rangle$ be the edge to be inserted in E, and $a = lca(i, j)$ be the least common ancestor of i and j in T_D. The subsequent Lemmas 16 and 17 prove that a vertex x changes its immediate dominator, due to the insertion of edge $\langle i, j \rangle$, if and only if x is reachable from j and $\text{IDOM}(x)$ lies on $P_{a, \text{IDOM}(j)}$, i.e., $\text{CHANGE}_i = \text{CHANGE} \cap \{x \in V \mid \text{IDOM}(x) \in P_{a, \text{IDOM}(j)}\}$. Moreover, Lemma 16 proves that the new immediate dominator of each vertex $x \in \text{CHANGE}_i$ is vertex a. Procedure T_D-Insert first labels each vertex y in $P_{r, \text{IDOM}(j)}$ with the length of $P_{r,y}$ (lines 3–4), and then computes $lca(i, j)$ (line 5). This can be accomplished in $O(n)$ worst case time. Using such labels, given a vertex $x \in V$, the procedure can test if $\text{IDOM}(x)$ belongs to $P_{a, \text{IDOM}(j)}$ in constant time simply checking if $\text{IDOM}(x)$ is labeled and its label is greater than the label of a. In this way, it is possible to check if $x \in \text{CHANGE}_i$ (line 7) in constant time.

Procedure T_D**-Insert** ($\langle i,j \rangle : edge$);
1.   **begin**
2. $E' := E \cup \{\langle i,j \rangle\}$
3.   **for each** vertex $x \in P_{r,\,\text{IDOM}(j)}$ **do**
4. $label(x) := length(P_{r,x})$
5.   starting from i visit $P_{r,i}$ until a labeled vertex a is found $\{a = lca(i,j)\}$
6.   **for each** $x \in V$ **do**
7.   **if** $(j,x) \in TC \wedge$ IDOM(x) is labeled $\wedge$ $label(\text{IDOM}(x)) > label(a)$
8.   **then** IDOM$'(x) := a$
9.   unlabel all the labeled vertices
10. **end**

Fig. 3. Insertion of edge $\langle i,j \rangle$

Lemma 16. *After the execution of Procedure T_D-Insert ($\langle i,j \rangle$), for each vertex $x \in$ CHANGE$_i$, IDOM$'(x) = a$ holds.*

Proof. Since $x \in$ CHANGE$_i$, then $(a, \text{IDOM}(x)) \in D$. Moreover, it is easy to see that $(a,x) \in D'$. By contradiction, let us suppose that there exists a vertex $z \in V - \{a,x\}$ such that $(a,z) \in D'$ and $(z,x) \in D'$. Due to the insertion of edge $\langle i,j \rangle$, vertex j can be reached from a using at least two disjoint paths, the first passing through IDOM(x), and the second passing through edge $\langle i,j \rangle$. The existence of such two paths imply that: if $x = j$ then vertex z cannot exist; if $x \neq j$ then (j,z) must belong to D'. In the latter case $(j,x) \in D$ must hold, contradicting the hypothesis that IDOM$(x) \in P_{a,\,\text{IDOM}(j)}$. $\qquad\square$

Lemma 17. *After the execution of Procedure T_D-Insert ($\langle i,j \rangle$), for each vertex $x \in V -$ CHANGE$_i$, IDOM$'(x) =$ IDOM(x) holds.*

Proof. Three possible cases may arise when $x \in V -$ CHANGE$_i$, according to which of the conditions tested in line 7 of Procedure T_D-Insert is not verified:
1. $(j,x) \notin TC$: by Observation 15, $x \notin$ CHANGE and then IDOM$'(x) =$ IDOM(x).
2. IDOM(x) is not labeled and $(j,x) \in TC$: if $(j,x) \notin D$ then Observation 15 guarantees that $x \notin$ CHANGE and then IDOM$'(x) =$ IDOM(x); if $(j,x) \in D$, the existence of a path from j to x in G contradicts the hypothesis that $(\text{IDOM}(x),x) \in D$.
3. IDOM(x) is labeled, $label(\text{IDOM}(x)) \leq label(a)$, and $(j,x) \in TC$: since $(\text{IDOM}(x),a) \in D'$ and $a =$ IDOM$'(j)$, then IDOM$'(x) =$ IDOM(x). $\qquad\square$

By Lemmas 16 and 17 and by discussion above, procedures T_D-Insert correctly updates T_D in $O(n)$ worst case time after each edge insertion using relationship TC. Then, the approach described in case 2 of Theorem 9 can be applied to maintain T_D during edge insertions, and the following theorem holds.

Theorem 18. *Let $G = (V, E; r)$ be a rooted dag with n vertices and m edges. The dominator tree of G can be incrementally maintained in $O(n(q + m))$ total time over each sequence of q edge insertions, and $O(1)$ time for each query.*

The decremental problem. Now we briefly explain how the approach described in case 2 of Theorem 9 can be used in order to update T_D after each edge deletion using relationships TC and D as building blocks.

Let us suppose that $\langle i, j \rangle$ is the edge to be deleted from E to obtain E'. We do not deal with the case in which the vertex j is not reachable from the root r in G', the extension being straightforward. The subsequent Lemmas 19 and 20 show that a vertex x changes its immediate dominator after the deletion of edge $\langle i, j \rangle$ if and only if it is reachable from j and $\text{IDOM}(x) = a$, where $a = \text{IDOM}(j)$, i.e., $\text{CHANGE}_d = \text{CHANGE} \cap \{x \in V \mid \text{IDOM}(x) = a\}$. Unlike the incremental case, in this situation it is not true that all the vertices that change the immediate dominator after the deletion of an edge have the same new immediate dominator. In fact the subsequent Lemma 20 shows that such new dominators lie on path $P_{a,z}$ in T_D, where z is the least common ancestor in T_D of the vertices in the set $Z = \{y \in V \mid \langle y, j \rangle \in E'\}$. Since we supposed that vertex j is still reachable from r after the deletion of edge $\langle i, j \rangle$, then $|Z| \geq 1$ holds.

    Procedure T_D-Delete $(\langle i, j \rangle : edge)$;
1.   **begin**
2. $E' := E - \{\langle i, j \rangle\}$
3. $a := \text{IDOM}(j)$
4.   select a vertex $v \in Z$
5. $l := length(P_{a,v})$
6.   **for each** vertex $x \in P_{a,v}$ **do**
7.   **if** $length(P_{x,v}) = k$
8.   **then** denote x as x_k $\{a = x_l$ and $v = x_0\}$
9.   **for each** $q \in V$ **do**
10.   **if** $\text{IDOM}(q) = a \wedge (j, q) \in TC$ **then**
11.   **begin**
12. $h := l \, \{(x_h, q) \in D'\}$
13.   **repeat** $h := h - 1$
14.   **until** $(x_h, q) \notin D' \vee x_h = z$
15.   **if** $(x_h, q) \in D'$ **then** $\text{IDOM}'(q) := x_h$
16.   **else** $\text{IDOM}'(q) := x_{h+1}$
17.   **end**
18. **end**

Fig. 4. Deletion of edge $\langle i, j \rangle$

Procedure T_D-**Delete**, shown in figure 4, first selects a vertex $v \in Z$, and then associates an integer subscript representing the length of path $P_{x,v}$ to each vertex x in $P_{a,v}$; in this way vertices a and v are referred to as x_l and x_0, respectively, where $l = length(P_{a,v})$. This can be trivially accomplished in $O(n)$ worst case time. The notation introduced above for the vertices in $P_{a,v}$ is used in order to determine the new immediate dominator of the vertices in CHANGE_d. The new immediate dominator of a vertex x in CHANGE_d is selected as the dominator of x in $P_{a,z}$ having the smallest subscript in lines 9–17 of procedure

T_D-Delete. The subsequent Lemma 21 shows that the above selection can be performed in $O(n^2)$ total time for all vertices that fall into CHANGE$_d$ during a sequence of edge deletions.

The correctness and the complexity of procedure T_D-Delete are formally stated by the following lemmas.

Lemma 19. *After the execution of Procedure T_D-Delete $(\langle i,j \rangle)$, for each $x \in V -$ CHANGE$_d$, IDOM$'(x) =$ IDOM(x) holds.*

Proof. Let us consider a vertex $x \in V -$ CHANGE$_d$. The following two cases may arise:

1. $(j,x) \notin TC$: by Observation 15, $x \notin$ CHANGE and then IDOM$'(x) =$ IDOM(x).
2. IDOM$(x) \neq a \wedge (j,x) \in TC$: we analize the various cases that may arise due to the position of IDOM(x) in T_D. We show that some cases are forbidden whereas others imply that IDOM$'(x) =$ IDOM(x).
   a. If j dominates IDOM(x) then j dominates x and hence by Observation 15 it follows that $x \notin$ CHANGE and then IDOM$'(x) =$ IDOM(x).
   b. If IDOM$(x) \in P_{a,y}$ for some $y \in Z$, the path from a to x passing through edge $\langle i,j \rangle$ gives a contradiction.
   c. If $(a,$ IDOM$(x)) \in D$ and IDOM$(x) \notin P_{a,y}$ for none $y \in Z$, then there exists a path between a and x passing through an edge $\langle y,j \rangle$ after the deletion, which contradicts the hypothesis that IDOM(x) was the immediate dominator of x before the deletion.
   d. If IDOM$(x) \in P_{r,a}$, then IDOM$'(x) =$ IDOM(x) after the deletion of edge $\langle i,j \rangle$. In fact, before the deletion there was a path from IDOM(x) to x not containing a. The deletion of edge $\langle i,j \rangle$ modifies neither this path nor paths from IDOM(x) to x passing through vertices in Z.
   e. The last case to be considered is that $lca($ IDOM$(x), a)$ differs from both IDOM(x) and a. Since there exists the path from $lca($ IDOM$(x), a)$ to x passing through a, this contradicts the hypothesis that IDOM(x) dominates x before the deletion. $\qquad\square$

Lemma 20. *After the deletion of edge $\langle i,j \rangle$, for each $x \in$ CHANGE$_d$, IDOM$'(x) \in P_{a,z}$ holds.*

Proof. Since $(a,i) \in D$ and $(a,j) \in D$ then after the deletion of edge $\langle i,j \rangle$ it follows that if $x \in$ CHANGE$_d$ then $(a,x) \in D'$. Hence a dominates IDOM$'(x)$.

By contradiction, let us suppose that there exists a vertex $x \in$ CHANGE$_d$ such that IDOM$'(x) \notin P_{a,z}$. Two possible cases may arise.

1. $(z,$ IDOM$'(x)) \in D'$. In this case, IDOM$'(x)$ belongs to the subtree of T'_D rooted in z, and it is different from z.
2. $(z,$ IDOM$'(x)) \notin D'$. In this case there is no path from IDOM$'(x)$ to x passing through j.

In both cases, by definition of vertex z, we derive a contradiction due to the existence of a path from z to x passing through j and not passing through IDOM$'(x)$. □

Lemma 21. *The total time required by procedure T_D-Delete in order to compute the new immediate dominators of vertices in* CHANGE$_d$ *during an arbitrary sequence of edge deletions, is $O(n^2)$.*

Proof. Let us consider a vertex q in CHANGE$_d$ and its history during a sequence of edge deletions. During the execution of procedure T_D-Delete after the deletion of an edge $\langle i, j \rangle$, all the edges in $P_{a,z}$ from $a =$ IDOM(q) to IDOM$'(q)$ are traversed exactly once, together with an edge outgoing from IDOM$'(q)$ (lines 9–17). During the next edge deletion in the sequence, procedure T_D-Delete looks for the new immediate dominator of vertex q starting from IDOM$'(q)$, possibly traversing the same edge outgoing from IDOM$'(q)$ traversed in the previous deletion. Since this reasoning also applies to the subsequent deletions, then each edge in the dominator tree is traversed at most twice during a sequence of edge deletions, while updating the immediate dominator of vertex q. Since the vertices are n the lemma follows. □

By Lemmas 19, 20 and 21, and by discussion above, procedure T_D-Delete correctly updates T_D in $O(nm)$ total time during an arbitrary sequence of edge deletions, using relationship TC and D. Then, the approach described in case 2 of Theorem 9 can be applied to maintain T_D during edge deletions, and the following theorem holds.

Theorem 22. *Let $G = (V, E; r)$ be a rooted dag with n vertices and m edges. The dominator tree of G can be decrementally maintained in $O(n(q + m))$ total time over each sequence of q edge deletions, and $O(1)$ time for each query.*

Maintenance of the nearest common dominator. Given a rooted dag $G = (V, E; r)$, a set $U \subseteq V$, and the dominator tree T_D of G, if we denote with $lca(U)$ the least common ancestor of the vertices in U in T_D and with $ncd(U)$ the nearest common dominator of vertices in U, it is easy to show that $lca(U) = ncd(U)$. Then it is straightforward to derive a procedure for the dynamic maintenance of $ncd(U)$. In fact, it is sufficient to update the dominator tree after each insertion (deletion), and finding the least common ancestor of vertices in U in the updated version of the dominator tree. Since we have showed in the previous section that the dominator tree can be updated in $O(n)$ amortized time after each insertion (deletion), and the least common ancestor of a given set of vertices in a tree can be found in $O(n)$ worst case time (see [10]), then the following theorem trivially holds.

Theorem 23. *Let $G = (V, E; r)$ be a rooted dag. The nearest common dominator of a given set $U \subseteq V$ can be maintained in $O(n(q + m))$ total time over each sequence of q edge insertions (deletions), and $O(1)$ time per query.*

References

1. A. V. Aho, R. Sethi, and J. D. Ullman. *Compilers, Principles, Techniques, and Tools.* Addison-Wesley, 1986.
2. G. Ausiello, G. F. Italiano, A. Marchetti Spaccamela, and U. Nanni. Incremental algorithms for minimal length paths. *Journal of Algorithms,* 12(4):615–638, 1991.
3. G. Ausiello and M. Scasso. Algoritmos para mantener dinamicamente clausura transitiva y arbol de dominadores en grafos aciclicos dirigidos. Technical report, Buenos Aires, November 1989.
4. M. D. Carrol and B. G. Ryder. Incremental dataflow update via attribute and dominator update. In *ACM Symposium on Principles of Programming Languages.,* pages 274–284, 1988.
5. J. Chu. Optimal algorithms for the nearest common dominator problem. *Journal of Algorithms,* 13:693–697, 1992.
6. M. J. Chung and M. S. Krishnamoorthy. Algorithms for placing recovery points. *Information Processing Letters,* 28:177–181, 1988.
7. R. Cytron, J. Ferrante, B. K. Rosen, M. N. Wegman, and K. Zadeck. Efficiently computing static single assignment form and the control dependence graph. *ACM Transactions on Programming Languages and Systems,* 13(4):451–490, 1991.
8. F. Harary. *Graph Theory.* Addison-Wesley, Reading, MA, 1969.
9. D. Harel. A linear time algorithm for finding dominators in flow graphs and related problems. In *ACM Symposium on Theory of Computing,* pages 185–194, 1985.
10. D. Harel and R. E. Tarjan. Fast algorithms for finding nearest common ancestors. *SIAM Journal on Computing,* 13(2):338–355, 1984.
11. G. F. Italiano. Amortized efficiency of a path retrieval data structure. *Theoretical Computer Science,* 48:273–281, 1986.
12. G. F. Italiano. Finding paths and deleting edges in directed acyclic graphs. *Information Processing Letters,* 28:5–11, 1988.
13. P. N. Klein, S. Rao, M. Rauch, and S. Subramanian. Faster shortest-path algorithms for planar graphs. In *ACM Symposium on Theory of Computing,* pages 27–37, Montreal, Quebec, Canada, 1994.
14. J. A. La Poutré and J. van Leeuwen. Maintenance of transitive closure and transitive reduction of graphs. In *International Workshop on Graph-Theoretic Concepts in Computer Science,* pages 106–120. Lect. Notes in Comp. Sci., 314, 1988.
15. T. Lengauer and R. E. Tarjan. A fast algorithm for finding dominators in a flowgraph. *ACM Transactions on Programming Languages and Systems,* 1(1):121–141, 1979.
16. S. R. Pawagi and I. V. Ramakrishnan P. S. Gopalakrishnan. Computing dominators in parallel. *Information Processing Letters,* 24:217–221, 1987.
17. G. Ramalingam and T. Reps. An incremental algorithm for maintaining the dominator tree of a reducible flowgraph. In *ACM Symposium on Principles of Programming Languages.,* pages 287–296, 1994.
18. S. Subramanian. A fully dynamic data structure for reachability in planar digraphs. In *Proc. of ESA '93,* pages 372–383. Lecture Notes in Comp. Sci., 726, 1993.
19. R. E. Tarjan. Finding dominators in directed graphs. *SIAM Journal on Computing,* 3(1):62–89, 1974.

Closure Properties of Context-Free Hyperedge Replacement Systems

Ornella Ciotti and Francesco Parisi-Presicce

Dipartimento di Scienze dell'Informazione,
Universita' di Roma 'La Sapienza', 00198-Roma (Italy)
parisi@dsi.uniroma1.it

Abstract. Hyperedge replacement systems are investigated in terms of closure properties with respect to language operators such as union, intersection and concatenation. It is shown that context-free HR systems are closed under union, coalesced sum and concatenation but not under intersection, giving an explicit counterexample.

1 Introduction

Graph grammars describe sets of graphs in a way that is analogous to the grammatical description of sets of strings. There are several types of graph grammars, that vary in such things as whether edges, nodes or handles serve as nonterminals, and how newly introduced subgraphs are glued to the current graph. The different approaches can be classified mainly as algorithmic, algebraic and categorical. A popular type of context-free graph grammar is the Hyperedge Replacement System, or HR grammar of [5] where hyperedges are replaced by multipointed graphs of the same type.

In this work we investigate operations on graph languages such as union, intersection, coalesced sum, concatenation and iteration and the closure properties of context free graph languages under these operations. Union and intersection are defined as usual by taking the set-theoretic union or intersection of the languages. The coalesced sum of two languages is the set of graphs resulting from the coalesced sum of pairs of graphs, one from each language where the coalesced sum of two multipointed graphs of the same type is obtained by taking the disjoint union and then identifying the BEGIN nodes of the two graphs and the END nodes of the two graphs, respectively. Concatenation of two languages is the set of graphs resulting from the concatenation of pairs of graphs, one from each language where concatenation of two graphs (not necessarily of the same type) is obtained from the disjoint union be identifying the END nodes of the first graph with the BEGIN nodes of the second one. Iteration is the usual Kleene star.

The main result is that the family of context free graph languages is closed under all the operations mentioned above with the exception of the intersection. Some of the results are already contained in [5] but with different proofs. The results are similar to those obtained for the union, concatenation, iteration and intersection of linear (string) languages, but they are not trivial, in particular the counterexample for the intersection of graph languages, since the well known example for string languages $L = \{(a^n b^n c^n) : n \geq 1\}$ is context free if viewed as a language of graphs [8].

2 Preliminaries

We briefly review the basic notions of edge-labelled hypergraphs and hyperedge replacement as in [5].

Definition 1. Let C be an arbitrary, but fixet set, called set of labels (or colors).

1. A (directed, hyperedge-labeled) hypergraph over C is a system (V, E, s, t, l) where
   - V is a finite set of nodes (or vertices),
   - E is a finite set of hyperedges,
   - $s : E \to V^*$ and $t : E \to V^*$ are two mappings assigning a sequence of sources s(e) and a sequence of targets t(e) to each $e \in E$, and
   - $l : E \to C$ is a mapping labeling each hyperedge.
2. Given a hypergraph (V, E, s, t, l), a hyperedge $e \in E$ is called an (m,n)-edge for some $m, n \in N$ if $length(s(e)) = m$, and $length(t(e)) = n$. The pair (m,n) is the type of e, denoted by type(e).
3. A multi-pointed hypergraph over C is a system $H = (V, E, s, t, l, begin, end)$ where (V, E, s, t, l) is a hypergraph over C and $begin, end \in V^*$. Components of H are denoted by V_H, E_H, s_H, t_H, l_H, $begin_H$, end_H, respectively. The set of a multi-pointed hypergraphs over C is denoted by H_C.
4. H is said to be an (m,n)-hypergraph, for some $m, n \in N$ if $length(begin_H) = m$, and $length(end_H) = n$. The pair (m,n) is the type of H, denoted by type(H).
5. For H an (m,n)-hypergraph, EXT_H denotes the set of nodes contained in the sequences $begin_H$ and end_H, called the set of external nodes of H. INT_H denotes the set of all other nodes, i.e., $INT_H = V_H - EXT_H$, called the set of internal nodes.

Example 1. In the following diagram, circles denote nodes and rectangles denote hyperedges; the hyperedge $h1$ has 2 source nodes and 1 target node. The graph $H1$ represented is of type (2,3).

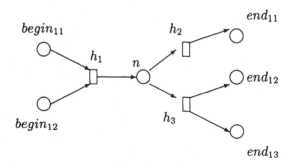

Definition 2. Let $H \in H_C$ be a multi-pointed hypergraph, and let $B \subseteq E_H$ be a set of hyperedge to be replaced. Let $repl : B \to H_C$ be a mapping with type(repl(b))=type(b) for all $b \in B$. Then the replacement of B in H through $repl$ yields a multi-pointed hypergraph X given by

- $V_X = V_H + \Sigma_{b \in B}(V_{repl(b)} - EXT_{repl(b)})$,
- $E_X = (E_H - B) + \Sigma_{b \in B}E_{repl(b)}$,
- each hyperedge keeps its labels,
- each hyperedge of $(E_H - B)$ keeps its sources and targets,
- each hyperedge of $E_{repl(b)}$ keeps its internal sources and targets and the external ones are handed over to the corresponding sources and targets of H
- $begin_X = begin_H$ and $end_X = end_H$

The multipointed hypergraph X is denoted by $REPLACE(H, repl)$.

The definition can be generalized to the case where $repl : B \to H_C$ is a mapping with $\pi_i(type(b)) \leq \pi_i(type(repl(b)))$ for $i = 1, 2$ and for all $b \in B$; in this case for each $b \in B$, either $begin_b$ corresponds (by default) to the initial subsequence of $begin_{repl(b)}$ or there is an explicit function $T_b : BEGIN_b \to BEGIN_{repl(b)}$, where $BEGIN_x$ is the set of nodes in the sequence $begin_x$. Similarly for end_b.

Definition 3. 1. Let $NT \subseteq C$ be a set of nonterminals. A production over NT is an ordered pair $p = (A, R)$ with $A \in NT$ and $R \in H_C$. A is called left hand side of p, R is called right hand side of p.
2. Let $H, H' \in H_C$, p=(A,R) a production, and $e \in E_H$ a hyperedge with $l_H(e) = A$ and type(e)=type(R). Then H directly derives H' through p (applied to e) if H' is isomorphic to $REPLACE(H, e \to R)$. We write $H \to_{p,e} H'$ or $H \to_p H'$.
3. A derivation (of length k) from H_0 to H_k is a sequence of direct derivations $H_0 \to_{p1,e1} H_1 \to_{p2,e2} \cdots_{pk,ek} H_k$

Definition 4. A Hyperedge Replacement Grammars is a system $HRG = (T, NT, S, P)$ where $T \subseteq C$ is a set of terminals, $NT \subseteq C$ is a set of nonterminals, P is a finite set of productions over NT and $S \in H_C$ is called axiom. The hypergraph language L(HRG) generated by HRG consist of all terminal labeled hypergraphs which can be derived from S by applying productions of P.

Hyperedge Replacement Systems can be seen as a special case of the general Graph Grammars defined in a categorical setting in [2]: it is enough to define each production $(L \leftarrow K \to R)$ so that the left hand side L is a single hyperedge, the interface graph K is a discrete graph whose nodes are the source and target nodes of L, and R is an arbitrary hypergraph containing K. This view of Hyperedge Replacement Systems allows the use of well known results from the classical theory based on double pushouts [2].

3 Some Positive Results

In this section, we analyze briefly the union of two context-free languages and then define the coalesced sum and prove that it too is an operation preserving the contex-freeness. We then define the composition of graphs and graph languages on three different levels of increasing generality, showing that context-freeness is preserved by each of them.

First the simplest result.

Let $G1 = < T_1, NT_1, S_1, P_1 >$ and $G2 = < T_2, NT_2, S_2, P_2 >$ be context-free graph grammars and define $G = G1 \cup G2$ to be

$< T_1 + T_2, NT_1 + NT_2 + \{A\}, S', P_1 + P_2 + \{(A, S_1), (A, S_2)\} >$

where $+$ indicates the disjoint set-theoretical union and S' is a new axiom labelled by the new nonterminal symbol A. It is immediate to check that G generates exactly the union of $L(G1)$ and $L(G2)$ and that the productions have the appropriate form.

Proposition 5. *The union of context-free graph languages is context-free.*

Notice that the union can be defined even if the types of the languages are different. If the languages have the same type, then the resulting language is of that type too; if not, the new axiom can be taken as a hyperedge whose begin and end sequences are the min of the corresponding of S_1 and S_2.

Unlike the union, the coalesced sum $\oplus$ is defined here only between languages of graphs of the same type.

Definition 6. – Given multi-pointed (hyper)graphs $H1$ and $H2$ of type (m,n), the coalesced sum $H1 \oplus H2$ is the multipointed (m,n)-hypergraph obtained by identifying, in the disjoint union $H1 + H2$, the nodes $begin_{1i}$ and $begin_{2i}$ for $i = 1, ..., m$ and the nodes end_{1j} and end_{2j} for $j = 1, ..., n$.
– Given languages $L(G1)$ and $L(G2)$ of type (m,n), the coalesced sum $L(G1) \oplus L(G2)$ is defined as $\{H1 \oplus H2 : Hi \in L(Gi)\ i = 1, 2\}$.

Example 2. Given the graph $H1$ of the previous example and following graph $H2$ of type (2,3)

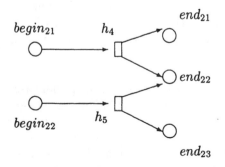

the coalesced sum $H1 \oplus H2$, of type (2,3), can be represented as follows

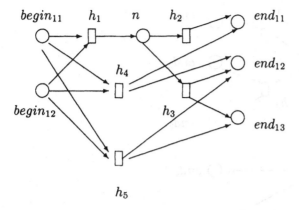

Proposition 7. *The coalesced sum of context-free graph languages is context-free.*

If $L(Gi)$ is generated by $Gi =< T_i, NT_i, S_i, P_i >$ for i=1,2, define $G =< T_1 + T_2, NT_1 + NT_2, S_1 \oplus S_2, P_1 + P_2 >$. It is not too difficult to check that $L(G) = L(G1) \oplus L(G2)$ and that the productions have the appropriate form.

At the opposite end of the spectrum is the *sum* of two graphs, which can be defined for any two graphs independently of their types. The *begin* nodes of the resulting graph comprise those of the operands (similarly for the *end* nodes. The operation is non commutative since the concatenation of the *begin* and *end* sequences is in general non commutative.

Definition 8. – Given multi-pointed (hyper)graphs $H1$ and $H2$ of type (m1,n1) and (m2,n2), respectively, the sum $H1 \pm H2$ is the multipointed (m1+m2,n1+n2)-hypergraph obtained by taking the disjoint union of the sets of nodes and the sets of edges, and by defining the nodes $begin_{H1\pm H2} = begin_{H1} \cdot begin_{H2}$ and the nodes $end_{H1\pm H2} = end_{H1} \cdot end_{H2}$.

– Given languages $L(G1)$ and $L(G2)$ of type (m1,n1) and (m2,n2), respectively, the sum $L(G1) \pm L(G2)$ is defined as $\{H1 \pm H2 : Hi \in L(Gi)\ i = 1, 2\}$.

Example 3. Given $H1$ and $H2$ as above, the sum $H1 \pm H2$ can be represented as follows

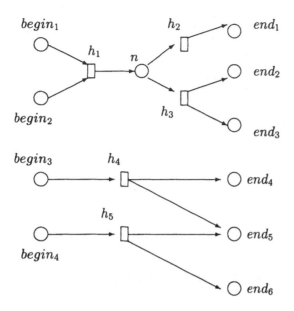

Proposition 9. *The sum of context-free graph languages is context-free.*

The proof is similar to that of the previous Proposition.

Definition 10.   – If $H1$ and $H2$ are hypergraphs with $type(H1) = (k, m)$ and
$type(H2) = (m, n)$, then their concatenation $H1 \star H2$ is the graph obtained
by first taking the disjoint union of $H1$ and $H2$ and then identifying the
i^{th} *end* node of $H1$ with the ith *begin* node of $H2$, for every i=1,...,m.
Moreover $begin(H1 \star H2) = begin(H1)$ and $end(H1 \star H2) = end(H2)$, and
so $type(H1 \star H2) = (k, n)$.
  – Given languages $L(G1)$ and $L(G2)$ of type (k,m) and (m,n), respectively, the
concatenation $L(G1) \star L(G2)$ is defined as $\{H1 \star H2 \ : \ Hi \in L(Gi) \ i = 1, 2\}$.

The concatenation $H1 \star H2$ can be defined even if $end(H1)$ and $begin(H2)$
do not have the same length. The concatenation can be parameterized by an
arbitrary partial function $j : end(H1) \rightarrow begin(H2)$ by "connecting" only the
end vertices for which j is defined with their image under j. The 'unused' (either
because not in the domain or not in the range of j) nodes contribute to the final
begin and *end* points. In this case $H1 \star_j H2$ can be defined as the disjoint
union of $H1$ and $H2$ and the subsequent identification of $x \in end(H1)$ with
$j(x) \in begin(H2)$ for all x for which j is defined. Then
$begin(H1 \star_j H2) = begin(H1) \cdot (begin(H2) - range(j))$ and
$end(H1 \star_j H2) = end(H2) \cdot (end(H1) - domain(j))$
with the obvious corresponding type.

Example 4. Given the hypergraphs $H1$ and $H2$ previously defined, let j be the
partial function with $j(end_{11}) = begin_{21}$ and $j(end_{12}) = begin_{22}$. Then the
parametrized concatenation $H1 \star_j H2$ is given by

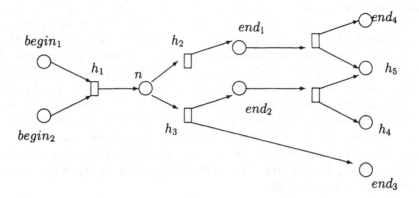

Given languages $L(G1)$ and $L(G2)$ of type (k,m) and (n,p), respectively, their concatenation via $j : [m] \rightharpoonup [n]$ is denoted by $L(G1) \star_j L(G2)$ and is defined as $\{H1 \star_j H2 \; : \; Hi \in L(Gi) \; i = 1,2\}$.

We adopt the convention that a 'missing' j indicates the more restrictive case of concatenation where j is the identity function.

The next result is similar to Lemma 4 in [3].

Proposition 11. *If a class of languages is closed under concatenation via partial functions, then it is closed under sum.*

We now establish the closure of the class of context free graph languages under concatenation.

Proposition 12. *The concatenation via a partial function j of context-free graph languages is context-free.*

The proof of this result is obtained, as usual, by defining the appropriate grammar to generate $L(G1) \star_j L(G2)$.

Let $L(Gi)$ be generated by $Gi =< T_i, NT_i, S_i, P_i >$, for i=1,2 and define $G3 = G1 \star_j G2$ by letting $G3 =< T_3, NT_3, P_3, S_3 >$ where:

- $T_3 = T_1 + T_2$
- $NT_3 = NT_1 + NT_2 + \{A\}$
- S_3 is a new axiom labelled by the new nonterminal A
- $P_3 =< P_1 + P_2 + \{(A, S_1 \star_j S_2)\} >$

To show that the concatenation $G1 \star_j G2$ of grammars generates the concatenation of languages, i.e., $L(G1 \star_j G2) = L(G1) \star_j L(G2)$ let $g1 \in L(G1)$ via the productions $(p1..pn)$ of $G1$ and let $g2 \in L(G2)$ via the productions $(q1..qm)$ of $G2$. Then $g1 \star_j g2 \in L(G1) \star_j L(G2)$ and $g1 \star_j g2 \in L(G1 \star_j G2)$ since the productions can be applied in the same order to S_3 without interference because they can overlap only in the "gluing" nodes which are preserved by each production, being the type of the non terminal hyperedge the same as that of the corresponding right hand side.

Conversely, let $g \in L(G1 \star_j G2)$ via the productions $(r1..rn)$ of $(G1 \star_j G2)$ so that $S_3 \Rightarrow_{r1} h1 \Rightarrow_{r2} h2 \ldots \Rightarrow_{rn} g$. By definition, $r1 = (A, S_1 \star_j S_2)$ while each ri is either in P_1 or in P_2. Notice now that by definition of $G3$, of HR production and of hyperedge replacement, any two consecutive applications of productions ri and $r(i+1)$ can overlap only on nodes which are at the same time (some of the) *end* nodes for $L(G1)$ and (some of the) *begin* nodes for $L(G2)$, both of which are preserved by each application of the productions of Gi. Hence the derivations are sequentially independent [2] and therefore, by the Church-Rosser property, their order can be reversed. Iterating the process, the derivation can be rewritten into an equivalent one $(t1...tn)$ where $ti \in \{r1, ..., rn\}$ and for some k, $\{t1, ..., tk\} \subseteq P_1$ and $\{t(k+1), ..., tn\} \subseteq P_2$. Then $S_3 \Rightarrow_{t1} \Rightarrow_{t2} \ldots \Rightarrow_{tk} g1 \star_j S_2 \Rightarrow_{t(k+1)} \ldots \Rightarrow_{tn} g1 \star_j g2$.

Definition 13. Given a context-free language $L(G)$ of type (m,m), the iteration $L(G)^*$ is defined as the union $\bigcup_{n \geq 0} L(G)^n$ where $L(G)^1 = L(G)$ and $L(G)^{n+1} = L(G) \star L(G)^n$

Proposition 14. *The iteration of context-free graph languages is context-free.*

It is sufficient to define, with S' a new axiom labelled by the new nonterminal A, $G^* = < T, NT + \{A\}, S', P + \{(A, S), (A, S \star S')\} >$ and show that $L(G^*) = L(G)^*$.

4 A negative result

In this section we sketch a proof that the set of context-free graph languages is not closed under intersection by giving a concrete counterexample. As mentioned in the Introduction, the classical example of $\{a^n b^n c^n : n \in N\}$ as the intersection of $\{a^m b^n c^n : m, n \in N\}$ and $\{a^n b^n c^m : m, n \in N\}$ cannot be used since it is a context-free graph language as shown in [8]

A necessary condition for a graph language to be context-free is given by the following lemma from [5].

Lemma 15. *If L is a hyperedge replacement language of order r, then there exist constants p and q depending only on L, such that for each multi-pointed hypergraph H in L with $size(H) \geq p$ there are a symbol X, natural numbers $m, n \in N$, a multi pointed hypergraph FIRST with a unique (m,n)-edge e labelled X, an (m,n)-hypergraph LINK with a unique (m,n)-edge f labelled X and an (m,n)-hypergraph LAST such that*

- *$H = FIRST \star LINK \star LAST$*
- *$size(LINK \star LAST) \leq q$*
- *$size(frame(LINK)) \leq r$*
- *LINK is non-trivial*
- *for $k \in N$, $FIRST \star LINK^k \star LAST$ is in L.*

This Pumping Lemma can be used to prove that a given graph language cannot be generated by a context free graph grammar. For example, a graph language whose underlying graphs are all square grids cannot be generated by context free grammar because it does not satisfy the Pumping Lemma no matter how the $LINK$ is chosen. We use this fact to construct, as the intersection of context-free languages, a language that contains 'almost' the set of all square grids.

Let $G1$ be the context free grammar of the form: $G1 =< T_1, NT_1, S_1, P_1 >$ where
$T_1 = \{a\}$
$NT_1 = \{u1, u2\}$
the initial axiom is

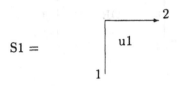

$S1 =$

and P_1 contains the following productions (3 for the nonterminal u_1 and 2 each for the nonterminals u_2 and u_3):

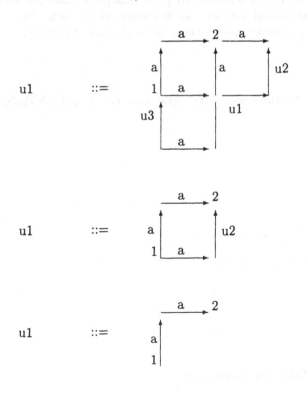

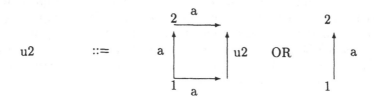

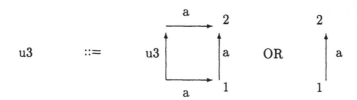

The language generated by $G1$ is formed by grids that grow along the horizontal dimension or along horizontal and vertical dimension at the same time, so that the vertical dimension is never greater than the horizontal dimension.

Similarly, let $G2$ be the context-free grammar $G2 =< T_2, NT_2, S_2, P_2 >$ where $T_2 = \{a\}$
$NT_2 = \{u1, u2\}$
the initial axiom is

S2 =

2 → u1 1

and P_2 contains the following productions:

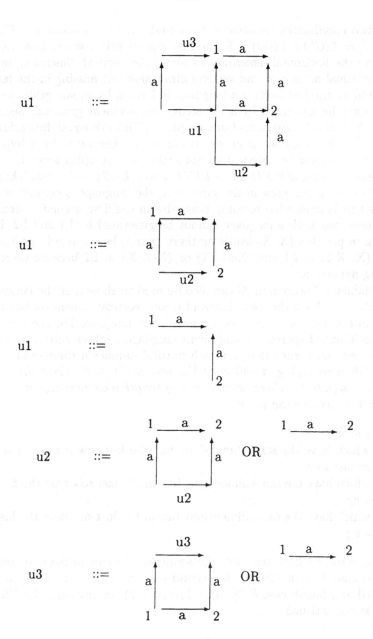

The language generated by $G2$ is formed by grids that grow along the vertical dimension or along the horizontal and the vertical dimensions at the same time.

Theorem 16. *The set of context-free graph languages is not closed under intersection*

The sketch of the proof will run the rest of the section.
Let us call four nodes and edges placed as single square an *item*; for each item

we have two coordinates (number of rows and number of columns). The new language $L = L(G1) \cap L(G2)$ is formed by square grids, which have as many items along the horizontal dimension as along the vertical dimension, or grids which have equal horizontal and vertical dimension but missing in the last row the first and/or the last item. If a language is formed by square grids, we know that it cannot be generated by a hyperedge replacement grammar because it grows too fast. If a language contains graphs which have equal dimensions but incomplete in the last row, as above, it cannot be generated by a hyperedge replacement grammar because it does not satisfy the pumping lemma.

If we select a subgraph $FIRST$, a $LINK$ and a $LAST$ in our graph language so that the dimensions grow in the same way, the language generated is not in L because this language has missing items also in the first, second ... etc. rows, and we know that such a language cannot be generated by L1 and L2. In fact if the item in position (X, X) ismissing there cannot be an item in position (X, X+1) or (X, X-1) in L1 and (X+1, X) or (X-1, X) in L2 because there is no supporting hyperedge.

The Diameter Theorem in [5] can also be used to show that the language L, with graphs that have the same horizontal and vertical dimension but are incomplete in the last row, cannot be a context free language. The theorem states that for an infinite hyperedge-replacement language L, there exist a constant K depending only on L such that, for each natural number n there exist a graph $N \in L$ with $card(V_N) \geq n$ and $card(V_N)/diam(N) \leq K$ where $diam(N) = max_{v,v' \in N} min\{n \in N : there\ exist\ a\ path\ of\ length\ n\ connecting\ v\ and\ v'\ in\ N\}$
In our language we can have:

- square grids,
- grids which have the same dimensions but the last row is missing the first and the last item,
- grids which have the same dimensions but in the last row only the first item is missing
- grids which have the same dimensions but in the last row only the last item is missing.

We can show, by induction over the sequence of items in the first row, that in the first case $K \geq n^2/2n$, in the second case $K \geq (n^2 - 2)/(2n - 3)$ while in the third and fourth case $K \geq (n^2 - 1)/(2n - 2)$. In any case, the Diameter Theorem is not satisfied.

5 Concluding Remarks

Languages generated by Hyperedge Replacement Systems can be manipulated by using standard set-theoretic operations. The two most common ones, union and intersection, behave differently with respect to the property of being context-free: it is easy to show that the union of context-free languages is still context-free while a counterexample has been constructed to show that the intersection of context-free languages is not a context-free language (although the typical example for string languages is, if seen as a graph language). It would be interesting to investigate whether the intersection of a 'linear' graph language (generated by productions with only one nonterminal on the right hand side) with a context-free language is still context-free.

The other two operations of coalesced sum and composition have been motivated by studies on visual languages and image recognition, where two images can be put together horizontally by matching part of the corresponding borders (discretized and represented by the *begin* and the *end* nodes) or can be integrated by superimposing them and then identifying the common border.

It may be of interest to investigate 'partial' coalesced sums by generalizing the definition to a coalesced sum when the types of the two graphs are not the same, by parameterizing it via partial functions, one between the respective *begin* nodes and another one between the respective *end* nodes of a pair of graphs and/or a pair of typed languages, in a manner similar to the parametrized concatenation. We have not encountered a meaningful application of such a generalized coalesced sum.

References

1. Courcelle, B.: Graph Rewriting: an algebraic and logical approach. Handbook of Theoret.Comp.Sci. vol.B, Elsevier 1990, 193–242
2. Ehrig, H.: Introduction to the Algebraic Theory of Graph Grammars. Springer Lect. Notes Comp. Sci. 73 (1979) 1–69
3. Engelfriet, J., Vereijken, J.J..: Concatenation of Graphs. partecipant edition of Proceedings of Fifth Internat.Wksp on Graph Grammars and their applications to Comp.Sci., USA nov 1994, 13–18
4. Golin, E.J., Reiss, S.P.: The Specification of Visual Language Syntax. Proc. IEEE Workshop on Visual Languages, Rome (Italy) (1989) 105–110
5. Habel,A.: Hyperedge Replacement: Grammars and Languages. Springer Lect. Notes Comp. Sci. 643 (1992)
6. Helm, R., Marriot, K.: Declarative Specification of Visual Languages. Proc. IEEE Workshop on Visual Languages, Skokie (Illinois, USA) (1990) 98–103
7. Nagl, M.: A tutorial and bibliographical survey on Graph Grammars. Springer Lect. Notes Comp. Sci. 73 (1979) 70–126
8. Nagl, M.: Formal languages of labelled graphs. Computing 16 (1976) 113–137

Upward Drawings of Search Trees
(Extended Abstract)

P. Crescenzi and P. Penna

Dipartimento di Scienze dell'Informazione
Via Salaria 113, 00198 Roma
e-mail: `piluc@dsi.uniroma1.it`

Abstract. We prove that any logarithmic binary tree admits a linear-area straight-line strictly-upward planar grid drawing (in short, upward drawing), that is, a drawing in which (a) each edge is mapped into a single straight-line segment, (b) each node is placed below its parent, (c) no two edges intersect, and (d) each node is mapped into a point with integer coordinates. Informally, a logarithmic tree has the property that the height of any (sufficiently high) subtree is logarithmic with respect to the number of nodes. As a consequence, we have that k-balanced trees, red-black trees, and BB[α]-trees admit linear-area upward drawings. We then generalize our results to logarithmic m-ary trees: as an application, we have that B-trees admit linear-area upward drawings.

1 Introduction

In several applications, information is better displayed by a graphical representation emphasizing its structure in a readable way. The automatic design of these graphical representations is one of the main motivations for the growing interest in the research area of *graph drawing* whose typical problem is the following: given a graph G, produce a geometric representation of G according to some graphic standards and optimization criteria.

Several graphic standards and optimization criteria have been proposed in the literature depending on the application at hand. The annotated bibliography maintained by Di Battista, Eades, Tamassia, and Tollis mentions most of them and refers to more than 300 papers in this research area [6]. In this paper we are interested in *straight-line strictly-upward planar grid drawings*, in short *upward drawings*, of rooted trees, that is, drawings in which each edge is mapped into a single straight-line segment, each node is placed below its parent, no two edges intersect, and each node is mapped into a point with integer coordinates. Each of these standards is naturally justified by "readability" considerations.

A natural and important optimization criterion for evaluating these drawings is that they take as little area as possible where the area of a drawing equals the area of the smallest isothetic rectangle bounding the drawing. This criterion belongs to the family of the so-called *aesthetic* criteria which are based on the fact that some drawings are better than others in conveying information regarding the tree.

1.1 Previous Results

The first $O(n \log n)$-area algorithm to produce an upward drawing of a binary tree appeared in [11] and recently [3] it has been proved that this algorithm is optimal (it is worth observing that, if we relax the upwardness requirement, then any binary tree of n nodes admits a linear-area planar grid drawing [12]). In [3] the authors also gave two algorithms producing a linear-area upward drawing of complete and Fibonacci binary trees, respectively. In [8], it has been proved that if we allow an edge to be represented by a chain of straight-line segments and a node to be on the same horizontal line as its parent, then *any* binary tree can be drawn in linear area. Subsequently, in [5] it has been shown that any AVL tree admits a linear-area upward drawing. Finally, in [1] the authors proved, among other things, that bounded-degree trees in some classes of balanced trees admit an $O(n \log \log n)$-area upward drawings: these classes include k-balanced trees, red-black trees, and BB[α]-trees.

1.2 Our Results

In Sect. 2 we show that, for any "logarithmic" binary tree t with n nodes, an upward drawing of t can be produced with area $O(n)$ in time $O(n)$. Informally, a logarithmic tree has the property that the height of any (sufficiently high) subtree is logarithmic with respect to the number of nodes. For example, several binary search trees such as k-balanced trees, red-black trees, and BB[α]-trees are logarithmic trees. In particular, we prove that, for any constant $\alpha > 1$, a constant κ exists such that any logarithmic binary tree with n nodes can be upward drawn in any rectangle whose shorter side is at least $\log^\alpha n$ and whose area is equal to κn. To this aim, we make use of the top-down approach developed in [5] suitably modified in order to deal with the case of logarithmic trees. Observe that the bound on the length of the shorter side allows a very great flexibility to applications that need to draw a binary tree in a prespecified rectangular region: indeed, the only requirement is essentially that the length of the shorter side is a little bit greater than the height of the tree.

In Sect. 3, we extend the previous result to the case of logarithmic m-ary trees. As a consequence of this generalization, we can show that, for any m, the class of m-ary B-trees admits linear-area upward drawings. In order to prove this result, we introduce the new notion of h-v-d^{m-2} drawing which is an extension of that of h-v drawing introduced in [3] and which is used as an intermediate drawing towards the upward one.

Due to the lack of space, most of the proofs will be only sketched.

1.3 Preliminaries

In this section we give preliminary definitions that will be used throughout the paper.

We refer to directed rooted trees. We denote by e the empty tree. Given m trees $t_1, \ldots, t_m$, we denote by $t_1 \oplus \ldots \oplus t_m$ the tree whose immediate subtrees are

$t_1, \ldots, t_m$. The definition of k-balanced trees, red-black trees, BB[α]-trees and B-trees can be found in any of the several textbooks on the design and analysis of algorithms such as [2, 10].

A *straight-line strictly-upward planar grid drawing*, in short *upward drawing*, of a tree t is a drawing of t such that:

1. Edges are straight-line segments.
2. Each node has an ordinate greater than that of its parent (we are thus assuming that the y-axis is downward oriented).
3. Edges do not intersect.
4. Nodes are points with integer coordinates.

The width (respectively, height) of a drawing is the width (respectively, height) of the smallest isothetic rectangle bounding the drawing. We adopt the convention that both the width and the height are measured by the number of grid points, so that any drawing of a nonempty tree has both width and height greater than zero. The *area* of a drawing is then defined as the product of the width and the height.

An *h-v drawing* of a binary tree is an upward drawing in which only rightward-horizontal and downward-vertical straight-line segments are allowed. More precisely, an h-v drawing of a non-empty binary tree $t = t_1 \oplus t_2$ is obtained by one of the two operations illustrated in Fig. 1 where δ_1 and δ_2 are two h-v drawings of t_1 and t_2, respectively. In the first operation, that is, the horizontal operation, δ_2 is translated to the right by as many grid points as the width of δ_1 and δ_1 is translated to the bottom by one grid point. The semantics of the second operation, that is, the vertical operation, is defined similarly.

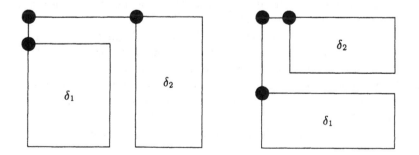

Fig. 1. The two operations of an h-v drawing

In [3], it has been shown that any h-v drawing of area A can be transformed into an upward drawing of area at most $2A$ (see also Lemma 9).

2 Binary Search Trees

In this section we give a sufficient condition to apply the techniques of [5] in order to obtain a linear-area upward drawing of a binary tree. This condition will then be applied to show that several classes of search trees admit upward drawings with area $O(n)$. These classes include k-balanced trees, red-black trees, and BB[α]-trees.

Let us first introduce the notion of a class of (h_0, β)-logarithmic trees. Informally, the trees belonging to this class have the property that the height of any (sufficiently high) subtree is logarithmic with respect to the number of nodes. Formally, we give the following definition.

Definition 1. A tree t of height h is (h_0, β)-*logarithmic* if, for any subtree t' of t whose height is greater than h_0,

$$n' \geq 2^{\beta h'}$$

where n' and h' denote the number of nodes and the height of t', respectively. A class of trees is said to be (h_0, β)-*logarithmic* if any of its members is an (h_0, β)-*logarithmic* tree.

It is well-known that the class of k-balanced trees is $(1, 1/k)$-logarithmic, that the class of red-black trees is $(1, 1/2)$-logarithmic, and that the class of BB[α]-trees is $(1, \log \frac{1}{1-\alpha})$-logarithmic.

In order to apply the techniques of [5], we have to define two functions which denote a lower bound on the smaller size of the rectangle in which the tree has to be drawn and the constant factor in the area function, respectively. Indeed, these two functions are the same as those used by Crescenzi and Piperno and are defined as follows:

$$l(h) = h^\alpha$$

where $\alpha > 1$ and

$$k(h+1) = \begin{cases} k_0 & \text{if } 1 \geq h < h_0, \\ k(h)\left(1 + \frac{1}{2h^\alpha}\right) & \text{otherwise} \end{cases}$$

where k_0 is a constant that will be specified later.

Clearly, for any h, $l(h+1) \geq l(h) + 1$. Moreover, the following two facts are already known.

Lemma 2 [5]. *A constant κ exists such that*

$$\lim_{h \to \infty} k(h) = \kappa.$$

Lemma 3 [5]. *For any h and for any n_1 and n_2 with $n_1 \leq n_2$,*

$$k(h+1)n \geq k(h)n_1 + k(h)n_2 + \frac{k(h+1)n - k(h)n_2}{l(h+1)}.$$

The following two results, whose proofs are here omitted, show that, for "almost any" tree in an (h_0, β)-logarithmic class of binary trees, two more properties are satisfied which are essential to prove that any tree in the class can be upward drawn in linear area.

Lemma 4. *Let C be an (h_0, β)-logarithmic class of binary trees. A constant $\bar{h}$ exists such that, for any $t \in C$ of height $h + 1 > \bar{h} + 1$ with n nodes,*

$$\frac{k(h+1)n - k(h)n_2}{\sqrt{k(h+1)n}} \geq l(h)$$

where n_2 is the number of nodes of the subtree of t with the bigger number of nodes.

Lemma 5. *Let C be an (h_0, β)-logarithmic class of binary trees. A constant $\hat{h}$ exists such that, for any $t \in C$ of height $h + 1 > \hat{h} + 1$ with n nodes,*

$$\frac{k(h)n_2}{\sqrt{k(h+1)n}} \geq l(h)$$

where n_2 is the number of nodes of the subtree of t with the bigger number of nodes.

We are now ready to prove the main result of this section. As stated before, the proof is based on the top-down approach used by Crescenzi and Piperno.

Theorem 6. *Let C be an (h_0, β)-logarithmic class of binary trees. For any $t \in C$ of height h with n nodes, t can be upward drawn within any rectangle R whose smaller size is at least $l(h)$ and whose area is at least $k(h)n$.*

Proof. The proof is by induction on h. For $h \leq \max\{h_0, \bar{h}, \hat{h}\}$ where $\bar{h}$ and $\hat{h}$ are the values specified by Lemmas 4 and 5, respectively, the proof is straightforward: indeed, since the number of trees with bounded height is finite, it is sufficient to choose k_0 big enough (that is why we did not specify the constant k_0).

Let $h \geq \max\{h_0, \bar{h}, \hat{h}\}$ and let us assume that the theorem is true for any height less than $h + 1$. Moreover, let l and L denote the smaller and the larger size of R, respectively. Given an (h_0, β)-logarithmic tree t of height $h + 1$ with n nodes, let us define

$$l_1 = L - \frac{k(h)n_2}{l} \qquad \text{and} \qquad l_2 = L - l_1$$

where n_2 denotes the number of nodes of the larger immediate subtree of t. Observe that if we denote by n_1 the number of nodes of the smaller immediate subtree of t, then $n = n_1 + n_2 + 1$.

The root of t is mapped into the grid point whose coordinate are (x, y) where x and y denote the coordinates of the top leftmost corner of R. Let us assume that the longer side of R is the vertical one (the other case can be proved in a similar way). We then isolate two rectangles R_1 and R_2 within the rectangle R

as follows. The top leftmost corners of R_1 and R_2 have coordinates $(x + 1, y)$ and $(x, y + \lfloor l_1 \rfloor)$, respectively. The vertical side and the horizontal side of R_1 have length l_1 and $l - 1$, respectively, while the vertical side and the horizontal side of R_2 have length l_2 and l, respectively (see Fig. 2 reproduced from [4]).

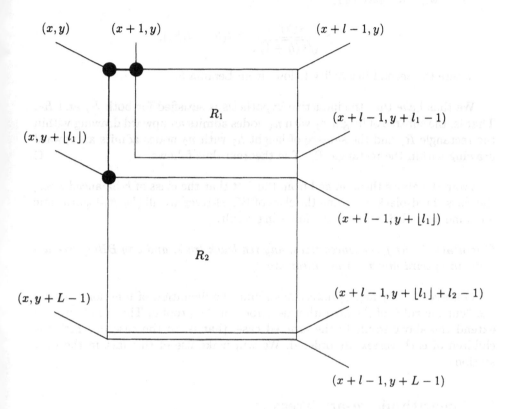

Fig. 2. The splitting of the rectangle R

Clearly, the area of R_2 is equal to $k(h)n_2$. Moreover, it is easy to see that Lemma 3 implies that the area of R_1 is at least $k(h)n_1$.

Let h_1 and h_2 denote the heights of the two subtrees with n_1 and n_2 nodes, respectively. We now shall prove that the shorter sides of R_1 and R_2 have length at least $l(h_1)$ and $l(h_2)$, respectively.

1. Rectangle R_1. If $l_1 \geq l - 1$, then

$$l - 1 \geq l(h + 1) - 1 \geq l(h) \geq l(h_1).$$

Otherwise,

$$l_1 \geq \frac{k(h + 1)n - k(h)n_2}{\sqrt{k(h + 1)n}} \geq l(h) \geq l(h_1),$$

where the second inequality follows from Lemma 4.

2. Rectangle R_2. If $l_2 \geq l$, then

$$l \geq l(h+1) \geq l(h) \geq l(h_2).$$

Otherwise, we have that

$$l_2 \geq \frac{k(h)n_2}{\sqrt{k(h+1)n}} \geq l(h) \geq l(h_2),$$

where the second inequality follows from Lemma 5.

We thus have that the inductive hypothesis is satisfied for both R_1 and R_2. That is, the subtree of height h_1 with n_1 nodes admits an upward drawing within the rectangle R_1 and the subtree of height h_2 with n_2 nodes admits an upward drawing within the rectangle R_2. The theorem thus follows. □

From the above theorem and from the fact that the class of k-balanced trees, the class of red-black trees, and the class of BB[α]-trees are all (h_0, β)-logarithmic for some h_0 and β, we have the following result.

Corollary 7. *Any k-balanced trees, any red-black trees, and any BB[α]-tree admits an upward drawing with linear area.*

Observe that by appropriately modifying the definition of function $k(\cdot)$ and the "cutting rule" of the algorithm described in the proof of Theorem 6, we can extend the above result to the ordered case, that is, to the case in which the children of each vertex are ordered. We will make use of this fact in the next section.

3 Logarithmic m-ary Trees

In this section we will extend the previous results to the case of m-ary trees. For the sake of clarity, we will describe our techniques in the case of ternary trees and we will only sketch how these techniques can be generalized for $m > 3$.

Basically, the linear-area upward drawing of a logarithmic ternary tree is obtained by the following four steps.

1. Transform the ternary tree into a binary tree.
2. h-v draw the binary tree.
3. Transform the h-v drawing of the binary tree into an 'h-v-d drawing' of the ternary tree.
4. Transform the h-v-d drawing into an upward drawing.

The second step will make use of the result of the previous section while the other three steps will be sketchily described in the following.

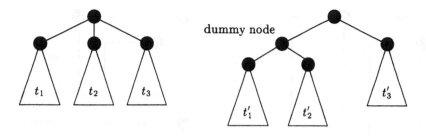

Fig. 3. From ternary to binary trees

3.1 From ternary to binary trees

The transformation from a ternary tree t to a binary tree t' is inductively defined as follows.

1. If the height of t is 1 then $t' = t$.
2. If $t = t_1 \oplus e$ (respectively, $t = t_1 \oplus t_2$), then $t' = t'_1 \oplus e$ (respectively, $t' = t'_1 \oplus t'_2$).
3. If $t = t_1 \oplus t_2 \oplus t_3$, then t' is obtained from t'_1, t'_2, and t'_3 by adding a 'dummy' node as shown in Fig. 3.

It is not hard to prove the following fact.

Lemma 8. *If t is an (h_0, β)-logarithmic ternary tree, then the corresponding binary tree t' is $(2h_0, \beta/2)$-logarithmic.*

3.2 From h-v drawings to h-v-d drawings

The h-v-d drawing standard is an extension of the h-v drawing one in the sense that we allow 'diagonal' rightward-downward segments between one node and one of its children. More formally (but still intuitively), if δ_1, δ_2, and δ_3 are the h-v-d drawings of three ternary trees t_1, t_2, and t_3, respectively, then an h-v-d drawing of $t = t_1 \oplus t_2 \oplus t_3$ can be obtained in one of the four ways shown in Fig. 4.

We are now interested in deriving an h-v-d drawing of a ternary tree t from the h-v drawing of the corresponding binary tree t'. To this aim, it is necessary that the drawing of t' satisfies an appropriate ordering of the subtrees of each node. This ordering depends on the existence of dummy nodes. In particular, assume that the root of t' has a dummy child as shown in Fig. 3. Then if the height (respectively, width) of the rectangle in which t' has to be drawn is greater than the width (respectively, height), then the h-v drawing of t' has to be performed as shown in Fig. 5(a) (respectively, 5(b)).

It is then easy to transform this h-v drawing of t' into an h-v-d drawing of t as shown in Fig. 6. Clearly, the area of the resulting h-v-d drawing is no more than the area of the original h-v drawing.

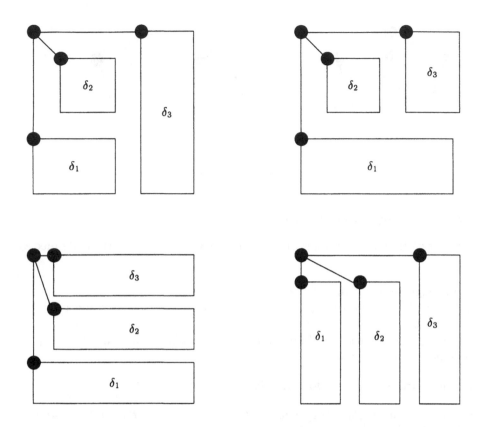

Fig. 4. The h-v-d operations

3.3 From h-v-d drawings to upward drawings

As in the case of h-v drawings, h-v-d drawings turn out to be a useful tool to produce intermediate drawings towards the final area-efficient upward drawings. Indeed, the following fact shows that transforming an h-v-d drawing into an upward drawing increases the area of the drawing by at most a factor of 2.

Lemma 9. *Any h-v-d drawing δ of height h and width w can be transformed in linear time into an upward drawing δ' of height $h + w$ and width w.*

Proof. We simply substitute each point (x, y) of δ with the point $(x, x + y)$. Clearly, the resulting drawing is upward and its height and width are equal to $h + w$ and w, respectively. It is not hard to see that this transformation preserves the planarity of the drawing. □

From the above construction and from Theorem 6, we have the following result.

Theorem 10. *Any (h_0, β)-logarithmic ternary tree admits a linear-area upward drawing.*

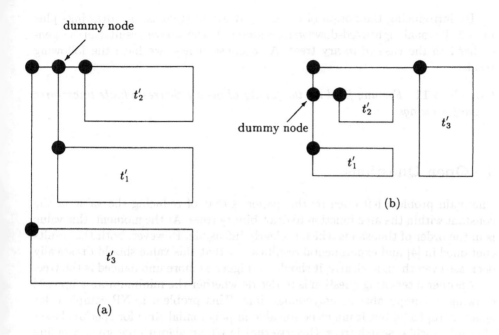

Fig. 5. The ordering of h-v drawing

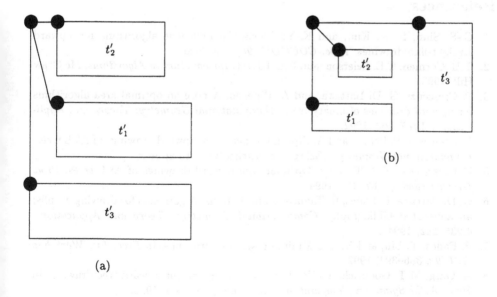

Fig. 6. The resulting h-v-d drawing

By introducing the notion of h-v-d^{m-2} drawing (that is, h-v drawings plus $m-2$ diagonal rightward-downward segments), the above result can be generalized to the case of m-ary trees. As a consequence, we have the following result.

Corollary 11. *For any fixed m, the family of m-ary B-trees admits linear-area upward drawings.*

4 Open Questions

The main problem left open by this paper is that of reducing the value of the constant within the area function to draw binary trees. At the moment, this value is in the order of thousands which is clearly infeasible. However, both the results contained in [4] and experimental results show that this value should drastically decrease even though, clearly, it should be bigger as more unbalanced is the tree.

Another interesting question is to decide whether the minimum-area upward drawing is computable in polynomial-time. This problem is NP-complete for general graphs [9] but it might be solvable in polynomial-time for special classes of graphs such as search trees. Observe that in [7], an algorithm is given yielding a minimum area h-v drawing of a binary tree with n nodes in time $O(n\sqrt{n}\log n)$.

References

1. C.-S. Shin, S. K. Kim, and K.-Y. Chwa. Area-efficient algorithms for upward straight-line drawings. *Proc. COCOON '96*, to appear.
2. T.H. Cormen, C.E. Leierson, and R.L. Rivest. *Introduction to Algorithms*, Mc Graw Hill, 1990.
3. P. Crescenzi, G. Di Battista, and A. Piperno. A note on optimal area algorithms for upward drawings of binary trees. *Computational Geometry: Theory and Applications*, 2:187–200, 1992.
4. P. Crescenzi, P. Penna, and A. Piperno. Linear area upward drawings of AVL trees. *Computational Geometry: Theory and Applications*, to appear.
5. P. Crescenzi and A. Piperno. Optimal-area upward drawings of AVL trees. *Proc. Graph Drawing*, 307–317, 1994.
6. G. Di Battista, P. Eades, R. Tamassia, and I. Tollis. Algorithms for drawing graphs: an annotated bibliography. *Computational Geometry: Theory and Applications*, 4:235–282, 1994.
7. P. Eades, T. Lin, and X. Lin. Minimum size h-v drawings. In *Proc. Int. Workshop AVI '92*, 386–394, 1992.
8. A. Garg, M.T. Goodrich, and R. Tamassia. Area-efficient upward tree drawing. In *Proc. ACM Symp. on Computational Geometry*, 359–368, 1993.
9. A. Garg and R. Tamassia. On the computational complexity of upward and rectilinear planarity testing. *Proc. Graph Drawing*, 286–297, 1994.

10. D.E. Knuth. *The Art of Computer Programming: Sorting and Searching*, Addison Wesley, 1975.
11. Y. Shiloach. Linear and planar arrangements of graphs. Ph.D. Thesis, Department of Applied Mathematics, Weizmann Institute of Science, Rehovot, Israel, 1976.
12. L. Valiant. Universality considerations in VLSI circuits.. *IEEE Trans. on Computers*, C-30(2):135–140, 1981.

More General Parallel Tree Contraction: Register Allocation and Broadcasting in a Tree

Krzysztof Diks[1]   Torben Hagerup[2]

[1] Instytut Informatyki, Uniwersytet Warszawski, Banacha 2, 02–097 Warszawa, Poland.
[2] Max-Planck-Institut für Informatik, D–66123 Saarbrücken, Germany.

Abstract. We extend the classic parallel tree-contraction technique of Miller and Reif to handle the evaluation of a class of expression trees that does not fit their original framework. We discuss applications to the following problems: (1) Register allocation, i.e., computing the number of registers needed to evaluate a given expression if all intermediate results must be kept in registers; and (2) Broadcasting in a tree, i.e., computing the number of steps needed to transmit a message from the root to all other nodes in a given tree if each node is a processor that can communicate with a single neighbor in each step. We show that on inputs of size n, both problems can be solved with optimal speedup in $O((\log n)^2)$ time on an EREW PRAM, in $O(\log n \log \log n)$ time on a CREW PRAM, and in $O(\log n)$ time on a CRCW PRAM.

1   Introduction

Register allocation for the evaluation of an expression is a central concern in code generation. Atomic operands (variables and constants) occurring in the expression are generally assumed to reside in main memory initially, while operators can take their operands from either registers or main memory. For reasons of speed, all intermediate results are to be kept in registers. Since registers are a scarce resource, the goal is to use as few registers as possible. Before an expression can be evaluated, its nonatomic immediate subexpressions must be evaluated and left in registers. The order in which subexpressions are evaluated, which the compiler is free to choose, is what determines the number of registers used.

The problem of *information dissemination* or *broadcasting* in a tree was introduced and motivated in [17]. At time 0, the root of the tree generates a message to be distributed to all other nodes in the tree. The goal is to minimize the *broadcast time*, the earliest instant in time at which all nodes have received the message. At each integral time instant after it receives the message, a node can pass it on to exactly one of its children. The order in which children are notified, which can be chosen freely, is what determines the broadcast time.

The problems of register allocation for an expression and broadcasting in a tree show strong similarities. These become most evident when one considers recursive characterizations of optimal solutions. Suppose that the immediate subexpressions of an expression E are $E_1, \ldots, E_k$ and that the number of registers needed to evaluate E_i is b_i, for $i = 1, \ldots, k$. If the subexpressions of E are evaluated in the order $E_1, \ldots, E_k$, the number of registers used in the evaluation of E will be $b = \max_{1 \le i \le k}(b_i + i - 1)$, the reasoning being that while E_i is under evaluation, $i - 1$ registers are needed to hold the values of $E_1, \ldots, E_{i-1}$. A moment's thought reveals that b is minimized, over all permutations of $E_1, \ldots, E_k$, if $b_1 \ge b_2 \ge \cdots \ge b_k$ (see [2, Fig. 9.23]). Similarly, it is shown in [17, Theorem 1] that if $T_1, \ldots, T_k$ are the maximal subtrees of a tree T rooted at the children of the root of T, if b_i is the optimal broadcast time of T_i, for $i = 1, \ldots, k$, and if $b_1 \ge b_2 \ge \cdots \ge b_k$, then the optimal broadcast time of T is $\max_{1 \le i \le k}(b_i + i)$. Thus the only formal difference between the two problems is an additive constant, plus different boundary conditions: The broadcast time of every 1-node tree is 0, whereas the register count assigned to a leaf in [2] is 1 if the leaf is the leftmost child of its parent, and 0 otherwise (see Fig. 1). Both problems can be solved in linear time; in the case of computing optimal broadcast times, this was shown in [17]. The register-allocation and broadcasting problems make sense in a more general setting: One can ask for the minimum number of registers needed to execute a straight-line program, represented as a directed acyclic graph, or for the optimal broadcast time of an arbitrary connected, undirected graph with a distinguished start node. However, (decision versions of) these more general problems are NP-hard [16, 17].

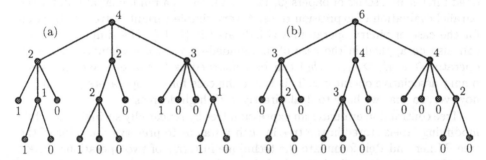

Fig. 1. A tree in which each node u is shown with (a) the register requirements (b) the optimal broadcast time of the maximal subtree rooted at u.

Let $\mathbb{N} = \{1, 2, \ldots\}$ and $\mathbb{N}_0 = \mathbb{N} \cup \{0\}$. Motivated by the applications described above, we take $\odot_\mu$, for $\mu \in \{0, 1\}$, to be the function from $\mathbb{N}_0^+$, the set of nonempty sequences of nonnegative integers, to $\mathbb{N}_0$ defined as follows: For all $k \in \mathbb{N}$ and all $a_1, \ldots, a_k \in \mathbb{N}_0$, $\odot_\mu(a_1, \ldots, a_k) = \max_{1 \le k \le k}(b_i + i - \mu)$, where $\{b_1, \ldots, b_k\}$ is the same multiset as $\{a_1, \ldots, a_k\}$ and $b_1 \ge b_2 \ge \cdots \ge b_k$. We define the $\odot$-*problem* as follows: An instance of the problem is given by a rooted tree T in which each leaf is labeled with either 0 or 1 and each internal node is labeled with one of the functions $\odot_0$ and $\odot_1$, and the task is to compute the *value* of T, defined as the value of its root; the value of a leaf is its associated label, and the value of an internal node u is $\odot_\mu(b_1, \ldots, b_k)$, where $\odot_\mu$ is the function labeling u and $b_1, \ldots, b_k$ are the values of its children.

While it is not difficult to show that the $\odot$-problem is in NC, obtaining our best results requires some care and rather intricate variants of tree contraction. Denoting the number of nodes in the input tree by n, we achieve the following time bounds, in all cases together with a linear time-processor product, i.e., with optimal speedup: $O((\log n)^2)$ on the EREW PRAM, $O(\log n \log \log n)$ on the CREW PRAM, and $O(\log n)$ on the CRCW PRAM. For the EREW PRAM, the actual time bound is $O(\log n \log(d+1))$, where d is the maximum degree of any node in the input tree. For $d = 2$, our (EREW PRAM) result was obtained previously by Miller and Teng [14].

2 Tree contraction

The $\odot$-problem is a special case of the more general problem of evaluating an expression tree over a domain D and a set Ω of operators from D^+ to D. An instance is here given by a rooted and ordered tree T in which each leaf is labeled with an element of D and each internal node is labeled with an operator in Ω. The goal again is to compute the value of the root, where the value of a leaf is its label, and the value of an internal node with label $\Diamond$, whose k children, in the order from left to right, have values $b_1, \ldots, b_k$, is $\Diamond(b_1, \ldots, b_k)$.

The *tree-contraction* technique, introduced by Miller and Reif [11] and developed further in a series of papers [9, 12, 13, 14, 15], is a fundamental tool for the parallel evaluation of expression trees. A very simple variant of tree contraction for the case of binary operators was indicated in [1, 10]. The simple algorithm can also be applied in the case of unbounded-degree trees whenever, for each operator $\Diamond \in \Omega$, $\Diamond(b_1, \ldots, b_k)$ can be construed as $b_1 \diamond b_2 \diamond \cdots \diamond b_k$ for some binary associative operator $\diamond$. This is not the case for our operators $\odot_0$ and $\odot_1$, however, so that we have to deal directly with high degrees.

Tree contraction evaluates an expression tree by repeatedly shrinking it, while modifying labels stored in the tree in such a way as to preserve the value of the tree. Miller and Reif formulate the technique in terms of two operations, RAKE and COMPRESS. Described at a high level that, in particular, ignores the calculations involving labels, RAKE removes a leaf from the (current) tree, while COMPRESS *merges* two adjacent degree-1 nodes. If RAKE and COMPRESS operations can be carried out in constant time (this depends on the label calculations), then tree contraction can reduce an n-node tree to a constant size in $O(\log n)$ time without changing its value, after which the value of the tree can be determined in constant time. In our case, COMPRESS operations can be executed in constant time, as required, but simultaneously raking k leaves with the same parent involves sorting k items, which cannot be done in constant time. Miller and Reif were faced with a similar problem in [13], where tree contraction is used to compute so-called canonical labelings of trees. In their case, if $k \geq 2$ children of the same node simultaneously become leaves, due to the execution of RAKE operations, it is necessary to spend $\Theta(\log k)$ time sorting labels associated with the k leaves, after which they can be removed. Miller and Reif showed that even with this complication, the tree contraction finishes within $O(\log n)$ time. What makes our problem more difficult is that even after any amount of processing, we cannot remove any leaf children of a node u as long as u has two or more nonleaf children; this is a reflection of the fact that even if all but one of the arguments of $\odot_\mu$ are known, the value of $\odot_\mu$, once the final argument is revealed, could still depend on all the other arguments individually. We show how to cope with such a situation within the framework of tree contraction.

The lemma below, which describes generic tree contraction, was essentially proved in a less general form by Miller *et al.* [12, 13, 9, 14]. Their description does not provide a clean interface between generic tree contraction and its various applications, however, so that the lemma can be extracted from their exposition only with some effort. For this reason we provide a self-contained proof of the lemma in an appendix. We restrict attention to operators $\diamond$ that are *commutative*, i.e., whose value is invariant under permutation of arguments. This is a genuine restriction, but the operators $\odot_0$ and $\odot_1$ of interest here are clearly commutative.

Let k and l be integers with $0 \leq l \leq k$, and let D be a set. An *l-dimensional projection* of a function $\diamond$ from D^k to D is the function from D^l to D obtained from $\diamond$ by fixing $k-l$ arguments to be constants in D, while leaving the remaining l arguments as indeterminates. E.g., the l-dimensional projection of $\diamond$ obtained by fixing the $k-l$ first arguments at the values $b_1, \ldots, b_{k-l}$ maps $(x_1, \ldots, x_l)$ to $\diamond(b_1, \ldots, b_{k-l}, x_1, \ldots, x_l)$ for all $(x_1, \ldots, x_l) \in D^l$. Given sets D and Ω, where each element of Ω is a function from D^+ to D, let $\mathcal{T}_{D,\Omega}$ be the set of all rooted trees in which each leaf is labeled with an element of D and each internal node is labeled with an element of Ω.

Lemma 1. *Let D be a set, Ω a set of commutative functions from D^+ to D, and $\mathcal{T}$ a subset of $\mathcal{T}_{D,\Omega}$. If there are τ, p and $\mathcal{F}$ satisfying Conditions (A)–(E) below, then, for all integers $n \geq 2$, every n-node tree $T \in \mathcal{T}$ can be evaluated in $O(\tau(T)\log n + \log(p(T)))$ time on a PRAM with $np(T)$ processors.*

(A) *$\mathcal{F}$ is a set of functions from D to D, and for all $f \in \mathcal{F}$ and all $x \in D$, $f(x)$ can be computed from f and x in constant time with one processor.*

(B) *For any two functions f and g in $\mathcal{F}$, $g \circ f$ belongs to $\mathcal{F}$ and can be computed from f and g in constant time with one processor.*

(C) *Every 1-dimensional projection of a function in Ω belongs to $\mathcal{F}$.*

(D) *τ and p are functions from $\mathcal{T}$ to $\mathbb{N}$ such that for all integers $n \geq 2$ and all n-node trees $T \in \mathcal{T}$, $\tau(T)$ and $p(T)$ can be computed from T in $O(\tau(T)\log n)$ time with $np(T)$ processors.*

(E) *Fix an input tree $T^* \in \mathcal{T}$ with $n \geq 2$ nodes, let $\tau^* = \tau(T^*)$ and $p^* = p(T^*)$ and consider the following setting: A sequence $a_1^*, \ldots, a_s^*$ of s integers, where s is a nonnegative integer bounded by the maximum degree of a node in T^* and where $1 \leq a_1^* \leq a_2^* \leq \cdots \leq a_s^* = O(\tau^* \log n)$, is available for preprocessing in $O(\tau^* \log n)$ time with s processors. Assume that the preprocessing finishes at time 0 (this is just a convention for fixing the origin of the time axis). Subsequently, at time a_i^*, for $i = 1, \ldots, s$, the value b_i^* of a node r_i in T^* becomes known, and $p^* n_i$ processors numbered $p^* \sum_{j=1}^{i-1} n_j + 1, \ldots, p^* \sum_{j=1}^{i} n_j$ become available, where n_i is the number of nodes in the maximal subtree*

T_i^* of T^* rooted at r_i, for $i = 1, \ldots, s$. *Thus every value that becomes known contributes new processors, and the available processors at all times are consecutively numbered. Moreover, the trees $T_1^*, \ldots, T_s^*$ are disjoint.*

In these circumstances, for some constant $C > 0$, the function $f : D \to D$ mapping x to $\Diamond(b_1^*, \ldots, b_s^*, x)$, for all $x \in D$, must be computable with the available processors to be ready by time $\max_{a_1^* \leq j \leq a_s^*}(j + C\tau^* \log(|F_j| + 2))$, where $F_j = \{i : 1 \leq i \leq s$ and $a_i^* = j\}$, for $j = a_1^*, \ldots, a_s^*$; for $s = 0$ we take this condition to mean that f must be computable in constant time with one processor.

The model of computation is specified in Lemma 1 only as a PRAM; the lemma holds for all variants of the PRAM at least as strong as the EREW PRAM. When there are no restrictions on concurrent reading, the term $\log(p(T))$ in the time bound can be removed. While Conditions (A)–(C) of Lemma 1 are readily interpreted in the context of [12, 13, 9, 14] and Condition (D) is a simple extension (allowing more time and processors), Condition (E) may look unfamiliar. Its closest analogue in the work of Miller and Reif is their requirement [13] that "the leaf children of a node with k leaf children can be raked in $O(\log k)$ time", but this formulation is neither precise nor general enough for our needs.

Since a major concern in the remainder of the paper will be to show that Condition (E) is satisfied in a number of situations, we introduce terminology that facilitates the discussion. First, the processing required in Condition (E), i.e., the computation of the function mapping x to $\Diamond(b_1^*, \ldots, b_s^*, x)$, for all $x \in D$, will be referred to as *raking*. The quantities $b_1^*, \ldots, b_s^*$ are called *keys* to distinguish them from other integers. For $1 \leq i < j \leq s$, we consider b_i^* and b_j^* to be distinct keys even if they have the same numerical value. For $i = 1, \ldots, s$, a_i^* is called the *arrival time* of the key b_i^*, and for $j = a_1^*, \ldots, a_s^*$, the set of keys with arrival time j will be called a *family*, also considered to have arrival time j. We distinguish between families with different arrival times even if they happen to be empty. The *lead* of a family with arrival time j is $t - j$, where t is the time at which the raking finishes. Finally, for any finite set S, we define the *log-size* of S as $\log(|S| + 2)$. Condition (E) can now be expressed by saying that the lead of some family must be within a constant factor of τ^* times its log-size.

3 The algorithms

In our application of tree contraction to the $\odot$-problem, we take $D = \mathbb{N}_0$ and $\Omega = \{\odot_0, \odot_1\}$, and we define $\mathcal{T}$ to be the set of those trees in $\mathcal{T}_{D,\Omega}$ whose leaf labels are all bounded by 1. For all $\alpha, \beta, \gamma \in \mathbb{N}_0$ with $\alpha < \beta$, denote by $\phi_{\alpha,\beta,\gamma}$ the function from $\mathbb{N}_0$ to $\mathbb{N}_0$ given by

$$\phi_{\alpha,\beta,\gamma}(x) = \begin{cases} \gamma, & \text{if } x < \alpha, \\ \gamma + 1, & \text{if } \alpha \leq x < \beta, \\ \gamma + 2 + x - \beta, & \text{if } x \geq \beta, \end{cases}$$

for all $x \in \mathbb{N}_0$, and take $\mathcal{F} = \{\phi_{\alpha,\beta,\gamma} \mid \alpha, \beta, \gamma \in \mathbb{N}_0$ and $\alpha < \beta\}$. Condition (A) of Lemma 1 is clearly satisfied, and Conditions (B) and (C) are expressed in the following lemmas.

Lemma 2. *For any two functions f and g in $\mathcal{F}$, $g \circ f$ belongs to $\mathcal{F}$ and can be computed from f and g in constant time with a single processor.*

Lemma 3. *For all $\mu \in \{0,1\}$, all $k \in \mathbb{N}_0$ and all $b_1, \ldots, b_k \in \mathbb{N}_0$, the function f_μ from $\mathbb{N}_0$ to $\mathbb{N}_0$ mapping x to $\odot_\mu(b_1, \ldots, b_k, x)$, for all $x \in \mathbb{N}_0$, belongs to $\mathcal{F}$.*

Proof sketch. For $k = 0$, $f_\mu = \phi_{1,2,1-\mu}$, so consider the case $k \geq 1$. Without loss of generality assume that $b_1 \geq \cdots \geq b_k$ and define i as the maximal integer with $1 \leq i \leq k$ such that $b_i + i = \odot_0(b_1, \ldots, b_k)$. One can observe that $f_0 = \phi_{\alpha,\beta,\gamma} \in \mathcal{F}$, where $\alpha = b_i$, $\beta = b_i + i + 1$ and $\gamma = b_i + i \geq 1$. This also implies that $f_1 = \phi_{\alpha,\beta,\gamma-1}$.

3.1 The EREW PRAM algorithm

For the EREW PRAM we take $\tau(T) = \lceil \log(d+1) \rceil$, where d is the maximum degree of a node in T, and $p(T) = 1$ for all $T \in \mathcal{T}$. Then Condition (D) of Lemma 1 is easily seen to be satisfied. What remains is to show how to satisfy Condition (E). But this is also easy: We simply wait until time a_s^*, i.e., until all of the keys $b_1^*, \ldots, b_s^*$ are available, and then process them as implicit in the proof of Lemma 3. For $\diamond = \odot_\mu$, this involves sorting $b_1^*, \ldots, b_s^*$ into nonincreasing order, adding $i - \mu$ to the ith element in the sorted sequence, for $i = 1, \ldots, s$ (we call this "adding appropriate offsets"), computing the maximum in the resulting sequence, and determining the largest position of an occurrence of the maximum. Since $s \leq d$, all of this can be done in $O(\tau^*)$ time on an EREW PRAM with s processors, the only nontrivial subroutine needed in fact being one for logarithmic-time sorting with optimal speedup [3, 6]. We have proved:

Lemma 4. *For all integers $n \geq 2$ and $d \geq 1$, $\odot$-problems can be solved on input trees with n nodes and maximum degree d in $O(\log n \log(d + 1))$ time on an EREW PRAM with n processors.*

3.2 The CRCW PRAM algorithm

For the CRCW PRAM we let $\tau(T) = 1$ and $p(T) = \lceil \log n \rceil^2$ for all $T \in \mathcal{T}$ with $n \geq 2$ nodes. Again Condition (D) of Lemma 1 is obviously satisfied. Consider Condition (E).

We cannot follow the strategy of the EREW PRAM algorithm of waiting until all keys $b_1^*, \ldots, b_s^*$ have become available before starting the raking, the reason being that a PRAM with a polynomial number of processors cannot sort in constant time. Instead we have to carry out the sorting incrementally, i.e., to sort keys as they arrive. More precisely, our raking algorithm proceeds as follows:

Let $\lambda \in \mathbb{N}$ be a constant to be fixed later. We divide the s keys into $q = \lceil (a_s^* - a_1^* + 1)/(2\lambda) \rceil$ *generations* $G_1, \ldots, G_q$. For $i = 1, \ldots, q$, a key belongs to the ith generation G_i if its arrival time lies in the set $\{a_s^* - i(2\lambda) + 1, \ldots, a_s^* - (i-1)(2\lambda)\}$, called the *span* of G_i. In other words, the generations are numbered "backwards", and except for G_q, which may be smaller, each generation spans exactly 2λ time units. We say that a family belongs to a generation

G if its arrival time lies in the span of G. The raking algorithm sorts the keys within each generation; we will use G_i also to denote the sorted sequence obtained by sorting the keys in the ith generation, for $i = 1, \ldots, q$, relying on context to resolve any ambiguity. In addition, with $H_q = G_q$, G_i is merged with H_{i+1} to create the sorted sequence H_i, for $i = q - 1, \ldots, 1$ (see Fig. 2). Each sorting and merging operation is started as soon as its input is available. Thus several generations may be sorted simultaneously, while the merges must happen in a strictly sequential fashion. When the final sorted sequence H_1 has been produced, we remember its maximum M and add appropriate offsets in the range $\{0, \ldots, s\}$. Since the maximum of the resulting sequence S obviously lies in the range $\{M, \ldots, M + s\}$, computing this quantity reduces to computing the maximum of the sequence S' obtained from S by subtracting M from each of its elements and replacing negative elements by zero. The elements of S' lie in the range $\{0, \ldots, s\}$, which means that their maximum can be computed in constant time with s processors [8, Theorem 1]. The final nontrivial step of the raking, the computation of the position in S of the last occurrence of the maximum, also reduces to computing the maximum of s integers in the range $\{0, \ldots, s\}$. Thus the raking finishes within a constant delay after the end of the last merge.

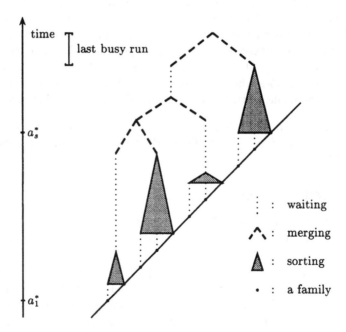

Fig. 2. Sorting and merging generations.

We still have to specify the routines used for sorting and merging. We employ a standard sorting algorithm using logarithmic time and a linear number of processors [3, 6] and a constant-time merging algorithm characterized in the lemma below, whose proof is based on techniques and results of [4, 5].

Lemma 5. *For all integers N and M with $2 \leq N \leq M$ and all fixed $\epsilon >$ 0, two sorted sequences, each containing N integers in the range $\{0, \ldots, M\}$, can be merged in constant time on a CRCW PRAM with $O(N(M/N)^\epsilon \log N)$ processors.*

It is easy to see by induction that the value of every N-node tree in $\mathcal{T}$ is bounded by N. Thus, whenever our raking algorithm merges two sequences involving $N \geq 2$ keys and a maximum key value of M, the number of processors available for the merge will be at least $p^* \max\{N, M\} \geq \max\{N(\log n)^2, M\}$. With $\epsilon = 1/2$, this is at least $N(M/N)^\epsilon \log N$. For if $(M/N)^\epsilon \leq \log n$, the claim is obvious; and if not, $M = N(M/N)^{1/2}(M/N)^{1/2} \geq N(M/N)^\epsilon \log N$. It now follows from Lemma 5 that every merge carried out by the raking algorithm can be executed in constant time. We choose the constant λ anticipated above as any constant upper bound on the number of time steps needed for a single merge.

Define the *lead* of a generation G as the number of time steps from the beginning of the sorting of G to the end of the raking. In order to show that the raking is sufficiently fast, we must find a generation whose lead is within a constant factor of its log-size; to see this, we use the facts that the log-size of each generation is within a constant factor of the log-size of the largest family in the generation (since there are at most 2λ such families), and that the lead of every family in a generation G is within a constant factor (in fact, within an additive term of 2λ) of the lead of G. We can clearly assume that $q \geq 1$, i.e., that at least one merge takes place.

Say that a time step is *busy* if some merge is in progress during the step. Note that sorting may take place during a nonbusy step—we care only about merges. Define a *busy run* as a maximal sequence of busy time steps, let k be the number of merges carried out during the last busy run and consider two cases:

Case 1: $k \leq 2$. What triggers the last busy run is the completion of the sorting of some generation G. Since the last busy run takes only constant time, certainly the lead of G is within a constant factor of its log-size.

Case 2: $k \geq 3$. Let t be the time at which the last busy run begins and note that the last merge completes no later than at time $t + \lambda k$. The fact that G_k and H_{k+1} cannot start merging before time t implies that for some l with $k \leq l \leq q$, the sorting of G_l has not been completed by time $t - 1$. Since generations begin only every 2λ steps, the sorting of G_l starts no later than $2\lambda(l - 1)$ steps before the beginning of the last merge, i.e., no later than at time $t + \lambda k - 2\lambda(l - 1) \leq t + \lambda k - 2\lambda(k - 1) = t - \lambda(k - 2)$. Thus the time r needed for the sorting of G_l is at least $\lambda(k - 2) \geq \lambda k/3$, and the lead of G_l is at most $r + \lambda k$. It follows that the lead of G_l is within a constant factor of its log-size.

Lemma 6. *For all integers $n \geq 2$, $\odot$-problems can be solved on n-node input trees in $O(\log n)$ time on a CRCW PRAM with $O(n(\log n)^2)$ processors.*

3.3 The CREW PRAM algorithm

For the CREW PRAM we take $\tau(T) = \lceil \log \log n \rceil$ and $p(T) = \lceil \log n \rceil^2$ for all $T \in \mathcal{T}$ with $n \geq 4$ nodes. The raking now has to perform not only the sorting, but also the maximum-finding in an incremental fashion, since a CREW PRAM cannot carry out either task in sublogarithmic time (even for inputs consisting of small integers). Our approach to incremental maximum-finding is to extend the sorted sequences employed by the CRCW PRAM algorithm by additional information, as expressed in the definition of an R-structure below. A *range query* about a sequence $a_1, \ldots, a_N$ of integers specifies two integers k and l with $1 \leq k \leq l \leq N$ and asks for the position of the last occurrence of $\max_{k \leq i \leq l} a_i$ in the subsequence $a_k, \ldots, a_l$.

Definition 7. Given a multiset S of N keys, an *R-structure* for S consists of

(1) The sorted sequence $b_1 \geq b_2 \geq \ldots \geq b_N$, where $S = \{b_1, \ldots, b_N\}$.
(2) A *range-query* structure Q that allows arbitrary range queries about the sequence $b_1 + 1, \ldots, b_N + N$ to be answered in $O(\log \log N)$ time with $O(\log N)$ processors.

It is easy to see that in the context of Condition (E) of Lemma 1, an R-structure for the full multiset $\{b_1^*, \ldots, b_s^*\}$ will allow us to finish the raking in $O(\log \log n)$ time. Except in the trivial case $s = 0$, we obtain such an R-structure by repeatedly combining R-structures for smaller multisets with the algorithm described in the following lemma, whose proof is omitted.

Lemma 8. *Given R-structures for two multisets X and Y with $N = |X|$ and $M = |Y|$ elements, where $2 \leq M \leq N$, an R-structure for $X \cup Y$ can be constructed in $O(\log \log N + \log M)$ time on a CREW PRAM with $O(N(\log N)^2)$ processors.*

We will use the phrase "R-merging X and Y" to denote the procedure described in Lemma 8. Except for the fact that R-merges take the place of usual merges, our CREW PRAM raking algorithm is similar to the CRCW PRAM algorithm described in the previous subsection. The only other difference is a new definition of the generations $G_1, \ldots, G_q$. Let λ be a positive integer with $\lambda = \Theta(\log \log n)$ such that λ steps suffice to (1) sort any sequence of at most $K = \lceil \log n \rceil^3$ keys, and (2) R-merge any sequence of at most K keys with any sequence of at most n keys. Let g be the function from $\{a_1^*, \ldots, a_s^*\}$ to $\mathbb{N}$ given by

$$g(x) = \left\lfloor \log \left(\left\lfloor \frac{a_s^* - x}{3\lambda} \right\rfloor + 1 \right) \right\rfloor + 1,$$

for all $x \in \{a_1^*, \ldots, a_s^*\}$, and define G_i, for $i = 1, \ldots, q = g(a_1)$, as the set of those keys whose arrival time is mapped to i by g. Thus G_1 comprises the last 3λ families, G_2 spans the previous 6λ families, and so on. In general, for $i = 2, \ldots, q$, G_i spans twice as many families as G_{i-1}, except that G_q may be smaller.

We must demonstrate the existence of a family whose lead is within $O(\log \log n)$ of its log-size. Recall that a busy run is a maximal sequence of steps in which (R-)merging takes place. Let t be the time at which the last busy run starts or, if $q = 1$ (in which case there are no busy runs), the time at which the sorting of G_1 ends. Take $k = 0$ if $q = 1$, and otherwise let k be the number of

R-merges in the last busy run. We define a *final-merger* as follows: The generations $G_1, \ldots, G_k$ are final-mergers. Additionally, if $k = 0$ or the sorting of G_{k+1} finishes after that of G_k, G_{k+1} is a final-merger. Let G_r be a largest final-merger and consider two cases:

Case 1: $|G_r| \leq K$. In this case each final-merger can be sorted in at most λ steps, and every R-merge in the last busy run, if any, takes at most λ steps. We can conclude that $k \leq 1$, since otherwise, when the R-merging of G_k with H_{k+1} completes, the sorting of G_{k-1} has not yet begun, contradicting the fact that the last busy run is in progress. But then any family in G_1 has a lead of $O(\log \log n)$.

Case 2: $|G_r| > K$. Let L be the log-size of G_r. It suffices to show that the lead of G_r is $O(L \log \log n)$. To see this, note the following: First, since the total number of families is $O(\log n \log \log n)$, the log-size of G_r is within a constant factor of the log-size of a largest family in G_r. And second, the lead of G_r is at most three times that of any of its families if $r > 1$, while it is at most an additive $3\lambda = O(\log \log n)$ larger if $r = 1$. In particular, this allows us to assume that $q > 1$ and hence $k \geq 1$.

Since the size of G_r dominates that of any final-merger, every R-merge that takes place during the last busy run can be executed in $O(L)$ time. Because the total number of R-merges is $O(\log \log n)$, the length of the final busy run is $O(L \log \log n)$. The last busy run is triggered by the completion of the sorting of a generation G_l whose size is dominated by that of G_r. Moreover, $l \geq r$, i.e., the sorting of G_r starts no earlier than that of G_l. Thus the sorting of G_r starts after time t or at time $t - O(L)$, which implies that the lead of G_r is $O(L \log \log n)$.

Lemma 9. *For all integers $n \geq 4$, $\odot$-problems can be solved on input trees with n nodes in $O(\log n \log \log n)$ time on a CREW PRAM with $O(n(\log n)^2)$ processors.*

A nontrivial preprocessing step, which we cannot describe here for lack of space, allows us to reduce the number of processors needed by the algorithms of Lemmas 4, 6 and 9 to the point of optimal speedup. This yields our main result:

Theorem 10. *For all integers $n \geq 4$ and $d \geq 1$, register-allocation and broadcasting problems can be solved on input trees with n nodes and maximum degree d using $O(n)$ operations and either $O(\log n \log(d + 1))$ time on an EREW PRAM, $O(\log n \log \log n)$ time on a CREW PRAM, or $O(\log n)$ time on a CRCW PRAM.*

Appendix: Proof of Lemma 1

We assume that the input tree is given according to a standard adjacency-list representation, where each node has an adjacency list with an entry for each of its children, and each nonroot node has a pointer to its entry in the adjacency list of its parent. When merging a node v with its parent u, we will identify the resulting node with the parent u—to emphasize this, we also denote the operation as "merging v into u". In terms of the representation, we simply remove v after appending its adjacency list to that of u.

Let $T^* = (V^*, E^*)$ be the input tree and take $n = |V^*|$, $\tau^* = \tau(T^*)$ and $p^* = p(T^*)$. We construct a new tree $T_0 = (V_0, E_0)$ as follows: For each node $v \in V^*$ with k children $v_1, \ldots, v_k$, we introduce k new nodes $u_1, \ldots, u_k$, make u_i the parent of v_i, for $i = 1, \ldots, k$, make u_{i-1} the parent of u_i, for $i = 2, \ldots, k$, and make v the parent of u_1. This is very easy to do; in fact, it can be viewed largely as adopting a new interpretation of the adjacency list of v. We will call the original nodes in V^* "black", while the new nodes in $V_0 \backslash V^*$ are "white".

We now associate a processor with each node in T_0 and *contract* T_0 in a sequence of *stages*, ending when only one node remains. We will use "T" to denote the evolving tree, reserving "T_0" for its original value. A stage ends, in general, with a number of nodes grouped into disjoint *clusters*; initially, there are no clusters. We will call a node *free* if it does not belong to a cluster. Each stage takes constant time and comprises the following four *phases*:

Phase 1: Cluster formation;
Phase 2: Cluster processing;
Phase 3: Cluster removal;
Phase 4: Leaves-cutting.

In Phase 1 (Cluster formation), certain free nodes are grouped into new clusters. Say that a *bond* exists between a node v and its parent u if they are both of degree ≤ 1 and both free and if either they are both white, or u is black and v does not have a white child (i.e., either it is a leaf, or it has a black child). Each node incident on one or two bonds becomes part of a new cluster, the other members of the cluster being those nodes that it can reach via one or more bonds. Note that no global information about a cluster is computed; each node merely determines whether it belongs to a cluster and, if so, which of its neighbors belong to the same cluster.

In Phase 2 (Cluster processing), each cluster carries out the task of repeatedly merging pairs of adjacent nodes in the cluster until a just a single node remains, at which point the cluster is removed in Phase 3 (Cluster removal). In each stage, however, only a constant number of steps of the task are executed, which is the reason why clusters may survive from one stage to the next. It is well-known and easy to see that the task of a cluster with k nodes can be carried out in $O(\log k)$ time, viewing the cluster as a linked list and performing repeated pointer doubling; as observed in [7, 9], concurrent reading during the pointer doubling can be avoided by letting each node maintain an indication of whether it points to the last element of the list. In Phase 4 (Leaves-cutting), every free nonroot leaf is merged into its parent.

Since only nodes of degree ≤ 1 are merged into their parents, it is easy to see that no node degree ever increases. Thus the degree of every white node remains bounded by 2, and the degree of every black node remains bounded by 1. We already used this implicitly above by assuming that all free leaves can be merged into their parents in constant time.

As an aid in analyzing the number of steps needed by the algorithm described above to contract T to a single node, we attach conceptual weights to nodes and clusters. Every free node is of weight 1. When k nodes form a cluster, their combined weight of k is transferred to the cluster. At the end of each phase, the weight of each cluster is reduced by a factor of θ, where θ is a constant with $0 < \theta < 1$ chosen so that the weight of no cluster ever drops below 1; this is possible because the initial weight of a cluster formed out of k nodes is $k \geq 2$, while the cluster survives for only $O(\log k)$ stages. When a cluster is removed, finally, the single node emerging from the cluster is restored to have a weight of 1; because the weight of the cluster was at least 1, this does not increase the total weight. Since the total weight is $2n - 1$ initially and clearly never drops below 1, the processing of T works in $O(\log n)$ time if we can show that each stage decreases the total weight by at least a constant factor. As the processing and removal of clusters certainly does not decrease the number of leaves, this is a consequence of the following claim, whose proof is omitted.

Claim 11. *At the beginning of Phase 2 (Cluster processing) in every stage, at least $1/20$ of the total weight is contributed by leaves and clusters.*

The procedure above, just shown to work in $O(\log n)$ time, serves exclusively to gather timing information for a second run, still to be described, which is augmented with computation that actually evaluates the input tree T^*. During the first run, whenever a black node is merged into its parent, it is marked with the time at which this happens, called its *completion time*, where we assume the origin of the time axis placed at the start of the procedure. We will actually slow the procedure down by a factor of $c\tau^*$, where $c \in \mathbb{N}$ is a suitably large constant to be chosen later; we here use Condition (D) of Lemma 1. This has the effect of multiplying all completion times by $c\tau^*$. After the first run, for each node u in T^*, we sort the children of u by their completion times [7, 19], after which we consider T^* to be an ordered tree. We use the Euler-tour technique as described in [18] to determine, for each node $u \in V^*$, the smallest and the largest preorder number in T^* of a descendant of u. For each node u in T^* with k children, the sorting provides us with an array $A_u[1 \mathinner{.\,.} k]$ of the sorted completion times of the children of u. Reinterpreting these as arrival times, in the sense of Condition (E) of Lemma 1, and ignoring the last entry, we have one half of the input to a raking problem of size $s = k - 1$ associated with u. We describe below how to obtain the other half of the input.

In addition to the processor associated with each node in T_0, the second run of the tree contraction employs np^* *special processors* numbered $1, \ldots, np^*$. We assume that the additional processing carried out in the second run is "hidden", by means of extra processors and/or the slowdown mentioned above, so that the timing information gathered for the first run will be accurate also for the second run. During the second run we maintain the following invariants: In every stage, at the end of Phase 2 and at the end of Phase 4,

(1) If a black node is a leaf in T, it is labeled with its value (in the input tree T^*);

(2) If a black node u has a black child v in T, u is labeled with a function in $\mathcal{F}$ that maps the value of v to its own value.

If we copy leaf labels from T^* to T_0, Invariant (1) is satisfied initially, and Invariant (2) holds vacuously. If Invariant (1) holds at the end of the second run, the root of T (and therefore of T^*) will be labeled with its value, and we are done. In order to maintain the invariants, we augment the first run with the following steps:

(a) Whenever a black leaf v merges into a black parent u, the function f labeling u (Invariant (2)) is applied to the value x labeling v (Invariant (1)), and $f(x)$ is attached as the new label of u, thus maintaining Invariant (1). By Condition (A) of Lemma 1, this can be done in constant time.

(b) Whenever a black nonleaf v merges into a black parent u, the rules for the formation of clusters imply that v has a black child. Thus u and v are labeled with functions f_u and f_v, respectively (Invariant (2)). We replace the label of u by $f_u \circ f_v$, maintaining Invariant (2). By Condition (B), this can be done in constant time.

(c) Whenever a black node v merges into a white parent, the rules for the formation of clusters imply that v is necessarily a leaf. Thus the value of v is known (Invariant (1)). We supply the value of v as an input key to an ongoing raking computation associated with the parent u of v in T^*. Moreover, if the smallest and largest preorder numbers of a descendant of v in T^* are l_1 and l_2, respectively, we dedicate the special processors numbered $(l_1 - 1)p^* + 1, \ldots, l_2 p^*$ to this task. When the raking finishes, we label u with the resulting function which, by Condition (C), belongs to $\mathcal{F}$.

In Step (c), suppose that v is the ith child of u. It is easy to see that if we take T_i^* to be the maximal subtree of T^* rooted at v, then the number of processors dedicated to the raking in Step (c) above is precisely as required by Condition (E). Furthermore, the trees $T_1^*, \ldots, T_s^*$ indeed are disjoint, and since they receive preorder numbers in the order in which their roots complete, the special processors allocated to the raking at all times are consecutively numbered, as required; they are not numbered starting at 1, but this is easy to take care of.

Steps (a)–(c) above specify no action in the case of a white node merging into another node. The operation may create a black leaf or a give a black node a black child, however, so that we must show the invariants to be satisfied. Say that a node u in T *contains* a node $v \in V_0$ if either $u = v$, or (recursively) a node containing v at some point merged into u. A simple yet useful observation, easy to prove by induction on the number of merges involved in deriving T from T_0, is that if u and v are nodes in T, then u is an ancestor of v in T if and only if it is an ancestor of v in T_0. Another observation of this kind is that every node u in T_0 is contained in the closest ancestor of u in T_0 that still belongs to T.

Suppose that at some time t, a white node merging into a black node u causes u to acquire a black child v. Let U be the set of proper descendants of u in T_0 that are not descendants of v, and let Z be the set of (black) children of u in T^* that are not ancestors of v. By the first observation made above, no node

w in U can still be present in T at time t, since it could not be a descendant of u without being a descendant of v. Since this holds for all nodes w in U, the second observation implies that at time t, all nodes in U are contained in u. A node in Z has no black proper ancestor that is also a proper descendant of u. Since it cannot merge into u (then u could never again acquire a white child, contrary to the fact that it has one at time t), it must merge into a white node in U, at which point its value is supplied to the raking problem associated with u. Suppose that this happens simultaneously in the step immediately before time j for a nonempty group F_j of nodes in Z. Since nodes merge only in disjoint pairs (or triples, if we allow two leaves to merge simultaneously with a common parent) and at least $|F_j|/2$ white nodes in U are still present in T at time j, we must have $t \geq j + c'c\tau^* \log(|F_j|+2)$, for some constant $c' > 0$ (recall that we slow down the contraction of T by a factor of $c\tau^*$). Since this holds for each integer j in the range $a_1^*, \ldots, a_s^*$, where a_1^* and a_s^* are the first and last completion times of a node in Z, Condition (E) guarantees that with the constant c chosen suitably (in dependence of C), the raking associated with u will finish before time t. The function computed by the raking is precisely as required by Invariant (2), which can therefore be satisfied.

Suppose now instead that at time t, a white node merging into a black node u makes u a leaf. An argument very similar to the one above, taking U as the set of proper descendants of u in T_0 and Z as the set of (black) children of u in T^*, shows that before time t, there will have been enough time to compute a function in $\mathcal{F}$ mapping the value of the last child v of u in T^* to the value of u. Invariant (1) and Condition (A) imply that when v becomes a leaf, constant time suffices to label u with its value, maintaining Invariant (1).

If a special processor is dedicated to the raking at two different (black) nodes u and w, then one of the two nodes is an ancestor in T^* of the other one. Suppose that u is an ancestor of w in T^* and let v be the child of u in T^* that is also an ancestor of w. As argued above, the raking at w finishes before w can acquire a black leaf or become a leaf itself. In particular, when the raking at w finishes, w still exists as a node in T. On the other hand, a special processor participating in the raking at w is not needed in the raking at u before v merges into a white parent, at which point v must be a leaf and w, if different from v, must have merged into a parent. Thus no special processor is simultaneously dedicated to different raking operations.

An issue that was ignored above is how to inform a special processor of the raking problem on which it is supposed to work at any given time. If concurrent reading is allowed, we can simply associate the special processors numbered $(i-1)p^* + 1, \ldots, ip^*$ with the node v_i of preorder i in T^*, for $i = 1, \ldots, n$, and let each of these processors keep track of the node in T containing v_i. We omit the description of a more complicated solution that works even for the EREW PRAM.

Acknowledgment. We thank Wojtek Plandowski for pointing out the similarity between register allocation and broadcasting.

References

1. K. Abrahamson, N. Dadoun, D. G. Kirkpatrick, and T. Przytycka, A simple parallel tree contraction algorithm, *J. Algorithms* **10** (1989), pp. 287–302.
2. A. V. Aho, R. Sethi, and J. D. Ullman, *Compilers: Principles, Techniques, and Tools*, Addison-Wesley, Reading, MA, 1986.
3. M. Ajtai, J. Komlós, and E. Szemerédi, An $O(n \log n)$ sorting network, *In* Proc. 15th Annual ACM Symposium on Theory of Computing (STOC 1983), pp. 1–9.
4. O. Berkman and U. Vishkin, On parallel integer merging, *Inform. and Comput.* **106** (1993), pp. 266–285.
5. S. Chaudhuri and T. Hagerup, Prefix graphs and their applications, *In* Proc. 20th International Workshop on Graph-Theoretic Concepts in Computer Science (WG 1994), Springer Lecture Notes in Computer Science, Vol. 903, pp. 206–218.
6. R. Cole, Parallel merge sort, *SIAM J. Comput.* **17** (1988), pp. 770–785.
7. R. Cole and U. Vishkin, Deterministic coin tossing with applications to optimal parallel list ranking, *Inform. and Control* **70** (1986), pp. 32–53.
8. F. E. Fich, P. Ragde, and A. Wigderson, Relations between concurrent-write models of parallel computation, *SIAM J. Comput.* **17** (1988), pp. 606–627.
9. H. Gazit, G. L. Miller, and S.-H. Teng, Optimal tree contraction in the EREW model, In *Concurrent Computations: Algorithms, Architecture, and Technology*, S. K. Tewksbury, B. W. Dickinson, and S. C. Schwartz (eds.), Chap. 9, pp. 139-156, Plenum Press, New York, 1988.
10. S. R. Kosaraju and A. L. Delcher, Optimal parallel evaluation of tree-structured computations by raking, *In* Proc. 3rd Aegean Workshop on Computing (AWOC 1988), Springer Lecture Notes in Computer Science, Vol. 319, pp. 101–110.
11. G. L. Miller and J. H. Reif, Parallel tree contraction and its application, *In* Proc. 26th Annual Symposium on Foundations of Computer Science (FOCS 1985), pp. 478–489.
12. G. L. Miller and J. H. Reif, Parallel tree contraction, Part 1: Fundamentals, preprint, 1987. The final version (not available to us) appeared in *Randomness and Computation*, Advances in Computing Research, Vol. 5, S. Micali (ed.), pp. 47–72, JAI Press, Greenwich, CT, 1989.
13. G. L. Miller and J. H. Reif, Parallel tree contraction, Part 2: Further applications, *SIAM J. Comput.* **20** (1991), pp. 1128-1147.
14. G. L. Miller and S.-H. Teng, Tree-based parallel algorithm design, manuscript. A preliminary version appeared in Proc. 2nd International Conference on Supercomputing (1987), pp. 392–403.
15. M. Reid-Miller, G. L. Miller, and F. Modugno, List ranking and parallel tree contraction, In *Synthesis of Parallel Algorithms*, J. H. Reif (ed.), Chap. 3, pp. 115–194, Morgan Kaufmann Publ., San Mateo, CA, 1993.
16. R. Sethi, Complete register allocation problems, *SIAM J. Comput.* **4** (1975), pp. 226–248.
17. P. J. Slater, E. J. Cockayne, and S. T. Hedetniemi, Information dissemination in trees, *SIAM J. Comput.* **10** (1981), pp. 692–701.
18. R. E. Tarjan and U. Vishkin, An efficient parallel biconnectivity algorithm, *SIAM J. Comput.* **14** (1985), pp. 862–874.
19. R. A. Wagner and Y. Han, Parallel algorithms for bucket sorting and the data dependent prefix problem, *In* Proc. International Conference on Parallel Processing (ICPP 1986), pp. 924–930.

System Diagnosis with Smallest Risk of Error

Krzysztof Diks[1] Andrzej Pelc[2]

[1] Instytut Informatyki, Uniwersytet Warszawski, Banacha 2, 02-097
Warszawa, Poland. E-mail: diks@mimuw.edu.pl

Research partly supported by a grant from KBN. This work was done during the
author's stay at the Université du Québec à Hull, supported by NSERC International
Fellowship.

[2] Département d'Informatique, Université du Québec à Hull, Hull,
Québec J8X 3X7, Canada. E-mail: pelc@uqah.uquebec.ca

Research supported in part by NSERC grant OGP 0008136.

Abstract. We consider the problem of fault diagnosis in multiprocessor systems. Every processor can test its neighbors; fault-free processors
correctly identify the fault status of tested neighbors, while faulty testers
can give arbitrary test results. Processors fail independently with constant probability $p < 1/2$ and the goal is to identify correctly the status
of all processors, based on the set of test results. We give fast diagnosis algorithms with the highest possible probability of correctness for
systems represented by complete bipartite graphs and by simple paths.
This is for the first time that the most reliable fault diagnosis is given
for these systems in a probabilistic model without any assumptions on
the behavior of faulty processors.

1 Introduction

With growing importance of large multiprocessor systems the issue of their reliability attracts increasing attention. One of the major problems in this area,
known as the fault diagnosis problem, is to locate all faulty processors in the
system. The classical approach to fault diagnosis was originated by Preparata,
Metze and Chien in their seminal paper [10]. They studied fault diagnosis in a
graph model in which processors are represented as nodes of the graph and links
along which tests can be performed are represented as edges. It was assumed
that fault-free processors always give correct test results, while tests conducted
by faulty processors are unpredictable: a faulty tester can output any test result
regardless of the status of the tested neighbor. Faults were assumed permanent,
i.e., the fault-status of a processor does not change during testing and diagnosis. In [10] a worst-case scenario was adopted: it was assumed that at most t
processors are faulty and that they are placed in a way most detrimental for diagnosis. This assumption precluded the possibility of diagnosis for t larger than
the number of neighbors of any processor.

This model and some of its variations have been thoroughly studied in literature (see the survey [6], where extensive bibliography can be found). It has
been argued that the worst case scenario often fails to reflect realistic diagnosis situations. As an alternative, various probabilistic models were proposed (cf.

[2, 3, 4, 5, 8, 9, 11, 12]). Instead of imposing an upper bound on the number of faulty processors and assuming their worst-case location, an a priori failure probability, independent for each processor, is assumed in these models. Diagnosis is then restricted to sets of faulty processors of sufficiently high a priori probability [8], in which case it can be performed unambiguously [5], or is done in general and has a high probability of correctness [1, 2, 3, 4, 9, 11, 12].

In this paper we work in the probabilistic model previously studied in [2, 12]. The assumptions concerning test results are the same as in the above described model of Preparata, Metze and Chien [10] and faults are also permanent. However, unlike in [10], it is assumed that processors fail with constant probability $p < 1/2$ and all faults are independent. It should be noted that this is the only probabilistic model in which no assumption is made on the behavior of faulty testers. Thus diagnosis algorithms working reliably under this model are very robust in that they produce correct diagnosis under any behavior of faulty processors.

In [2] this model is formalized and the probability of correctness of a diagnosis is rigorously defined. The authors show a simple diagnosis strategy based on majority vote that works for a class of systems called *tester graphs* which includes the complete graph. They estimate the probability of correctness of this strategy without computing it precisely and show that it converges to 1 as the system grows.

The goal of the present paper is to construct efficient diagnosis algorithms that maximize the probability of correctness. We call such diagnosis strategies *optimal*. We give them for two classes of multiprocessor systems: those represented by complete bipartite graphs and those represented by (simple) paths. In both cases the running time of the diagnosis algorithm is linear in the number of tests performed in the system. An easy modification of our diagnosis for paths yields optimal diagnosis for systems represented by rings. In an upcoming paper [7] we study tester graphs from [2] and show that the majority strategy proposed in [2] is in fact optimal fro these graphs. This is for the first time that optimal diagnosis is given under a very general probabilistic model, in which no assumptions are made on the behavior of faulty processors.

The paper is organized as follows. In section 2 we establish terminology used throughout the paper and formalize the model. In sections 3 and 4 we give algorithms for optimal diagnosis in complete bipartite graphs and paths, respectively. Section 5 contains conclusions and open problems.

2 Terminology and model description

A system is modeled as an undirected graph whose set of nodes $U = \{u_1, ..., u_n\}$ represents processors, some pairs of which are connected by direct communication channels (links) represented by edges of the graph. Neighboring processors perform tests on each other. A *test assignment* in a system is represented by a symmetric directed graph $G = (U, E)$ where $(u, v) \in E$ means that processor u tests processor v. Thus $(u, v) \in E$ whenever u and v are connected by a link in the system. We fix this directed graph G till the end of this section. The outcome of a test $(u, v) \in E$ is 1 (0) if u evaluates v as faulty (fault-free). A complete collection of test results is called a *syndrome*. Formally, a syndrome is any function $S : E \to \{0, 1\}$. The set of all possible syndromes is denoted by Σ. The set of all faulty processors in the system is called a *fault set*. This can be any subset of U. A syndrome S is said to be *compatible* with a fault set F if, for any $(u, v) \in E$, such that $u \in U \setminus F$, $S(u, v) = 1$ iff $v \in F$. This corresponds to the assumption that fault-free processors always give correct test results. Since faulty testers can give arbitrary test results, any syndrome compatible with a fault set F can occur when faulty processors in the system are exactly those in F. The set of all syndromes compatible with a fault set F is denoted by $\sigma(F)$. Fault sets F_1 and F_2 are called *associated* if $\sigma(F_1) \cap \sigma(F_2) \neq \emptyset$.

We consider only deterministic diagnosis algorithms. The input of such an algorithm is a syndrome and the output is the set of processors that the algorithm diagnoses as faulty (all other processors are implicitly diagnosed as fault-free). Thus a *diagnosis* is any function $D : \Sigma \to \mathcal{P}(U)$.

We now define formally the probability of correctness of any diagnosis. The sample space is the set of all fault sets, i.e.,

$$\Omega = \{F : F \subseteq U\}.$$

The probability function P is defined for all subsets of Ω by the formula

$$P(X) = \sum_{F \in X} p^{|F|} (1 - p)^{n - |F|},$$

for any $X \subseteq \Omega$. If D is a diagnosis, $\mathrm{Cor}(D)$ is the event consisting of those fault sets F for which D returns F on any syndrome compatible with F, i.e. the event that diagnosis D is correct regardless of faulty processors' behavior. More precisely,

$$\mathrm{Cor}(D) = \{F \subseteq U : \forall_{S \in \sigma(F)} D(S) = F\}.$$

Now the performance of diagnosis D is measured by its probability of correctness which is simply $P(\mathrm{Cor}(D))$. A diagnosis D is *optimal* if $P(\mathrm{Cor}(D)) \geq P(\mathrm{Cor}(D'))$ for every diagnosis D'.

It is shown in [7] that the above definition of the probability of correctness of a diagnosis D is equivalent to the slightly more complicated definition given in [2].

If two subsets are associated then at most one of them can belong to $\mathrm{Cor}(D)$. Since every set F and its complement are associated (the common compatible syndrome is the one that gives result 0 for u and v both in or both outside of F and result 1 otherwise), we have the following observation.

Proposition 1. *For any diagnosis D and any fault set F, F and $U \setminus F$ cannot both belong to $\mathrm{Cor}(D)$.*

3 Complete bipartite graphs

In this section we consider test assignments in which all processors are partitioned into two non-empty subsets A and B, all processors from A testing all processors from B and vice-versa. Such assignments are modeled by *complete bipartite graphs* $G = (U, E)$ for which $U = \{u_1, ..., u_k\}$ is a disjoint union of sets A and B and $E = \{(u, v) : (u \in A \text{ and } v \in B) \text{ or } (u \in B \text{ and } v \in A)\}$.

We assign weight $W(u)$ to every element $u \in U$. W.l.o.g. we can assume that $|A| \geq |B|$. If $|U|$ is odd then $W(u) = 1$ for any u. If $|U|$ is even then choose any $u_0 \in A$ and let $W(u) = 1$ for all $u \in U \setminus \{u_0\}$ and $W(u_0) = 2$. (This is done to break tie in the vote in case of an even number of voters.) Call a set $F \subseteq U$ *winning* if $\sum_{v \in F} W(v) < \sum_{v \in U \setminus F} W(v)$. Thus B is winning. Also, if F is winning then $|F| \leq |U \setminus F|$. Hence winning fault sets are at least as probable as their complements. Call F *normal* if both F and $U \setminus F$ intersect both A and B.

We now define the diagnosis Bip for complete bipartite graphs. This diagnosis will be proved optimal. Given a syndrome S define the undirected graph $H = (U, \tilde{E})$, where $\{u, v\} \in \tilde{E}$ iff $S(u, v) = S(v, u) = 0$. Next, for any $u \in U$, let $N(u) = \{v \in U : \{u, v\} \in \tilde{E}\}$ and let $C(u) = \{u\} \cup N(u) \cup N(u_i)$, where u_i is the element of $N(u)$ with lowest index. Let C be the largest among all sets $C(u)$.

$$\text{Bip}(S) = \begin{cases} U \setminus C & \text{if } |C| > 1 \\ B & \text{otherwise.} \end{cases}$$

Thus, if there are at least two processors that consider each other fault-free, processors in the largest set $C(u)$ are diagnosed as fault-free and all other processors as faulty. If no such two processors exist, all processors in A are diagnosed as fault-free and all those in B as faulty. It is easy to see that the function Bip can be computed in time $O(|E|)$.

Theorem 2. *Diagnosis Bip for complete bipartite graphs is optimal.*

Proof: We first show that the following fault sets belong to Cor(Bip): normal winning sets, subsets of B and proper subsets of A. Let F be a fault set and S a syndrome compatible with F. First suppose that F is a normal winning set. By normality there are fault-free processors both in A and in B. For any fault-free processor u, the set $C(u)$ consists of all fault-free processors. Since F is winning, this set $C(u)$ must be equal to C and since $|C| > 1$, we have $\text{Bip}(S) = U \setminus C = F$. Next, suppose that F is a subset of B. If $F = B$ then there ara no edges in the graph H and thus $|C| = 1$. By definition $\text{Bip}(S) = B$ in this case. If F is a proper subset of B then the only set $C(u)$ containing more than one element is $U \setminus F$ and consequently $\text{Bip}(S) = F$. Finally, suppose that F is a proper subset of A. Again, the only set $C(u)$ containing more than one element is $U \setminus F$ and thus $\text{Bip}(S) = F$.

In order to show that diagnosis Bip is optimal, note that all subsets of U belong to one of the following categories:
1. normal,

1. normal,
2. A,
3. B,
4. proper subset of A,
5. proper subset of B,
6. $A \cup X$ with non-empty $X \subseteq B$,
7. $B \cup X$ with non-empty $X \subseteq A$.

Normal sets are associated in pairs: a winning normal set with its (normal) complement. Sets in category 4 are associated with their complements in category 7 and those in category 5 with their complements in category 6. Finally, all sets in categories 2,3,6 and 7 are mutually associated: the syndrome constantly equal 1 is compatible with all of them.

Let D be any diagnosis. Since all normal winning sets are in Cor(Bip), the sum of probabilities of normal sets in Cor(D) cannot exceed the sum of probabilities of normal sets in Cor(Bip), in view of proposition 1. Consider two cases. If $B \in$ Cor(D) then no subset of categories 2, 6 or 7 can belong to Cor(D). Since all sets of categories 3, 4 and 5 belong to Cor(Bip), it follows that probability of Cor(D) does not exceed probability of Cor(Bip). If $B \notin$ Cor(D), at most one set of categories 2, 6 or 7 can belong to Cor(D). Every such set has probability at most equal to that of B. Thus exchanging B for one of these sets cannot increase the probability of correctness of a diagnosis. Since all sets of categories 3, 4 and 5 belong to Cor(Bip), it follows, in this case as well, that the probability of Cor(D) does nos not exceed the probability of Cor(Bip). Hence diagnosis Bip is optimal. □

Example 1. Consider the complete bipartite graph on the set $U = \{u_1, ..., u_7\}$ of 7 nodes, with sets $A = \{u_1, u_2, u_3, u_4, u_5\}$ and $B = \{u_6, u_7\}$. The set Cor(Bip) consists of the empty set, all sets of size 1, all sets of size 2, all sets of size 3 except $B \cup \{u_i\}$, for $i = 1, 2, 3, 4, 5$, and 5 subsets of A of size 4. Thus the probability of diagnosis Bip is

$$(1-p)^7 + 7p(1-p)^6 + 21p^2(1-p)^5 + 30p^3(1-p)^4 + 5p^4(1-p)^3.$$

4 Paths

In this section we give optimal diagnosis for systems represented by (simple) paths. Thus the test assignment is a graph $P_k = (U, E)$ where $U = \{u_1, ..., u_k\}$ and $E = \{(u_i, u_{i+1}) : 1 \le i < k\} \cup \{(u_{i+1}, u_i) : 1 < i \le k\}$. It will be convenient to represent fault sets as sequences of pluses and minuses called *configurations*, the set of minuses corresponding to the fault set, e.g., $+++--+$ corresponds to the fault set $\{u_4, u_5\}$ in P_6. Thus a configuration is a function $\delta : U \to \{+, -\}$. Terminology defined for fault sets (e.g. "associated fault sets" or "syndrome compatible with a fault set") will be used for corresponding configurations. A *block* in a configuration δ is any sequence $(u_i, u_{i+1}, ..., u_j)$, $j \ge i$, of consecutive processors, satisfying one of the following conditions:

- $\delta(u_l) = +$, for all $i \le l \le j$, and $\delta(u_{j+1}) = -$, if $j < k$, and $\delta(u_{i-1}) = -$, if $i > 1$. (In this case the block is called *fault-free*.
- $\delta(u_l) = -$, for all $i \le l \le j$, and $\delta(u_{j+1}) = +$, if $j < k$, and $\delta(u_{i-1}) = +$, if $i > 1$. (In this case the block is called *faulty*.

Processors u_{i-1} and u_{j+1} are called the *borders* of the block.

Thus a block is a maximal sequence of consecutive processors having the same value. A block containing at least 3 processors is called *hard*. Blocks B_1 and B_2 are said to *touch* each other if the last processor of one of them is a neighbor of the first processor of the other. In this case blocks B_1 and B_2 are also called *consecutive*. A *chain* is a maximal sequence of consecutive two-element blocks.

Configurations δ_1 and δ_2 are called *adjacent* if the following conditions are satisfied:

- each fault-free block in δ_1 is either a fault-free block in δ_2 or is contained in a faulty block in δ_2,
- each fault-free block in δ_2 is either a fault-free block in δ_1 or is contained in a faulty block in δ_1.

Lemma 3. *Two configurations are adjacent if and only if they are associated.*

Proof: Omitted. $\qquad\square$

We now construct the set $\mathcal{R}$ of *regular* configurations. Let $[i_1, j_1], ..., [i_n, j_n]$ be segments in $(1, ..., k)$, such that $j_l > i_l$, for every $l = 1, ..., n$ and $j_l < i_{l+1} - 1$, for every $l = 1, ..., n-1$. Every such (possibly empty) sequence of segments yields a regular configuration in the following way.

1. Processors with indices in a segment form a fault-free block.
2. For every segment $[i_l, j_l]$ processors u_{i_l-1} (if $i_l > 1$) and u_{j_l+1} (if $j_l < k$) are faulty (they are borders of the respective fault-free block).
3. A nonempty sequence of processors between consecutive borders defined in 2. has alternating values, starting with value $+$ (e.g., $(++++-+-+-+++))$. In this example, the first 4 processors form a segment, the 5th processor is a border, processors 6,7 and 8 are in between borders, the 9th processor is a border and processors 10,11 and 12 form a segment. Note that two consecutive values $-$ can result (e.g., in $(++++-+--+++))$.
4. Processors preceding the first border defined in 2. have alternating values starting with $+$ (e.g., $(+-+-+++)$ or $(+-+--+++))$. Again two consecutive minuses are possible.
5. Processors following the last border defined in 2. have alternating values starting with $+$ (e.g., $(+++-+-+-)$ or $(+++-+-+-+))$.
6. Processors of the entire line, in case of the empty sequence of segments, have alternating values starting with $+$ (e.g., $(+-+-+-)$ or $(+-+-+-+))$.

Example 2. The family of regular configurations for the 5-node path P_5 is:

$$\{(+-+-+), (++-+-), (-++-+), (+-++-), (+--++),$$

$$(+ + + - +), (- + + + -), (+ - + + +),$$

$$(+ + + + -), (- + + + +), (+ + + + +), (+ + - + +)\}.$$

Let S be any syndrome. A *pseudo-block* is any sequence $(u_i, u_{i+1}, ...u_j), j \geq i$, of consecutive processors, satisfying the following conditions:

- $S(u_m, u_{m+1}) = 0$ for all $i \leq m < j$,
- $S(u_m, u_{m-1}) = 0$ for all $i < m \leq j$,
- $S(u_j, u_{j+1}) = 1$ or $S(u_{j+1}, u_j) = 1$ or $j = k$,
- $S(u_i, u_{i-1}) = 1$ or $S(u_{i-1}, u_i) = 1$ or $i = 1$.

Thus a pseudo-block is a maximal sequence of consecutive processors in which neighbors consider each other fault-free. A pseudo-block containing at least 3 processors is called *hard*. Pseudo-blocks B_1 and B_2 are said to *touch* each other if the last processor of one of them is a neighbor of the first processor of the other. In this case pseudo-blocks are also called *consecutive*. A *pseudo-chain* is a maximal sequence of consecutive two-element pseudo-blocks.

We now define the diagnosis Path-Diag which will be proved optimal for test assignments represented by paths. Let S be any syndrome for the path P_k. Consider the following algorithm to construct the configuration Path-Diag(S):

Algorithm Path-Diag

1. Divide P_k into pseudo-blocks.
2. Assign value $+$ to all processors in hard pseudo-blocks.
3. Assign alternating values to pseudo-blocks in pseudo-chains starting as follows:
   - if the last pseudo-block B of the pseudo-chain touches a hard pseudo-block then assign value $-$ to processors in B.
   - otherwise, assign value $+$ to processors in B.
4. If a processor u in a hard pseudo-block or in a pseudo-block in a pseudo-chain has a neighbor v without an assigned value then assign to v a different value than that of u (v will be a guard of the block containing u).
5. For every maximal sequence of consecutive processors without assigned values assign alternating values to consecutive processors, always starting with value $+$.

The above algorithm runs in linear time and hence function Path-Diag can also be computed in time $O(k)$.

Example 3. Consider the path P_{16} and a syndrome yielding the following pseudo-blocks:

$$(u_1, u_2), (u_3), (u_4), (u_5), (u_6), (u_7, u_8, u_9),$$

$$(u_{10}, u_{11}), (u_{12}, u_{13}), (u_{14}, u_{15}), (u_{16}).$$

In the first step of algorithm Path-Diag these pseudo-blocks are identified. In the second step value $+$ is assigned to u_7, u_8, u_9. In the third step pseudo-chains

(u_1, u_2) and (u_{10}, u_{11}), (u_{12}, u_{13}), (u_{14}, u_{15}) are considered: value $+$ is assigned to pseudo-block (u_1, u_2) and processors u_{10}, u_{11}, u_{12}, u_{13}, u_{14}, u_{15} in the second pseudo-chain get values $-, -, +, +, -, -$, respectively. In the fourth step u_3 and u_6 get value $-$ and u_{16} gets value $+$. In the fifth step processors u_4, u_5 get values $+, -$, respectively.

Theorem 4. *Diagnosis Path-Diag for the path P_k is optimal.*

Proof: We first prove that every regular configuration belongs to Cor(Path-Diag). Let δ be a regular configuration and let S be a syndrome compatible with δ. Every hard block is fault-free and corresponds to a hard pseudo-block, hence all processors in such blocks are correctly diagnosed in step 2. Next consider any chain C. It may correspond to a sequence of several pseudo-chains with splits occurring whenever processors in a faulty block do not diagnose each other as fault-free. The last block in the chain is faulty only if it touches a (fault-free) hard block. Otherwise it is fault-free. This last block is diagnosed correctly in step 3 and hence the last pseudo-chain yielded by C is diagnosed correctly. All other pseudo-chains yielded by C correspond to sequences of blocks starting and ending with fault-free blocks because splits could occur only between faulty processors. Consequently these pseudo-chains are also diagnosed correctly in step 3. Faulty processors between which splits occurred are diagnosed correctly in step 4. Thus all processors in C are diagnosed correctly. Also neighbors of the first and last processor of hard blocks and of the first and last processor of every chain are diagnosed correctly in step 4. Finally, consider any maximal sequence s of processors that are not in blocks of size greater than 1. These processors must have alternating values. If the first processor of s is a neighbor of a processor in fault-free block of size greater than 1, then it has value $-$ and it is correctly diagnosed in step 4. Hence the second processor in s has value $+$ and the entire sequence s is correctly diagnosed in step 5. Otherwise, the first processor of s has value $+$ and the entire sequence s is correctly diagnosed in step 5. This shows that all processors are diagnosed correctly by Path-Diag and hence Path-Diag$(S) = \delta$ which proves that δ belongs to Cor(Path-Diag).

Let D be any diagnosis for the path P_k. In order to finish the proof, it is enough to show a one-to-one function $f : \mathrm{Cor}(D) \to \mathcal{R}$, such that for any configuration $\delta \in \mathrm{Cor}(D)$ we have

$$p^{|F_\delta|}(1-p)^{k-|F_\delta|} \le p^{|F_{f(\delta)}|}(1-p)^{k-|F_{f(\delta)}|}, \tag{1}$$

where F_δ and $F_{f(\delta)}$ are fault sets corresponding to configurations δ and $f(\delta)$, respectively. Since $\mathcal{R} \subseteq \mathrm{Cor}(\text{Path-Diag})$, this will imply that the probability of Cor(D) does not exceed that of Cor(Path-Diag).

The function f is constructed as follows. Let $\delta \in \mathrm{Cor}(D)$. Every fault-free block of at least two processors in δ becomes a fault-free block in $f(\delta)$. Borders of these blocks in δ become borders of corresponding blocks in $f(\delta)$. Every sequence of processors between consecutive borders, as well as the sequence before the first border and the sequence following the last border get alternating values, always starting with $+$ (e.g. configuration $(--++--+-+++---)$ is transformed into $(+ - + + - + - - + + + - + -))$. The resulting configuration is regular. It is clear that the number of values $+$ cannot decrease and hence condition 1 is satisfied.

Using lemma 3 it can be shown that the function f is one-to-one. □

5 Conclusion

We considered the problem of constructing algorithms for optimal diagnosis of multiprocessor systems, i.e., diagnosis with the highest possible probability of correctness. We provided linear algorithms to perform such optimal diagnosis for two examples of systems: complete bipartite graphs and paths. This is for the first time that optimal diagnosis is given for these systems in a probabilistic model without any assumptions on the behavior of faulty processors. Our results also permit to precisely compute the probability of correctness of a diagnosis under this general scenario. This quantitative measure of performance can be used to give a meaningful comparison of various diagnostic strategies. If an optimal diagnosis running in linear time can be found for a given test assignment, as in the cases mentioned above, it is a natural choice, as it combines best possible diagnostic quality with speed. Otherwise, a simple heuristic strategy should be sought and its probability of correctness could be evaluated using the definition of section 2. We do not know what is the complexity of the problem of finding an optimal diagnosis for any test assignment (more precisely, the problem: given a directed graph G and a syndrome S, find a fault set $D(S)$ such that diagnosis D is optimal for G). We conjecture that it is NP-hard.

The most challenging open problems yielded by our research are those of finding fast optimal diagnosis strategies for test assignments in other important networks, such as grids, tori or hypercubes. It would be also interesting to know whether the general problem of finding an optimal diagnosis for any test assignment can be solved in polynomial time, and, if not, to find heuristics combining speed with good performance.

References

1. D.M. Blough and A. Pelc, Complexity of fault diagnosis in comparison models, IEEE Transactions on Computers 41 (1992), 318-324.
2. D.M. Blough, G.F. Sullivan and G.M. Masson, Efficient diagnosis of multiprocessor systems under probabilistic models, IEEE Transactions on Computers 41 (1992), 1126-1136.
3. D.M. Blough, G.F. Sullivan and G.M. Masson,Intermittent fault diagnosis in multiprocessor systems, IEEE Transactions on Computers 41 (1992), 1430-1441.
4. M. Blount, Probabilistic treatment of diagnosis in digital systems, Dig. 7th Int. Symp. Fault-Tolerant Computing, IEEE Computer Society Press, (1977), 72-77.
5. A.T. Dahbura, An efficient algorithm for identifying the most likely fault set in a probabilistically diagnosable system, IEEE Transactions on Computers 35 (1986), 354-356.

6. A.T. Dahbura, System-level diagnosis: A perspective for the third decade, Concurrent Computation: Algorithms, Architectures, Technologies, Plenum Press, New York (1988).

7. K. Diks and A. Pelc, Globally optimal diagnosis in systems with random faults, IEEE Transactions on Computers, to appear.

8. S.N. Maheshwari and S.L. Hakimi, On models for diagnosable systems and probabilistic fault diagnosis, IEEE Transactions on Computers 25 (1976), 228-236.

9. A. Pelc, Undirected graph models for system-level fault diagnosis, IEEE Transactions on Computers 40 (1991), 1271-1276.

10. F. Preparata, G. Metze and R. Chien, On the connection assignment problem of diagnosable systems, IEEE Transactions on Electron. Computers 16 (1967), 848-854.

11. S. Rangarajan and D. Fussell, A probabilistic method for fault diagnosis of multiprocessor systems, Dig. 18th Int. Symp. Fault-Tolerant Computing, IEEE Computer Society Press, (1988), 278-283.

12. E. Scheinerman, Almost sure fault-tolerance in random graphs, SIAM Journal on Computing 16 (1987), 1124-1134.

Efficient Algorithms for Shortest Path Queries in Planar Digraphs [*]

Hristo N. Djidjev

Department of Computer Science, Rice University
P.O. Box 1892, Houston, TX 77251, USA
email: hristo@cs.rice.edu

Abstract. This paper describes algorithms for answering shortest path queries in digraphs with small separators and, in particular, in planar digraphs. In this version of the problem, one has to preprocess the input graph so that, given an arbitrary pair of query vertices v and w, the shortest-path distance between v and w can be computed in a short time. The goal is to achieve balance between the preprocessing time and space and the time for answering a distance query. Previously, efficient algorithms for that problem were known only for the class of outerplanar digraphs and for the class of digraphs of constant treewidth. We describe efficient algorithms for this problem for any class of digraphs for which an $O(\sqrt{n})$ separator theorem holds. For such graphs our algorithm uses $O(S)$ space and answers queries in $O(n^2/S)$ time, for any previously chosen $S \in [n, n^2]$. For the class of planar digraphs improved algorithms are described.

1 Introduction

Let G be a digraph with weights on the edges and let p be a path in G with endpoints v and w. The *length* of p is the sum of the weights of all edges of p. The *shortest-path distance* between v and w is the minimum length of a path from v to w. The path of minimum length is called a *shortest path* between v and w.

Finding shortest path information in a graph is a very important and intensively studied problem with applications in communication systems, transportation, scheduling, computation of network flows, etc. [1]. For the single-source shortest path (SSSP) problem with non-negative edge weights, Dijkstra's algorithm takes $O(n^2)$ time [5]. An implementation of Dijkstra's algorithm that uses Fibonacci heaps [10] reduces the time to $O(n \log n + m)$.

Many recent papers address shortest path problems for special classes of graphs. Johnson proposes an $O(nm + n^2 \log n)$-time algorithm for the all-pairs shortest path (APSP) problem for sparse graphs [12]. Frederickson and Janardan construct an $O(n)$ algorithm for the SSSP problem for outerplanar graphs [8]. Using planar separator theorems, Frederickson develops an $O(n\sqrt{\log n})$-time

[*] This work was partially supported by the NSF grant No. CCR-9409191.

algorithm for the single-source and an $O(n^2)$-time algorithm for the all-pairs shortest path problem for planar digraphs [9]. The time for the single-source problem for planar digraphs was later improved to $O(n)$ in [13]. More generally, given a class of graphs satisfying an $O(n^{1-\delta})$-separator theorem where $0 < \delta < 1$, the SSSP problem can be solved in $O(n)$ time for any n-vertex graph of the class, not including the time to find a recursive separator decomposition [13].

In this paper we consider the so-called on-line version of the shortest path problem, where one has to construct in a preprocessing phase a data structure such that, given an arbitrary pair of query vertices v and w, the shortest-path distance between v and w can be found fast. This version of the shortest path problem arises in applications where shortest-path distances have to be computed only between a small and previously unknown set of vertex pairs. Such applications include facility allocation, transmission in communication systems, circuit and VLSI design. An efficient solution to the on-line version of the shortest path problem is given in [6] for the class of outerplanar digraphs, where the preprocessing algorithms uses $O(n)$ time and space and a distance query is answered in $O(\log n)$ time. This result implies an algorithm for the class of planar digraphs with face-on-vertex cover q with preprocessing time and space $O(n \log n + q^2)$ and query time $O(\log n)$. That algorithm is, however, not efficient for arbitrary planar digraphs, since in the worst case $q = \Theta(n)$ and the preprocessing space is $\Theta(n^2)$, the same as in the APSP algorithm [9] for planar digraphs. An algorithm with $O(\alpha(n))$ query time was described for the class of digraphs of constant treewidth in [3, 4].

We study here the on-line version of the shortest path problem for the class of all planar digraphs and, more generally, for any class of digraphs satisfying an $O(\sqrt{n})$ separator theorem. Our goal is to construct algorithms for which the preprocessing-space $\times$ query-time product is not greater compared to the best algorithms for the SSSP and APSP problems. Specifically, for digraphs with $O(\sqrt{n})$ separators, if S denotes the preprocessing-space and Q denotes the query time, then for the SSSP problem $S = O(n)$ and $Q = O(n)$ and for the APSP problem $S = O(n^2)$ and $Q = O(1)$. In both cases $SQ = O(n^2)$. We will show in this paper that for the same class of graphs we can construct algorithms whose preprocessing spaces and query times yield the same $SQ = O(n^2)$ tradeoff, for S in the range $[\Theta(n), \Theta(n^2)]$. Surprisingly, if the input graph is planar, we can use the topology of planar graphs in order to achieve even better space-query time tradeoff (but for a smaller range for S). We construct an algorithm for planar digraphs with continuous space-query time tradeoff $SQ = O(n\sqrt{S} \log n)$ for $S \in [\Theta(n^{4/3}), \Theta(n^{3/2}))$. This algorithm is most efficient for $S = \Theta(n^{4/3})$, for which value $SQ = O(n^{5/3} \log n)$.

This paper is organized as follows. In Section 2, we describe a simple divide-and-conquer algorithm for digraphs with $O(\sqrt{n})$ separators which works for $S = \Theta(n^{3/2})$. That algorithm will be used in the next two sections, where algorithms for $S > n^{3/2}$ and $S < n^{3/2}$ will be described, respectively, for the same class of graphs. More efficient algorithms for the class of planar digraphs are described in Section 5.

2 The basic algorithm

A class of graphs satisfies an $f(n)$ *separator theorem*, if for any n-vertex graph G from the class there exists a set of $O(f(n))$ vertices (called a *separator*) whose removal leaves no component of size exceeding αn for some previously fixed $\alpha < 1$. In the next three sections, by $\mathcal{G}_{\sqrt{n}}$ we will denote a class of graphs for which an $O(\sqrt{n})$ separator theorem holds and G will denote a digraph from $\mathcal{G}_{\sqrt{n}}$ with non-negative edge weights. It is assumed that a recursive separator decomposition can be found in $O(n)$ time (so that the linear time SSSP algorithm from [13] can be used). Given a set C of vertices of G, let $G - C$ denote the subgraph of G induced by $V(G) \setminus C$.

The preprocessing algorithm

Algorithm PREPROCESS–BASIC

1. Construct a balanced decomposition tree T_G of G as follows.
   (a) If G contains a single vertex, then return a tree T_G of a single vertex labeled by $V(G)$.
   (b) Construct a set C of $O(\sqrt{n})$ vertices whose removal divides G into components with no more than αn vertices each, where $\alpha < 1$.
   (c) Define a new vertex labeled by C. Find the connected components of $G - C$. For each component K of $G - C$ construct a decomposition tree T_K by running the algorithm recursively on K and make the root of T_K a child of C. Let T_G be the resulting tree rooted at C.
2. For each vertex (set) M of T_G compute and store the shortest-path distances from any vertex of M to all vertices belonging to a descendant of M in T_G (including M).

Note that in Step 2 we compute the distances between any vertex of C and any vertex of G, for any component K of $G - C$ we compute the distances between any vertex of the separator of K and any vertex of K, etc.

Lemma 2.1 *Algorithm* PREPROCESS–BASIC *runs in* $O(n^{3/2})$ *time and uses* $O(n^{3/2})$ *space.*

Proof: For all steps of the algorithm the time and the space requirements are the same. Hence we shall analyze only the time complexity. Constructing the separator and finding the connected components in Step 1 requires $O(n)$ time for one iteration. Since T_G is balanced, the total time needed for Step 1 is $O(n \log n)$.

Let us now analyze Step 2. Finding the shortest-path distances from a vertex of C to all vertices of G takes $O(n)$ time using the linear SSSP algorithm from [13]. Thus, computing all distances associated with the vertices of C requires $|C|O(n) = O(n^{3/2})$ time.

Similarly, for any component K of $G - C$ (i.e. for any child K of C in T_G) and any vertex v of the separator of K, we would like to compute the distances (in G) between v and all vertices of K in $O(|K|)$ total time. We can not just run a SSSP algorithm in K and ignore the rest of G, since a shortest path may use vertices which are not in K. For this end we modify K by adding edges joining v to any vertex of C with weight equal to the length of the corresponding shortest path. Since K is a component of $G - C$, the shortest-path distance in G between v and any vertex of K is the same as the shortest-path distance between these vertices in the modified graph K'. Moreover, any separator of K corresponds to a separator of K' with at most one additional vertex (vertex v). Thus, the shortest path distances from v to all vertices in K can be computed using the SSSP algorithm from [13] in $O(|K'|) = O(|K|)$ time. For all vertices of the separator C_K of K the time to compute all shortest-path distances is $O(|K||C_K|) = O(|K|^{3/2})$.

Thus, if $T(n)$ is the maximum time needed for Step 2 for any n-vertex digraph from $\mathcal{G}_{\sqrt{n}}$, then $T(n)$ satisfies the recurrence

$$T(n) \leq \max\{\sum_i T(n_i) + O(n^{3/2}) \mid n_i \leq \alpha n, \sum_i n_i = n\},$$

where $\alpha < 1$, whose solution is $T(n) = O(n^{3/2})$. □

The query algorithm

Algorithm QUERY–BASIC
{Finds the shortest-path distance between vertices v and w}

1. Find vertices (sets) M_v and M_w of T_G such that $v \in M_v$ and $w \in M_w$.
2. Find the nearest common ancestor (NCA) M in T_G of M_v and M_w.
3. Compute the shortest-path distance between v and w using the formula

$$distance(v, w) := \min\{dist(v, m) + dist(m, w) \mid m \in M\},$$

where distances $dist$ have been computed during the preprocessing phase.

Lemma 2.2 *Algorithm* QUERY–BASIC *computes the shortest-path distance between v and w in $O(\sqrt{n})$ time.*

Proof: *Correctness:* Let U be the union of the sets of vertices corresponding to M and all its descendants in T_G and let H be the subgraph of G induced by U. By the construction of T_G, M is the separator of H found in Step 1(b) of Algorithm PREPROCESS–BASIC and, since M_v and M_w are descendants of M, v and w both belong to H. Thus Step 3 of Algorithm QUERY–BASIC correctly computes the shortest-path distance between v and w, if v and w belong to different components of $H - M$.

Assume for the sake of contradiction that v and w belong to some component H' of $H - M$. The subtree $T_{H'}$ of T_G corresponding to H' contains both v and w and is rooted at a proper descendant of M. Then the nearest common ancestor of M_v and M_w is a proper descendant of M which contradicts the choice of M.

Time analysis: Searching T_G and finding the NCA of M_v and M_w in Step 1 takes $O(\log n)$ time since T_G is balanced. Step 3 requires finding the minimum of $|M|$ numbers and thus takes $O(|M|) = O(\sqrt{n})$ time. □

We summarize the above results in the following theorem.

Theorem 1 *Given an n-vertex digraph G from $\mathcal{G}_{\sqrt{n}}$, a data structure of size $O(n^{3/2})$ can be constructed in $O(n^{3/2})$ time so that, given any two vertices v and w, the shortest-path distance from v to w can be computed in $O(\sqrt{n})$ time.*

Theorem 1 shows that we can reduce the space and the preprocessing time for the shortest path problem by a factor of $\Theta(\sqrt{n})$, which will lead to an increase of the query time by a factor of $\Theta(\sqrt{n})$ (from $\Theta(1)$ to $\Theta(\sqrt{n})$). (A similar result is independently obtained in [2].) A natural question arises whether it is possible to construct an algorithm that allows faster queries, possibly at the expense of more preprocessing time or space, or alternatively, algorithms that are more efficient in terms of preprocessing resources, but slower in answering queries. In the next sections we will show that the answer to both questions is positive.

3   Algorithms with improved query times

By repeated application of a separator theorem, one can divide a graph G from $\mathcal{G}_{\sqrt{n}}$ into multiple components of roughly the same size. In addition to the requirement that the separator must be small in size, we will also require that each component into which G is divided is adjacent to a roughly the same (small) number of separator vertices. In order to formally specify this requirement we need some definitions. The vertices of G will be divided into sets called *regions*, and vertices will be classified as either *internal* or *boundary* vertices. Internal vertices belong to only one region and boundary vertices belong to more than one region. Moreover, any pair of adjacent internal vertices belong to the same region.

For any internal vertex v, by $R(v)$ we denote the region containing v, by $C(v)$ we denote the component of the graph induced by $R(v)$ that contains v, and by $B(v)$ we denote the set of boundary vertices of $R(v)$. $B(v)$ will be called the *boundary* of $R(v)$. If v is a boundary vertex we define $R(v) = C(v) = B(v) = \{v\}$.

We will use the following result of Frederickson [9].

Theorem 2 *For any n-vertex graph G from $\mathcal{G}_{\sqrt{n}}$ and any $\varepsilon \in (0,1)$ the vertices of G can be divided into $O(1/\varepsilon)$ regions with $O(\varepsilon n)$ internal vertices and $O(\sqrt{\varepsilon n})$ boundary vertices each. Moreover, each boundary vertex is contained in no more than 3 regions. Such a division (called an ε-division) can be found in $O(n \log n)$ time and $O(n)$ space.*

The preprocessing algorithm

The algorithms described in the following sections compute in a preprocessing step shortest path distances between the boundary vertices of some ε-division and the other vertices of the graph. We illustrate that storing an appropriate subset of that information and choosing a suitable value of ε can result in various shortest path algorithms with improved performance. The following algorithm gives the typical structure of such a preprocessing algorithm with details to be given in the concrete implementation.

Algorithm PREPROCESS–GENERIC

1. Let $\varepsilon \in (0, 1)$. Construct an ε-division of the input graph G as in Theorem 2.
2. Compute the connected components of the subgraph of G induced by the set of internal vertices of the division. Keep a pointer from each internal vertex u of G to the component $C(u)$ and to the region $R(u)$ containing u.
3. For any boundary vertex b compute the shortest-path distances from b to each vertex of G and from each vertex of G to b.
4. **Store a subset** (to be specified in the concrete implementation) of the shortest path information computed in the previous step.
5. Do **additional preprocessing** (to be specified in the concrete implementation), if necessary.

Lemma 3.1 *Steps 1–3 of Algorithm* PREPROCESS–GENERIC *require $O(n^{3/2}\varepsilon^{-1/2})$ time and $O(n)$ space.*

Proof: Step 1 requires $O(n \log n)$ time and $O(n)$ space according to Theorem 2. Step 2 takes $O(n)$ time and space. The total number of boundary vertices according to Theorem 2 is $O(\sqrt{n/\varepsilon})$. Thus Step 3 takes $O(\sqrt{n/\varepsilon}\, n) = O(n^{3/2}\varepsilon^{-1/2})$ time. The space for any iteration of Step 3 is $O(n)$ and that space can be reused; thus the total space for that step is $O(n)$. □

Next we give the main preprocessing and query algorithms of this section.

Algorithm PREPROCESS–I

Steps 1–3 are the same as in Algorithm PREPROCESS–GENERIC.

4. *Store:* the shortest-path distances from (boundary vertex) b to each vertex of G and from each vertex of G to b.
5. *Additional preprocessing:* apply Algorithm PREPROCESS–BASIC on any component of internal vertices found in Step 2.

Lemma 3.2 *Algorithm* PREPROCESS–I *runs in $O(n^{3/2}\varepsilon^{-1/2})$ time and uses $O(n^{3/2}\varepsilon^{-1/2})$ space.*

Proof: The total number of boundary vertices according to Theorem 2 is $O(\sqrt{n/\varepsilon})$. Thus Step 4 takes $O(\sqrt{n/\varepsilon}\, n) = O(n^{3/2}\varepsilon^{-1/2})$ time and uses $O(n^{3/2}\varepsilon^{-1/2})$ space. By Lemma 2.1 the time and space for Step 5 is $O((n\varepsilon)^{3/2})O(1/\varepsilon) = O(n^{3/2}\varepsilon^{1/2}) = O(n^{3/2}\varepsilon^{-1/2})$ (since $\varepsilon < 1$). □

The query algorithm

Algorithm QUERY-I
{Finds the shortest-path distance between vertices v and w}

1. If v or w is a boundary vertex, then find the shortest-path distance between v and w by using the information computed in Step 3 of Algorithm PREPROCESS-I and halt.
2. Determine the components $C(v)$ and $C(w)$ and the boundary $B(v)$.
3. If $C(v) \neq C(w)$, then compute the shortest-path distance between v and w by the formula

$$distance(v, w) := \min\{dist(v, b) + dist(b, w) \mid b \in B(v)\},$$

   where *dist* are the shortest-path distances computed by the preprocessing algorithm.
4. If $C(v) = C(w)$, then compute the shortest-path distance between v and w by applying Algorithm QUERY-BASIC on $C(v)$ with a query pair v, w.

Lemma 3.3 *Algorithm* QUERY-I *computes the shortest-path distance between v and w in $O(\sqrt{\varepsilon n})$ time.*

Proof: *Correctness:* If $C(v) \neq C(w)$, then any path from v to w must contain a boundary vertex from $R(v)$. Thus Step 3 correctly computes the shortest-path distance between v and w in that case. If $C(v) = C(w)$, then the correctness of Algorithm QUERY-I follows from the correctness of Algorithm QUERY-BASIC (Lemma 2.2).

   Time analysis: Steps 1 and 2 take $O(1)$ time. Step 3 takes time proportional to the maximum size of a boundary of a region, which is $O(\sqrt{\varepsilon n})$. Finally, Step 4 requires $O(\sqrt{|C(v)|}) = O(\sqrt{|R(v)|}) = O(\sqrt{\varepsilon n})$ time. $\square$

We have the following theorem that summarizes the results from this section.

Theorem 3 *Given an n-vertex digraph G from $\mathcal{G}_{\sqrt{n}}$ and any number $S \in [n^{3/2}, n^2]$, a data structure of size $O(S)$ can be constructed in $O(S)$ time so that, for any two vertices v and w, the shortest-path distance from v to w can be computed in $O(n^2/S)$ time.*

Proof: Follows from Lemma 3.2 and Lemma 3.3 by setting $\varepsilon = n^3/S^2$. $\square$

   Theorem 3 yields a continuous trade-off $SQ = \Theta(n^2)$ between the area S and the query time Q, for S in the range $[n^{3/2}, n^2]$. For $S = n^2$, Theorem 3 gives the same time as the optimal algorithm for the all pairs shortest paths problem for digraphs with $O(\sqrt{n})$ separator [9], which uses $O(n^2)$ space and answers queries in $O(1)$ time.

   Next, we will extend the above result by describing an algorithm achieving the same space–query time tradeoff $SQ = \Theta(n^2)$ as in Theorem 3, but for S in the range $[n, n^{3/2}]$.

4 Space–efficient algorithms

The preprocessing algorithm

The algorithm is similar to Algorithm PREPROCESS–I except that, for any boundary vertex b, we store only the shortest-path distances between b and the other boundary vertices. (In Algorithm PREPROCESS–I we stored the shortest-path distances between any boundary vertex b and all vertices of G.) A similar preprocessing for a dynamic version of the shortest path problem was used in [7].

Algorithm PREPROCESS–II

Steps 1–3 are the same as in Algorithm PREPROCESS–GENERIC.

4. *Store:* the shortest-path distances between (boundary vertex) b and all boundary vertices of G (in both directions).
5. *Additional preprocessing:* None.

Lemma 4.1 *Algorithm* PREPROCESS–II *runs in* $O(n^{3/2}\varepsilon^{-1/2})$ *time and uses* $O(n/\varepsilon)$ *space.*

Proof: The time bound follows from Lemma 3.1. The space needed to store the shortest-path distances from any boundary vertex b is $O((n/\varepsilon)^{1/2})$. The space bound follows, since the number of all boundary vertices is $O((n/\varepsilon)^{1/2})$. □

The query algorithm

Algorithm QUERY–II
{Finds the shortest-path distance between vertices v and w}

1. If both v and w are boundary vertices, then find the shortest-path distance between v and w by using the distance table computed in Step 3 of Algorithm PREPROCESS–II and stop.
2. If $C(v) \neq C(w)$, then
   (a) compute the shortest-path distances from v to all vertices of $B(v)$ by using the linear time SSSP algorithm applied on graph $R(v)$;
   (b) compute the shortest-path distances from all vertices of $B(w)$ to w by using the SSSP algorithm applied on graph $R(w)$;
   (c) compute the shortest-path distance D from v to w by the formula

   $$D := \min\{d_{query}(v,b_1)+dist(b_1,b_2)+d_{query}(b_2,w) \mid b_1 \in B(v), b_2 \in B(w)\},$$

   where *dist* denotes the distance computed by the preprocessing algorithm, and d_{query} denotes the distance computed in Steps 2 (a) and (b) of this algorithm.

3. If $C(v) = C(w)$, then compute the shortest-path distance between v and w by adding in $R(v)$ edges between v and all vertices of $B(v)$ with weights equal to the lengths of the corresponding shortest paths and applying the SSSP algorithm with source v on the resulting graph.

Lemma 4.2 *Algorithm* QUERY-II *computes the shortest-path distance between* v *and* w *in* $O(\varepsilon n)$ *time.*

Proof: The correctness of the algorithm follows from the definition of regions and boundary vertices. Step 1 requires $O(1)$ time. The time for Steps 2(a)and (b) is $O(|R(v)| + |R(w)|) = O(\varepsilon n)$. Step 2(c) needs $O(|B(v)| |B(w)|) = O(\sqrt{\varepsilon n}\sqrt{\varepsilon n}) = O(\varepsilon n)$ time. Step 3 requires $O(|R(v)| + |B(v)|) = O(\varepsilon n)$ time. $\qquad\square$

We combine Lemma 4.1 and Lemma 4.2 the following theorem.

Theorem 4 *Given an n-vertex digraph G from $\mathcal{G}_{\sqrt{n}}$ and any $S \in [n, n^{3/2}]$, a data structure of size $O(S)$ can be constructed in $O(nS^{1/2})$ time so that, for any two vertices v and w, the shortest-path distance from v to w can be computed in $O(n^2/S)$ time.*

Proof: Follows from Lemma 4.1 and Lemma 4.2 by setting $\varepsilon = n/S$. $\qquad\square$

Note that the space and the query time from Theorem 4 satisfy the same relationship $SQ = O(n^2)$ as the best known algorithm for the SSSP problem [13] (where $S = O(n)$ and $Q = O(n)$).

In the next section we will show that improvements to Theorem 3 and Theorem 4 are possible in the special case of planar input graphs.

5 Faster algorithms for the class of planar digraphs

The results from the previous sections apply also to planar graphs, since the class of planar graphs has an $O(\sqrt{n})$-separator theorem and recursive separator decomposition can be found in $O(n)$ time [11]. In order to speed-up our algorithms, we will additionally use some topological properties of planar graphs. In particular, we will make use of the Jordan Curve Theorem, which states that any simple closed curve in the plane divides it into exactly two connected regions "inside" and "outside".

The preprocessing algorithm

In the preprocessing algorithm we will compute, as before, the shortest-path distances between selected pairs of vertices. In this algorithm, however, we will not store information (will ignore) paths that cross the boundaries of some regions multiple times. For this end, we will modify the shortest path trees rooted at the boundary vertices in the following manner.

Let v be any vertex of G. In our algorithms bellow we will need in some cases shortest paths that avoid vertices of $B(v)$. For given $b \in B(v)$, we call a *reduced shortest path tree* from b with respect to $B(v)$ the shortest path tree in

$G - \{B(v) \setminus \{b\}\}$ rooted at b. Note that since there are at most three regions adjacent to any boundary vertex b, we need to compute and store no more than three reduces shortest path trees from b.

Algorithm PREPROCESS–PLANAR

Steps 1–3 are the same as in Algorithm PREPROCESS–GENERIC.

4. *Store:* the shortest-path distances between (boundary vertex) b and all boundary vertices of G (in both directions).
5. *Additional preprocessing:*
   (a) Compute the reduced shortest path trees from any boundary vertex x with respect to the boundaries of all regions adjacent to x and store the corresponding shortest-path distances between x and all boundary vertices of G (in both directions).
   (b) Apply Algorithm PREPROCESS–BASIC on any component found in Step 2.

Lemma 5.1 *Algorithm* PREPROCESS–PLANAR *runs in* $O(n^{3/2}\varepsilon^{-1/2})$ *time and uses* $O(n/\varepsilon + n^{3/2}\varepsilon^{1/2})$ *space.*

Proof: The time and space for Steps 1–4 are the same as in Algorithm PREPROCESS-II, i.e., $O(n^{3/2}\varepsilon^{-1/2})$ time and $O(n/\varepsilon)$ space. The time and space for Step 5 (b) are $O((n\varepsilon)^{3/2})$ per region, according to Lemma 2.1. Hence, the total time and the total space for Step 5 are $O((n\varepsilon)^{3/2})O(1/\varepsilon) = O(n^{3/2}\varepsilon^{1/2})$. $\qquad\square$

The query algorithm

The idea of the query algorithm is the following. We have already computed during the preprocessing phase the shortest-path distances between any two boundary vertices of G. Using a similar type of information, in Algorithm QUERY–II we found the distance between query vertices v and w by computing the length of any shortest path from v to w containing intermediate vertices $x \in B(v)$ and $y \in B(w)$, for all possible choices of x and y. Thus the query time was proportional to $|B(v)||B(w)|$. If we use the planarity of G, we will be able to reduce the number of considered pairs (x, y) to $|B(w)| \log |B(v)|$. For example, assume that c' and c'' are cycles of vertices of $B(v)$ and $B(w)$, and the paths p_1 and p_2 illustrated on Figure 3 are shortest paths from x_1 and x_2 to w containing vertices y_1 and y_2, respectively. Then, if x is a vertex on the clockwise portion of c' from x_1 to x_2, a shortest path will exist from x to w that contains a vertex y from the clockwise portion of c'' from y_1 to y_2. Thus, we can limit our search to only a portion of c''.

We assume w.l.o.g. that G is a maximal planar digraph. (If G is not maximal, we can add new edges with infinite lengths.) In a maximal planar graph, each face is a triangle and the union of any number of triangles results in a region whose boundary is a union of edge-disjoint cycles. Thus, the boundary of any region of the ε-partition of G is a union of simple edge-disjoint cycles.

By $dist(x,y)$ we denote the distance between vertices x and y computed in Step 5 of Algorithm PREPROCESS–PLANAR. Assume that v and w are 2 vertices

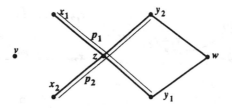

Fig. 1. Intersecting paths p_1 and p_2.

such that $C(v) \neq C(w)$. Moreover, let $x_1 \neq x_2$ be 2 vertices from $B(v)$ and y_1 and y_2 be 2 vertices from $B(w)$ such that, for $i = 1, 2$, the sum

$$dist(x_i, b_i) + dist(b_i, w) \text{ for } b_i \in B(w)$$

is minimized for $b_i = y_i$. Then $dist(x_i, y_i) + dist(y_i, w)$ will be the shortest-path distance from x_i to w. Let p_1 and p_2 be shortest paths from x_1 to y_1 and from x_2 to y_2, respectively. We will show that such paths p_1 and p_2 can be chosen that do not "cross", i.e., have the structure illustrated on Figure 2.

Lemma 5.2 *If* $dist(x_2, y_2) + dist(y_2, w) < dist(x_2, y_1) + dist(y_1, w)$, *then* p_1 *and* p_2 *are vertex disjoint.*

Proof: Assume that p_1 and p_2 have a common vertex, say z (Figure 1). Since both p_1 and p_2 are shortest paths, then $dist(z, y_2) + dist(y_2, w) = dist(z, y_1) + dist(y_1, w)$. Hence $dist(x_2, y_2) + dist(y_2, w) = dist(x_2, z) + dist(z, y_1) + dist(y_1, w) \geq dist(x_2, y_1) + dist(y_1, w)$ – a contradiction. Thus p_1 and p_2 are vertex disjoint. $\square$

If the condition of Lemma 5.2 is not satisfied, i.e., if $dist(x_2, y_2) + dist(y_2, w) \geq dist(x_2, y_1) + dist(y_1, w)$, then we can replace p_2 by another shortest path by choosing $y_2 := y_1$. We can, therefore, make the assumption that in this case p_1 and p_2 consist of two vertex-disjoint paths p_1' and p_2', respectively, followed by a common subpath with an endvertex on $B(w)$ (Figure 2 (a)). If the conditions of Lemma 5.2 are satisfied, then p_1 and p_2 are vertex disjoint (Figure 2 (b)).

Suppose that c' and c'' are the two cycles from the boundaries of v and w, respectively, that separate v and w (Figure 3). Then, any path from v to w will contain at least one vertex from c' and at least one vertex from c''. Let x_1 and

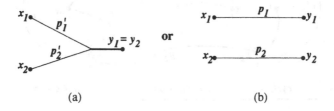

(a)   or   (b)

Fig. 2. Possible structure of the paths p_1 and p_2.

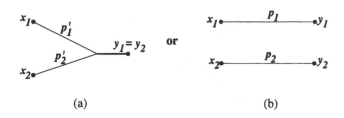

(a)   (b)

Fig. 3. Separating cycles of boundary vertices.

x_2 be chosen from c' and y_1 and y_2 be chosen from c'', where x_1, x_2, y_1, and y_2 are as defined for Lemma 5.2.

Denote, for any cycle c and two vertices x and y on c, by $Cchain(x, y, c)$ the subpath of c between vertices x and y in clockwise direction. Similarly, denote by $CCchain(x, y, c)$ the subpath of c between vertices x and y in counterclockwise direction. We will write simply $Cchain(x, y)$ and $CCchain(x, y)$ if the cycle c is clear from the context.

Lemma 5.3 *If x is a vertex from $Cchain(x_1, x_2)$ such that there is a shortest path between x and w containing no vertices from $B(v)$, then the sum $dist(x, b) + dist(b, w)$ for $b \in c''$ will be minimized for some y from $CCchain(y_1, y_2)$ (Figure 3).*

Proof: Let p_1 and p_2 be shortest paths from x_1 and x_2 to y_1 and y_2 that have no other vertices from $B(v)$ except for their start vertices.

First, consider the case when p_1 and p_2 are vertex disjoint. Define a cycle C consisting of the following paths: a shortest path from v to x_1, p_1, a shortest path from y_1 to w, a shortest path from w to y_2, p_2, and a shortest path from x_2 to v (Figure 4). Call the *inside* of C the region that contains a vertex from the clockwise chain on c' from x_1 to x_2.

Let p be a shortest path from x to w. We have the following two cases (Figure 3).

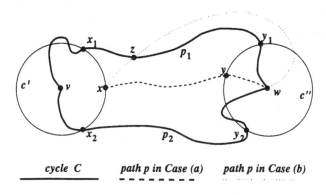

cycle C path p in Case (a)   path p in Case (b)

Fig. 4. Illustration to the proof of Lemma 5.3.

(a) p has no other vertex in common with $p_1 \cup p_2$ except w. Then p will contain a vertex from $CCchain(y_1, y_2)$. Hence p will contain a vertex from $CCchain(y_1, y_2)$.

(b) p contains a vertex $z \neq w$ from $p_1 \cup p_2$. Assume that $z \in p_1$. Then the path consisting from the subpath of p from x to z plus the subpath of p_1 from z to w is a shortest path from x to w which includes y_1. The claim follows.

Now consider the case when p_1 and p_2 have a common vertex. Then, by the assumption about the structure of p_1 and p_2, $y_1 = y_2$ (see Lemma 5.2 and Figure 2) and any path from x to w that contains no other vertices from $B(v)$ except x will intersect p_1 or p_2. Hence, a shortest path between x and w that contains y_1 can be constructed as in Case (b) above. □

The following query algorithm is based on Lemma 5.3.

Algorithm QUERY–PLANAR
{Finds the shortest-path distance between vertices v and w}

A. If $C(v) \neq C(w)$ then do

1. Modify the graph $C(v)$ by adding an edge between any pair of boundary vertices with weight equal to the shortest-path distance in G between the vertices as computed by the preprocessing algorithm. (Note that the number of additional edges is $O(|C(v)|)$.) Let $C'(v)$ denote the resulting (possibly non-planar) graph. Compute the shortest-path distances from v to all vertices from $B(v)$ by running Dijkstra's SSSP algorithm in $C'(v)$. Similarly, compute the shortest-path distances from all vertices from $B(w)$ to w.

2. Choose any two different vertices x_1 and x_2 from $B(v)$. Using the reduced shortest-path trees and Lemma 5.2, find a pair of vertices y_1 and y_2 such that there are shortest paths q_1 from x_1 to y_1 and q_2 from x_2 to y_2 with the properties (i) q_1 and q_2 are vertex disjoint or $y_1 = y_2$; and (ii) q_1 and q_2 contain no other vertices from $B(v)$ except x_1 and x_2.

3. For any vertex $x \in Cchain(x_1, x_2)$ find the minimum length of a path from x to w that does not intersect c' and contains a vertex from $CCchain(y_1, y_2)$ as follows.

   (a) Find a vertex x_3 on $Cchain(x_1, x_2)$ that divides $Cchain(x_1, x_2)$ into equal (within difference in size at most one) parts. Find a vertex $y_3 \in CCchain(y_1, y_2)$ such that the distance

   $$D(x_3) = dist(x_3, b) + d_{query}(b, w) \mid b \in CCchain(y_1, y_2)$$

   is minimized for $b = y_3$, where $dist$ denotes the distance computed by the preprocessing algorithm and d_{query} denotes the distance computed in Step 1 of this algorithm. Store $D(x_3)$ as the shortest-path distance from x_3 to w (avoiding vertices from c').

   (b) Find, recursively, for any vertex $x \in Cchain(x_1, x_3)$, the minimum length of a path from x to w that does not intersect c' and that contains a vertex from $CCchain(y_1, y_3)$.

(c) Find, recursively, for any vertex $x \in Cchain(x_3, x_2)$, the minimum length of a path from x to w that does not intersect c' and that contains a vertex from $CCchain(y_3, y_2)$.

4. Similarly, find the lengths of the shortest paths from all vertices from $CCchain(x_1, x_2)$ to w that do not intersect c' and contain a vertex from $Cchain(y_1, y_2)$.

5. Compute the shortest-path distance from v to w as

$$distance(v, w) := \min\{d_{query}(v, b) + d_{query}(b, w) \mid b \in B(v)\},$$

where distances $d_{query}(v, b)$ are computed in Step 1 and distances $d_{query}(b, w)$ are computed in Steps 3 and 4 of this algorithm.

endif

B. **If $C(v) = C(w)$ then** compute the shortest-path distance between v and w by applying Algorithm QUERY–BASIC on $C(v)$ with a query pair v, w.

Lemma 5.4 *Algorithm* QUERY–PLANAR *computes the shortest-path distance between v and w in $O(\sqrt{\varepsilon n} \log(\varepsilon n))$ time.*

Proof (sketch): The correctness of the algorithm follows from the definition of reduced shortest path trees and Lemma 5.3. Dijkstra's algorithm runs in $O(k \log k)$ time on a graph with k vertices. Thus Step 1 takes $O(\varepsilon n \log(\varepsilon n))$ time, since $\max\{|C(v)|, |C(w)|\} = O(\varepsilon n)$ and the number of new edges added in Step 1 is $O(|B(v)|^2 + |B(w)|^2) = O((\sqrt{\varepsilon n})^2) = O(\varepsilon n)$. Step 2 takes $O(|B(w)|) = O(\sqrt{\varepsilon n})$ time. For the time analysis of Step 3, denote by k and m, respectively, the number of edges of $Cchain(x_1, x_2)$ and $CCchain(y_1, y_2)$. Finding the shortest-path distance between x_3 and $CCchain(y_1, y_2)$ in Step 3 (a) takes $O(m)$ time. Thus the time $T(k, m)$ for Step 3 satisfies the inequality

$$T(k, m) \leq \max\{T(\lfloor k/2 \rfloor, m_1) + T(\lceil k/2 \rceil, m_2) \mid m_1 + m_2 = m\},$$

where m_1 and m_2 correspond to the sizes of the parts into which $CCchain(y_1, y_2)$ is divided by y_3 (see Steps 2 (b) and 2 (c)). The solution to this recurrence is $T(k, m) = O(m \log k)$. Similarly, Step 4 takes $O(m' \log k')$ time, where k' and m' are the lengths of $CCchain(x_1, x_2)$ and $Cchain(y_1, y_2)$. Step 5 takes $O(|B(w)|) = O(\sqrt{\varepsilon n})$ time. The lemma follows since $k + k' = O(\sqrt{\varepsilon n})$ and $m + m' = O(\sqrt{\varepsilon n})$. $\square$

We can summarize the results from the last section in the following theorem, which significantly improves Theorem 4 in the case of planar digraphs.

Theorem 5 *Given an n-vertex planar digraph G and any $S \in [n^{4/3}, n^{3/2}]$, a data structure of size $O(S)$ can be constructed in $O(n\sqrt{S})$ time so that, for any two vertices v and w, the shortest-path distance from v to w can be computed in $O(n \log n / \sqrt{S})$ time.*

Proof: Follows from Lemma 5.1 and Lemma 5.4 by setting $\varepsilon = n/S$. $\square$

For comparison, the time bound to compute the shortest-path distance from v to w in Theorem 4 is $O(n^2/S)$, which is proportional to the square of the corresponding time bound in Theorem 5 (ignoring the logarithmic factor).

For graphs of bounded genus the Jordan Curve Theorem does not hold and the method of proof of Theorem 5 can not be directly applied. It is an interesting open problem to generalize Theorem 5 for such class of graphs.

References

1. Ravindra K. Ahuja, Thomas L. Magnanti, and James B. Orlin. *Network flows : theory, algorithms, and applications.* Prentice Hall, 1993.
2. S. Arikati, D.Z. Chen, L.P. Chew, G. Das, M. Smid, and C.D. Zaroliagis. Planar spanners and approximate shortest path queries among obstacles in the plane. In *Proceedings of ESA'96*, to appear.
3. H. Bodlaender. Dynamic algorithms for graphs with treewidth 2. In *WG'93, Lecture Notes in Computer Science, vol. 790*, pages 112–124. Springer-Verlag, Berlin, Heidelberg, New York, Tokio, 1994.
4. Shiva Chaudhuri and Christos D. Zaroliagis. Shortest path queries in digraphs of small treewidth. In *ICALP'95, Lecture Notes in Computer Science, vol. 944*, pages 244–255. Springer-Verlag, Berlin, Heidelberg, New York, Tokio, 1995.
5. E.W. Dijkstra. A note on two problems in connection with graphs. *Numer. Math*, 1:269–271, 1959.
6. H. Djidjev, G. Pantziou, and C. Zaroliagis. Computing shortest paths and distances in planar graphs. *Proc. of 18th International Colloquium on Automata Languages and Programming*, pages 327–339, 1991.
7. E. Feuerstein and A.M. Spaccamela. Dynamic algorithms for shortest paths in planar graphs. In *WG'91, Lecture Notes in Computer Science, vol 570*, pages 187–197. Springer-Verlag, Berlin, Heidelberg, New York, Tokio, 1991.
8. G. N. Frederickson and R. Janardan. Designing networks with compact routing tables. *Algorithmica*, 3:171–190, 1988.
9. G.N. Frederickson. Fast algorithms for shortest paths in planar graphs, with applications. *SIAM Journal on Computing*, 16:1004–1022, 1987.
10. Michael L. Fredman and Robert E. Tarjan. Fibonacci heaps and their uses in improved network optimization algorithms. *Journal of the ACM*, 34:596–615, 1987.
11. Michael T. Goodrich. Planar separators and parallel polygon triangulation. *Proceedings of 24th Symp. on Theory of Computing*, pages 507–516, 1992.
12. Donald B. Johnson. Efficient algorithms for shortest paths in sparse networks. *Journal of the ACM*, 24:1–13, 1977.
13. P. Klein, S. Rao, M. Rauch, and S. Subramanian. Faster shortest-path algorithms for planar graphs. In *26th ACM Symp. Theory of Computing*, pages 27–37, 1994.

LexBFS–Orderings and Powers of Graphs*

Feodor F. Dragan[1], Falk Nicolai[2], Andreas Brandstädt[3]

[1] Department of Mathematics and Cybernetics, Moldova State University,
A. Mateevici str. 60, Chişinău 277009, Moldova
e–mail : dragan@cinf.usm.md
[2] Gerhard-Mercator-Universität –GH– Duisburg, FB Mathematik, FG Informatik,
D 47048 Duisburg, Germany
e–mail : nicolai@informatik.uni-duisburg.de
[3] Universität Rostock, FB Informatik, Lehrstuhl für Theoretische Informatik,
D 18051 Rostock, Germany
e–mail : ab@informatik.uni-rostock.de

Abstract. For an undirected graph G the k-th power G^k of G is the graph with the same vertex set as G where two vertices are adjacent iff their distance is at most k in G. In this paper we consider LexBFS–orderings of chordal, distance–hereditary and HHD–free graphs (the graphs where each cycle of length at least five has two chords) with respect to their powers. We show that any LexBFS–ordering of a chordal graph is a common perfect elimination ordering of all odd powers of this graph, and any LexBFS–ordering of a distance–hereditary graph is a common perfect elimination ordering of all its even powers. It is well-known that any LexBFS–ordering of a HHD–free graph is a so–called semi–simplicial ordering. We show, that any LexBFS–ordering of a HHD–free graph is a common semi–simplicial ordering of all its odd powers. Moreover we characterize those chordal, distance–hereditary and HHD–free graphs by forbidden isometric subgraphs for which any LexBFS–ordering of the graph is a common perfect elimination ordering of all its nontrivial powers. As an application we get a linear time approximation of the diameter for weak bipolarizable graphs, a subclass of HHD–free graphs containing all chordal graphs, and an algorithm which computes the diameter and a diametral pair of vertices of a distance–hereditary graph in linear time.

1 Introduction

Powers of graphs play an important role for solving certain problems related to distances in graphs : p–center and q–dispersion (cf. [7, 3]), k–domination and k–stability (cf. [8, 3]), diameter (cf. [13]), k–colouring (cf. [26, 20]) and approximation of bandwidth (cf. [27]). For instance, consider the k–colouring problem. The vertices of a graph have to be coloured by a minimal number of colours such that no two vertices at distance at most k have the same colour. Obviously, k–colouring a graph is equivalent to colour (in the classical sense)

---

* First author supported by DAAD, second author supported by DFG.

its k-th power. It is well–known that the colouring problem is NP–complete in general. On the other hand, there are a lot of special graph classes with certain structural properties for which the colouring problem is efficiently solvable. One of the most popular class is the one of chordal graphs. Here we have a linear time colouring algorithm by stepping through a certain dismantling scheme — the so–called perfect elimination ordering — of the graph. So it is quite natural to consider graph classes for which certain powers are chordal.

In the last years some papers investigating powers of chordal graphs were published. One of the first results in this field is due to DUCHET ([18]) : If G^k is chordal then G^{k+2} is so. In particular, odd powers of chordal graphs are chordal, whereas even powers of chordal graphs are in general not chordal. Chordal graphs with chordal square were characterized by forbidden configurations in [28].

It is well–known that any chordal graph has a perfect elimination ordering which can be computed in linear time by Lexicographic Breadth–First–Search (LexBFS, [32]) or Maximum Cardinality Search (MCS, [33]). Thus each chordal power of an arbitrary graph has a perfect elimination ordering. A natural question is whether there is a common perfect elimination ordering of all (or some) chordal powers of a given graph. The first result in this direction using minimal separators is given in [17] : If both G and G^2 are chordal then there is a common perfect elimination ordering of these graphs (see also [4]). The existence of a common perfect elimination ordering of all chordal powers of an arbitrary given graph was proved in [3]. Such a common ordering can be computed in time $O(|V||E|)$ using a generalized version of Maximum Cardinality Search which simultaneously uses chordality of these powers.

Here we consider the question whether LexBFS, working only on an initial graph G, produces a common perfect elimination ordering of chordal powers of G. Hereby we consider chordal, distance–hereditary and HHD–free graphs as initial graphs. Recall, that in chordal graphs every cycle of length at least four has a chord and in distance–hereditary graphs each cycle of length at least five has two crossing chords. HHD–free graphs can be defined as the graphs in which every cycle of length at least five has two chords. Analogously to chordal graphs, HHD–free graphs can be dismantled via a so–called semi–simplicial ordering which can be produced in linear time by LexBFS (cf. [25]). Since a semi–simplicial ordering in reverse order is a perfect ordering (in sense of CHVATAL), HHD–free graphs are perfectly orderable, and hence they can be coloured in linear time (cf. [10]).

2 Preliminaries

Throughout this paper all graphs $G = (V, E)$ are finite, undirected, simple (i.e. loop–free and without multiple edges) and connected.

A *path* is a sequence of vertices $v_0, \ldots, v_k$ such that $v_i v_{i+1} \in E$ for $i = 0, \ldots, k-1$; its *length* is k. As usual, an induced path of k vertices is denoted by P_k. A graph G is *connected* iff for any pair of vertices of G there is a path in G joining both vertices.

The *distance* $d_G(u, v)$ of vertices u, v is the minimal length of any path connecting these vertices. Obviously, d_G is a metric on G. If no confusion can arise we will omit the index G. An induced subgraph H of G is an *isometric* subgraph of G iff the distances within H are the same as in G, i.e.

$$\forall x, y \in V(H) \; : \; d_H(x, y) = d_G(x, y).$$

The *k–th neighbourhood* $N^k(v)$ of a vertex v of G is the set of all vertices of distance k to v, i.e.

$$N^k(v) := \{u \in V : d_G(u, v) = k\},$$

whereas the *disk* of radius k centered at v is the set of all vertices of distance at most k to v :

$$D_G(v, k) := \{u \in V : d_G(u, v) \leq k\} = \bigcup_{i=0}^{k} N^i(v).$$

For convenience we will write $N(v)$ instead of $N^1(v)$. Again, if no confusion can arise we will omit the index G. The *k-th power* G^k of G is the graph with the same vertex set V where two vertices are adjacent iff their distance is at most k. If $k \geq 2$ then G^k is called *nontrivial power*.

The *eccentricity* $e(v)$ of a vertex $v \in V$ is the maximum over $d(v, x)$, $x \in V$. The minimum over the eccentricities of all vertices of G is the *radius* $rad(G)$ of G, whereas the maximum is the *diameter* $diam(G)$ of G. A pair x, y of vertices of G is called *diametral* iff $d(x, y) = diam(G)$.

Next we recall the definition and some characterizations of chordal graphs. An *induced cycle* is a sequence of vertices $v_0, \ldots, v_k$ such that $v_0 = v_k$ and $v_i v_j \in E$ iff $|i - j| = 1$ (modulo k). The *length* $|C|$ of a cycle C is its number of vertices. A graph G is *chordal* iff any induced cycle of G is of length at most three. One of the first results on chordal graphs is the characterization via dismantling schemes. A vertex v of G is called *simplicial* iff $D(v, 1)$ induces a complete subgraph of G. A *perfect elimination ordering* is an ordering of G such that v_i is simplicial in $G_i := G(\{v_i, \ldots, v_n\})$ for each $i = 1, \ldots, n$. It is well–known that a graph is chordal if and only if it has a perfect elimination ordering (cf. [21]). Moreover, computing a perfect elimination ordering of a chordal graph can be done in linear time by Lexicographic Breadth–First–Search (LexBFS, [21]). To make the paper self–contained we present the rules of this algorithm.

Let $s_1 = (a_1, \ldots, a_k)$ and $s_2 = (b_1, \ldots, b_l)$ be vectors of positive integers. Then s_1 is *lexicographically smaller* than s_2 ($s_1 < s_2$) iff

1. there is an index $i \leq \min\{k, l\}$ such that $a_i < b_i$ and $a_j = b_j$ for all $j = 1, \ldots, i - 1$, or
2. $k < l$ and $a_i = b_i$ for all $i = 1, \ldots, k$.

If $s = (a_1, \ldots, a_k)$ is a vector and a is some positive integer then $s + a$ denotes the vector $(a_1, \ldots, a_k, a)$.

procedure LexBFS
Input   : A graph $G = (V, E)$.
Output : A LexBFS–ordering $\sigma = (v_1, \ldots, v_n)$ of V.

begin forall $v \in V$ **do** $l(v) := ()$;
    for $n := |V|$ **downto** 1 **do**
        choose a vertex $v \in V$ with lexicographically maximal label $l(v)$;
        define $\sigma(n) := v$;
        forall $u \in V \cap N(v)$ **do** $l(u) := l(u) + n$;
        $V := V \setminus \{v\}$;
    endfor;
end.

In the sequel we will write $x < y$ whenever in a given ordering of the vertex set of a graph G vertex x has a smaller number than vertex y. Moreover, $x < \{y_1, \ldots, y_k\}$ is an abbreviation for $x < y_i$, $i = 1, \ldots, k$.

In what follows we will often use the following property (cf. [25]) :

(P1)   If $a < b < c$ and $ac \in E$ and $bc \notin E$ then there exists a vertex d such that $c < d$, $db \in E$ and $da \notin E$.

Lemma 1. (1) *Any LexBFS–ordering has property (P1).*
(2) *Any ordering fulfilling (P1) can be generated by LexBFS.*

Proof. (1) We refer to the well–known proof in [21].
(2) Let $\sigma = (v_1, \ldots, v_n)$ be an ordering fulfilling (P1) and suppose that $(v_{i+1}, \ldots, v_n)$, $i \leq n - 1$, can be produced by LexBFS but not $(v_i, \ldots, v_n)$, i.e. v_i cannot be chosen via LexBFS. Let u be the vertex chosen next by LexBFS. Then there must be a vertex $w > v_i$ adjacent to u but not to v_i. We can choose w rightmost in σ. Thus in σ we have $u < v_i < w$, $uw \in E$ and $wv_i \notin E$. Now (P1) implies the existence of a vertex $z > w$ adjacent to v_i but not to u. Since w is chosen rightmost all vertices with a greater number than w which are adjacent to u are adjacent to v_i too. Hence the LexBFS–label of v_i is greater than that of u, a contradiction. $\qquad\square$

3   Chordal Graphs

A set $S \subseteq V$ is *m–convex* (monophonically convex) iff for all pairs of vertices x, y of S each vertex of any induced path connecting x and y is contained in S too.

Lemma 2 [19]. *If G is a chordal graph and $(v_1, \ldots, v_n)$ is a perfect elimination ordering of G then $V(G_i)$ is m–convex in G and, in particular, G_i is an isometric subgraph of G, for every $i = 1, \ldots, n$.*

Using property (P1), m–convexity and isometricity of G_i in G we can prove

Theorem 3. *For a chordal graph G every LexBFS-ordering of G is a perfect elimination ordering of each odd power G^{2k+1} of G.*

Since we do not use chordality of odd powers in the proof of the above theorem we reproved that odd powers of chordal graphs are again chordal.

Theorem 4. *If G is a chordal graph which does not contain the graphs of Figure 1 as isometric subgraphs then every LexBFS-ordering of G is a perfect elimination ordering of each even power G^{2k}, $k \geq 1$, of G.*

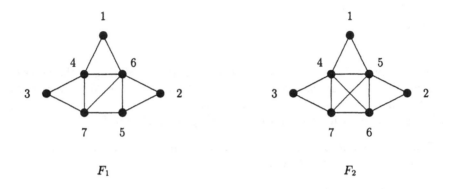

Fig. 1. Chordal graphs labeled by a LexBFS-ordering such that vertex 1 is not simplicial in G^2.

Corollary 5. *If G is chordal and does not contain the graphs of Figure 1 as isometric subgraphs then all powers of G are chordal.*

Ptolemaic graphs (cf. [9, 24]) are the graphs fulfilling the ptolemaic inequality, i.e. for any four vertices u, v, w, x it holds

$$d(u, v)d(w, x) \leq d(u, w)d(v, x) + d(u, x)d(v, w).$$

In [24] it was shown that the ptolemaic graphs are exactly the chordal graphs without a 3-fan (cf. Figure 4), i.e. the distance-hereditary chordal graphs (cf. [2]). For the well-known class of interval graphs we refer to [21].

Corollary 6. *If G is a ptolemaic or interval graph then any LexBFS-ordering of G is a common perfect elimination ordering of all powers of G.*

Corollary 7. *If G is a ptolemaic or interval graph and v is the first vertex of a LexBFS-ordering of G, then $e(v) = diam(G)$.*

Proof. Let σ be a LexBFS-ordering of G, v be the first vertex of σ and k its eccentricity. By Corollary 6 σ is a perfect elimination ordering of the power G^k of G. In particular, v is simplicial in G^k. Thus G^k is complete. □

Hence the diameter and a diametral pair of vertices of a ptolemaic or interval graph can be computed in linear time by only using a LexBFS-ordering.

4 HHD–free Graphs

Note that a vertex is simplicial if and only if it is not midpoint of a P_3. In [25] this notion was relaxed : A vertex is *semi–simplicial* iff it is not a midpoint of a P_4. An ordering $(v_1, \ldots, v_n)$ is a *semi–simplicial ordering* iff v_i is semi–simplicial in G_i for all $i = 1, \ldots, n$. In [25] the authors characterized the graphs for which every LexBFS–ordering is a semi–simplicial ordering as the HHD–free graphs, i.e. the graphs which do not contain a house, hole or domino as induced subgraph (cf. Figure 2).

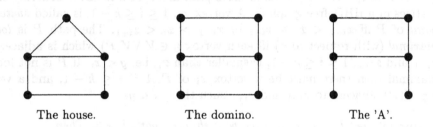

The house.   The domino.   The 'A'.

Fig. 2. The house, the domino and the 'A'.

If a HHD–free graph does not contain the 'A' of Figure 2 as induced subgraph then this graph is called *weak bipolarizable* (HHDA–free) [31].

In [16] we investigated powers of HHD–free graphs. We proved that odd powers of HHD–free graphs are again HHD–free. Furthermore, an odd power G^{2k+1} of a HHD–free graph G is chordal if and only if G does not contain a $C_4^{(k)}$ as an isometric subgraph (cf. [1] and [5] for the role of $C_4^{(k)}$ in distance–hereditary graphs and hole–free graphs). Hereby, a $C_4^{(k)}$ is a graph induced by a C_4 with pendant paths of length k attached to the vertices of the C_4, see Figure 3.

Fig. 3. A $C_4^{(k)}$ and the $C_4^{(1)}$ minus a pendant vertex.

As a relaxation of m–convexity in chordal graphs we introduced the notion of m^3–convexity in [15] : A subset $S \subseteq V$ is called m^3*–convex* iff for any pair of vertices x, y of S each induced path of length at least 3 connecting x and y is completely contained in S.

Lemma 8 [15]. *An ordering $(v_1, \ldots, v_n)$ of the vertices of a graph G is semi-simplicial if and only if $V(G_i)$ is m^3-convex in G for all $i = 1, \ldots, n$.*

The above lemma implies that the minimum (with respect to a semi–simplicial ordering) of an induced path of length at least three must be one of its endpoints.

The proofs of our results are based on nice properties of shortest paths in HHD–free graphs with respect to a given LexBFS–ordering.

Let $P = x_0 - \ldots - x_k$ be an induced path and σ be a LexBFS–ordering of the vertices of a HHD-free graph G. A vertex x_i, $1 \leq i \leq k - 1$, is called *switching point* of P iff $x_{i-1} < x_i > x_{i+1}$ or $x_{i-1} > x_i < x_{i+1}$. The path P is *locally maximal* (with respect to σ) iff each vertex $y \in V \setminus V(P)$ which is adjacent to x_{i-1} and x_{i+1}, $1 \leq i \leq k - 1$, is smaller than x_i, i.e. $y < x_i$. If P is not locally maximal then there must be a vertex x_i of P, $1 \leq i \leq k - 1$, and a vertex $y \notin V(P)$ adjacent to x_{i-1} and x_{i+1} such that $x_i < y$.

Lemma 9. *Let $P = x_0 - \ldots - x_k$ be a shortest path, $k \geq 3$. Then*

1. *The number s of switching points of P is at most three.*
2. *The switching points of P induce a subpath of P.*
3. *If P is locally maximal then $s \leq 1$.*

Lemma 10. *Let $P = x_0 - \ldots - x_k$, $k \geq 3$, be a shortest path which is locally maximal. Furthermore let $x_0 < x_k$ and let x_i, $1 \leq i \leq k - 1$, be the switching point of P. Then*

1. *$d(x_0, x_i) \geq d(x_i, x_k)$ and*
2. *if $d(x_0, x_i) = d(x_i, x_k)$, i.e. $k = 2i$, then $x_0 < x_k < \ldots < x_j < x_{k-j} < \ldots < x_{i-1} < x_{i+1} < x_i$.*

Using property $(P1)$, m^3–convexity and the above path properties we can show

Theorem 11. *Any LexBFS–ordering of a HHD–free graph G is a common semi-simplicial ordering of all odd powers of G.*

Theorem 12. *Any LexBFS–ordering of a HHD–free graph G is a common perfect elimination ordering of all nontrivial odd powers of G if and only if G does not contain a $C_4^{(1)}$ minus a pendant vertex (cf. Figure 3) as isometric subgraph.*

Corollary 13. *Any LexBFS–ordering of a weak bipolarizable graph is a common perfect elimination ordering of all its nontrivial odd powers.*

Corollary 14. *Let v be the first vertex of a LexBFS–ordering of a weak bipolarizable graph G. Then $diam(G) - 1 \leq e(v) \leq diam(G)$.*

Proof. First note that for $e(v) = 1$ there is nothing to show. If $e(v) = 2k + 1$, $k \geq 1$, then G^{2k+1} is complete and hence $diam(G) = e(v)$. For $e(v) = 2k$ the odd power G^{2k+1} is complete implying $diam(G) \leq 2k + 1 = e(v) + 1$. □

Theorem 15. *Any LexBFS–ordering of a HHD–free graph G is a common perfect elimination ordering of all even powers of G if and only if G does not contain one of the graphs of Figure 1 as isometric subgraph.*

Similar to Corollary 14 we can prove

Corollary 16. *Let v be the first vertex of a LexBFS–ordering of a HHD–free graph G which does not contain a graph of Figure 1 as isometric subgraph. Then $diam(G) - 1 \leq e(v) \leq diam(G)$.*

5 Distance–Hereditary Graphs

A graph G is *distance–hereditary* ([23]) iff each connected induced subgraph of G is isometric. Distance–hereditary graphs were extensively studied in [2], [22], [11], [1] and [29]. For proving our results we used the following property :

Theorem 17 (The four–point condition [2]). *Let G be a distance–hereditary graph. Then, for any four vertices u, v, w, x at least two of the distance sums*
$$d(u, v) + d(w, x), \quad d(u, w) + d(v, x), \quad d(u, x) + d(w, v)$$
are equal, and, if the two smaller sums are equal then the larger one exceeds this by at most two.

Furthermore, distance–hereditary graphs can be characterized by forbidden subgraphs ([2], [22]) : A graph is distance–hereditary if and only if it does not contain a hole, a house, a domino and a 3–fan as induced subgraph (see Figure 4).

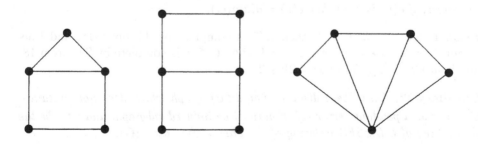

Fig. 4. A house, a domino and a 3–fan.

Thus distance–hereditary graphs are HHD–free, and each LexBFS–ordering of G is a semi–simplicial ordering of G.

Using the four–point condition, m^3–convexity and property $(P1)$ we can show

Theorem 18. *Each LexBFS–ordering σ of a distance–hereditary graph G is a perfect elimination ordering of each even power G^{2k}, $k \geq 1$.*

Thus we reproved that even powers of distance–hereditary graphs are chordal (cf. [1]). In [1] it was proved that all odd powers of a distance–hereditary graph are HHD–free. Moreover, an odd power G^{2k+1} is chordal if and only if G does not contain an induced subgraph isomorphic to the $C_4^{(k)}$, cf. Figure 3.

Theorem 19. *Any LexBFS–ordering σ of a given distance–hereditary graph G is a common perfect elimination ordering of all its nontrivial powers if and only if G does not contain a $C_4^{(1)}$ minus a pendant vertex (cf. Figure 3) as induced subgraph.*

Theorem 20. *Any LexBFS–ordering σ of a distance–hereditary graph G is a common semi–simplicial ordering of all its powers.*

Computing a diametral pair of vertices

In [12] a linear time algorithm for computing the diameter of a distance–hereditary graph was presented, but that approach is not usable for finding a diametral pair of vertices. As an application of the preceding results we present a simpler algorithm which computes both the diameter and a diametral pair of vertices of a distance–hereditary graph in linear time. This points out once more the importance of considering chordal powers of graphs and perfect elimination orderings of them.

Lemma 21. *Let v be the first vertex of a LexBFS–ordering of a distance–hereditary graph G. Then*

$$diam(G) - 1 \leq e(v) \leq diam(G).$$

Moreover, if $e(v)$ is even then $e(v) = diam(G)$.

Proof. If $e(v) = 2k$, $k \geq 1$, then G^{2k} is complete by Theorem 18, and thus $diam(G) = 2k$. If $e(v) = 2k + 1$, $k \geq 1$, then G^{2k+2} is complete by Theorem 18, and hence $2k + 1 \leq diam(G) \leq 2k + 2$. □

Corollary 22. *Let G be a distance–hereditary graph which does not contain a $C_4^{(1)}$ minus a pendant vertex (cf. Figure 3) as induced subgraph, and let v be the first vertex of a LexBFS–ordering of G. Then $e(v) = diam(G)$.*

Recall that the ptolemaic graphs are exactly the chordal distance–hereditary graphs. Thus they do not contain a $C_4^{(1)}$ minus a pendant vertex. Therefore, any LexBFS–ordering of a ptolemaic graph is a diametral ordering. In [30] such an ordering is used to check the Hamiltonicity of a ptolemaic graph in linear time.

For the sequel we may assume that G is not complete for otherwise there is nothing to do. In what follows we describe the steps of the algorithm.

At first we compute a LexBFS–ordering σ of a given distance–hereditary graph G. Let v be the first vertex of σ. If $e(v) = 2k$, $k \geq 1$, then, by Lemma 21, $e(v) = diam(G)$, and the vertices v and $w \in N^{e(v)}(v)$ form a diametral pair of G. So let $e(v) = 2k + 1$. Now we start LexBFS at vertex v yielding a LexBFS–ordering τ with first vertex u. If $e(u) = 2k + 2$ then, by Lemma 21, $diam(G) = 2k + 2$ and the vertices u and $w \in N^{e(u)}(u)$ form a diametral pair of G. Otherwise ($e(v) = e(u) = 2k + 1$) we choose a vertex z at distance k to u and at distance $k + 1$ to v.

Lemma 23. $k + 1 \leq e(z) \leq k + 2$.

Proof. Since $d(z, v) = k + 1$ we immediately have $e(z) \geq k + 1$. So let w be a vertex of V such that $d(z, w) \geq k + 2$. We obtain the following distance sums :

$$
\begin{aligned}
d(u, v) + d(z, w) &= 2k + 1 + d(z, w) \geq 3k + 3 \\
d(u, z) + d(v, w) &= k + d(v, w) \qquad\quad \leq 3k + 1 \\
d(u, w) + d(v, z) &= k + 1 + d(u, w) \quad\;\; \leq 3k + 2
\end{aligned}
$$

Now the four–point condition gives

$$
d(v, w) = 2k + 1, \quad d(u, w) = 2k, \quad \text{and} \quad d(z, w) = k + 2.
$$

This settles the proof. $\qquad\qquad\qquad\qquad\qquad\qquad\qquad\qquad\qquad\qquad\square$

For every vertex w of $V \setminus D(z, k)$ we store in $track(w)$ the second edge of an arbitrary shortest path from z to w. Define $F := \{track(w) : w \in V \setminus D(z, k)\}$. We will say that two edges in a graph are *independent* iff the vertices of this edges induce a $2K_2$ in G.

Lemma 24. $diam(G) = 2k+2$ if and only if the set F contains two independent edges.

Proof. Let $diam(G) = 2k + 2$ and let x, y be vertices of G such that $d(x, y) = 2k + 2$. Since both u and v (as first vertices of LexBFS–orderings) are simplicial in G^{2k} we get

$$
d(u, x) = d(u, y) = d(v, x) = d(v, y) = 2k + 1.
$$

With $d(z, u) = k$ this implies $d(z, x) \geq k + 1$. So we obtain the following distance sums :

$$
\begin{aligned}
d(u, v) + d(z, x) &= 2k + 1 + d(z, x) \geq 3k + 2 \\
d(u, z) + d(v, x) &= k + 2k + 1 \qquad\quad = 3k + 1 \\
d(u, x) + d(v, z) &= 2k + 1 + k + 1 \quad\;\; = 3k + 2
\end{aligned}
$$

Now the four–point condition gives $d(z, x) = k+1$. By symmetry, $d(z, y) = k+1$. Thus z lies on a shortest path joining x and y. Obviously, $track(x)$ and $track(y)$ are independent edges due to $d(x, y) = 2k + 2$ and $d(x, z) = d(y, z) = k + 1$.

Now let $s_1 s_2$ and $t_1 t_2$ be independent edges in F. Let $z - s_1 - s_2 - \ldots - w_1$ and $z - t_1 - t_2 - \ldots - w_2$ be shortest paths of length at least $k + 1$. We will prove

$d(w_1, w_2) = 2k + 2$. Since $s_2 - s_1 - z - t_1 - t_2$ is induced we get $d(s_2, t_2) = 4$. Using $k + 1 \leq e(z) \leq k + 2$ we obtain the following distance sums :

$$
\begin{aligned}
d(w_1, z) \; + d(s_2, t_2) &= 4 + d(w_1, z) \in \{k + 5, k + 6\} \\
d(w_1, s_2) + d(z, t_2) &= 2 + d(w_1, s_2) \in \{k + 1, k + 2\} \\
d(w_1, t_2) + d(z, s_2) &= 2 + d(w_1, t_2)
\end{aligned}
$$

Since the difference between the first and second distance sum is at least three the four–point condition implies that the larger two sums must be equal, i.e. the first and third one. So we get

$$
k + 3 \leq d(w_1, t_2) \leq k + 4 \quad \text{and} \quad k + 3 \leq d(w_2, s_2) \leq k + 4
$$

by symmetry. Together with $d(s_2, t_2) = 4$ this implies

$$
\begin{aligned}
d(w_1, w_2) + d(s_2, t_2) &= 4 + d(w_1, w_2) \\
d(w_1, s_2) \; + d(w_2, t_2) &\in \{2k - 2, 2k - 1, 2k\} \\
d(w_1, t_2) \; + d(w_2, s_2) &\in \{2k + 6, 2k + 7, 2k + 8\}
\end{aligned}
$$

By the same argument as above the four–point condition implies that the first and the third distance sum must be equal, i.e. $d(w_1, w_2) \geq 2k + 2$. $\qquad\qquad\square$

Therefore the following algorithm correctly computes the diameter and a diametral pair of a distance–hereditary graph :

Algorithm DHGDiam.

Input : A connected distance–hereditary graph G.

Output : $diam(G)$ and a diametral pair of vertices of G.

(1)   **begin** $\sigma :=$ LexBFS(G, s) for some $s \in V(G)$.

(2)   Let v be the first vertex of σ.

(3)   **if** $e(v)$ is even **then return**$(e(v), (v, w))$ where $w \in N^{e(v)}(v)$.

(4)   **else** $\tau :=$ LexBFS(G, v).

(5)   Let u be the first vertex of τ.

(6)   **if** $e(u) = e(v) + 1$ **then return**$(e(u), (u, w))$ where $w \in N^{e(u)}(u)$.

(7)   **else** Let $k \in \mathbb{N}$ such that $e(v) = e(u) = 2k + 1$.

(8)   Choose a vertex z from $D(u, k) \cap D(v, k + 1)$.

(9) $F := \{track(w) : w \in V \setminus D(z, k)\}$.

(10)   **if** F contains a pair e_1, e_2 of independent edges

(11)   **then return**$(2k + 2, (x, y))$

                     where $x, y \in V$ such that $track(x) = e_1$ and $track(y) = e_2$.

(12)   **else return**$(2k + 1, (v, u))$

(13) **end.**

Before going into the implementation details consider the examples of Figure 5. In the first one, a $C_4^{(1)}$ minus a pendant vertex, the algorithm correctly stops in step (6). In the second one both first vertices of both LexBFS–orderings have odd eccentricity. Thus we must compute the $track$–values and the set F.

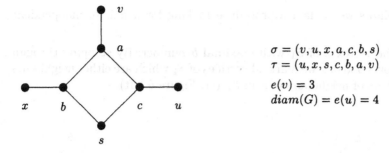

$$\sigma = (v, u, x, a, c, b, s)$$
$$\tau = (u, x, s, c, b, a, v)$$

$$e(v) = 3$$
$$diam(G) = e(u) = 4$$

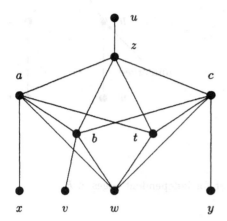

$$\sigma = (v, x, y, w, b, a, c, t, z, u)$$
$$\tau = (u, x, y, t, z, a, c, w, b, v)$$

$$e(v) = e(u) = 3$$
$$diam(G) = d(x, y) = 4$$

$$F = \{xa, yc, vb, wb\}$$
$$xa, yc \text{ independent}$$

Fig. 5. Algorithm DHGDiam — Examples.

It remains to show that the above algorithm can be implemented to run in linear time. It is well–known that LexBFS and BFS run in linear time. So it is sufficient to consider steps (9) and (10).

Step (9). At first we build a BFS–tree rooted at z yielding the set of neighbourhoods $N^i(z)$, $i = 0, \ldots, e(z)$ of z. For any vertex $x \in V \setminus \{z\}$ let $f(x)$ denote the father of x in the BFS–tree.

We compute the *track*–values levelwise : For all vertices w in $N^2(z)$ define $track(w) := wy$ where $y = f(w)$. Recursively we compute $track(w) := track(f(w))$ for $w \in N^i(z)$, $i = 3, \ldots, e(z)$.

Now we can compute F by collecting all *track*–edges of the vertices of the set $V \setminus D(z, k)$. Obviously the above procedure runs in linear time.

Step (10). We use the BFS–tree rooted at z which was already computed in step (9). Let $b : V \to \mathbb{N}$ be the numbering of the vertices of G produced by BFS where $b(z) = 1$. Let S_1 (S_2) be the vertices of $N(z)$ ($N^2(z)$) which are endpoints of edges of F.

In what follows we explain a procedure looking for a pair of independent edges :

Consider the vertex x of S_1 with maximal b–number. By stepping through the neighbourhood of x we mark all vertices of S_1 which are either neighbours of x or fathers of neighbours of x in S_2 (cf. Figure 6 left).

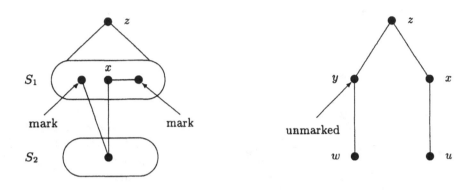

Fig. 6. Algorithm DHGDiam — Test for independent edges in F.

If there is an unmarked vertex $y \in S_1$ then there must be a neighbour w of y in S_2. We claim that the edges yw and xu, for some neighbour u of x in S_2, are independent (cf. Figure 6 right).

Indeed, since y is unmarked we must have $xw \notin E$ and $xy \notin E$. Since $b(x) > b(y)$, $x = f(u)$ and $y = f(w)$ the rules of BFS imply $uy \notin E$ (if $uy \in E$ then $f(u) = y$). Now $uw \notin E$ for otherwise the set $\{z, x, y, w, u\}$ induces a cycle of length five. Therefore, edges yw and xu are independent.

Now assume that all vertices of S_1 are marked. Then x cannot be an endpoint of a pair of independent edges. So we delete x from S_1 and all neighbours of x of S_2. We repeat the above procedure until we get a pair of independent edges or S_1 is empty.

Since processing a vertex x of S_1 takes $O(deg(x))$ the total running time of step (10) is linear.

Summarizing the above we get

Theorem 25. *For distance–hereditary graphs the diameter and a diametral pair of vertices can be computed in linear time.*

References

1. H.J. BANDELT, A. HENKMANN and F. NICOLAI, Powers of distance–hereditary graphs, *Discr. Math.* 145 (1995), 37-60.
2. H.-J. BANDELT and H.M. MULDER, Distance–hereditary graphs, *Journal of Combin. Theory (B)* 41 (1986), 182-208.
3. A. BRANDSTÄDT, V.D. CHEPOI and F.F. DRAGAN, Perfect elimination orderings of chordal powers of graphs, *Technical Report* Gerhard–Mercator–Universität – Gesamthochschule Duisburg SM–DU–252, 1994 (to appear in *Discr. Math.*).
4. A. BRANDSTÄDT, F.F. DRAGAN, V.D. CHEPOI and V.I. VOLOSHIN, Dually chordal graphs, *Proc. of WG'93, Springer, Lecture Notes in Computer Science* 790 (1994), 237-251.
5. A. BRANDSTÄDT, F.F. DRAGAN and V.B. LE, Induced cycles and odd powers of graphs, *Technical Report* Universität Rostock CS-09-95, 1995.
6. A. BRANDSTÄDT, F.F. DRAGAN and F. NICOLAI, LexBFS–orderings and powers of chordal graphs, *Technical Report* Gerhard–Mercator–Universität – Gesamthochschule Duisburg SM–DU–287, 1995 (to appear in *Discr. Math.*).
7. R. CHANDRASEKARAN and A. DOUGHETY, Location on tree networks: p–center and q–dispersion problems, *Math. Oper. Res.* 6 (1981), No. 1, 50–57.
8. G.J. CHANG and G.L. NEMHAUSER, The k–domination and k–stability problems on sun-free chordal graphs, *SIAM J. Algebraic and Discrete Methods*, 5 (1984), 332–345.
9. G. CHARTRAND and D.C. KAY, A characterization of certain ptolemaic graphs, *Canad. Journal Math.* 17 (1965), 342-346.
10. V. CHVATAL, Perfectly orderable graphs, *Annals of Discrete Math.* 21 (1984), 63–65.
11. A. D'ATRI and M. MOSCARINI, Distance–hereditary graphs, Steiner trees and connected domination, *SIAM J. Computing* 17 (1988), 521–538.
12. F.F. DRAGAN, Dominating cliques in distance–hereditary graphs, Proceedings of SWAT'94, *Springer, Lecture Notes in Computer Science* 824, 370–381.
13. F.F. DRAGAN and F. NICOLAI, LexBFS–orderings of distance–hereditary graphs, *Technical Report* Gerhard–Mercator–Universität – Gesamthochschule Duisburg SM–DU–303, 1995.
14. F.F. DRAGAN and F. NICOLAI, LexBFS–orderings and powers of HHD–free graphs, *Technical Report* Gerhard–Mercator–Universität – Gesamthochschule Duisburg SM–DU–322, 1996.
15. F.F. DRAGAN, F. NICOLAI and A. BRANDSTÄDT, Convexity and HHD–free graphs, *Technical Report* Gerhard–Mercator–Universität – Gesamthochschule Duisburg SM–DU–290, 1995.
16. F.F. DRAGAN, F. NICOLAI and A. BRANDSTÄDT, Powers of HHD–free graphs, *Technical Report* Gerhard–Mercator–Universität – Gesamthochschule Duisburg SM–DU–315, 1995.
17. F.F. DRAGAN, C.F. PRISACARU and V.D. CHEPOI, Location problems in graphs and the Helly property (in Russian), *Discrete Mathematics, Moscow*, 4 (1992), 67–73.
18. P. DUCHET, Classical perfect graphs, *Annals of Discr. Math.* 21 (1984), 67–96.
19. M. FARBER and R.E. JAMISON, Convexity in graphs and hypergraphs, *SIAM Journal Alg. Discrete Meth.* 7, 3 (1986), 433–444.
20. M. GIONFRIDDO, A short survey on some generalized colourings of graphs, *Ars Comb.* 30 (1986), 275–284.

21. M.C. GOLUMBIC, Algorithmic Graph Theory and Perfect Graphs, *Academic Press*, *New York* 1980.
22. P.L. HAMMER and F. MAFFRAY, Completely separable graphs, *Discr. Appl. Math.* 27 (1990), 85–99.
23. E. HOWORKA, A characterization of distance–hereditary graphs, *Quart. J. Math. Oxford* Ser. 2, 28 (1977), 417-420.
24. E. HOWORKA, A characterization of ptolemaic graphs, *Journal of Graph Theory* 5 (1981), 323-331.
25. B. JAMISON and S. OLARIU, On the semi–perfect elimination, *Advances in Applied Math.* 9 (1988), 364–376.
26. T.R. JENSEN and B. TOFT, Graph coloring problems, *Wiley* 1995.
27. T. KLOKS, D. KRATSCH and H. MÜLLER, Approximating the bandwidth for AT–free graphs, *Proceedings of European Symposium on Algorithms ESA'95, Springer, Lecture Notes in Computer Science* 979 (1995), 434–447.
28. R. LASKAR and D.R. SHIER, On powers and centers of chordal graphs, *Discr. Appl. Math.* 6 (1983), 139–147.
29. F. NICOLAI, A hypertree characterization of distance–hereditary graphs, *Technical Report* Gerhard–Mercator–Universität – Gesamthochschule Duisburg SM–DU–255 1994.
30. F. NICOLAI, Hamiltonian problems on distance–hereditary graphs, *Technical Report* Gerhard–Mercator–Universität – Gesamthochschule Duisburg SM–DU–264, 1994.
31. S. OLARIU, Weak bipolarizable graphs, *Discr. Math.* 74 (1989), 159–171.
32. D. ROSE, R.E. TARJAN and G. LUEKER, Algorithmic aspects on vertex elimination on graphs, *SIAM J. Computing* 5 (1976), 266–283.
33. R.E. TARJAN and M. YANNAKAKIS, Simple linear time algorithms to test chordality of graphs, test acyclicity of hypergraphs, and selectively reduce acyclic hypergraphs, *SIAM J. Computing* 13, 3 (1984), 566–579.

Efficient Union-Find for Planar Graphs and Other Sparse Graph Classes

(Extended Abstract)

Jens Gustedt*

TU Berlin, Sekr. MA 6-1, D-10623 Berlin – Germany

Abstract. We solve the Union-Find problem (UF) efficiently for the case the input is restricted to several graph classes, namely partial k-trees for any fixed k, d-dimensional grids for any fixed dimension d and for planar graphs. For the later we develop a technique of decomposing such a graph into small subgraphs, patching, that might be useful for other algorithmic problems on planar graphs, too.

By efficiency we do not only mean "linear time" in a theoretical setting but also a practical reorganization of memory such that a dynamic data structures for UF is allocated consecutively and thus to reduce the amount of page fault produced by UF implementations drastically.

1 Introduction and Overview

In this paper we present a new idea for the Union-Find Problem, *UF* for short, that extends the approach of Gabow & Tarjan given in [1]. In this paper the authors efficiently solve the UF problem where the Unions that are allowed form a tree: in linear time on a **RAM** .

Here we attack this in a more general way. Given is a graph G. The objective is to perform Unions and Finds on the vertices $V(G)$ such that for every step the actual subsets that are obtained form connected subgraphs of G.

Clearly that, if $G = K_n$ we are back to the usual UF problem with no restrictions at all; at least if we ignore complexity issues for a moment. In general we will have some family $\mathcal{G}$ of graphs and we will call the problem that the graph G might be arbitrarily chosen from $\mathcal{G}$ the $\mathcal{G}$-**graphical** Union-Find **Problem**, $\mathcal{G}UF$. Then the idea is, that when assuming that such an instance G is given explicitly[1], we might avoid the lower bounds of the complexity of the UF problem, see [2,3].

The method we use (and develop in Sections 2 and 3) is an extension of the one given by Gabow & Tarjan, namely by

(i) solving the problem for "small" sets in a preprocessing,
(ii) then dividing the instance by a "neglectable" portion of the graph, a **skeleton**, into such small sets, the **clusters**, and

---

* Supported by the IFP "Digitale Filter".

[1] So the input has at least a size as large as the number of edges of G.

(iii) giving a technique how to perform UF as a combination of local information in a cluster and global information in the skeleton.

This technique can also be seen as a method of **reorganizing memory** in order to reduce memory faults or delay and thus improve the real time behavior of UF data structures.

The real world bottleneck for Union-Find is the use of a dynamic data structure. The usual path compression data structure makes no guarantee at all where in memory the next item to fetch might be located. So if we are doing Unions more or less arbitrary the **real** processing time of the algorithm is dominated by loading data, either from memory into cache or —even worse— from disk into memory.

Since for such practical considerations the distinction between random access machines (RAM) and pointer machines (PM) is rather academic we restrict ourselves to the RAM. This aspect is briefly explained in Section 7.

The progress made here in this work for this class of problems is summarized by the following theorem.

Main Theorem. $\mathcal{G}UF$ *is solvable on a RAM in time proportional to the number of* Find*s for* $\mathcal{G}$ *any of the following classes of graphs:*

(i) Trees and partial k-trees, for any fixed parameter k.
(ii) d-dimensional grids for fixed d and 8-neighborhood graphs of a 2-dimensional grids.
(iii) Planar graphs.

Just to give an example of an application of this theorem consider the problem of computing minimum spanning trees. Suppose we are in a situation that we have graph with given edge weights such that in addition the sort order of the edges is known (or easy to compute). With Kruskal's algorithm our Main Theorem then immediately gives linear time bounds if we restrict ourselves to the graph classes in question.

For the proof of the Main Theorem the three parts are handled in Sections 4, 5 and 6. For the scope of this introduction we briefly describe the parts of the Main Theorem in the following three paragraphs.

Partial k-trees The result obtained for partial k-trees is a straight forward generalization of the work of Gabow and Tarjan. It is mainly chosen as a first illustration of the power of this approach.

Grids and UF for Image Segmentation An important application will be that the underlying graph G is e.g a grid as it appears in image segmentation. In image segmentation the goal is to group a digital image into homogeneous connected regions, so called segments. An important technique to do this is region growing: starting from one-point segments and gluing together neighboring segments if appropriate.

Up to now there are only two special cases where the complexity of this approach is known to be linear, see [4,5]. These special cases strongly restrict the order in which Unions may be performed (called **scanning order**, [4]) and the neighborhood definition that is used for the connectivity property. One of our goals here is to extend this to *arbitrary scanning orders*, to *other neighborhood definitions* on digital images and even to 3-dimensional images.

Planar Graphs For planar graphs we develop a technique, called patching, of decomposing such a graph into small subgraphs that might be useful for other algorithmic problems on planar graphs, too. A patching is a separator of the graph of negligible size, i.e smaller than $n/\text{slow}(n)$ for some growing function slow, that separates the graph into small components.

2 Basic Definitions and Facts

2.1 Notations

Graphs are simple and without loops or isolated vertices. For a graph G, $V(G)$ and $E(G)$ denote the vertex and edge sets. The degree of a vertex v is denoted by $\deg_G(v)$, $\deg_G$ is the maximum degree over all vertices. $d_G(u, w)$ denotes the distance between two vertices u and w, i.e the number of edges on a shortest path between u and w. The notation $d_G(U, W)$ is the obvious extension to arbitrary vertex sets U and W. For some vertex set U and some value p we define the p-**neighborhood** of U as

$$N_G^{\text{p}}(U) = \{v \in V \mid d_G(v, U) \le \text{p}\}. \tag{1}$$

We will always assume that graphs that are given as instances are given **explicitly** as lists of edges, say. So in particular we always have an input size that is proportional to the number of edges.

The problem we are dealing with, is given by the following specification.

Problem 1 (Graphical Union-Find *Problem, GUF).*
Instance: A graph G.
Task: Perform a sequence of $n < |V(G)|$ Unions and $m \ge |V(G)|$ Finds on $V(G)$ that respect G.

Here a sequence of Unions and Finds **respects** a graph G if after each Union every subset created induces a connected subgraph of G. This is equivalent of saying that every Union can be realized as an edge contraction in G and thus that the actual state of the UF is always represented by a **minor** of G. So we easily obtain the following remark.

Remark 2. Let G^- be a minor of G^+ such that a sequence of edge contractions and deletions that lead from G^+ to G^- is explicitly given. Suppose GUF is solvable for G^+ in time $t(n, m)$ then any sequence of $n^- < |V(G^-)|$ Unions and $m^- \ge |V(G^-)|$ Finds on $V(G^-)$ that respects G^- is solvable in time $t(n^- + |V(G^+)|, m^-)$.

Observe that this means in particular that if the size of G^+ is linear in the size of G^- GUF may be solved on G^- in the same complexity as on G^+.

We will investigate the GUF problem restricted to certain graph classes. If $\mathcal{G}$ is a class of graphs $\mathcal{G}$UF refers to GUF restricted to $\mathcal{G}$. To warm up let us consider the problem $\mathcal{G}^3$UF, where $\mathcal{G}^3$ is the class of graphs of degree bounded by 3.

Remark 3. GUF is solvable in linear time iff $\mathcal{G}^3$UF is solvable in linear time.

Proof. "$\Longrightarrow$" is trivial. For "$\Longleftarrow$" replace in instance G any vertex of degree higher than 3 by an appropriate binary tree. By that we easily obtain a graph G' that fulfills the requirements of Remark 2 and is at most twice as large as G. $\square$

To simplify the discussion in the rest of the paper a bit we will always assume that $n = |V(G)| - 1$. We will also assume that the demand for a Union is presented by pointing out an edge of G for which the components/subsets of the endpoints should be glued into one; *the possible question whether or not two current subsets may be united or not is not part of the problem specification.*

2.2 Slowly Growing Functions

We will reduce the problem so that the number of edges in the underlying graph must not be too large compared to the number of vertices. Therefore throughout the following we use the notation of slow(n) for a *slowly* growing function. By that we mean a monotone function that at least fulfills

$$\text{slow}(n) \le (1/18) \log \log(n) \tag{2}$$

and is dominating α the inverse of the Ackerman function. We use a definition for that function α that turns out to be basically (in O-notation) the same as the traditional one but is a bit simpler to handle:

$$\alpha(m, n) = \min \left\{ x \mid A(x, \lceil m/n \rceil) > n \right\}, \tag{3}$$

where A is the Ackermann function given by the usual recursion.

Observe that α is increasing in the second argument but decreasing in the first.

We require for slow that there is a constant c such that $A(c, \lceil \text{slow}(n) \rceil) > n$ for all n. In particular this means that

$$\alpha(\text{slow}(n)n, n) \le c. \tag{4}$$

Observe that there are many commonly used functions that fulfill these requirements, e.g (almost[2]) any iterated log-function or log*. For these functions choosing c to be 2 can be easily seen to do the job. In the rest of the paper we will

---

[2] almost for the magic constant 1/18 in (2) that is needed for an estimation later on

assume that such a function slow with corresponding constant c is chosen and fixed.

Recall that by the work of Tarjan, see [6], we know that any UF problem can be solved in time proportional to $\alpha(m, n)m$. As a corollary from that we easily obtain:

Corollary 4. *Let G be an instance of GUF that is given explicitly as input and $n = |V(G)|$ with $|E(G)| > \alpha(n, n)n$. Then any sequence of Unions and Finds that respects G can be performed in linear time.*

Together with Remark 3 this shows in particular that it makes sense to restrict our discussion to classes of sparse graphs.

2.3 Micro-Encoding of Small Sets

A basic method will be to encode UF problems on small subsets of the ground set by bit-vectors, so called **micro-encoding**. Here by "small" we mean sets that are smaller than a function ℓ with[3]

$$\ell(n) \leq \sqrt[3]{\log n}. \tag{5}$$

Such micro-encodings will then be used to do UF on small subsets of the groundset. Therefore let V with $|V| = n$ be a set and $V_1, V_2, \ldots$ a partition of V. We say that a sequence of Unions and Finds on V **respects** the partition if for any subset U produced by those Unions there is an i such that $U \subseteq V_i$. Our aim is the following lemma.

Lemma 5. *Let V with $|V| = n$ be a set and $V_1, V_2, \ldots$ a partition of V such that $|V_i| \leq \ell(n)$ for all i. Then any sequence of Unions and Finds on V that respects the partition can be implemented in linear time.*

Proof. To have all operations in constant time we maintain tables for the Union and the Find operation. For the first the input needed is a current state and two elements, and the output is the resulting state.

Here a **state** represents all information needed for such problems of sets of size less than $\ell(n)$. This can be done by representing each state by $\ell(n)$ numbers of size $\ell(n)$. Thus each such state may then be encoded in $\ell(n) \log \ell(n)$ bits which can be bounded as follows:

$$\ell(n) \log \ell(n) \leq 1/3 \log \log(n) \sqrt[3]{\log(n)} \leq 1/3 \log(n), \tag{6}$$

if n is large enough.

It is then easy to encode such a UF tree in a bit vector. To obtain the right complexity (i.e constant time per Union and Find) observe that the number of possible states is less than $2^{1/3 \log(n)} \leq \sqrt[3]{n}$. So clearly the size of such tables for Union and Find can be bounded by $(\sqrt[3]{n})^2 \log^2 \ell(n)$ which in turn is dominated by n if n is suitable large. The preprocessing to build up all tables consistently is also easily seen to lay in that bound. $\qquad\square$

---

[3] ℓ stands for large, since it will be much larger than slow.

Observe that a concrete choice of ℓ has not been necessary for the proof, we only needed the upper bound (5).

3 Union-Find by Clustering

For a graph $G = (V, E)$ and $n = |V(G)|$ a **skeleton**, skel $=$ skel(G), is a vertex set with cardinality bounded by $n/\text{slow}(n)$. A **cluster** C of G with respect to skel is then a component of $G \setminus$ skel. The **boundary** $\partial(C)$ of C are those $v \in$ skel that have an edge to a vertex in C. The **closure** $\Delta(C)$ of C is $\Delta(C) = C \cup \partial(C) = N_G^1(C)$.

Denote by $\mathcal{C}_{\text{skel}}$ the **clustering** w.r.t skel, i.e the family of components of $G \setminus$ skel. A clustering $\mathcal{C}$ is **valid** if in addition $|\Delta(C)| \le \ell(n)$ for all $C \in \mathcal{C}$.

Because every edge either joins two elements of skel or is inside exactly one $\Delta(C)$ for some cluster C and every such $\Delta(C)$ is connected we have that

$$\sum_{C \in \mathcal{C}} |\Delta(C)| \le |E(G)| \tag{7}$$

and thus we may encode any clustering with an additional overhead that is linear in $|E(G)|$.

We are now able to formulate a UF data structure for that context, see Algorithms 1 and 2. We assume that for each $\Delta(C)$ we maintain a local (micro-encoded) UF data structure and a global (path compression) one on skel. For the later we assume that if $v \ne w$ are the representatives of their sets and we perform Union(v, w) the representative of the newly created set will be v, i.e the first argument to Union.

Algorithm 1 (Find).

Input: Vertex $v \in V(G)$.
Output: Root $r \in V(G)$ a unique identifier for the current subset of v.
1. **if** $v \notin$ skel **then** $v := \text{Find}_{\Delta(C(v))} v$
2. **else return** $\text{Find}_{\text{skel}} v$
3. **if** $v \in$ skel **then return** $\text{Find}_{\text{skel}} v$
4. **else return** v

First observe that Find for some subset S indeed gives a unique identifier $r = r(S) \in S$. Because of the interchange of v and w in line **swap** of Algorithm 2 it is also easy to see that the identifier of a subset is an element of skel whenever this is possible.

Remark 6. For any subset S created by Unions with $S \cap$ skel $\ne \emptyset$ we have that $r(S) \in$ skel.

Lemma 7. *Let $G = (V, E)$ be an instance of GUF s.t a valid clustering $\mathcal{C}$ of G is given. Then the above data structure solves this problem for any valid sequence of $n < |V|$ Unions and $m \ge |V|$ Finds in time $O(n + m)$ on a RAM.*

Proof. Correctness follows directly from Remark 6. It remains to show the complexity. We have $O(n + m)$ calls to the operations of the local UF data structure

Algorithm 2 (Union).

Input: Edge $\{v, w\} \in E(G)$.
local:
  1. **if** $v \notin$ skel **then** $v := \text{Find}_{\Delta(\mathcal{C}(v))} v$
  2. **if** $w \notin$ skel **then** $w := \text{Find}_{\Delta(\mathcal{C}(w))} w$
global:
  3. **if** $v \in$ skel **then** $v := \text{Find}_{\text{skel}} v$
  4. **if** $w \in$ skel **then** $w := \text{Find}_{\text{skel}} w$
do it:
  5. **if** $v \neq w$ **then**
  6. **if** $w \notin$ skel **then** $\text{Union}_{\Delta(\mathcal{C}(w))}(v, w)$
swap: else if $v \notin$ skel **then** $\text{Union}_{\Delta(\mathcal{C}(v))}(w, v)$
  7. **else** $\text{Union}_{\text{skel}}(v, w)$

and an over all $O(n)$ preprocessing time, so the local data structures poses no problems.

In addition we have $m + 2n$ Finds on the global data structure. This global data structure has at most $n' = n/\text{slow}(n)$ elements and in particular we can't perform more than n' Unions here. We find that

$$m \geq n > n' = n/\text{slow}(n) \tag{8}$$
$$m/n' \geq n/n' = \text{slow}(n) \geq \text{slow}(n'). \tag{9}$$

So the complexity of the global Unions and Finds is bounded by

$$\alpha(m + 2n, n')m \leq \alpha(m, n')m \leq \alpha\left(\text{slow}(n')n', n'\right)m. \tag{10}$$

Since $\text{slow}(n')$ is suitably growing, by (4) we have that the right hand side is bounded by $c \cdot m$, and we are done. $\qquad\square$

4 Partial k-Trees

Let us now prove part (i) of the Main Theorem. Since the ideas are just straight forward extensions of the ideas of Gabow & Tarjan, we give the proof in an informal way.

Let us first concentrate on the case that is settled by Gabow & Tarjan, namely trees. The key observation is to restrict ourselves to the case where the degree of the instances is bounded by 3, say. This can easily be achieved as described for Remark 3. If we now root such an instance at an arbitrary vertex r we may easily collect subtrees as clusters of size s with $\text{slow}(n) \leq s < 2\,\text{slow}(n)$ from leaves to r and cut of such subtrees by their root. So for a choice of $\ell(n) = 2\,\text{slow}(n)$ Lemma 7 easily applies to that case.

To prove the theorem for partial k-trees, recall that k is assumed to be a fixed constant. From Bodlaender, [7], we know that a tree decomposition of width k for such a graph can be found in linear time and it is also well known that such a decomposition may be easily adapted such that the underlying tree is binary. Since every vertex of the decomposition tree represents a separator of size at most $k + 1$ (and k is fixed) the ideas for trees immediately extend: collect subgraphs of size s with $\text{slow}(n) \leq s < 2\,\text{slow}(n) + k$ corresponding to subtrees of the decomposition tree and cut them of the graph by the separator corresponding to the root of the subtree.

5 d-dimensional Grids and Relatives

Recall from the introduction that one of our goals is to provide UF algorithms for image segmentation. This in fact is easily modeled as graphical UF problem restricted to 2 or 3-dimensional grids. Recall also that we want to be able to attack arbitrary sequences of UF operations (as long as they respect the grid) with generalized neighborhood relations inside the grid.

Besides the usual neighborhood definition on grids another one that also is widely used is the so called 8-neighborhood which connects a pixel to all 8 pixels surrounding it: left, right, up, down, left-up, right-up, left-down, right-down. The arguments used in [5,4] to show linear time complexity heavily rely on the fact that the underlying graph is planar which is not the case for 8-neighborhood any more. So there is no hope to extend this to the 8-neighborhood.

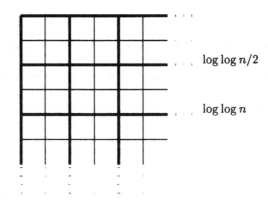

$\log \log n/2$

$\log \log n$

Fig. 1. The skeleton in a 2-dimensional grid

Proof (Main Theorem, part (ii)). Let d be some fixed integer and the instance of the GUF G be a d-dimensional grid. Assume that the vertices are identified with their position $(p_1, \dots, p_d)$ in the grid. An appropriate skeleton of G can easily be defined.

$$\text{skel}(G) = \{(p_1, \dots, p_d) \in V(G) \mid \text{there is } i \text{ s.t } p_i = 0 \bmod \log \log(n)/d\} \quad (11)$$

We easily obtain that $|\mathrm{skel}(G)|$ is bounded by $n/\log\log(n)$ and that for every connected component K of $G \setminus \mathrm{skel}(G)$

$$|\Delta K| \le (1 + \log\log(n)/d)^d. \tag{12}$$

So if we choose $\ell(n)$ to be $(1 + \log\log(n)/d)^d$ the part of the statement for grids follows easily with Lemma 7. To solve the problem on the 8-neighborhood in a 2-dimensional grid (or a higher dimensional analogue) observe that the skeleton given above still is a separator and so there is nothing new to prove. $\square$

6 Planar Graphs

A natural approach to apply our technique to planar graphs would be to use the Planar Separator Theorem, see [8], by dividing the graph into two halves and going on recursively in each of the halves until we remain with parts that are small enough. But for two reasons a straight forward application of that theorem doesn't lead to the desired result. First the running time guaranteed by that approach would only lead to $O(n \log n)$ which is much too bad for our purposes. Second, it is not easy to establish an appropriate bound on the size of the skeleton; by a single application of that theorem we obtain a separator of size $\sqrt{n}$ and a partition into unequal halves. So we are only able to show that each level l of recursion in total contributes $\sqrt{D^l}\sqrt{n}$ for some $D > 2$ to the skeleton. But this is only leads to a super-linear bound on the size of the skeleton.

So we follow another way to solve GUF if the input graph is planar. What we would like to do is to proceed analogously as for 2-dimensional grids. But this is also not as straight forward as one could hope. We have to introduce some technical definitions (p-patches) that will play the (key) role that the tiling squares of size $\log\log(n)/2$ played for the grids.

6.1 p-Patches

For some positive value p a p-**patch** is a planar graph $G = (V, E)$ together with a designated face $F = F_G$, the **boundary**, and an interval $I_G \subset F$, the **lower boundary**, such that

 (i) all faces but F_G are triangles,
 (ii) $\deg_G \le 8$,
 (iii) $d_G(v, I) \le p$ for all $v \in V$ and
 (iv) $|F \setminus I| \le 3p \cdot 5^p$.

The constants 3, 5 and 8 appearing in this definition are more or less arbitrary and chosen to make the estimations easy. We call $\Upsilon_G = F_G \setminus I_G$ the **upper boundary** of G. An p-patch is **narrow** if (iv) is enforced to $\Upsilon_G \le 3p$. We will omit the subscript G at F, I and Υ whenever possible.

Let G and G' be p-patches. G' is a p-**subpatch** of G if it is a subgraph of G and if $I_{G'} = I_G|_{V(G)}$. Observe that if $G' \ne G$ then also $F_{G'} \ne F_G$. Important properties of p-patches are reflected by the following lemmas, see Figure 2 for an illustration.

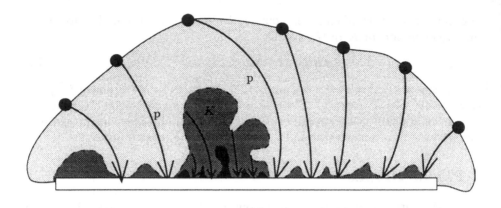

Fig. 2. The p-neighborhood of Υ

Lemma 8. *Let G be an p-patch then $|V(G \setminus N_G^p(\Upsilon))| \leq 15p \cdot 5^{2p}$.*

Proof. This is easy to see because of properties (i) and (ii). The (not so important) constant 5 comes from the fact that when coming from Υ at least one edges is needed as back edge and at least two are needed as side edges because of the triangulation, enforced by (i). □

Lemma 9. *Let G be an p-patch and K be a connected component of $G \setminus N_G^p(\Upsilon)$ and $G' = N_G^1(K)$. Then G' with $I_{G'} = I_G|_{V(G)}$ is p-subpatch of G.*

Proof. Consider Figure 2. Here the vertical bar represents I_G, the filled circles represent elements of Υ_G and the dashed line represents the boundary of the connected components of $G \setminus N_G^p(\Upsilon)$.

We have to show that $\Upsilon_{G'}$ is not too large. Because (iii) of the definition we have that for every $v \in \Upsilon$ a path P from v to I is included in $N_G^p(\Upsilon)$. Since G is planar, such a path separates G and thus $G \setminus N_G^p(\Upsilon)$ into two different components. But now the component K under investigation can be cut of by two such paths P, one on the left and one on the right. Thus there is $V' \subseteq \Upsilon$ with $|V'| \leq 2$ such that K can already be found as $G \setminus N_G^p(V')$.

But this means also that $\Upsilon_{G'} \subseteq N_G^p(V')$ and thus because of the degree constraints that $|\Upsilon_{G'}| \leq 2 \cdot 5^p$. □

Lemma 10. *There is an algorithm that, given an integer p and a narrow p-patch G as input, runs in linear time and finds a skeleton $\mathrm{skel}\, G$ of size $\leq |V(G)|/p$ such that every cluster is of size at most $15p \cdot 5^{2p}$.*

Proof. Compute a BFS-tree growing from Υ and calculate the cardinalities of each level i of that tree. For each $j = 0, \dots, p-1$ sum up the cardinalities of all levels i with $i \equiv j \mod p$. There exists a j_0 such that this sum of cardinalities is $\leq |V(G)|/p$. The union of all corresponding levels gives our desired skeleton.

Because G is narrow, level j_0 itself has at most $3p \cdot 5^{j_0} \leq 3p \cdot 5^p$ elements and thus $G \setminus N_G^{j_0}(\Upsilon)$ is an p-patch and $N_G^{j_0}(\Upsilon)$ is not too large. Now applying Lemma 9 iteratively shows that $G \setminus N_G^{j_0}(\Upsilon)$ breaks down into p-patches and Lemma 8 shows that these have the appropriate sizes. □

Observe also, that the multiplicative constant hidden in the linear time bound does not depend on p.

6.2   Patching an Arbitrary Planar Graph

To apply these methods to an arbitrary planar graph we first have to ensure that we always may find a major that is triangulated and has bounded degree, see Algorithm 3.

Algorithm 3 (Bound Degree).

Input: arbitrary planar graph G.
Output: Triangulated planar graph G' of degree bounded by 8 such that G is
   a minor of G' and such that $|V(G')| \leq 10\,|V(G)|$.
   1. Triangulate G.
   2. Isolate vertices of high degree, see Fig. 3(a)
   3. Replace the neighborhood of high degree vertices, see Fig. 3(b).

Clearly triangulation poses no problem at all. **Isolation** of vertices of high degree is done in such a way that afterwards we have

(i) all vertices v with $\deg(v) > 8$ only have neighbors of degree 6 or less
(ii) every vertex w has at most 2 neighbors v with $\deg(v) > 8$.

This can easily be achieved by dividing each edge by a new vertex and joining all these new vertices to the other 4 new vertices on the neighboring facets. It is clear that now

(i) every old vertex only has new vertices as neighbors,
(ii) every new vertex has degree 6,
(iii) every such new vertex has 2 old vertices as neighbors and
(iv) the resulting graph is triangulated again, see Fig. 3(a).

Replacement at vertices of high degree is done as follows. Let v be a vertex of degree larger than 8 and w and w' two neighbors that form a triangle together with v. Then

(i) delete the edges vw and vw' from G
(ii) introduce a new vertex v' and join it to v, w and w'.

Do so for all adequate pairs of edges of v. These are $\lfloor \deg(v)/2 \rfloor$. If $\deg(v)$ is odd divide the remaining edge by yet another new vertex. Now join all new vertices obtained in this step to a cycle, see Fig. 3(b). Up to now all new vertices

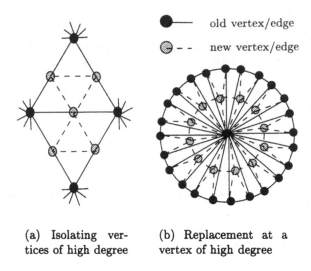

old vertex/edge

new vertex/edge

(a) Isolating vertices of high degree

(b) Replacement at a vertex of high degree

Fig. 3. Bounding the Degree

have degree at most 5 and all old vertices (but v) have their degree unchanged. To triangulate this newly obtained graph observe that the only facets that are not triangles are those with two new vertices and two former neighbors of v. These squares can be triangulated by diagonals that all go in the same direction, to the right, say. Now all new vertices have degree 6 and all former neighbors of v have their degree increased by one.

We iterate this procedure until $deg(v) \leq 8$. Then all new vertices introduce have degree at most 7.

We then loop for all vertices of high degree. Any vertex w that originally had degree at most 6 is involved into such a replacement for at most 2 vertices of high degree. So the new degree of w is at most 8. To summarize we state

Proposition 11. *Let G be a planar graph then there is a triangulated planar graph G' of degree not larger than 8 such that*

(i) G is a minor of G'
(ii) $|V(G')| \leq 10\,|V(G)|$

and such a graph G' can be found in linear time.

Proof (Main Theorem, part (iii)). If we chose $p = p(n) = slow(n)$ and $\ell(n) = 15\,slow(n)5^{2\,slow(n)}$ we have by (2) that

$$\ell(n) \leq \frac{5}{6} \cdot \log\log(n)5^{\frac{1}{9}\cdot\log\log(n)} \leq \frac{5}{6} \cdot \log\log(n)\log(n)^{\frac{\log 5}{9}} < \sqrt[3]{\log(n)} \qquad (13)$$

and thus Lemma 7 and 10 show the claim for narrow $p(n)$-patches. With Proposition 11 it remains to show how to obtain narrow p-patches from a triangulated

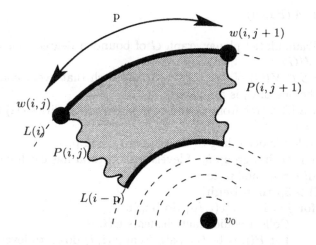

Fig. 4. Patch on Level i

planar graph of bounded degree 8. We mainly use the same idea that was used to split narrow p-patches themselves, see Algorithm 4.

Now every connected component of $G \setminus S$ (but the one of v_0) is a narrow p-patch, see Figure 4. The only non-p-patch maybe the component of v_0 but this is of size at most 5^P, so it is already an appropriate cluster.

By line (5) we ensure that a level is only split into multiple patches if this is necessary. Observe that otherwise we possibly would collect too much vertices into the paths $P(i, j)$. So S as chosen above is not too large

$$|S| \leq \sum_{i=i_0 \bmod p} |L(i)| + \sum_{\substack{i=i_0 \bmod p \\ j=1,\ldots\lfloor c(i)/p \rfloor}} |P(i, j)|$$

$$\leq |V(G)|/p + \sum_{\substack{i=i_0 \bmod p \\ j=1\ldots\lfloor c(i)/p \rfloor}} p. \qquad (14)$$

Since this in turn is bounded by $\leq 2 |V(G)|/p$ we are done. $\qquad \square$

Algorithm 4 (Patch).

Input: Triangulated planar graph G of bounded degree 8, $n = |V(G)|$, vertex $v_0 \in V(G)$.

Output: $S \subset V(G)$ of size $|S| \le n/p$ and such that every component of $G \setminus S$ but the one of v_0 is a p-patch.

1. Grow a BFS-tree from v_0 and collect the cardinalities $c(i)$ of all levels $L(i)$, for $i = 1, \ldots,$.
2. Group levels according to their index $i \mod p$.
3. There is such a group i_0 of levels that in total has cardinality $\le |V(G)|/p$.
4. **for all** $i = i_0 \mod p$ **do**
5. **if** $c(i) > 3p$ **then begin**
6.   **for** $j = 1, \ldots, \lfloor c(i)/p \rfloor$ **do begin**
7.   Collect equidistant elements $w(i,j)$
8.   Let $P(i,j)$ be the path from $w(i,j)$ down to level $L(i - p)$.
9.   **end end**
10.

$$S = \bigcup_{i=i_0 \bmod p} L(i) \cup \bigcup_{\substack{i=i_0 \bmod p \\ j=1,\ldots \lfloor c(i)/p \rfloor}} P(i,j)$$

7 Perspectives

Practical Considerations: Reorganization of Memory

Clearly one disadvantage of our approach is the intensive use of a RAM as computational model. It is often claimed that using microencoding to handle "small" subproblems would only pay off in cases where the over all problem size is so huge, that we never expect it to occur in our limited universe. Therefore there have been intensive studies to avoid such use of the RAM and to restrict the computational model to a *pointer machine*, PM, see e.g [3].

On the other hand it is also often argued that α is such a slowly growing function that the factor is completely neglectable for practical purposes. So some people claim, that avoiding α-factors is a useless theoretical game.

Both criticisms go to short, since the real world bottleneck for doing UF on large data as it occurs e.g in image processing is the **von Neumann bottleneck**, i.e the different speed main memory and CPU have in modern computers. Thus the **bad thing** for practical purposes is creating programs that really do **random access** on the memory; in running such programs the CPU is idle most of the time and waiting for the memory to serve it with data.

Clearly that the inability to reflect this real world restriction applies to both, to the PM as well as to the RAM. So from that point of view, none of these models is preferable. On the other hand the approach of clusters as presented in this paper easily allows a **reorganization of memory** since it may cooperate with the strategies of modern machines to reduce the effect of the von Neumann bottleneck: its effect much less dramatic if memory access goes onto the same page several times.

For a conventional UF data structure, it is very unlikely that dereferencing a parent pointer of a UF element stays on the same page or even on a small group of pages. But if we are given a valid clustering and allocate each cluster continuously on subsequent pages of memory and do the same for the skeleton as a whole this behavior changes drastically. Then a Find may either

(i) stay inside a cluster and thus on the pages allocated for it or
(ii) lead into the skeleton.

Thus only a sublinear number of pages are potentially accessed. This is so, even if we do not impose any other restriction for the sequence of UF than to respect the underlying graph. If we even ensure that the sequence always proceeds with a neighbor of the current element the behavior will improve even further.

So we reduce the bottleneck of data location by a substantial factor, ie by the page size divided by the size of a UF element. On todays machines with a page size of 1KB we expect this factor to be more than 100.

Possible Generalizations

For graph algorithms people have in many cases been able to avoid the α-bottleneck of UF by looking closely a the graphical structure occurring in their particular problem. We have the impression, that there must be a common pattern behind all these attempts. In regard of Remark 3 and by hoping on the best of all worlds we conclude with the following guess.

Guess. *GUF is solvable in linear time on a RAM.*

References

1. Harold N. Gabow and Robert E. Tarjan. A linear-time algorithm for a special case of disjoint set union. *J. Comput. System Sci.*, 30:209–221, 1984.
2. Robert E. Tarjan. A class of algorithms which require non-linear time to maintain disjoint sets. *J. Comput. System Sci.*, 18:110–127, 1979.
3. J. A. La Poutré. Lower bounds for the union-find and the split-find problem on pointer machines (extended abstract). 1990.
4. Michael B. Dillencourt, Hanan Samet, and Markku Tamminen. A general approach to connected-component labeling for arbitrary image representations. *J. Assoc. Comput. Mach.*, 39(2):253–280, April 1992. Corr. p. 985-986.
5. Christophe Fiorio and Jens Gustedt. Two linear time Union-Find strategies for image processing. *Theoret. Comput. Sci.*, 154(2):165–181, February 1996.
6. Robert E. Tarjan. Efficiency of a good but not linear set union algorithm. *J. Assoc. Comput. Mach.*, 22:215–225, 1975.
7. Hans L. Bodlaender. A linear time algorithm for finding tree-decompositions of small treewidth. *in [9]*, 1993.
8. Richard J. Lipton and Robert Endre Tarjan. A separator theorem for planar graphs. *SIAM J. Appl. Math.*, 36:177–189, 1979.
9. In *Proceedings of the Twenty Fifth Anual ACM Symposion on Theory of Computing.* ACM, Assoc. for Comp. Machinery, 1993.

Switchbox Routing in VLSI Design: Closing the Complexity Gap

Extended Abstract

Stephan Hartmann, Markus W. Schäffter and Andreas S. Schulz

Technische Universität Berlin, Fachbereich Mathematik (MA 6–1)
Straße des 17. Juni 136, D–10623 Berlin, Germany
hartmann, shefta, schulz@math.TU-Berlin.DE

Abstract. The design of integrated circuits has achieved a great deal of attention in the last decade. In the routing phase, there have survived two open layout problems which are important from both the theoretical and the practical point of view. Up to now, switchbox routing has been known to be solvable in polynomial time when there are only 2–terminal nets, and to be NP–complete in case there exist nets involving at least five terminals. Our main result is that this problem is NP–complete even if no net has more than three terminals. Hence, from the theoretical perspective, the switchbox routing problem is completely settled.
The NP–completeness proof is based on a reduction from a special kind of the satisfiability problem. It is also possible to adopt our construction to channel routing which shows that this problem is NP–complete, even if each net does not consist of more than four terminals. This improves upon a result of Sarrafzadeh who showed the NP–completeness in case of nets with no more than five terminals.

1 Introduction

Very large scale integrated circuit layout (VLSI) is one of the amazingly growing areas in discrete mathematics and computing science of the last years, due to both its practical relevance and its importance as a trove of combinatorial problems. Usually, in VLSI design one distinguishes between the phase of placing physical components and the subsequent routing phase realizing the conducting connections between them.

The routing phase itself consists of the layout problem and the corresponding layer assignment. We refer the reader to the book of Lengauer [Len90] and to the survey of Möhring, Wagner, and Wagner [MWW95] for a detailed description of this process as well as for comprehensive surveys of the use of combinatorial and graph–theoretic methods in VLSI design. Here, we concentrate on the layout problem where the course of the wires to connect the cells in a single plane has to be determined.

Most generally, the problem is to find an edge–disjoint packing of Steiner trees in a given planar graph. To be more precise, we are given a graph $G = (V, E)$, the so–called *routing graph*, and k sets $N_1, \ldots, N_k \subseteq V$ called *nets*. In this context, the elements of the nets are referred to as *terminals*. The task is to find k pairwise edge–disjoint Steiner trees $T_1, \ldots, T_k \subseteq E$ such that T_i connects the terminals of net N_i, if they exist. A solution of the Steiner tree packing problem is called *layout*. For planar graphs, Kramer and van Leeuwen [KvL84] showed that the Steiner tree packing problem is NP–complete even if there are only two–terminal nets. Korte, Prömel and Steger [KPS90] complemented their result by proving the NP–completeness of the problem if there are only two multi–terminal nets. If all terminals are assigned to the outer face of the routing graph, Okamura and Seymour [OS81] gave sufficient conditions for instances that can be solved in polynomial time.

The routing graphs arising in VLSI design are actually very special planar graphs. Most frequently, they are rectangular grids, corresponding to the usual shape of the physical layout areas. Such routing problems have been attacked by quite different methods ranging from purely bottom–up methods over floor–planning techniques up to polyhedral combinatorics (see, e.g., [Len90,GMW93]). There are two types of problems on a grid which are of particular importance, namely *switchbox routing* and *channel routing*. In both cases, all terminals are placed on the boundary of the grid. In switchbox routing the terminals may be placed on all four sides. Channel routing is a special case of switchbox routing where the terminals are only placed on the lower and the upper side of the grid.

For the switchbox problem, Preparata and Mehlhorn [MP86] gave a polynomial time algorithm that constructs a layout, if all nets contain only two terminals. For the channel routing problem, Sarrafzadeh [Sar87] proved the NP–completeness if some of the nets involved have six or more terminals. He also claimed (without giving a proof) the NP–completeness of problems involving nets with at least five terminals. This implies the same result for switchbox routing. We show that switchbox routing is NP–complete even if all nets have at most three terminals. Hence, this paper closes the gap between the algorithm of Preparata and Mehlhorn on one side, and the NP–completeness result of Sarrafzadeh on the other side. As a consequence, heuristic algorithms are of interest for all instances of the switchbox routing problem that contain nets with more than 2 terminals. An overview of different heuristics can be found in [MS92]. It is also possible to transfer our construction to channel routing. This results in an NP–completeness proof for the case that every net has at most four terminals, see [Har96]. We would like to mention that our reduction is partially based on refinements of some of the ideas in [Szy85] and [Sar87].

The paper is organized as follows. In Section 2, we define the 3–bounded 3–SAT problem and the switchbox routing problem and give a first, introductory description of the transformation. The following sections discuss all the details of the transformation. In Section 5, we prove the correctness of our result. We conclude with some remarks in Section 6.

Due to space limitations some details are omitted from this paper. A complete version can be obtained from the authors.

2   A First Description of the Reduction

An instance of the switchbox routing problem consists of a *routing region* and a set of *nets*. The routing region is assumed to be a rectangular grid, called *switchbox*, with n vertical lines and m horizontal lines, also called *tracks*. The set of nets consists of k nets $N_1, .., N_k$, where each net is a set of so–called *terminals* which here are intersection points at the boundary of the grid.

A solution of the switchbox routing problem, called a *layout* or a *routing*, is given by pairwise edge–disjoint Steiner trees $T_1, ..., T_k$ embedded in the grid such that T_i connects the terminals of net N_i, $i = 1, ..., k$. In the layout, all induced paths must have disjoint edges but they may meet at the intersection points of the grid. In VLSI design, this is called the *knock–knee model* since at an intersection point, two induced paths may cross or both may change their direction (forming a double–bend, called knock–knee). In contrast to the knock–knee model, there is the *Manhattan model* where only crossings but no knock–knees are allowed. For the Manhattan model, Szymanski [Szy85] showed that channel routing with 4–terminal nets is NP–complete and hence so is switchbox routing. This result is extended by Middendorf [Mid93] who showed that even the 2–terminal channel routing problem in the Manhattan model is NP–complete. In the following, we consider the knock–knee model.

The 3–terminal switchbox routing problem

Instance: A rectangular routing region consisting of n vertical and m horizontal lines with terminals assigned to the intersection points at the boundary of the grid. A collection $\{N_1, .., N_k\}$ of nets, each net consists of at most 3 terminals.
Question: Is there an edge-disjoint knock-knee routing for the nets in the given routing region?

Before explaining the basics of our reduction, we introduce some notions which prove useful in the discussions to follow. We number the horizontal lines of the grid top–down and the vertical lines from the left to the right. The segment of the grid between the vertical lines with indices i and $i + 1$ is called a *column* and is denoted by $\vec{i}$. The *(local) horizontal density* $d_h(i)$ of column $\vec{i}$ is defined as the number of nets that have to cross column $\vec{i}$, i.e., it is the number of nets that contain terminals at the left–hand side as well as terminals at the right–hand side of the column $\vec{i}$. We call the vertical line with index i *density–increasing* if $d_h(i) > d_h(i - 1)$ and *density–decreasing* if $d_h(i) < d_h(i - 1)$. Otherwise, we call the vertical line *density–preserving*. We call the number m of horizontal lines the *horizontal capacity* and the number n of vertical lines the *vertical capacity* of the switchbox. The *free horizontal capacity* of a column is the difference of the capacity minus the density of the column.

We interchangeably use the term net for a set of terminals and for the realization of its Steiner tree in the layout. The respective meaning should always be clear from the context.

The following is our main theorem.

Theorem 1. *The 3-terminal switchbox routing problem is NP–complete.*

The rest of the paper is devoted to prove this result. It is clear that 3–terminal switchbox routing is in NP. The reduction is from a special case of the 3–SAT problem.

The 3–bounded 3–SAT problem

Instance: A set $\mathcal{X} = \{x_1, \ldots, x_N\}$ of Boolean variables, and a collection $\mathcal{C} = \{C_1, \ldots, C_M\}$ of clauses over $\mathcal{X}$. Thereby, for each variable x_i there are at most three clauses in $\mathcal{C}$ that contain either x_i or $\overline{x}_i$. Moreover, each clause contains at most three literals.

Question: Is there an assignment for the variables in $\mathcal{X}$ such that every clause in $\mathcal{C}$ is satisfied?

The 3–bounded 3–SAT problem is NP–complete (see [GJ79, p. 259]). Without loss of generality, we assume that all clauses contain strictly more than one literal and that every Boolean variable occurs negated as well as unnegated.

The main ideas of the transformation are as follows.

– For each Boolean variable $x_i \in \mathcal{X}$ we introduce two nets X_i and $\overline{X}_i$, called *real–nets*. If x_i occurs in two clauses, X_i and $\overline{X}_i$ consist of two terminals each, otherwise they have three terminals.

– A variable x_i is meant to be TRUE if and only if net X_i is routed below net $\overline{X}_i$ in the layout, and FALSE otherwise.

– In order to maintain the relative vertical ordering of the real–nets X_i and $\overline{X}_i$, we aim to keep the horizontal density high throughout the switchbox. Therefore, we introduce so–called *extension–nets*. Each of these nets has exactly one terminal on the left or the right boundary of the grid. Their functionality is discussed in Section 3.

– We also fix the real– and the extension–nets to certain tracks. This is the function of *sandwich–nets*. The details are given in Subsection 4.4.

– Each clause C_j is modelled as a block B_j of consecutive vertical lines. The constructed switchbox consists of these clause blocks chained from the left to the right (see Figure 1).

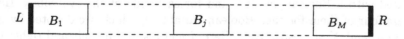

Fig. 1. The coarse–structure of the resulting switchbox instance.

– The link of the Boolean variables to the clauses in which they occur is essentially captured as follows. If the Boolean variable x_i appears in the clause C_j, both nets X_i and $\overline{X}_i$ have exactly one terminal on the same vertical line (called the *variable–line*) of the clause block B_j.

- This variable–line is surrounded by a certain collection of vertical lines (and nets placed on these lines) to guarantee that the terminals of X_i and $\overline{X}_i$ can be connected to its corresponding nets, respectively. This makes sure that each literal can be TRUE or FALSE. The precise structure of clause blocks is described in Section 4.
- Throughout the paper, repeatedly occuring collections of vertical lines and nets of the same topological structure are called modules. The collection used in the previous item (which provides the free capacity needed) is called a *detour module* and is explained in Subsection 4.2.
- We use *gate modules* (see Subsection 4.3) in order to ensure that more than the available capacity is needed, in the case that all literals of a clause are FALSE.

The resulting instance of the switchbox routing problem is quite complex. To be precise, a 3–bounded 3–SAT instance consisting of N Boolean variables and M clauses is transformed into a switchbox routing problem instance with $n = 8N + 16M_2 + 31M_3$ vertical lines, $m = 10N + 10M_2 + 20M_3 + 4$ horizontal lines, and $k = 14N + 16M_2 + 32M_3 + 4$ nets, where M_2 denotes the number of 2–literal and M_3 the number of 3–literal clauses of the 3–bounded 3–SAT instance, respectively.

3 The Sides of the Switchbox

In this section, we introduce all the tracks of the switchbox instance, explain to which nets they are dedicated to, and finally assign terminals to their left and right endpoints. Roughly, we can distinguish between three different areas of tracks. The top area is dedicated to the real– and extension–nets, the middle one to the gate–nets (see Subsection 4.3), and the bottom one to the detour–nets (see Subsection 4.2).

For each Boolean variable x_i we insert two horizontal lines, called *variable–tracks*. They are designated for the associated real–nets. Since the terminals of the real–nets only appear within clause blocks, we need to keep the variable–tracks occupied between the left–hand side (right–hand side, respectively) of the switchbox and the left–most (right–most) terminals of the real–nets. This is the task of the extension–nets. There is one of extension–net for each real–net associated with a Boolean variable and for each side of the switchbox. Hence, there are four of them for each Boolean variable x_i which are denoted by $X_{i,lt}$, $X_{i,lb}$, $X_{i,rt}$, and $X_{i,rb}$, respectively. Here, l and r stand for left and right, and t and b stand for top and bottom, respectively.

Each extension–net has exactly three terminals. One is directly placed on the left (right) boundary of the grid and two of them are in the clause block which contains the left–most (right–most) terminal of the associated real–net. The terminals in the clause blocks are combined with a clamp module which is discussed in detail in Subsection 4.4.

To explain the precise dedication of the introduced tracks and the arrangement of the terminals on the sides, we fix an arbitrary ordering of the clauses and re–number the Boolean variables correspondingly. The variables are numbered from 1 to N in order of their first occurence with respect to the chosen clause ordering (from left to right).

Starting from the top, we assign the terminals of the extension–nets in the order $1, \ldots, N$ to the sides of the switchbox. Thereby, the associated terminals of the upper and the lower extension–nets alternate. Each of these terminals is surrounded by a certain number of sandwich–nets. Sandwich–nets are also assigned to gate– and detour–nets. They have a special structure: they consist of three terminals; two of them are placed at the left–hand and at the right–hand side of the switchbox, respectively, both on the same track. Because of lack of capacity, we will see that sandwich–nets are forced to occupy the whole track to which their outer terminals are assigned to.

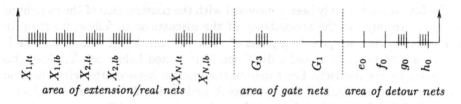

Fig. 2. The left side L of the switchbox (in landscape mode).

Figure 2 depicts the terminal assignment at the left side of the switchbox. The terminals of extension–, gate– and detour–nets are represented by long dashes while the terminals of sandwich–nets, assigned to these nets, are represented by short dashes. The terminal assignment of the right side is similar. Note that all intersection points at the left and the right boundary of the grid are occupied by terminals.

4 The Clause Block

In this section, we give the precise structure of a clause block. Every clause block consists of vertical variable–lines for every Boolean variable of the clause. The variable–lines are surrounded by detour modules which are themselves embedded into gate modules. We present these concepts as well as the clamp modules which serve to maintain the vertical ordering of the nets, in the following subsections.

4.1 The Variable–Line

For a given clause C, we introduce a variable–line for each variable occurring in C. If a variable x appears unnegated in C, we place a terminal of net X at the lower position of the variable–line of x and a terminal of net $\overline{X}$ at the upper position. If x appears negated in C, the assignment is the other way around (see Figure 3).

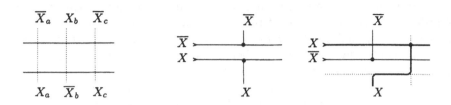

Fig. 3. Var.–lines for $C = x_a \vee \overline{x}_b \vee x_c$ **Fig. 4.** a) TRUE–routing b) FALSE–routing.

So far, we have mainly been concerned with the construction of the switchbox instance. To improve the accessibility of the discussions to follow, we turn for a moment to the interpretation point of view. As mentioned before, a variable x is meant to be TRUE if and only if net X is routed below net $\overline{X}$. Now, this is rendered more precisely. Let I denote the interval between the left–most and the right–most terminals of the real–nets X and $\overline{X}$. Let I^* be the maximal sub–interval of I such that the left– and the right–most vertical line is at maximum density. The variable x is defined to be TRUE if and only if net X is routed below net $\overline{X}$ in I^*, and FALSE otherwise.

Given the terminal assignment at the variable–lines and the interpretation of the routing of the real–nets, we can distinguish two possible routing types at each variable–line. If a variable x is TRUE (FALSE, respectively) and appears negated (unnegated) in the clause under consideration, then not both terminals of the associated variable–line can directly be connected to their dedicated track. Consequently, additional horizontal and vertical capacity is needed to route the necessary detour (see Figure 4 b). Such a routing is called a FALSE–routing. Every other kind of local routing (see, for instance, Figure 4 a) is called a TRUE–routing. It corresponds to a literal with value TRUE.

In order to allow a TRUE– or a FALSE–routing at every variable–line, the detour modules provide the required horizontal and vertical capacity.

4.2 The Detour Module

Detour modules appear around each variable–line. They are intended to provide the capacity needed for a FALSE–routing and, at the same time, to keep the horizontal and vertical density high enough to prevent a change of the vertical ordering of the nets. We distinguish between two types of detour modules, namely *right–handed* and *left–handed* ones. Since they are symmetric about the associated variable–line, we restrict ourselves to the description of right–handed detour modules.

Consider, say, the ith detour module. This module consists of four vertical lines, of two terminals of sandwich–nets, and of six terminals of four so–called *detour–nets* e_{i-1}, e_i, g_{i-1}, and g_i. These detour–nets serve to keep the capacity low which is caused by the special terminal assignment at the introduced vertical lines. The first terminal of the net e_{i-1} and the first two terminals of the net g_{i-1} are located in the previous detour module. The third terminal of g_{i-1} is placed on the upper endpoint of the first vertical line of the ith detour module. We put this line itself to the left of the variable–line surrounded by the ith detour module. The other three vertical lines of this detour module are placed to its right (see Figure 5). Terminals of the detour–net g_i are assigned to the bottom endpoints of the second and the fourth of these vertical lines whereas the bottom endpoints of the first and the third vertical line are used for the nets e_{i-1} and e_i, respectively. The remaining upper endpoints of the vertical lines to the right of the variable–line are designated for terminals of the nets S_x, e_{i-1}, and S_g, respectively. Here, S_x denotes a sandwich–net whose remaining two terminals are placed directly above the associated extension–nets of the Boolean variable x at both sides of the switchbox instance (see Figure 2). The other two terminals of the sandwich–net S_g are directly placed above the associated detour–net at both sides of the switchbox instance (see again Figure 2). The exact terminal assignment within the detour module can be taken from Figure 5.

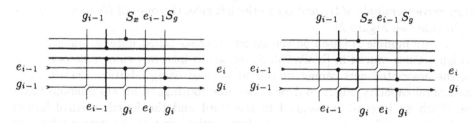

Fig. 5. A layout for the detour module: a TRUE– and a FALSE–routing (left/right).

The vertical line immediately to the right of the variable–line is called the *detour–line*. The described detour module is called right–handed since the detour–line is on the right side of the variable–line. A left–handed detour module is obtained by reversing the terminal assignment from left–to–right. The detour–nets of a left–handed module are denoted by f_i and h_i instead of e_i and g_i, respectively.

The name detour–line is caused by its functionality. The detour caused by a FALSE–routing cannot be realized without using the vertical capacity of this line. For the same reason, the detour module provides free horizontal capacity of one track between its first vertical line, which is density–decreasing, and its detour–line which is density–increasing. The remaining vertical lines of the detour module, however, are density–preserving. The capacity provided by the detour module can be used to realize a TRUE– as well as a FALSE–routing. Notice that the free horizontal capacity of one track must be used in both cases (see Figure 5).

Finally, we should mention that, due to their exposed location, the detour–nets e_0, f_0, g_0, h_0 as well as $e_{M_2+2M_3}$, $f_{M_2+2M_3}$, $g_{M_2+2M_3}$, and $h_{M_2+2M_3}$ need a special treatment. First, the terminals of e_0 f_0, g_0, and h_0 which have not been assigned so far are placed on the left side of the switchbox, as already indicated in Figure 2. Similarly, the remaining terminals of the detour–nets $e_{M_2+2M_3}$, $f_{M_2+2M_3}$, $g_{M_2+2M_3}$, and $h_{M_2+2M_3}$ are assigned to the right side of the switchbox. Second, in contrast to the other detour–nets, the detour–nets g_0, h_0, $e_{M_2+2M_3}$, and $f_{M_2+2M_3}$ have two terminals only.

4.3 The Gate Module

In order to ensure that each clause will be satisfied by the variable assignment deduced from a layout, we have to guarantee that not at every variable–line within a clause a FALSE–routing can be realized. For this purpose, we introduce gate modules.

Each gate module consists of six vertical lines, four sandwich–nets and four *gate-nets* which consist of three terminals each. To simplify notation, we focus on one gate module and denote its corresponding gate–nets by G_i, $i = 1, .., 4$. One terminal of each of these nets is assigned to the sides of the switchbox, the respective terminals of G_1 and G_3 to the left side, the ones of G_2 and G_4 to the right side (see Figure 2).

A gate module is based on the structure of so–called *autonomous intervals* which are introduced in [FWW93]. Autonomous intervals enforce detours, that is, the associated nets cannot be routed without using additional vertical and horizontal capacity. Figure 6 depicts possible routings of an autonomous interval. Such an interval is assigned to the third and the fourth vertical line of each gate module. The first and the last vertical line of the gate module are density–preserving. The second vertical line is density–decreasing whereas the fifth vertical line is density–increasing. Consequently, the free horizontal capacity is one between these vertical lines.

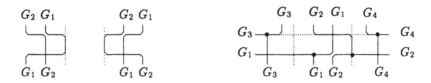

Fig. 6. Layouts for an auton. interval.   **Fig. 7.** Structure of the gate module.

The precise terminal assignment of the involved gate– and sandwich–nets to these vertical lines is given in Figure 10, after the introduction of the clamp modules. Figure 7 depicts the pure structure of a gate module.

Two detour modules are embedded into one gate module. Their detour–lines are placed between the second and the third vertical line of the gate–module and between the fourth and fifth vertical line, respectively. In Figure 7, the embedded detour–lines are represented by dotted lines. For 2–literal clauses, the combination of gate– and detour–modules is as follows: we assign a right–handed detour module to the left variable–line and a left–handed detour module to the right variable–line; a gate module lies inbetween. For 3–literal clauses, we first assign two detour modules to the variable–line in the middle: a left–handed one and a right–handed one. The left variable–line of the corresponding clause block is embedded into a right–handed detour module whereas the right one is embedded into a left–handed detour module. Afterwards, two gate modules are placed between the first two and the last two detour–lines. The interaction between the detour and the gate modules is illustrated in Figure 8.

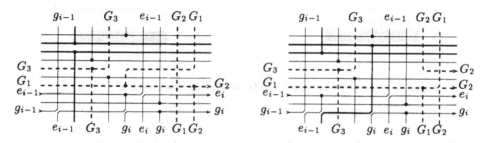

Fig. 8. The interplay between gate and detour modules: a TRUE– and a FALSE–routing

In order to route the detour enforced by a gate module, the free horizontal capacity provided by the gate module itself and the capacity of at least one embedded detour–line must be used. This guarantees that not at every variable–line a FALSE–routing can be realized. In our interpretation, this means that at least one literal of the corresponding clause has to be TRUE. This will be proved in Lemma 7.

4.4   The Clamp Module

The crux of the transformation is to guarantee a certain vertical ordering of the nets. Therefore, we force nets to dedicated tracks. This is the task of *clamp modules*. A clamp module consists of two consecutive vertical lines, of two middle terminals of sandwich–nets, and of two terminals of a net which either starts or ends within a clause block and which should be forced to a dedicated track. Clamp modules are assigned to the gate–nets G_3 and G_4, and to each of the extension–nets.

Consider, say, a net Y whose left–most terminal is placed on the left side of the switchbox. The terminals of the associated sandwich–nets S^a and S^b (where *a* stands for above and *b* for below) are directly placed above and below the left–most terminal of Y at the left side and are assigned to the same positions at the right side of the switchbox. The first vertical line of the clamp module is density–preserving and the second is density–decreasing. The terminals of S^a and Y are assigned to the top, and the terminals of Y and S^b are placed on the bottom side of these lines, in this order. If a clamp module is assigned to a starting net whose right–most terminal is assigned to the right side of the switchbox, the first vertical line of the clamp module is density–increasing and the second is density–preserving. For the exact terminal assignment of a clamp module, we refer the reader to Figure 9.

Fig. 9. A clamp module assigned to an ending/starting net (left/right).

We will show in Proposition 3 that an ending net (starting net, respectively) is fixed at the left (right) side to its dedicated track by its terminal and at the right (left) side by a clamp module. At this point, all ingredients of the construction are described in detail. Consequently, we are now able to combine the different modules to a clause block.

4.5 Combining the Different Modules

In the previous subsections we have introduced a bunch of different nets which have been combined to different modules of certain functionality. In this subsection, we give a brief description how the presented modules are combined to obtain a clause block for a certain clause $C \in \mathcal{C}$. The structure of a clause block is quite complex since the different modules overlap. Nevertheless, the main principle is that most of the nets have the same routing in every layout, if one exists. This "skeleton" consists of all sandwich–, detour–, and extension–nets. Within this skeleton, the routings of the real– and gate–nets have to be realized.

Figure 10 depicts the complete clause block for the 2–literal clause $\{\bar{x}_1, x_2\}$. It contains the routing of all nets of the "skeleton", i.e., nets that have the same routing in every layout. Note that only one pair of extension–nets occurs since we assume that only the real–nets of variable x_2 start in this clause block. The remaining nets marked with an arrow may have different routings depending on different variable assignments.

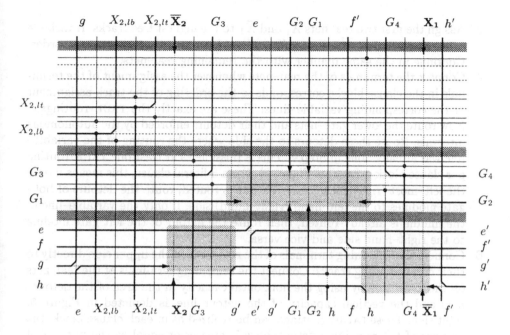

Fig. 10. An example clause block for $\{\bar{x}_1, x_2\}$.

5 The Proof

At this point, the reduction from an instance of the 3–bounded 3–SAT problem to an instance of the 3–terminal switchbox routing problem is completely described. The NP–completeness proof has the following structure: Lemma 2 shows that a satisfying variable assignment for an instance of the 3–bounded 3–SAT problem induces a layout for the resulting switchbox routing problem. For the reverse direction, it is most important to show that the determination of the values of the Boolean variables from the layout is well defined. This is captured by Lemma 6. Once this lemma is proved, it remains to show that the assignment deduced from the layout satisfies all clauses of the underlying instance of the 3–bounded 3–SAT problem. This is done in Lemma 7.

Lemma 2 (Existence of a Layout). *Every feasible solution of the 3–bounded 3–SAT problem induces a layout for the resulting switchbox routing problem.*

Proof. Given an instance of the 3–bounded 3–SAT problem and a satisfying variable assignment, we now describe how to obtain a layout for the resulting switchbox routing problem instance.

1. Route all sandwich–nets along the tracks where their left and right terminals are assigned to and connect their middle terminals.
2. Route all extension– and upper gate–nets only along their dedicated tracks.

3. Assign the first two real nets X_1 and $\overline{X}_1$ to the highest free tracks. If variable x_1 is TRUE, route $\overline{X}_1$ above X_1, otherwise route them in reverse order. Continue in the same manner with the remaining real–nets.

4. Connect the terminals of the real–nets whenever the assignment of the terminals in the clause blocks corresponds to the ordering of the track assignment at the corresponding variable–lines. These are exactly the TRUE–routings.

5. Now route the lower gate–nets and their detours enforced by the autonomous intervals. Route these detours to the side where the routing of the variable–line was already done in Step 4. This is possible as at least one TRUE–routing per clause block exists (see Figure 8). For a 3-literal clause, this means if only the left–most (right–most) variable–line is routed, route the detours of both gate modules to the left–hand (right–hand) side. If only the variable–line in the middle is already routed, then route the detour of the left gate module to the right–hand side and vice versa.

6. Connect the remaining terminals of the real–nets on the upper side directly to their corresponding tracks. Connect the remaining terminals of the real–nets on the lower side by using the horizontal capacity in the area of the detour–nets and the vertical capacity of the detour–line as depicted in Figure 5, right side. These FALSE–routings can be realized as in each clause block, the number of the remaining detour–lines is greater or equal as the number of FALSE–routings.

7. Route the remaining detour–nets in a canonical way (see Figure 5).

This procedure yields a layout for the resulting switchbox routing problem. □

For the reasons of brevity, the proofs of the following proposition and lemmas are omitted

Proposition 3 (Clamp Module). *For every layout of the switchbox routing problem, all starting (ending) nets that are combined with a clamp module are routed along their dedicated tracks.*

Lemma 4 (Detour Module). *For every layout of the resulting switchbox routing problem, the track assignment of the nets does not change inside a detour module. That is, the two new inserted detour–nets are assigned to the tracks of the corresponding ending detour–nets.*

Lemma 5 (Gate Module). *For every layout of the resulting switchbox routing problem, the track assignment of the nets does not change inside a gate module.*

Lemma 6 (Ordering Lemma). *For every layout of an instance of the switchbox routing problem, the upper variable-track of a variable $x_i \in \mathcal{X}$ is either occupied by the real–net X_i or by the real–net $\overline{X}_i$ inside the associated interval I^*. The vertical ordering of the real–nets X_i and $\overline{X}_i$ is the same within this interval.*

Lemma 7 (Gate Lemma). *Consider a layout for the resulting instance of the switchbox routing problem. Then, in every clause block, a TRUE–routing is realized for at least one variable of the associated clause.*

209

Proof. As mentioned before, a clause block corresponding to a 3–literal (2–literal) clause consists of 4(2) detour–lines. At the detour–lines, vertical capacity is available in order to route a detour of a gate module or to realize a FALSE–routing. A detour of a gate module and a FALSE–routing cannot be realized at the same detour–line since FALSE–routings occupy the vertical capacity of the detour–lines from the area of the detour–nets up to the area of the extension/real–nets. Hence, it covers the area where the detours of the gate modules have to be routed (see Figure 8).

Suppose a layout of the switchbox routing problem implies a variable assignment which does not satisfy a clause C. Hence, at each variable–line of the clause block corresponding to C a FALSE–routing is realized. This means that for a 3–literal(2–literal) clause, 3(2) FALSE–routings plus 2(1) detours of the gate modules have to be realized. Thus, the demand of vertical capacity is strictly greater than the number of detour–lines which contradicts the existence of a layout. □

Proof (of Theorem 1). Simple counting shows that all introduced nets consist of at most three terminals. Then, the theorem follows directly from the above lemmas. Lemma 2 states that a satisfying variable assignment induces a layout of the corresponding instance for the resulting switchbox routing problem. Such a layout can be constructed by the algorithm which is given in the proof of Lemma 2. By Lemma 6, the variable assignment is shown to be well defined while Lemma 7 guarantees that every clause of the 3–bounded 3–SAT instance is satisfied by the obtained variable assignment. Hence, the 3-terminal switchbox routing problem is NP–complete. □

6 Concluding Remarks

In this paper, we determined the border between the polynomial time solvable and the NP-complete instances of the switchbox routing problem. The transition between the 2–terminal case (which can be solved efficiently) and the k–terminal case, $k \geq 3$, (which is NP–complete) corresponds directly to the transition between path embeddings and Steiner tree embeddings in grid graphs. The techniques introduced in this paper also apply to the channel routing problem. A proof similar to the one above shows that the 4–terminal channel routing problem is NP–complete [Har96]. This improves upon the result of [Sar87] for 5–terminal nets. Only the complexity status of the 3–terminal channel routing problem remains open.

Acknowledgements

The authors are grateful to Dorothea Wagner for helpful comments and discussions.

References

[FWW93]   M. Formann, D. Wagner, and F. Wagner. Routing through a dense channel with minimum total wire length. *Journal of Algorithms*, 15:267 – 283, 1993.

[GJ79]   M. R. Garey and D. S. Johnson. *Computers and Intractability - A Guide to the Theory of NP-Completeness*. W. H. Freeman and Company, New York, 1979.

[GMW93]   M. Grötschel, A. Martin, and R. Weismantel. Routing in grid graphs by cutting planes. In G. Rinaldi and L. Wolsey, editors, *Third IPCO Conference*, pages 447 – 461, 1993.

[Har96]   S. Hartmann. Channel routing with 4-terminal nets is NP–complete. Preprint, Fachbereich Mathematik, Technische Universität Berlin, Berlin, Germany, 1996.

[KPS90]   B. Korte, H.-J. Prömel, and A. Steger. Steiner trees in VLSI–layout. In B. Korte, L. Lovasz, H. J. Prömel, and A. Schrijver, editors, *Paths, Flows and VLSI-Layout*, pages 185–214. Springer Verlag, 1990.

[KvL84]   M. R. Kramer and J. van Leeuwen. The complexity of wire routing and finding minimum area layouts for arbitrary VLSI circuits. In F. P. Preparata, editor, *Advances in Computing Research, Vol. 2 VLSI theory*, pages 129–146. JAI Press, Reading, MA, 1984.

[Len90]   Th. Lengauer. *Combinatorical Algorithms for Integrated Circuit Layout*. Teubner/Wiley&Sons, 1990.

[Mid93]   M. Middendorf. Manhattan channel routing is $\mathcal{NP}$–complete under truly restricted settings. Preprint, Universität Karlruhe, 1993. To appear in the Chicago Journal of Theoretical Computer Science.

[MP86]   K. Mehlhorn and F. P. Preparata. Routing through a rectangle. *Journal of the Association for Computing Machinery*, 33(1):60 – 85, 1986.

[MS92]   M. Marek-Sadowska. Switch box routing: a retrospective. *Integration, the VLSI Journal*, 13:39 – 65, 1992.

[MWW95]   R. H. Möhring, D. Wagner, and F. Wagner. *Network Routing*, chapter VLSI Network Design, pages 625 – 712. Handbooks in Operations Research and Management Science. Elsevier, 1995.

[OS81]   H. Okamura and P. D. Seymour. Multicommodity flows in planar graphs. *Journal of Computer Theory*, pages 75 – 81, 1981.

[Sar87]   M. Sarrafzadeh. Channel-routing problem in the knock-knee mode is NP–complete. *IEEE Transaction on Computer-Aided Design*, 6(4):503 – 506, 1987.

[Szy85]   T. G. Szymanski. Dogleg channel–routing is NP–complete. *IEEE Transaction on Computer-Aided Design*, 4(1):31 – 41, 1985.

Detecting Diamond Necklaces in Labeled Dags (A Problem from Distributed Debugging)

Michel Hurfin, Michel Raynal

IRISA
Campus de Beaulieu – 35042 RENNES Cedex – FRANCE
{name}@irisa.fr
Fax: +33 99 84 71 71

Abstract. The problem tackled in this paper originates from the debugging of distributed applications. Execution of such an application can be modeled as a partially ordered set of process states. The debugging of control flows (sequences of process states) of these executions is based on the satisfaction of predicates by process states. A process state that satisfies a predicate inherits its label. It follows that, in this context, a distributed execution is a labeled directed acyclic graph (dag for short). Debug or determine if control flows of a distributed execution satisfy some property amounts to test if the labeled dag includes some pattern defined on predicate labels.

This paper first introduces a general pattern (called *diamond necklace*) which includes classical patterns encountered in distributed debugging. Then an efficient polynomial time algorithm detecting such patterns in a labeled dag is presented. To be easily adapted to an on-the-fly detection of the pattern in distributed executions, the algorithm visits the nodes of the graph according to a topological sort strategy.

1   Introduction

This paper presents an algorithm to detect a sophisticated pattern (called *diamond necklace*) in a labeled directed acyclic graph. The problem solved by this algorithm originated from the detection of properties of distributed computations in our current effort to design and implement a facility for debugging distributed programs [9]. These programs are composed of a finite set of sequential processes cooperating by the only means of message passing. The concurrent execution of all the processes on a network of processors is called a distributed computation. The computation is asynchronous: each process evolves at his own speed and messages are exchanged through communication channels, whose transmission delays are finite but arbitrary. During a computation, each process executes a sequence of actions. At a given time, the local state of a process is defined by the values of the local variables managed by this process. From an initial state a process produces a sequence of process states according to its program text. In the context of the debugging of distributed programs, a distributed execution is usually modeled as a partially ordered set of process states [7]. Due to the asynchronous nature of a distributed computation, programmers refer to a classical causal precedence relation between states rather than to a non-available global clock. Informally, process state s_1 precedes s_2 if both have been produced by the same process with s_1 first, or if s_1 has been produced by some process before it sent a message to another process and the receiver process produced s_2 after receiving this message; this causal precedence relation is nothing else than Lamport's "happened before" relation expressed on process states [12]. A directed path of process states starting from an initial process state is usually called a control flow.

The ability to specify and verify properties which refer to the evolution of the distributed computation state, is essential. Many problems (assertion checking, error reporting, process control, decentralized coordination, workload characterization, bottleneck identification, ...) can be solved provided that an appropriate behavioral property is able to be specified and detected at the proper time. For instance, a programmer will check whether an expected or unexpected behavior occurs during the execution to ensure that the distributed program does what it is supposed to do. Since several years, we design and implement different distributed algorithms that on-the-fly detect properties on control flows of distributed computations [10, 5, 4]. Basically a property is defined as a language on an alphabet of predicates (a predicate being a boolean expression in which appear variables of a single process); a pattern is a word of this language. If a local state satisfies a given predicate, it inherits its label: so words can be associated with each control flow. Finally a control flow satisfies a property if one of its words belongs to the language defining the property, i.e., if it matches some pattern. Such an approach has been formalized in [1]. These properties are fundamentally sequential in the sense they consider each control flow separately.

Sequential properties are not powerful enough to express patterns which are on several control flows. An example of such a property is the following one: "there is a process state s_1 satisfying a predicate P_1 causally preceding a process state s_2 satisfying a predicate P_2 and all paths of process states starting at s_1 and ending at s_2 satisfy some sequential property". A logic able to express such non-sequential properties has been introduced in [6].

Here we abstract from distributed executions and consider labeled directed acyclic graphs (dags), in which vertices represent local states and edges represent dependent relation over states. We first define (Section 2) a general type of patterns (diamond necklace) for labeled dags which includes as particular cases sequential and non-sequential patterns useful in distributed debugging, and then (Section 3) we present an algorithm to detect these patterns. In order to be adaptable to on-the-fly distributed detection in the context of distributed debugging, it is required that the algorithm visits the nodes of the dag according to a topological sort strategy.

So this paper solves a new problem (to our knowledge), namely deciding if a labeled dag includes some specific pattern, that we met in designing and implementing a distributed debugging facility.

2   Diamond Necklaces

2.1   Labeled Dags

Let $G = (V, E)$ be a finite dag with n vertices. Notations v, v', v_i, v^i are used to represent elements of V. Let v_i and v_j be two vertices of V; $\mathcal{P}(v_i, v_j)$ is the set of all the paths in G from v_i to v_j.

$$
\mathcal{P}(v_i, v_j) \;=\; \left\{\; (v^1, v^2, \cdots, v^u) \;\middle|\;
\begin{array}{l}
v^1 = v_i \\
\wedge \\
\forall i,\; 1 \le i < u,\; (v^i, v^{i+1}) \in E \\
\wedge \\
v^u = v_j
\end{array}
\;\right\}
$$

In order to facilitate the explanation of the algorithm we suppose that G has a source vertex and a sink vertex denoted v_1 and v_n, respectively. By definition:

$$
\forall v_i \in V, \quad
\left\{
\begin{array}{l}
\mathcal{P}(v_i, v_1) = \emptyset \\
\wedge \\
\mathcal{P}(v_n, v_i) = \emptyset \\
\wedge \\
v_i \neq v_1 \iff \mathcal{P}(v_1, v_i) \neq \emptyset \\
\wedge \\
v_i \neq v_n \iff \mathcal{P}(v_i, v_n) \neq \emptyset
\end{array}
\right.
$$

Let Σ be a finite set of l labels: $\Sigma = \{a_1, a_2, \cdots, a_l\}$. The set of all strings over the alphabet Σ is denoted by Σ^*. λ is a labeling function that maps edges of G to sets of labels. If $(v_i, v_j) \in E$, $\lambda(v_i, v_j)$ denotes the set of labels associated with the edge (v_i, v_j). We assume the "empty" label ϵ is implicitly associated with every edge for which the labeling function defines no label. G^λ denotes the dag G with labeling λ.[1]

For each pair of vertices (v_i, v_j) of the graph, $\mathcal{L}(v_i, v_j)$ represents the set of words defined by considering all possible labeling of all paths starting at v_i and ending at v_j. More formally, let $a_1 a_2 \cdots a_u$ be an element of Σ^*:

$$
a_1 a_2 \cdots a_u \;\in\; \mathcal{L}(v_i, v_j)
$$
$$
\Longleftrightarrow
$$
$$
\exists (v^1, v^2, \cdots, v^u, v^{u+1}) \in \mathcal{P}(v_i, v_j) \text{ such that } \forall i,\; 1 \le i \le u,\; a_i \in \lambda(v^i, v^{i+1})
$$

Let R^k be the name of a property defined as a set of words (language $\mathcal{L}(R^k)$) on the alphabet Σ.

---

[1] We assign labels to each arc of the graph rather than to each vertex. When the goal is to detect properties of distributed computations, each vertex represents a local state and, in that case, the labels of all the predicates satisfied by a local state v are assigned to all incoming arcs of vertex v.

2.2 The Primitive Pattern SOME

Let v_i and v_j be two vertices of G and R^k be a property. The pair (v_i, v_j) satisfies the pattern $SOME(R^k)$ if there is a path from v_i to v_j such that at least one of the labelings of the path is a word of $\mathcal{L}(R^k)$. More formally:

$$((v_i, v_j) \models SOME(R^k)) \equiv \mathcal{L}(v_i, v_j) \cap \mathcal{L}(R^k) \neq \emptyset$$

2.3 The Primitive Pattern ALL

The pair (v_i, v_j) satisfies the pattern $ALL(R^k)$ if all labelings of all paths from v_i to v_j belong to $\mathcal{L}(R^k)$. More formally:

$$((v_i, v_j) \models ALL(R^k)) \equiv (\mathcal{P}(v_i, v_j) \neq \emptyset) \wedge (\mathcal{L}(v_i, v_j) \subseteq \mathcal{L}(R^k))$$

Note that this condition can also be expressed as follow:

$$(\mathcal{P}(v_i, v_j) \neq \emptyset) \wedge (\mathcal{L}(v_i, v_j) \cap (\Sigma^* - \mathcal{L}(R^k)) = \emptyset)$$

2.4 The General Pattern

This pattern is an alternating sequence of primitive patterns SOME and ALL. An ALL pattern reminds of a diamond and two consecutive diamonds are connected by a link, *i.e.*, a pattern SOME, the whole pattern forming a necklace of diamonds. The alternating sequence is denoted $R^1 \, R^2 \, R^3 \, \cdots \, R^m$.

A sequence of $m + 1$ vertices $(v^1, v^2, v^3, v^4, \cdots, v^m, v^{m+1})$ is a solution of the general pattern (*i.e.*, $(v^1, v^2, v^3, v^4, \cdots, v^m, v^{m+1}) \models R^1 \, R^2 \, R^3 \, \cdots \, R^m$), if these vertices satisfy the following constraints:

- $(v^1 = v_1) \wedge (v^{m+1} = v_n)$
- $\forall k, \ 1 \leq 2k+1 \leq m, \ (v^{2k+1}, v^{2k+2}) \models SOME(R^{2k+1})$
- $\forall k, \ 2 \leq 2k \leq m, \ (v^{2k}, v^{2k+1}) \models ALL(R^{2k})$

Figure 1 gives a pictorial representation of a diamond necklace. A line from v^{2k-1} to v^{2k} represents a path satisfying a pattern SOME, and a diamond-shaped plane figure represents a diamond starting at v^{2k} and terminating at v^{2k+1}.

The following prefix notation will be used in what follows. Let us consider the subgraph of G whose v_1 and v^k are the source and the sink vertices (*i.e.*, all maximal paths of this subgraph start at v_1 and terminate at v^k). If $(v^1, v^2, \cdots, v^k) \models R^1 \, R^2 \, \cdots \, R^{k-1}$ then we say the sequence $(v^1, v^2, \cdots, v^k)$ is a solution of the prefix $R^1 \, R^2 \, \cdots \, R^{k-1}$ of the pattern.

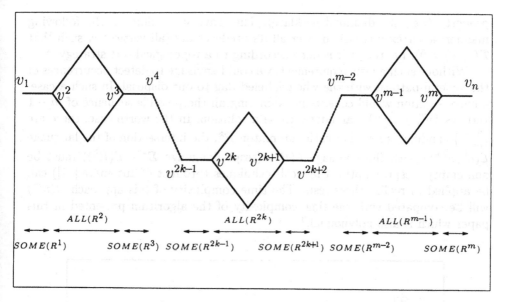

Fig. 1. A diamond necklace pattern

3 A Detection Algorithm

3.1 Regular Properties

For the sake of simplicity, in what follows, we consider only properties R^k whose corresponding languages $\mathcal{L}(R^k)$ are regular [8]. Moreover, these properties are sufficient to solve practical problems encountered in distributed debugging.

Let $\mathcal{A}^k$ ($1 \leq k \leq m$) be the finite automaton recognizing $\mathcal{L}(R^k)$. Formally, an automaton is a tuple $\mathcal{A}^k = (Q^k, \Sigma, \delta^k, q_0^k, F^k)$, where Q^k is a finite set of states (q_x^k is one of these states), Σ is a finite alphabet (equal to the set of l labels associated with edges of the graph), q_0^k is the initial state, F^k is the set of final states and δ^k its transition function mapping $Q^k \times \Sigma$ to 2^{Q^k}.

We extend δ to map $Q^k \times \Sigma^*$ to 2^{Q^k} by reflecting sequences of inputs as follows: Let w be a sequence of inputs, a_u be a basic predicate, and q be an state. Then,

1. $\delta(q, \epsilon) = \{q\}$, and
2. $\delta(q, w \cdot a_u) = \{q_i \mid q_i \in \delta(q_j, a_u) \text{ where } q_j \in \delta(q, w)\}$

All automata $\mathcal{A}^{2k}$ ($2 \leq 2k \leq m$) are supposed to be deterministic and complete; automata $\mathcal{A}^{2k+1}$ ($1 \leq 2k+1 \leq m$) can be non-deterministic.

3.2 Visiting the Graph

The algorithm proposed in this paper visits the vertices of G, starting from v_1. When it visits a new vertex v, it computes information necessary to detect the property (*i.e.*, the diamond necklace). The traversal is done in the following manner: a vertex v is visited after all its predecessors (all vertices v_i such that $\mathcal{P}(v_i, v) \neq \emptyset$); *i.e.* the visit is done according to a topological sort strategy. [2]

Without such a visit requirement, we could envisage to detect occurrences of the general pattern with the whole labeled dag to our disposal. In such a case, a naive solution would consists in examining all the possible sequence of $m - 1$ vertices $\{v^2, \cdots, v^m\}$, candidates to be a solution. In the worth case, there are $\binom{n-2}{m-1}$ candidates sets. For each automaton R^k, the intersection of the language $\mathcal{L}(v^k, v^{k+1})$ with the language $\mathcal{L}(R^k)$ (respt. language $\Sigma^* - \mathcal{L}(R^k)$) must be non-empty (respt. empty). Classical techniques (product of automata [11]) can be applied to realize these tests. The time complexity of this approach $O(n^m)$ will be compared with the time complexity of the algorithm presented in this paper which is also polynomial.[3]

```
begin
    states(v, k)  :=  ∅;
    foreach  v_p  such that  ((v_p, v) ∈ E) :
        foreach  q_x^k ∈ states(v_p, k) :
            foreach  a ∈ λ(v_p, v) :
                states(v, k)  :=  states(v, k) ∪ {δ^k(q_x^k, a)};
            endfor
        endfor
    endfor
end
```

Fig. 2. Visit of a vertex $v \neq v_1$

3.3   Detecting $(v_1, v) \models SOME(R^k)$

To facilitate the understanding of the general algorithm (Section 3.5), we first present simpler algorithms which constitute building blocks of the general one.

A variable $state(v, k)$ is associated with each vertex v; its definition is the following one:

---

[2] This visit strategy is particularly interesting in the context of on-the-fly detection of properties of distributed executions. Actually, in that case, the partially ordered set of local states is generated on-the-fly by the execution itself: due to this visit strategy of the vertices (local states) of the graph, the detection algorithm can be easily superimposed [2] on such an execution.

[3] Therefore, this algorithm can also be used to detect efficiently diamond necklaces in dags such as the lattice of global states of distributed computations which may be constructed at a designated process [3].

$$states(v, k) = \{\ q_x^k\ |\ \exists\, w \in \mathcal{L}(v_1, v)\ \text{such that}\ q_x^k \in \delta^k(q_0^k, w)\ \}$$

By visiting the vertices of G, starting from v_1 and using the traversal strategy explained above, the value of $states(v, k)$ is computed as indicated by Figure 2 (Initially: $states(v_1, k) = \{q_0^k\}$).

It follows that answering to the question "$(v_1, v) \models SOME(R^k)$" is equivalent to test the following predicate:

$$\exists\, q_x^k \in states(v, k)\ :\ q_x^k \in F^k$$

3.4 Detecting $(v_1, v) \models ALL(R^k)$

The previous discussion about the computation of $states(v, k)$ is still valid. Only the predicate to decide "$(v_1, v) \models ALL(R^k)$" has to be defined. As indicated we constrain all the automata recognizing a language whose property appears in an ALL pattern to be deterministic. With such a constraint the decision test becomes:

$$\forall\, q_x^k \in states(v, k)\ :\ q_x^k \in F^k$$

3.5 Detecting Diamond Necklace Patterns

Determine the set of solutions $(v^1, v^2, \cdots, v^m, v^{m+1})$ requires to analyze all the words associated with all possible labelings of all the paths. This demands to keep information related to word analyses and to launch the next automaton each time a prefix of the pattern has been recognized. As indicated previously, it is supposed that vertices of G are visited according to the strategy explained in Section 3.2, v_1 being the first vertex visited.

Launching automata. When v is visited, if $(v^1, v^2, \cdots, v^{2k}, v)$ is solution of the prefix $R^1\ R^2\ \cdots\ R^{2k}$ then a copy of the automaton $\mathcal{A}^{2k+1}$ has to be launched in order to start the search for a vertex v' such that $(v, v') \models SOME(R^{2k+1})$.

Similarly, if $(v^1, v^2, \cdots, v^{2k+1}, v)$ is solution of the prefix $R^1\ R^2\ \cdots\ R^{2k+1}$ then a copy of $\mathcal{A}^{2k+2}$ has to be launched to search for a vertex v' such that $(v, v') \models ALL(R^{2k+2})$.

Data structure to record past word analyses. As an automaton $\mathcal{A}^k$ can be launched from any vertex, the data structure $states(v, k)$ has to be enriched to record the vertices in which copies of $\mathcal{A}^k$ have been started. An array of m variables $start\_states$ is associated with each vertex v; $start\_states(v, k)$ is a set of pairs (v_i, q_x^k) whose first component is a vertex of G and second component is a state of Q^k. Its semantics is the following one:

$$(v_i, q_x^k) \in start\_states(v, k) \iff \left\{ \begin{array}{l} \text{a copy of } \mathcal{A}^k \text{ has been started in } v_i \\ \wedge \\ \exists w \in \mathcal{L}(v_i, v) \text{ such that } q_x^k \in \delta^k(q_0^k, w) \end{array} \right.$$

These data structures keep a record of all the word analyses done in the past of the vertex v that is currently visited.

Procedure *Visit* (v : vertex);
begin
/* Recognition */
  if ($v = v_1$) **then**
    $start\_states(v_1, 1) := \{ (v_1, q_0^1) \}$;
    for $k := 2$ **to** m :
      $start\_states(v_1, k) := \emptyset$;
    endfor
  else
    /* v is not the least vertex of G */
    /* All predecessor of v have been already visited */
    for $k := 1$ **to** m :
(1) $start\_states(v, k) := \emptyset$;
(2)   **foreach** v_p such that $((v_p, v) \in E)$:
(3)   **foreach** $(v_i, q_x^k) \in start\_states(v_p, k)$:
(4)   **foreach** $a \in \lambda(v_p, v)$:
(5) $start\_states(v, k) := start\_states(v, k) \cup \{(v_i, \delta^k(q_x^k, a))\}$;
          endfor
        endfor
      endfor
    endfor
  endif
/* Launching a copy of the automaton $\mathcal{A}^{k+1}$ */
  for $k := 1$ **to** $m - 1$:
    if ($k \bmod 2 = 0$) **then**
(6)   **if** ($\exists v_i$ such that: ($\forall (v_i, q_x^k) \in start\_states(v, k)$: $q_x^k \in F^k$)) **then**
        /* A pattern ALL has been recognized: $(v_i, v) \models ALL(R^k)$ */
(7) $start\_states(v, k + 1) := start\_states(v, k + 1) \cup \{(v, q_0^{k+1})\}$;
      endif
    else
(8)   **if** ($\exists (v_i, q_x^k) \in start\_states(v, k)$ such that: $q_x^k \in F^k$) **then**
        /* A pattern SOME has been recognized: $(v_i, v) \models SOME(R^k)$ */
(9) $start\_states(v, k + 1) := start\_states(v, k + 1) \cup \{(v, q_0^{k+1})\}$;
      endif
    endif
  endfor
  if ($v = v_n$) **then**
    $output\_solutions(m + 1, v_n)$;
  endif
(10) /* if necessary, call the procedure *reduction* (See Section 3.6) */
end

Fig. 3. General algorithm

The algorithm. The procedure described in Figure 3 specify the set of actions executed when a vertex v is visited. Two tasks have to be done.

1. All the copies of all the automata previously launched, in the past of v, have to progress in word recognition (lines 1–5).

   If $(v_p, v) \in E$ and $start\_states(v_p, k) \neq \emptyset$ then copies of $\mathcal{A}^k$ have been previously launched. From $(v_i, q_x^k) \in start\_states(v_p, k)$, we conclude first that a copy of $\mathcal{A}^k$ has been launched in v_i and second that there is at least one word $w \in \mathcal{L}(v_i, v_p)$ such that $q_x^k \in \delta^k(q_0^k, w)$. So the algorithm makes this copy of $\mathcal{A}^k$ progress according to labelings of the edge (v_p, v).

   Note that all the copies of the automata previously launched continue their analysis till the vertex v_n is visited. This is necessary as we do not known in advance if a partial solution will give rise to a solution.

2. When a prefix of the general pattern has been recognized, a new copy of the next automaton has to be launched (lines 6–9).

   If there is an automaton $\mathcal{A}^k$ such that a copy of $\mathcal{A}^k$ has been launched when visiting vertex v_i and $(v_i, v) \models ALL(R^k)$ (when k is an even number), or $(v_i, v) \models SOME(R^k)$ (when k is an odd number), then a copy of the automaton $\mathcal{A}^{k+1}$ has to be launched from v.

The set of all the solutions is obtained with the procedure described in Figure 4 by calling $output\_solutions(m + 1, v_n)$. If the whole pattern has not been recognized at the end of the computation, it is also possible to find out the longest prefix of the diamond necklace for which partial solution exist.

It is important to note that actions executed when visiting a vertex v only depend on values of variables $start\_states$ of v's immediate predecessors (This allows to adapt the algorithm to an on the fly detection when used in debugging distributed applications [4]).

It is also important to note that if we are only interested in the simpler problem which consists in deciding if an a priori given set of $m + 1$ vertices is a solution, then the data structures and the algorithm can be greatly simplified.

3.6 Deciding if There Exists a Solution

The previous algorithm finds all the solutions, *i.e.*, all sets of $m + 1$ vertices $(v^1, v^2, \cdots, v^m, v^{m+1})$ satisfying the pattern in G^λ. If we are only interested in knowing if there is a solution, the contents of variables $start\_states(v, k)$ can be

---

[4] [5] presents such an algorithm which detects on the fly the simple primitive pattern: $(v_1, v_n) \models SOME(R_1)$. In that case, there is only one pair of vertices that can be a solution; moreover this pair is defined a priori. For this very simple pattern, variables needed for the detection reduce to a boolean array whose size is equal to the number of states of the corresponding automaton. Each process of the distributed application which is debugged, manages a copy of this array and each application message piggybacks the value of the sender process array. In the general case, every process has to manage an array $start\_states[1..m]$ and messages have to carry the value of this array.

```
Solution: array[1..m + 1] of vertex;

Procedure output_solutions (k : integer;  v : vertex);
begin
    Solution[k] := v;
    if (k = 1) then
        print(Solution);
    else
        if (k mod 2 = 0) then
            /* Continue with all vᵢ such that (vᵢ, v) ⊨ ALL(Rᵏ) */
            foreach  vᵢ  such that: (∀ (vᵢ, qₓᵏ) ∈ start_states(v, k) : qₓᵏ ∈ Fᵏ))
                output_solutions(k − 1, vᵢ);
            endfor
        else
            /* Continue with all vᵢ such that (vᵢ, v) ⊨ SOME(Rᵏ) */
            foreach  vᵢ  such that: (∃ (vᵢ, qₓᵏ) ∈ start_states(v, k) : qₓᵏ ∈ Fᵏ))
                output_solutions(k − 1, vᵢ);
            endfor
        endif
    endif
end
```

Fig. 4. Enumerating the set of solutions

reduced in the following way. The procedure *reduction* (line 10) decreases the size of variables $start\_states$. This procedure does the following actions. At the end of the visit of vertex v, a pair (v_i, q_x^k) belonging to the set $start\_states(v, k)$ is suppressed if one of the two following predicates is true.

1. **Predicate P1**: k is an odd number and there exists another pair (v_j, q_x^k) in $start\_states(v, k)$.
   From $(v_i, q_x^k) \in start\_states(v, k)$, we deduce that there exists at least one word $w1$ such that $w1 \in \mathcal{L}(v_i, v)$ and $q_x^k \in \delta^k(q_0^k, w1)$. Similarly, we conclude that there exists also a word $w2$ such that $w2 \in \mathcal{L}(v_j, v)$ and $q_x^k \in \delta^k(q_0^k, w2)$. Thus, if a word $w3$ is such that $F^k \cap \delta^k(q_x^k, w3) \neq \emptyset$, we can conclude that both words $w1.w3$ and $w2.w3$ belong to $\mathcal{L}(R^k)$. So, if we are not interested in computing all solutions, it is sufficient to indicate that q_x^k is a state in which a copy of automaton $\mathcal{A}^k$ arrived after the vertex v has been visited.
2. **Predicate P2**: k is an even number, $\mathcal{L}(R^k)$ is a suffix language and there exist another pair (v_j, q_y^k) such that there is path from v_i to v_j (*i.e.* $\mathcal{P}(v_i, v_j) \neq \emptyset$).
   For each word $w3 \in \mathcal{L}(v_j, v)$, there exists at least one word $w1 \in \mathcal{L}(v_i, v)$ such that $w1 = w2.w3$.

If $\mathcal{L}(R^k)$ is a suffix language, $\forall w \in \Sigma^*, w1.w \in \mathcal{L}(R^k) \Rightarrow w3.w \in \mathcal{L}(R^k)$. Therefore, if v' is a vertex such that $\mathcal{P}(v, v') \neq \emptyset$ then: $(v_i, v') \models ALL(R^k) \Rightarrow (v_j, v') \models ALL(R^k)$. It follows that only (v_j, q_y^k) has to be memorized if we are not interested in computing all solutions [5].

3.7 Complexity

During an on-the-fly detection and when one try to find all the solutions, the storage complexity of this algorithm is $O(m.n^2.r)$ where m is the number of automata (i.e., the length of the diamond necklace), n is the number of vertices in the graph and r is the maximal number of states of an automaton (i.e. $r = max\{r^k \mid 1 \leq k \leq m\}$ with $r^k = \mid Q^k \mid$). Note that automaton $\mathcal{A}^1$ is launched only once when vertex v_1 is visited. Therefore, the size of the structure $start\_states(v, 1)$ is bounded by r^1 whereas the size of $start\_states(v, k)$ is bounded by $(p_v.r^k + 1)$ if $2 \leq k \leq m$ (where p_v is the number of immediate predecessors of vertex v).

Let $t_x^k(a) = \mid \delta^k(q_x^k, a) \mid$ and let $t^k = max\{max\{t_x^k(a) \mid a \in \Sigma\} \mid q_x^k \in Q^k\}$. Note that $t^k = 1$ if automaton $\mathcal{A}^k$ is deterministic. Let $t = max\{t^k \mid 1 \leq k \leq m\}$. Assume that elements (v_i, q_x^k) of $start\_states(v, k)$ are sorted according to the first component. The time complexity of the general algorithm is

$$O(m.n^3.r.t.l)$$

where l is the number of labels in Σ.

If $k \geq 2$, computation of the set $start\_states(v, k)$ requires less than $p_v^2.r^k.t^k.l$ insertions of elements.

The time complexity of this algorithm is cubic whereas the complexity of the naive approach described in Section 3.2 is $O(n^m)$. Note that, when $m = 1$, the naive approach consists in determining the product of two automata.

When the two reduction rules (described in Section 3.6) are applied, the size of the structure $start\_states(v, k)$ is bounded by $(s.r^k)$ where s is the width of the partial order (i.e., the size of the largest antichain). In the dag corresponding to the execution of a distributed application, the value of s is bounded by the number of processes observed during the debugging activity. In this case, the storage and time complexities of the algorithm also decrease .

4 Conclusion

The problem tackled in this paper originated from the debugging of distributed applications. Execution of such an application can be modeled as a partially

---

[5] In [10], a particular simple kind of diamond necklaces called atomic sequences is defined. The language associated to each diamond contains all the words built with all the symbols of an alphabet except those containing a particular forbidden symbol. Such a language is a suffix language. Consequently, the second reduction rule explained above can by applied in this particular case.

ordered set of process states. The debugging of control flows (sequences of process states) of these executions is based on the satisfaction of predicates by process states. A process state that satisfies a predicate inherits its label. It follows that, in this context, a distributed execution is a labeled directed acyclic graph. Debug or determine if control flows of a distributed execution satisfies some property amounts to test if the labeled acyclic graph includes some pattern defined on predicate labels.

This paper first introduced a general pattern (called *diamond necklace*) which includes classical patterns encountered in distributed debugging. Then an algorithm detecting such patterns in a labeled acyclic graph has been presented. To be easily adapted to an on-the-fly detection of the pattern in distributed executions, the algorithm has been based on a visit of the nodes of the graph according to a topological sort. Its time complexity is polynomial.

Acknowledgments

The authors would like to thank Didier Caucal and Jean-Xavier Rampon whose comments greatly improved both the content and the presentation of the paper.

References

[1] . Babaoğlu, E. Fromentin, and M. Raynal: "A Unified Framework for the Specification and Run-time Detection of Dynamic Properties in Distributed Computations", *The Journal of Systems and Software*, special issue on Software Engineering for Distributed Computing, 33:(3), pp. 287–298, 1996.

[2] L. Bougé, and N. Francez: "A compositional approach to superimposition", In *Proc. of the 15th ACM SIGACT-SIGPLAN Symposium on Principle of Programming Languages*, pp. 240–249, San Diego, California, January 1988.

[3] R. Cooper, and K. Marzullo: "Consistent Detection of Global Predicates", In *Proc. of the ACM/ONR Workshop on Parallel and Distributed Debugging*, pp. 163–173, Santa Cruz, California, May 1991.

[4] E. Fromentin, C. Jard, G. Jourdan, and M. Raynal: "On-the-fly Analysis of Distributed Computations", *Information Processing Letters 54*, pp. 267–274, 1995.

[5] E. Fromentin, M. Raynal, V.K. Garg, and A.I. Tomlinson: "On the fly testing of regular patterns in distributed computations", In *Proc. of the the 23rd International Conference on Parallel Processing*, pp. 73–76, St. Charles, IL, August 1994.

[6] V.K. Garg, A.I. Tomlinson, E. Fromentin, and M. Raynal: "Expressing and Detecting General Control Flow Properties of Distributed Computations", In *Proc. of the 7th IEEE Symposium on Parallel and Distributed Processing*, pp. 432–438, San-Antonio (USA), October 1995.

[7] V.K. Garg and B. Waldecker: "Detection of Unstable Predicates in Distributed Programs", In *Proc. of the 12th International Conference on Foundations of Software Technology and Theoretical Computer Science*, Springer Verlag, LNCS 652, pp. 253–264, New Delhi, India, December 1992.

[8] M.A. Harrison: "Introduction to Formal Language Theory", Addison-Wesley series in computer science, 1978.

[9] M. Hurfin, N. Plouzeau, and M. Raynal: "A Debugging Tool for Estelle Distributed Programs", *Journal of Computer Communications*, 28(5), pp. 328–333, May 1993.

[10] M. Hurfin, N. Plouzeau, and M. Raynal: "Detecting atomic sequences of predicates in distributed computations", In *Proc. ACM workshop on Parallel and Distributed Debugging*, pp. 32–42, San Diego, May 1993. (Reprinted in SIGPLAN Notices, Dec. 1993).

[11] J.E. Hopcroft and J.D. Ullman: "Introduction to Automata Theory, Languages, and Computation", Addison-Weslay Publishing Company, 418 pages, 1979.

[12] L. Lamport: "Time, clocks and the ordering of events in a distributed system", *Communications of the ACM*, 21(7), pp. 558–565, July 1978.

Algebraic Graph Derivations for Graphical Calculi

WOLFRAM KAHL

Department of Computing Science, German Armed Forces University Munich
e-mail: kahl@informatik.unibw-muenchen.de

1 Introduction

Relational formalisations can be very concise and precise and can allow short, calculational proofs under certain circumstances. Examples are can be found in [SS93], and also in the formalisation of second-order term graph rewriting in [Kah95b, Kah96]; for further applications of relational methods see also the book [BS96].

In situations corresponding to the simultaneous use of many variables in predicate logic, however, either a style using predicate logic with point variables has to be adopted or impractical and clumsy manipulations of tuples have to be employed inside relation calculus. In the application of relational formalisation to term graphs with bound variables [Kah95b, Kah96] we have been forced to employ both methods extensively, and, independently of other approaches, have been driven to develop a *graphical calculus* for making complex relation algebraic proofs more accessible.

It turns out that, although our approach shares many common points with those presented in the literature [BH94, CL95], it still is more general and more flexible than those approaches since we draw heavily on additional background in algebraic graph rewriting (see [EKL90] for a tutorial overview).

The part of the structure of relation algebra that can readily be exploited in graphical calculi is that of a *unitary pretabular allegory* (UPA, introduced in [FS90]). Allegories are a generalisation of categories to cope with relation-like structures; we shall not need any allegory theory in this paper, but only refer to it for comparison with one of the main streams of related work in the literature. In [BH94], an approach to transformations of expressions in UPAs via transformations of graphs has been presented and proven correct. The approach has been developed with a bias towards VLSI circuit development and the formalisation and drawings reflect this.

More or less building on the approach of [BH94], another approach to graphical calculi has been presented in [CL95], where a gentler introduction is given and an attempt is made to somewhat generalise beyond UPAs.

Both approaches, however, present the transformation rules as low-level graph manipulation rules and do not resort to any established graph transformation mechanism. As a result, there is only a fixed set of transformation rules that correspond to the basic axioms of the calculus, but no general mechanism to formulate new rules corresponding to proven theorems or special definitions.

In this paper we start from a slightly more general definition of *diagram* as basic data structure for our graphical calculus, and we proceed to give algebraic definitions of rule application and derivation. We cleanly separate the syntax and the semantics of our diagrams and we define correctness of rules on a high level.

For reasons of space we do not present any proofs, but concentrate on giving ample motivation and at least a few examples. I gratefully acknowledge the comments of an anonymous referee.

2 Type and Relation Terms

The structure we are going to exploit in our diagram proofs is that of a locally complete unitary pretabular allegory (LCUPA) [FS90], which is essentially an abstract relation algebra in the sense of [SS93] (without negation), equipped with all direct products (which are understood to be formed in the underlying category of total functions throughout this paper).

For improving understandability we shall use the more widespread nomenclature of abstract relation algebra (rather than that of LCUPAs) and also its notation as agreed upon in [BS96]. So we call the morphisms **relations**; **composition** of two relations $R : A \leftrightarrow B$ and $S : B \leftrightarrow C$ is written $R;S$; the **converse** of a relation $R : A \leftrightarrow B$ is $R^\smile : B \leftrightarrow A$; **intersection** of two relations $R, S : A \leftrightarrow B$ is $R \sqcap S$, and their **union** is $R \sqcup B$; for any object A, the **identity relation** is $\mathbb{I}_A$; for two objects A and B the **universal relation** is $\mathbb{T}_{A,B}$, and the **empty relation** is $\perp\!\!\!\perp_{A,B}$. Inclusion of $R : A \leftrightarrow B$ in $S : A \leftrightarrow B$, i.e. the fact that $R \sqcap S = R$, is denoted by $R \sqsubseteq S$.

Among the binary operators, relational compositon ";" has higher priority than union "$\sqcup$" and intersection "$\sqcap$".

For the **direct product** $A \times B$ of two objects A and B, $\pi_{A \times B}$ and $\rho_{A \times B}$ denote the first resp. second **projection mapping**. For two products $A \times B$ and $C \times D$ and two relations $R : A \leftrightarrow C$ and $S : B \leftrightarrow D$, the product $(R\|S) : (A \times B) \leftrightarrow (C \times D)$ of R and S is defined as $(R\|S) = \pi_{A \times B};R;\pi^\smile_{C \times D} \sqcap \rho_{A \times B};S;\rho^\smile_{C \times D}$.

The laws that are required to hold are the usual laws of relation calculus which we do not restate here.

For a set A we denote the set of **finite sequences** of elements of A with A^*. Two sequences s and t can be *concatenated* to form the sequence $s^\smallfrown t$. For a function $f : X \to Y$, we denote the mapping of f to sequences by $f^* : X^* \to Y^*$.

A sequence of objects of a LCUPA is understood to be the corresponding finite product.

We write set-comprehensions according to the Z-notation [Spi89], which uses the pattern "{ *signature* | *predicate* • *term* }" instead of the otherwise frequently observed pattern "{ *term* | *predicate* }". So we can write, as an example, the set containing the first four square numbers as $\{n : \mathbb{N} \mid n < 4 \bullet n^2\} = \{0, 1, 4, 9\}$.

We now introduce type terms and relation terms as the syntactic basis of our calculus. We reuse the operator symbols introduced above, but we shall employ "≡" for syntactical equality of terms.

Definition 2.1 A **type term** can be
- a *type constant*, including $\mathbb{1}$ for the unit type,
- a *type variable* $(\alpha, \beta, \gamma, \ldots)$,
- a *product type* $T \times U$ of two type terms T and U, or
- a *constructor type* $C(T_1, \ldots, T_n)$ created from n type terms $T_1, \ldots, T_n$ by application of an n-ary *type term constructor* C. ☐

Obviously, the product type could be considered as just another constructor type, but since it has a special status in LCUPAs, we rather treat it separately.

A **type substitution** is a partial function with finite domain from type variables to type terms. Application of a type substitution is defined as usual.

Definition 2.2 A **relation term** of type $A \leftrightarrow B$ for two type terms A and B can be
- a *relation constant*, including $\mathbb{I}$ (if $A \equiv B$), $\top$, $\bot$, π (if there is a type term C such that $A \equiv B \times C$), and ρ (if there is a type term C such that $A \equiv C \times B$),
- a *relation variable*,
- the *converse* $R^{\smile}$ of a relation term R of type $B \leftrightarrow A$ (also written $R : B \leftrightarrow A$),
- the *composition* $R;S$ of two relation terms $R : A \leftrightarrow C$ and $S : C \leftrightarrow B$,
- the *intersection* $R \sqcap S$ or the *union* $R \sqcup S$ of two relation terms $R : A \leftrightarrow B$ and $S : A \leftrightarrow B$,
- a *constructor term* $c(R_1, \ldots, R_n)$ created from n relation terms R_i by application of an n-ary *relation term constructor* c, with type constraints on the R_i depending on c.

Additionally, in any composite term, all occurrences of a relation variable must be of the same type. ☐

A **relation substitution** is a partial function with finite domain from relation variables to relation terms. Application of a substitution is again defined as usual.

Finally, an atomic **relational formula** is either an equality $R = S$ or an inclusion $R \sqsubseteq S$ for two relational terms R and S of the same type.

3 Relational Diagrams

3.1 Syntax

We now introduce *relational diagrams* as a special kind of labelled graphs or hypergraphs. Although hypergraphs are of course more general, we include the graph case for offering the reader a smoother access:

Definition 3.1 A **relational diagram** is a labelled directed (hyper-)graph $(\mathcal{N}, \mathcal{E}, \mathbf{s}, \mathbf{t}, \mathbf{n}, \mathbf{e})$ with $\mathcal{N}$ its node sets, $\mathcal{E}$ its edge set, $\mathbf{s} : \mathcal{E} \to \mathcal{N}$ (resp. $\mathbf{s} : \mathcal{E} \to \mathcal{N}^*$) the source mapping, $\mathbf{t} : \mathcal{E} \to \mathcal{N}$ (resp. $\mathbf{t} : \mathcal{E} \to \mathcal{N}^*$) the target mapping, $\mathbf{n}$ is the node labelling, assigning every node a type term, and $\mathbf{e}$ is the edge labelling, assigning every edge a relation term of type $\mathbf{n}(\mathbf{s}(e)) \leftrightarrow \mathbf{n}(\mathbf{t}(e))$ (resp. $\mathbf{n}^*(\mathbf{s}(e)) \leftrightarrow \mathbf{n}^*(\mathbf{t}(e))$). $\qquad\square$

Homomorphisms between relational diagrams are defined as usual:

Definition 3.2 A **relational diagram homomorphism** f from one relational diagram G_1 to another G_2 is a pair $(f\mathbf{n}, f\mathbf{e})$ of functions, with
- $f\mathbf{n} : \mathcal{N}_1 \to \mathcal{N}_2$, $f\mathbf{e} : \mathcal{E}_1 \to \mathcal{E}_2$,
- $\mathbf{s}_2(f\mathbf{e}(a)) = f\mathbf{n}(\mathbf{s}_1(a))$, $\mathbf{t}_2(f\mathbf{e}(a)) = f\mathbf{n}(\mathbf{t}_1(a))$, or, in the case of hypergraphs, $\mathbf{s}_2(f\mathbf{e}(a)) = f\mathbf{n}^*(\mathbf{s}_1(a))$, $\mathbf{t}_2(f\mathbf{e}(a)) = f\mathbf{n}^*(\mathbf{t}_1(a))$,
- there is a type substitution τ such that for all nodes v we have

$$\mathbf{n}_2(f\mathbf{n}(v)) \equiv \tau(\mathbf{n}_1(v)) \ ,$$

- there is a relation substitution σ such that for all edges a we have

$$\mathbf{e}_2(f\mathbf{e}(a)) \equiv \sigma(\mathbf{e}_1(a)) \ .$$

A homomorphism is called **plain** if τ and σ can be set to empty substitutions. $\square$

For any relational diagram $G = (\mathcal{N}, \mathcal{E}, \mathbf{s}, \mathbf{t}, \mathbf{n}, \mathbf{e})$ we define its **emptied diagram** as the corresponding discrete graph: $G^0 := (\mathcal{N}, \emptyset, \emptyset, \emptyset, \mathbf{n}, \emptyset)$.

In contrast with the graphs of [CL95], which are equipped with a designated source and a designated target node, and with the pictures and networks of [BH94], which use a connection mechanism that is also based on a source-target view, but includes the possibility of considering multiple ports, our diagrams are not equipped with any such indication of direction.

Therefore, when considering the semantics of a diagram, a direction has to be imposed from the outside, and we use *interfaces* for this purpose.

In the simple graph case, an interface essentially is just a graph with two nodes and one variable-labelled edge inbetween together with a homomorphism that essentially just serves to flag a source node and a target node in the diagram in question. The general definition that also copes with hyperedges has a more complicated formulation:

Definition 3.3 An **interface** (I, j) for a relational diagram G consists of a relational diagram I and a homomorphism j from I^0 to G, where I has only one edge ($|\mathcal{E}_I| = 1$) that is in addition labelled with a variable, and I does not have any isolated nodes. $\qquad\square$

j could be considered as a restricted kind of partial morphism from I to G. For graph rules, we have an inclusion semantics "$L \sqsubseteq R$" in mind. Since it is usually advisable to preserve previously established information, we can arrive at a useful rule concept by just asking for a homomorphism between the rule sides. This homomorphism has to be plain since otherwise there could be clashes in the instantiation of variables:

Definition 3.4 A rule $(L \xrightarrow{r} R)$ consists of two relational diagrams L and R together with a plain homomorphism r from L to R. ☐

An example rule is draw in the following diagram; here α and β are type variables, Θ syntactically is a constant (an arbitrary equivalence relation) and P and Q are relation variables; the rule then reads, that if $\Theta : \beta \leftrightarrow \beta$, for all $A : \alpha \leftrightarrow \beta$ and $B : \alpha \leftrightarrow \beta$ the following inclusion holds:[1]

$$(A;\Theta \sqcap B);\Theta \sqsubseteq (A \sqcap B;\Theta);\Theta$$

According to the definition of rules as single homomorphisms, rewriting will be defined by a single pushout construction[2] — the difference to the single-pushout approach of [Ken90, Löw90] is that here we still consider *total* homomorphisms:

Definition 3.5 A **rewrite step** for a rule $(L \xrightarrow{r} R)$ and a relational diagram G together with a homomorphism f from L to G is the pushout

$$
\begin{array}{ccc}
L & \xrightarrow{\ r\ } & R \\
{\scriptstyle f}\downarrow & & \downarrow{\scriptstyle g} \\
G & \xrightarrow{\ s\ } & H
\end{array}
$$

of r and f; the **result diagram** is the pushout object H.

A **derivation** of $H := G_n$ from $G := G_0$ is a sequence of rewrite steps

$$
\begin{array}{ccc}
L_i & \xrightarrow{\ r_i\ } & R_i \\
{\scriptstyle f_i}\downarrow & & \downarrow{\scriptstyle g_i} \\
G_{i-1} & \xrightarrow{\ s_i\ } & G_i
\end{array}
$$

and we let the **derivation morphism** be $s_1; \ldots; s_n$. ☐

---

[1] Actually, the diagram given here is a little bit stronger than the original inclusion, in that we did not draw an independent second "B"-edge, but this stronger version follows easily from the symmetry of the equivalence relation Θ and is easier to draw.

[2] In a category, for three objects A, B, and C and two arrows $f : A \to B$ and $g : A \to C$, a *pushout* is an object D together with two arrows $h : B \to D$ and $j : C \to D$ with $f;h = g;j$, such that for everyobject D' together with two arrows $h' : B \to D'$ and $j' : C \to D'$ with $f;h' = g;j'$ there is a unique arrow $u : D \to D'$ such that $h;u = h'$ and $j;u = j'$. For an introduction to the use of the pushout concept in graph rewriting see [EKL90].

With the example rule above, we can obtain the following rewrite step:

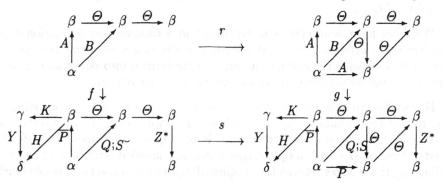

3.2 Semantics

For abbreviating the formal treatment, we informally treat the product type constructor as associative, and we consider a sequence $\langle T_1, \ldots, T_n \rangle$ of type terms as denoting the product term $T_1 \times \cdots \times T_n$. For the image of a set $S : \mathbb{P}(A)$ under a function $f : A \to B$ we just write $f(S) : \mathbb{P}(B)$.

Definition 3.6 (Readout) Let a relational diagram G and an interface $\mathcal{I} = (I, j)$ for G with the one hyperedge $a_{\mathcal{I}}$ be given. Let $\langle m_1, \ldots, m_{\text{ex}} \rangle$ be a sequentialisation of the nodes of G outside the range of j (this could be made uniquely determined by for example demanding a total ordering on the node set).

Then let $\mathcal{T}_{G,\mathcal{I}}$ be the following type term representing all nodes of the interface together with all other nodes of G:

$$\mathcal{T}_{G,\mathcal{I}} \equiv \mathbf{n}^*(j^*(\mathbf{s}(a_{\mathcal{I}}))) \times \mathbf{n}^*(j^*(\mathbf{t}(a_{\mathcal{I}}))) \times \mathbf{n}^*(\langle m_1, \ldots, m_{\text{ex}} \rangle)$$

Furthermore, let $\pi_{\text{in}} : \mathcal{T}_{G,\mathcal{I}} \to \mathbf{n}^*(j^*(\mathbf{s}(a_{\mathcal{I}})))$ and $\pi_{\text{out}} : \mathcal{T}_{G,\mathcal{I}} \to \mathbf{n}^*(j^*(\mathbf{t}(a_{\mathcal{I}})))$ be the projections onto the source and target of the interface image. Let for every node $x \in \mathcal{N}_I$ of the interface diagram $\pi_x : \mathcal{T}_{G,\mathcal{I}} \to \mathbf{n}(j(x))$ be the projection onto its component, and for every node sequence $s \in \mathcal{N}^*$ let $\pi_s : \mathcal{T}_{G,\mathcal{I}} \to \mathbf{n}^*(s)$ be a projection onto the components of $\mathcal{T}_{G,\mathcal{I}}$ corresponding to s. (Since j need not be injective, there can be a choice of projections for π_s, but as we shall see by the construction below, this does not influence the result significantly. We can choose a canonic construction that assigns to $z \in j(\mathcal{N}_I)$ the projection corresponding to that node $x \in \mathcal{N}_I$ that is the first one in the sequence $\mathbf{s}(a_{\mathcal{I}})\char`\^\mathbf{t}(a_{\mathcal{I}})$, for which $z = j(x)$.)

The **readout** of G via $\mathcal{I}$ is now defined to be the relation term

$$G_{\lceil \mathcal{I} \rceil} : \mathbf{n}^*(j^*(\mathbf{s}(a_{\mathcal{I}}))) \leftrightarrow \mathbf{n}^*(j^*(\mathbf{t}(a_{\mathcal{I}})))$$

with (see example and explanation below):

$$
\begin{aligned}
G_{\lceil \mathcal{I} \rceil} \equiv \pi_{\text{in}}^{\smile} ; (&\bigsqcap \{ a : \mathcal{E} \bullet (\pi_{\mathbf{s}(a_{\mathcal{I}})}; e(a) \sqcap \pi_{\mathbf{t}(a_{\mathcal{I}})}); \mathbb{T} \} \\
&\sqcap \bigsqcap \{ x, y : \mathcal{N}_I \mid x \neq y \wedge j(x) = j(y) \bullet (\pi_x \sqcap \pi_y); \mathbb{T} \} \\
&\sqcap \pi_{\text{out}})
\end{aligned}
$$

$\square$

(The intersections over sets strictly speaking also present choices wrt. the construction of $G_{\mathcal{I}}$.)

What we have done here is to construct in a canonic way a relational expression that corresponds to that encoded in the graph — using the laws of relational calculus with products it can be transformed into equivalent expressions that may be more appealing for one or the other reason.

Below, for an example, to the left a relational diagram G consisting of four hyperedges is shown together with a three-input, one-output interface $\mathcal{I}$. To the right, we have drawn an intermediate diagram that should clarify the construction of $G_{\mathcal{I}}$. All simple edges there are labelled with the identity $\mathbb{I}$. Collapsing those edges returns the original diagram G, so the two are obviously equivalent.

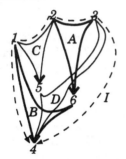 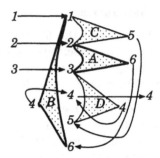

On the other hand, the three main layers of nodes in the right diagram obviously correspond to the input type $\mathbf{n}^*(j^*(\mathbf{s}(a_{\mathcal{I}})))$, $\mathcal{T}_{G,\mathcal{I}}$ and the output type $\mathbf{n}^*(j^*(\mathbf{t}(a_{\mathcal{I}})))$ respectively. The nodes ending the hyperedges can be regarded as the input of $\mathbb{T}$ in the first set component of $G_{\mathcal{I}}$; the readout is

$$\pi_{123}^{\smile};((\pi_{23};A \sqcap \pi_6);\mathbb{T} \sqcap (\pi_{16};B \sqcap \pi_4);\mathbb{T} \sqcap (\pi_{12};C \sqcap \pi_5);\mathbb{T} \sqcap (\pi_{53};D \sqcap \pi_3);\mathbb{T} \sqcap \pi_4)$$

This is equivalent to

$$\pi_{123}^{\smile};((\pi_1 \sqcap \pi_{23};A);B \sqcap (\pi_{12};C \sqcap \pi_3);D);\pi_4$$

and again (under the assumption of an appropriately nested input product and using the isomorphism $\mathsf{assPA} : (\alpha \times \beta) \times \gamma \to \alpha \times (\beta \times \gamma)$) with:

$$(\mathbb{I}\|A);B \sqcap \mathsf{assPA};(C\|\mathbb{I});D \ ,$$

which is easy to relate to the original diagram G.

The second set component from the readout definition is empty here, since the interface is injective; otherwise there would be additional $\mathbb{I}$-edges between nodes of the middle layer.

Unlike [CL95], we did not switch to predicate logic formulae, so we could stay inside the language of relation calculus extended with direct products (i.e., the language of LCUPAs).

Unlike [BH94], we started from a graph without additional hierarchic structure, so we had to construct $G_{\overline{\tau}}$ by "brute force". But since we consider $G_{\overline{\tau}}$ to be only an intermediary result anyway, the artificialness of its structure does not hurt our approach at all.

A different approach would have been to use the algorithm proposed in [VH91] for transforming any predicate logic formula into a relational expression, but that algorithm has to be capable to deal with more general situations than those reflected in relational diagrams, and the result would have been similarly artificial in its nature anyway.

We now start considering an arbitrary model $\mathcal{R}$ for our formalism, that is, a relation algebra (or LCUPA) together with interpretations for all constants and term constructors. When we write $\mathcal{R} \models F$ for some relational formula F then that has to be taken to mean that for every valuation of type variables with objects of $\mathcal{R}$ and every valuation of relation variables with relations from $\mathcal{R}$ the semantics of F in $\mathcal{R}$ is true.

The central result about the readout construction then is that all the choices encountered there do not influence the semantics of the result:

Proposition 3.7 For every relational term X resulting from changing any choices made while constructing $G_{\overline{\tau}}$, we have $\mathcal{R} \models G_{\overline{\tau}} = X$. □

We also obtain the fact that plain homomorphisms can only decrease the semantics:

Lemma 3.8 For every interface (I, j) for K and every plain homomorphism k from K to G, the following holds: $\mathcal{R} \models G_{\lceil(I,j;k)\rceil} \sqsubseteq K_{\lceil(I,j)\rceil}$.

Accordingly, for every interface (I, j) for G and every subgraph K of G with natural injection k, whenever $j(I^0) \subseteq k(K)$ then $\mathcal{R} \models G_{\lceil(I,j)\rceil} \sqsubseteq K_{\lceil(I,j;k^{-1})\rceil}$. □

Based on the semantics, we can now form a general concept of admissibility of rules:

Definition 3.9 A rule $(L \xrightarrow{r} R)$ is **correct** if for all interfaces (I, j) for L,

$$\mathcal{R} \models L_{\lceil(I,j)\rceil} = R_{\lceil(I,j;r)\rceil} \ .$$

□

It is not difficult to construct concrete rules where this equality holds for some interfaces, but not for others, and where application of these rules leads to invalid proofs. This equality is, however, guaranteed to hold for all interfaces for L whenever it holds for any interface (I, j) for L where j is surjective on the nodes.

For all the specific rules listed in [BH94, CL95] corresponding diagram rules can be formulated, and the correctness proofs carry over for any $\mathcal{R}$.

Application of correct rules yields derivations with a useful semantics:

Proposition 3.10 Let an interface (I,j) for a relational diagram G and a rule $(L\xrightarrow{r}R)$ with a matching homomorphism f from L to G be given, and consider the rewriting step yielding the pushout object H and the homomorphism s from G to H, then we have $\mathcal{R} \models G_{\lceil(I,j)\rceil} = H_{\lceil(I,j;s)\rceil}$.

Accordingly, for every interface (I,j) for the starting diagram G of a derivation from G to H with derivation morphism σ, we have $\mathcal{R} \models G_{\lceil(I,j)\rceil} = H_{\lceil(I,j;\sigma)\rceil}$. $\square$

With all this, we can formulate a strategy for finding a proof of the inclusion formula $R \sqsubseteq S$ as a graph derivation on relational diagrams:

i) Construct a diagram G together with an interface (I,j) such that $\mathcal{R} \models R = G_{\lceil(I,j)\rceil}$.
ii) Perform a suitable graph derivation on G, yielding H and the derivation morphism σ.
iii) Factorise σ into σ' from G to a suitable diagram H' and k from H' to H
iv) Recognise H' as a diagram with $\mathcal{R} \models S = H'_{\lceil(I,j;\sigma')\rceil}$.

Only in rare cases the full derivation result H will be needed (yielding an equality), usually only an inclusion is required anyway.

3.3 Examples without Hyperedges

For our first example, a part of the proof of [Kah95b, Lemma 4.2.3], let us assume an object O and relations C, V and W in the underlying relation algebra $\mathcal{R}$ for which the following two rules are correct:

$$W; C^\smile \sqsubseteq C; W^\smile$$

$$C; V^\smile \sqsubseteq V^\smile; C$$

For both rules we have drawn the right hand side; the left hand side is the subgraph induced by the boldened edges — as long as the rule morphism is injective, this abbreviating method of representation is possible. (The different layout of the two rules has been chosen for better fitting to the application below. Furthermore note that the first rule could also have been read $C; W^\smile \sqsubseteq W^\smile; C$ — the rules are valid no matter wich interface into the left-hand side is considered.)

Now, for a derivation of $\mathbb{T};(C \sqcap V;W) \sqsubseteq \mathbb{T};(C \sqcap V;W);C$, first these two rules are applied in order and then Lemma 3.8:

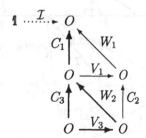

A standard inclusion chain for the same proof would be the following:

$$
\begin{aligned}
&\mathbb{T};(C \sqcap V;W) \\
&\sqsubseteq \mathbb{T};V;(W \sqcap V^{\smile};C) && \text{modal rule} \\
&\sqsubseteq \mathbb{T};(V^{\smile} \sqcap W;C^{\smile});C && \mathbb{T};V \sqsubseteq \mathbb{T}, \text{modal rule} \\
&\sqsubseteq \mathbb{T};(V^{\smile} \sqcap C;W^{\smile});C && W;C^{\smile} \sqsubseteq C;W^{\smile} \\
&\sqsubseteq \mathbb{T};C^{\smile};(W \sqcap C;V^{\smile});C && \text{modal rule} \\
&\sqsubseteq \mathbb{T};(W \sqcap V^{\smile};C);C && \mathbb{T};C^{\smile} \sqsubseteq \mathbb{T}, C;V^{\smile} \sqsubseteq V^{\smile};C \\
&\sqsubseteq \mathbb{T};V^{\smile};(C \sqcap V;W);C && \text{modal rule} \\
&\sqsubseteq \mathbb{T};(C \sqcap V;W);C && \mathbb{T};V^{\smile} \sqsubseteq \mathbb{T}
\end{aligned}
$$

A different way to present the diagrammatic proof could be via one graph with additional annotation of the edges with their "generation" (in addition, the edges needed in the result have been boldened):

Obviously, the diagram proof is simpler and more intuitive than the linear (term) proof. The main reason for this are the frequent "changes of point of view" that are reflected in applications of modal rules or of the Dedekind rule. Not every proof, however, exhibits such a behaviour. For our second example consider a part of the proof of [Kah95b, Lemma 3.5.9]: ("$\overline{R}$" is the complement of "R"; since it does not play any part in the graphical part of the calculus, we omitted it in the introduction. From the point of view of Def. 2.2, the complement operator is just a unary term constructor.)

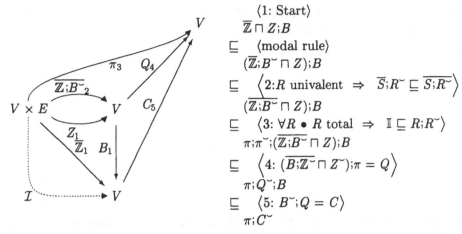

$$\langle 1: \text{Start} \rangle$$
$$\overline{\mathbb{Z}} \sqcap Z;B$$
$$\sqsubseteq \quad \langle \text{modal rule} \rangle$$
$$(\overline{\mathbb{Z};B^\smile} \sqcap Z);B$$
$$\sqsubseteq \quad \langle 2:R \text{ univalent} \;\Rightarrow\; \overline{S};R^\smile \sqsubseteq \overline{S;R^\smile} \rangle$$
$$(\overline{\mathbb{Z};B^\smile} \sqcap Z);B$$
$$\sqsubseteq \quad \langle 3: \forall R \bullet R \text{ total} \;\Rightarrow\; \mathbb{I} \sqsubseteq R;R^\smile \rangle$$
$$\pi;\pi^\smile;(\overline{\mathbb{Z};B^\smile} \sqcap Z);B$$
$$\sqsubseteq \quad \langle 4: (\overline{B;\mathbb{Z}^\smile} \sqcap Z^\smile);\pi = Q \rangle$$
$$\pi;Q^\smile;B$$
$$\sqsubseteq \quad \langle 5: B^\smile;Q = C \rangle$$
$$\pi;C^\smile$$

Here, only one "change of the point of view" was necessary. Therefore, the diagram proof has almost the same length as the linear proof.

When, however, in addition to changes of the point of view there are also references to many previously introduced nodes, then the linear proof would have to resort to heavy use of tuple constructions and manipulations, and the proof would become unreadable.

One example is a part of a proof that would be pretty hard to understand even in a mixed style using point variables in a predicate logic argument; it has been taken from [Kah95b, page 160] and uses quite a few special symbols and laws from the context there; these are however irrelevant for getting a general impression of the bandwith and graph size involved here, since it is these factors that make other approaches extremely hard to handle on such a problem:

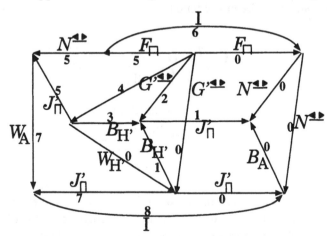

3.4 Example with Hyperedges

To see the beneficial effect the introduction of hyperedges can have, consider the following law (valid in the context of [Kah95b, Kah96]) which we want to use as a rule:

$$(\mathbb{I} \,\|\, \mathsf{conc});Z \sqsubseteq \mathsf{assPA};(Z \,\|\, \mathbb{I});Z$$

The middle diagram below directly depicts that rule:

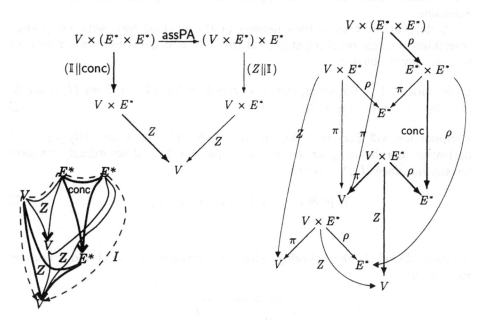

On the left we have drawn a hypergraph diagram denoting the same law, and it obviously is far simpler and more intuitive.

Although the term rule and its immediate rendering as a diagram rule still look simple enough, the problem is that for applicability of this rule usually lots of auxiliary laws for tuple manipulation would be necessary. The same applies to making use of the transformation results.

When one tries to avoid this by expanding the definitions of $(\_\|\_)$ and assPA, the result is the diagram rule on the right, which is far more complicated than the hypergraph rule, although it presents exactly the same information.

4 Extension to Branching Derivations

So far, the only operations that have been reflected in the calculus are composition and intersection.

Although these are already the most useful, we still can extend our approach without too much effort to cover laws that have a union as the outermost constructor of the right-hand side. We call the corresponding rules *branching rules*, since they give rise to several branches inside one derivation.

Definition 4.1   A **branching rule** $\mathbf{R} = (L, (r_i, R_i)_{i \in \Gamma_{\mathbf{R}}})$ is a relational diagram L together with a $\Gamma_{\mathbf{R}}$-family of diagrams R_i with respective plain homomorphisms r_i from R_i to L.   □

Every rule according to Def. 3.4 can be considered as a branching rule with a one-element index set, so our approach so far integrates smoothly with the extension.

A **derivation tree** is then defined in the obvious way, with every edge resembling a single rewriting step from Def. 3.5, and the semantics carries over without any problems:

Definition 4.2 A branching rule is **correct**, if for all interfaces (I, j) for L,
$$\mathcal{R} \models L_{[(I,j)]} = \bigsqcup \{i : \Gamma_{\mathbf{R}} \bullet R_{\hat{i}[(I,j;r_i)]}\}.$$ $\square$

Proposition 4.3 For every derivation tree for G with leaves $(H_j)_{j \in \Gamma^\bullet}$ and derivation morphisms s_j for the respective paths j defined accordingly, we have for any interface (I, j) of G that
$$\mathcal{R} \models K_{[(I,j)]} = \bigsqcup \{j : \Gamma^* \bullet H_{\hat{j}[(I,j;s_j)]}\}$$ $\square$

An example rule is the following, with type variables α and β and a relation variable R:

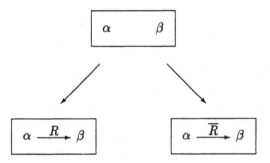

It corresponds, of course, to $\top \sqsubseteq R \sqcup \overline{R}$.

Even *joining rules* could be imagined, with union as outermost operator on the left-hand side, giving rise to derivations in the form of *directed acyclic graphs* (DAGs), but so far I have not yet encountered any useful examples for this.

5 Outlook and Conclusion

The rules we have considered could merge nodes through non-injective rule morphisms, but they could not delete any nodes or edges.

For this, the single-pushout approach we have presented here (albeit with total morphisms!) would have to be replaced with a double-pushout approach.

A nice application of this could be the restriction of derivations to rules that never increased the total number of nodes — the *DANGLING* condition in the gluing condition (see [Ehr78]) guarantees that with the applicability of such a rule the necessary nodes can in fact be deleted.

Restriction to three nodes would correspond to what is possible in the conventional relation calculus without resorting to direct products, and Roger Maddux has expressed interest in the class of theorems derivable with a *"bandwidth"* of at most four or five.

Summarising, we have seen that we can express

- **intersection** and **composition** inside relational diagrams,
- **products** via hyperedges and multiple nodes,
- **Dedekind** and **modal rules** via "changing point of view",
- **transitivity of inclusion** in derivations and
- **union** in branching rules.

Therefore, relational diagram proofs are useful whenever these concepts have to be used heavily, either explicitly in relational formulae or implicitly, as for example for products, that are used implicitly in predicate logic formulations with many variables.

The most important difference of our approach to those of [BH94] and [CL95] is that we have introduced an independent rule concept that can be used for arbitrary applications, and that we have exploited the categorical concepts of the algebraic approach to graph rewriting as the driving mechanism behind our derivation concept.

At the same time, we have created a rule mechanism that supports rule parameters and genericity in the typing of the rules at the same time — this is of course motivated by our work on typed term graph rewriting [Kah95a] in the context of the graphically interactive functional programming and program transformation system HOPS [ZSB86, Kah94, BK94].

References

[BH94]   Carolyn Brown and Graham Hutton. Categories, allegories and circuit design. In *Proceedings, Ninth Annual IEEE Symposium on Logic in Computer Science*, pages 372–381, Paris, France, 4–7 July 1994. IEEE Computer Society Press.

[BK94]   Arne Bayer and Wolfram Kahl. The **Higher-Object Programming** System "HOPS". In Bettina Buth and Rudolf Berghammer, editors, *Systems for Computer-Aided Specification, Development and Verification*, Bericht Nr. 9416, pages 154–171. Universität Kiel, 1994. URL: http://inf2-www.informatik.unibw-muenchen.de/HOPS/papers/Bayer-Kahl-94.ps.gz.

[BS96]   Chris Brink and Gunther Schmidt, editors. *Relational Methods in Computer Science*. Springer-Verlag, Wien, 1996. to appear.

[CL95]   Sharon Curtis and Gavin Lowe. A graphical calculus. In Bernhard Möller, editor, *Mathematics of Program Construction, Third International Conference, MPC '95, Kloster Irsee, Germany, July 1995*, volume 947 of *LNCS*, pages 214–231. Springer Verlag, 1995.

[Ehr78] Hartmut Ehrig. Introduction to the algebraic theory of graph grammars. In Volker Claus, Hartmut Ehrig, and Grzegorz Rozenberg, editors, *Graph-Grammars and Their Application to Computer Science and Biology, International Workshop*, volume 73 of *Lecture Notes in Computer Science*, pages 1–69, Bad Honnef, November 1978. Springer-Verlag.

[EKL90] Hartmut Ehrig, Martin Korff, and Michael Löwe. Tutorial introduction to the algebraic approach of graph grammars based on double and single pushouts. In Ehrig et al. [EKR90], pages 24–37.

[EKR90] Hartmut Ehrig, Hans-Jörg Kreowski, and Grzegorz Rozenberg, editors. *Graph-Grammars and Their Application to Computer Science, 4th International Workshop*, volume 532 of *Lecture Notes in Computer Science*, Bremen, Germany, March 1990. Springer-Verlag.

[FS90] Peter J. Freyd and Andre Scedrov. *Categories, Allegories*, volume 39 of *North-Holland Mathematical Library*. North-Holland, Amsterdam, 1990.

[Kah94] Wolfram Kahl. Can functional programming be liberated from the applicative style? In Bjørn Pehrson and Imre Simon, editors, *Technology and Foundations, Information Processing '94, Proceedings of the IFIP 13th World Computer Congress, Hamburg, Germany, 28 August – 2 September 1994, Volume I*, volume A-51 of *IFIP Transactions*, pages 330–335. IFIP, North-Holland, 1994.

[Kah95a] Wolfram Kahl. Aspects of typed term graphs. In Tiziana Margaria, editor, *Kolloquium Programmiersprachen und Grundlagen der Programmierung, Adalbert Stifter Haus, Alt Reichenau, 11.-13. Oktober 1995*, Bericht MIP-9519, pages 104–109. Universität Passau, Fakultät für Mathematik und Informatik, December 1995.

[Kah95b] Wolfram Kahl. Kategorien von Termgraphen mit gebundenen Variablen. Technischer Bericht 9503, Fakultät für Informatik, Universität der Bundeswehr München, September 1995.

[Kah96] Wolfram Kahl. *Algebraische Termgraphersetzung mit gebundenen Variablen*. Reihe Informatik. Herbert Utz Verlag Wissenschaft, München, 1996. ISBN 3-931327-60-4; also doctoral dissertation at Fakultät für Informatik, Universität der Bundeswehr München.

[Ken90] Richard Kennaway. Graph rewriting in some categories of partial morphisms. In Ehrig et al. [EKR90], pages 490–504.

[Löw90] Michael Löwe. Algebraic approach to graph transformation based on single pushout derivations. Technical Report 90/05, TU Berlin, 1990.

[Spi89] J. M. Spivey. *The Z Notation: A Reference Manual*. Prentice Hall International Series in Computer Science. Prentice Hall, 1989.

[SS93] Gunther Schmidt and Thomas Ströhlein. *Relations and Graphs, Discrete Mathematics for Computer Scientists*. EATCS-Monographs on Theoretical Computer Science. Springer Verlag, 1993.

[VH91] Paulo A. S. Veloso and Armando M. Haeberer. A finitary relational algebra for classical first-order logic. *Bulletin of the Section on Logic of the Polish Academy of Sciences*, 20(2):52–62, 1991.

[ZSB86] Hans Zierer, Gunther Schmidt, and Rudolf Berghammer. An interactive graphical manipulation system for higher objects based on relational algebra. In Gottfried Tinhofer and Gunther Schmidt, editors, *Proc. 12th International Workshop on Graph-Theoretic Concepts in Computer Science*, LNCS 246, pages 68–81, Bernried, Starnberger See, June 1986. Springer-Verlag.

Definability Equals Recognizability
of Partial 3-Trees

Damon Kaller

School of Computing Science
Simon Fraser University
Burnaby, B.C., Canada, V5A 1S6
E-mail: kaller@cs.sfu.ca

Abstract. We show that a graph decision problem can be *defined* in the Counting Monadic Second-order logic if the partial 3-trees that are yes-instances can be *recognized* by a finite-state tree automaton. The proof generalizes to also give this result for k-connected partial k-trees. The converse—definability implies recognizability—is known to hold over all partial k-trees. It has been conjectured that recognizability implies definability over partial k-trees; but a proof was previously known only for $k \leq 2$. This paper proves the conjecture—and hence the equivalence of definability and recognizability—over partial 3-trees and k-connected partial k-trees.

1 Introduction

Many NP-hard graph problems are known to have linear-time algorithms over the class of partial k-trees (see Arnborg [1] for a survey of early results). The partial k-trees encompass many important graph families including trees and forests ($k = 1$), series-parallel and outerplanar graphs ($k = 2$) and Halin graphs ($k = 3$). Takamizawa, Nishizeki and Saito [18] described a general technique to construct linear-time algorithms for a large group of problems over series-parallel graphs. A number of more general formalisms were later developed [4, 6, 7, 15], including several based on the Monadic Second-order (or MS) logic [3, 10, 11]: These capture large groups of problems with linear time-complexity over partial k-trees. Often, it is very easy to define a given graph decision problem with an MS logical statement; such a statement can be translated into a dynamic-programming algorithm automatically [3, 10, 11]. In this paper, we show that a decision problem can be solved by such an algorithm *only* if it can be defined within this formalism over partial 3-trees and k-connected partial k-trees. We use general techniques which may lead to a proof of this equivalence over all partial k-trees.

Any partial k-tree can be decomposed into a tree-like hierarchy of small *basic* graphs—each on $k + 1$ or fewer vertices. Arnborg, Corneil and Proskurowski [2] showed how to construct such a *tree decomposition* in polynomial time. Later, Bodlaender [8] gave a linear-time algorithm to do this. Tree decompositions are executed by a *tree automaton* [13] in much the same way as strings are executed by a conventional finite-state automaton. Such a machine provides a model of a dynamic-programming algorithm that decides, in linear time, whether or not to accept a given tree decomposition. We say that a subclass of the partial k-trees is *recognizable* if there exists a tree automaton accepting exactly the tree decompositions of graphs in that subclass. An MS logical *definition* of such a subclass is known to be sufficient for it to be recognizable: This was shown independently by Arnborg, Lagergran and Seese [3], by Borie, Parker and Tovie [10], and by Courcelle [11]—who also showed that a statement of the *Counting MS* (or *CMS*) logic necessarily defines any recognizable subclass of partial 1-trees [11] or partial 2-trees [12].

A tree decomposition can be interpreted as a rooted tree T for which each node has one of a constant number of labels. A finite-state tree automaton then computes the state of each node as a function of its label and the states of its children; T is accepted iff its root is thus assigned to an *accepting state*. Courcelle [11] describes how CMS predicates encoding the structure of T can be used to encode whether or not T is accepted by a given tree automaton. To prove that definability equals recognizability, then, we need only show that CMS logic can encode the structure of a fixed tree decomposition of any partial k-tree. This is easy for $k = 1$ because a tree is (roughly speaking) its own tree decomposition. The class of partial 2-trees can be characterized by one forbidden *minor*: namely, the clique on four vertices. (A minor of a graph G is a graph that can be obtained from a subgraph of G by a series of edge contractions.) This characterization imposes a regular structure on partial 2-trees; and Courcelle [12] used that structure to encode tree decompositions in CMS over partial 2-trees. Robertson and Seymour [16] have shown that partial k-trees (for any k) can be characterized with some finite number of forbidden minors. Unfortunately this number explodes with increasing k; and the forbidden minors are known only for $k \leq 3$ [5]. So it is difficult to use this characterization in general to obtain fixed tree decompositions of partial k-trees.

In this paper, we begin with an arbitrary tree decomposition of a *2-connected* partial 3-tree G. We then modify the tree decomposition to satisfy several special properties. We partition the modified tree decomposition into *trunks*, thereby decomposing G into a collection of *trunk-graphs*. We show that CMS logic can independently encode a fixed path decomposition for each trunk-graph; and that the collection of these path decompositions can be assembled into a fixed tree decomposition of G. For an arbitrary partial 3-tree, each of its 2-connected *blocks* can be decomposed separately in this way; and the resulting collection of tree decompositions can then be assembled together. Thus, CMS logic can encode a fixed tree decomposition of any partial 3-tree.

The rest of this paper is organized as follows: In Section 2 we review notation and preliminaries relating to tree decompositions and CMS logic. In Section 3 we give a formal definition of a tree automaton, and an outline of the approach used to derive a CMS formula encoding its behavior. In Section 4 we discuss *simple* tree decompositions, and describe how *trunk-graphs* are derived from them. In Section 5 we decompose a 2-connected partial 3-tree G into a tree-like hierarchy of these trunk-graphs; and we develop several important properties of the hierarchy. In Section 6 we show how those properties enable a CMS formula to deduce the structure of each trunk-graph, and to encode a *canonical* tree decomposition of G. In Section 7, we make concluding remarks, obtaining the result that definability equals recognizability of partial 3-trees. The generalization to k-connected partial k-trees is not difficult, and can be found in [14].

2 Preliminaries

The graphs in this paper are finite, simple and undirected. If G is a graph, then $V(G)$ is its vertex set and $E(G)$ is its edge set. We write $G' \sqsubseteq G$ to indicate that G' is a subgraph of G. If V' is a set (but not necessarily a subset of $V(G)$) then $G \backslash V'$ is the subgraph of G that is induced by $V(G) - V'$. For a singleton set $\{v\}$, we may write simply $G \backslash v$ for $G \backslash \{v\}$. For a subgraph G' of G, we may write simply $G \backslash G'$ for $G \backslash V(G')$.

If b is a vertex of a rooted tree T, then T_b is the subtree of T that is rooted at b. In this paper, a *leaf* of a rooted tree is either a degree-1 vertex other than the root, or a degree-0 vertex (which is the root). This definition is for the convenience of having a unique leaf in any path that is rooted at one of its endpoints.

A *cut-set* (or *cut-vertex*) of a graph G is a vertex subset (respectively, vertex) V' for which G has fewer components than $G \backslash V'$. V' is said to *separate* any pair of vertices that are in the same component of G, but in different components of $G \backslash V'$. An *ℓ-connected* graph (for $\ell \in \mathbf{Z}^+$) is a graph with ℓ vertex-disjoint paths between each pair of nonadjacent vertices. Note that this definition is somewhat unusual, but differs from the more standard definition [9] only for small graphs (on ℓ or fewer vertices): We consider such a small graph to be ℓ-connected iff it is a clique; by the standard definition, no such graph would be ℓ-connected.

2.1 Partial k-Trees and Tree Decompositions

A *k-tree* is either the clique on k vertices, or a graph that can be obtained (recursively) from a k-tree G by adding a new vertex and making it adjacent to any k distinct vertices that induce a clique in G. A *partial k-tree* is a subgraph of a k-tree. For example, a graph is a partial 0-tree iff its edge set is empty; a graph is a partial 1-tree iff it is a forest. Series-parallel graphs and outerplanar graphs are subclasses of the partial 2-trees; Halin graphs form a subclass of the partial 3-trees. The partial k-trees can be characterized as those graphs that admit a width-k *tree decomposition* [17]:

Definition 1. A *tree decomposition* of a graph G is a pair $(T, \mathcal{X})$ where T is a tree and $\mathcal{X} = \{X_a\}_{a \in V(T)}$ is a collection of subsets of $V(G)$, indexed by the nodes of T, for which

- $\bigcup_{a \in V(T)} X_a = V(G)$, and
- each edge of G has both endpoints in some $X_a \in \mathcal{X}$, and

- if $a, b, c \in V(T)$, and b lies on the path between a and c, then $X_a \cap X_c \subseteq X_b$.

We refer to the elements of $V(T)$ as *nodes*, so as not to confuse them with the vertices of G. The set X_b is called the *bag* indexed by $b \in V(T)$. A *width-k* tree decomposition is a tree decomposition for which no bag has more than $k + 1$ vertices. If T' is a subgraph of T, then $\mathcal{X}_{T'}$ consists of the bags indexed by nodes of T', and $X_{T'}$ is the the union of those bags:

$$\mathcal{X}_{T'} = \{X_a \in \mathcal{X} \mid a \in V(T')\} \qquad\qquad X_{T'} = \bigcup_{a \in V(T')} X_a$$

We refer to the subgraph of G induced by $X_{T'}$ as the subgraph *underlying T'*.

Definition 2. A *terminal set* of a partial k-tree G is a proper subset V' of $V(G)$ with cardinality $|V'| \le k$, such that G admits a width-k tree decomposition in which V' is a subset of some bag.

We assume that any graph G has a specially-designated (possibly empty) terminal set, denoted by $V_{\text{term}}(G)$. A *rooted* tree decomposition of G is a tree decomposition $(T, \mathcal{X})$ for which T is a rooted tree, and $V_{\text{term}}(G)$ is a subset of the root bag.

Definition 3. Suppose $(T, \mathcal{X})$ is a rooted tree decomposition of a graph G; and let $b \in V(T)$. If b is the root of T, then each vertex of $X_b - V_{\text{term}}(G)$ is a *drop vertex* of X_b; otherwise, each vertex of $X_b - X_a$ is a *drop vertex* of X_b (where $a \in V(T)$ is the parent of b).

2.2 Counting Monadic Second-Order Logic

A graph $G = (V, E)$ can be interpreted as a logical structure over a universe consisting of a set V of vertices and a set E of edges. G is described by the predicate $Edge(e, v)$, which holds whenever $v \in V$ is an endpoint of $e \in E$. The CMS logic [11] is a predicate calculus which can encode properties of such structures by using individual variables to represent vertices or edges; set variables to represent sets of vertices or edges; the equality ($=$) and membership ($\in$) symbols; existential ($\exists$) and universal ($\forall$) quantifiers; the logical operators $\wedge$ ("and"), $\vee$ ("or"), $\neg$ ("not"), $\Rightarrow$ ("implies") $\Leftrightarrow$ ("if and only if"); the *Edge* predicate; and unary predicates $\mathbf{card}_{\ell,c}$ for nonnegative integer constants ℓ, c (with $\ell < c$). If S is a set, then $\mathbf{card}_{\ell,c}(S)$ is true iff S has cardinality ℓ (mod c). Quantification is allowed over both individual and set variables. Predicates can be defined with any constant number of arguments, by constructing logical expressions in which the arguments are free variables. A *CMS statement* is a CMS predicate Ψ with zero arguments, and we write $G \models \Psi$ to indicate that Ψ is true when evaluated over a given graph G.

Lemma 4 is given by Courcelle [11, Lemma 3.7]. To encode the predicates described by Lemmas 5 and 6, we need only "color" the vertices and edges of a partial k-tree.

Lemma 4. *If a CMS predicate can encode a binary relation on the vertex set of a graph, then a CMS predicate can encode the transitive closure of that relation.*

Lemma 5. *A CMS predicate can encode a direction for each edge of an (undirected) partial k-tree.*

Lemma 6. *A CMS predicate can encode a constant-length string of bits for each vertex and each edge.*

3 Tree Automata

We are interested in decision problems for which an instance is a single graph (in particular, a partial k-tree). Often, the yes-instances of such a problem can be *defined* with a CMS statement Ψ. The problem is then to *recognize* whether a given partial k-tree belongs the class $\{G \mid G \models \Psi\}$. A *tree automaton* [13] can perform this recognition; the input to such a machine is obtained from a tree decomposition of the instance graph.

Definition 7. A *tree automaton* over an alphabet Σ_k is a quadruple $(\mathcal{S}, S_0, \mathcal{S}_A, f)$ where $\mathcal{S}$ is a finite set of *states*; $S_0 \in \mathcal{S}$ is the *initial* state; $\mathcal{S}_A \subseteq \mathcal{S}$ is the set of *accepting* states; and $f : \mathcal{S} \times \mathcal{S} \times \Sigma_k \to \mathcal{S}$ is the *transition function*.

For our purposes, the alphabet Σ_k is the set of graphs on $k + 1$ or fewer (labeled) vertices: These are the *basic* graphs induced by the bags of any width-k tree decomposition. Such a tree decomposition $(T, \mathcal{X})$ is interpreted as a rooted tree T in which each node b is labeled with the basic graph (denoted by $\sigma(b) \in \Sigma_k$) that is isomorphic to the (vertex-labeled) subgraph of G induced by X_b. We assume that any such input tree T is binary: This is no loss of generality, because any width-k tree decomposition can be quite easily modified into one with a binary tree. Each leaf b of T is then assigned to the state $f(S_0, S_0, \sigma(b))$; and each other node b of T is assigned to the state $f(S, S', \sigma(b))$, where $S, S' \in \mathcal{S}$ are the states to which the children of b are (recursively) assigned. The tree T is *accepted* by the tree automaton iff its root is thus assigned to an accepting state.

Theorem 8. [11] *If Ψ is a CMS statement, then (for each $k \in \mathbf{N}$) there exists a tree automaton $\mathcal{A}$ for which the input is a width-k (binary) tree decomposition $(T, \mathcal{X})$ of a graph G such that: $\mathcal{A}$ accepts $(T, \mathcal{X})$ iff $G \models \Psi$.*

Furthermore, a CMS statement Ψ can be automatically translated into a suitable tree automaton that *recognizes* the class $\{G \mid G \models \Psi\}$. This tree automaton performs only constant work for each node of the decomposition tree. Since tree decompositions can be generated in linear time [8], it follows that any CMS-definable graph decision problem has a linear-time algorithm over the class of partial k-trees.

To prove the converse of Theorem 8 for $k = 3$, we suppose there exists a tree automaton $\mathcal{A}$ that accepts a width-3 (binary) tree decomposition iff the underlying graph has some (unknown) property Π. Our goal is to derive a CMS statement defining Π. We proceed by showing that any partial 3-tree admits a fixed (or *canonical*) tree decomposition—whose structure can be encoded by a CMS formula. The tree of the canonical tree decomposition is not necessarily binary; so it cannot serve as the input to a tree automaton (as defined above). However, using a standard construction, we can transform the canonical tree decomposition into a binary form after adding some "dummy" nodes. Then, a CMS formula can explicitly represent the state computed by $\mathcal{A}$ for each original ("non-dummy") node of the tree decomposition. Using the $\mathbf{card}_{\ell,c}$ predicates, a CMS formula can verify that the state of each such node is consistent with the states of its children (in the canonical tree decomposition). Hence, a CMS predicate defining Π need only test whether the root of T is thus assigned to an accepting state. Details of this construction can be found in [11, 14].

4 Simple Tree Decompositions

To develop a canonical tree decomposition, we begin with a *simple* tree decomposition, and then modify it. Each node of a simple tree decomposition must satisfy three special properties. It is not hard to show that any connected partial k-tree admits a simple tree decomposition.

Definition 9. A *simple* tree decomposition is a rooted tree decomposition $(T, \mathcal{X})$ for which each node b of T satisfies the following properties:

P1: There is exactly one drop vertex of X_b.
P2: The subgraph underlying T_b is connected.
P3: If V' is a subset of the non-drop vertices of X_b, then V' is not a cut-set of the subgraph underlying T_b.

A tree decomposition $(P, \mathcal{X})$ for which P is a path is sometimes called a *path decomposition*. If P is, furthermore, rooted at one of its endpoints; and if each node of P satisfies **P1**, **P2** and **P3**, then we say that $(P, \mathcal{X})$ is a *simple path decomposition*. A *simple partial k-path* is any graph that admits a width-k simple path decomposition. In Section 5, we will decompose a partial k-tree into a collection of simple partial k-paths, by recursively choosing *trunks* in a simple tree decomposition.

Definition 10. Suppose $(T, \mathcal{X})$ is a simple width-k tree decomposition of a graph G. A *trunk* is a path P between some node $b \in V(T)$ and a leaf of T_b. The *trunk-graph* of P is then obtained from the subgraph of G underlying P as follows: Add an edge between each pair of vertices contained in the intersection $X_p \cap X_c$, for each child $c \in V(T \backslash P)$ of each node $p \in V(P)$. The terminal set of this trunk-graph consists of the non-drop vertices of X_b.

Although $(P, \mathcal{X}_P)$ is not necessarily a *simple* path decomposition of the subgraph (of G) underlying P, it can be shown that each node of P satisfies **P1**, **P2** and **P3** relative to the corresponding trunk-graph. Hence, a trunk-graph is a simple partial k-path.

Lemma 11. *If P is a trunk of a simple width-k tree decomposition $(T, \mathcal{X})$, then $(P, \mathcal{X}_P)$ is a simple path decomposition of the trunk-graph of P.*

It can be shown that any simple partial k-path contains a structure called a *pyramid*. We use the pair $(A, \rightarrow)$ to denote a sequence on a set A, where "$x \rightarrow y$" means that $y \in A$ immediate follows $x \in A$. We use "$\rightarrow^+$" to denote the transitive closure of "$\rightarrow$"; and we use "$\rightarrow^*$" to denote its reflexive, transitive closure.

Definition 12. Suppose R is a simple partial k-path. A *pyramid* in R consists of a vertex $v_1 \in V(R)$ and k vertex sequences $(A_1, \rightarrow_1), (A_2, \rightarrow_2), \ldots, (A_k, \rightarrow_k)$ with the following properties:

D1: $\{A_1, A_2, \ldots, A_k\}$ is a partition of $V(R) - \{v_1\}$.

D2: For $1 \leq i \leq k$: $v \in A_i$ is adjacent to v_1 (*i.e.* $\{v, v_1\} \in E(R)$) only if v is the first vertex of $(A_i, \rightarrow_i)$.

D3: For $1 \leq i \leq k$: $v \in A_i \cap V_{\text{term}}(R)$ only if v is the last vertex of $(A_i, \rightarrow_i)$.

D4: For $1 \leq i \leq k$: two vertices $u, v \in A_i$ are adjacent (*i.e.* $\{u, v\} \in E(R)$) only if $u \rightarrow_i v$.

D5: For $2 \leq \ell \leq k$: if $i_1, i_2, \ldots, i_\ell$ are distinct indices between 1 and k, and each A_{i_j} $(1 \leq j \leq \ell)$ contains two distinct vertices (say $u_{i_j} \rightarrow_{i_j}^+ u'_{i_j}$), then not all of the following are edges of R: $\{u_{i_1}, u'_{i_2}\}, \{u_{i_2}, u'_{i_3}\}, \ldots, \{u_{i_{\ell-1}}, u'_{i_\ell}\}, \{u_{i_\ell}, u'_{i_1}\}$.

The vertex v_1 is called the *apex* of the pyramid; and each sequence $(A_i, \rightarrow_i)$ is called an *axis* of the pyramid. An edge $e \in E(R)$ is an *apical* edge if one of its endpoints is the apex; e is an *axial* edge if both endpoints belong to the same axis; otherwise e is a *cross* edge.

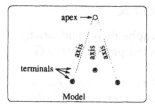
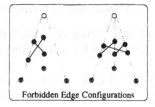

Fig. 1. A pyramid in a simple partial 3-path

A pyramid can be constructed by considering the bags of the simple path decomposition $(P, \mathcal{X}_P)$ in order from the leaf to the root, and inductively placing any newly-seen vertices at the ends of the axes. Figure 1 illustrates the structure of a pyramid in a simple partial 3-path with three terminals. If there are fewer terminals, then not every axis ends with one. The apex may be adjacent only to the first vertex of each axis. Axial edges exist only between consecutive vertices. Property **D5** says that pairs and triples of cross edges are forbidden to "criss-cross" as illustrated. In Section 6 we will use this pyramid structure to develop a *canonical* path decomposition of a trunk-graph.

5 A Trunk Hierarchy

In this section, we decompose a 2-connected partial 3-tree G into a tree-like hierarchy of trunk-graphs (Def. 10). We begin with a simple width-3 tree decomposition $(T, \mathcal{X})$ of G, and recursively select trunks partitioning the nodes of T. The first trunk P is a path between the root and a leaf of T. For each child $c \in V(T \backslash P)$ of each $p \in V(P)$, we then recursively select a trunk between c and some leaf of T_c. As the trunks are selected, we enforce special properties that will be used in Section 6 to obtain a CMS formula encoding a canonical tree decomposition of G.

Throughout this section, $(T, \mathcal{X})$ is a simple width-k tree decomposition of a 2-connected partial 3-tree G; $\mathcal{P}$ is a partition of T into trunks; and $\mathcal{R}$ is the corresponding *trunk hierarchy*:

Definition 13. A *trunk hierarchy* of a simple tree decomposition $(T, \mathcal{X})$ is the collection of trunk-graphs corresponding to a collection $\mathcal{P}$ of trunks that partitions the nodes of T.

Let T' be the tree (the minor of T) obtained by contracting each trunk in $\mathcal{P}$ into a single node. The trunk-graphs in $\mathcal{R}$ have an obvious one-to-one correspondence with the nodes of T'. We will use the terms "root", "child", "parent" *etc.*, with implied reference to T', when referring to these trunk-graphs.

Remark. The pair $(T', \mathcal{R})$ describes a structure similar to a tree decomposition: A trunk-graph R_b with k or fewer terminals corresponds to each $b \in V(T')$; each vertex of G belongs to at least one of these trunk-graphs; and each edge of G has both endpoints in some trunk-graph. Furthermore, if $R_p, R_c \in \mathcal{R}$ where p is the parent of c, then $V_{\text{term}}(R_c) = V(R_p) \cap V(R_c)$; and R_p has an edge (possibly not an edge of G though) between each pair of terminals of R_c.

We will make several claims concerning the trunk-graphs in $\mathcal{R}$, and briefly indicate how to enforce those properties for any 2-connected partial 3-tree G.

Property 14. *The vertices of each $R \in \mathcal{R}$ can be ordered $v_1, v_2, \ldots, v_{|V(R)|}$ such that, for each $i = 2, 3, \ldots, |V(R)|$, there is a non-terminal vertex v_j of R, where $j \leq i - 1$, such that at least one of the following conditions is satisfied:*

C1: v_i and v_j are adjacent (in G).
C2: R has a child $R' \in \mathcal{R}$ for which $V_{\text{term}}(R') = \{v_i, v_j\}$.
C3: R has a child $R' \in \mathcal{R}$ for which $V_{\text{term}}(R') = \{v_i, v_j, v_{j'}\}$, where $j' \leq i - 1$; and there is a path in $G \backslash \{v_j, v_{j'}\}$ between v_i and some terminal in $V_{\text{term}}(G)$.

It is not hard to show that Property 14 is satisfied if each node p of each trunk $P \in \mathcal{P}$ satisfies the following two properties (in addition to properties **P1**, **P2** and **P3** of Def. 9).

P4: If p has children $c \in V(P)$ and $c' \in V(T \backslash P)$, then $X_{c'}$ contains at most one vertex of $X_p - X_c$.

P5: If $v \in X_p - X_c$ (where $c \in V(P)$ is the child of p), then either $v \in V_{\text{term}}(G)$, or v is adjacent to some vertex of $G \backslash X_{T_p}$.

Each trunk P is recursively chosen between a given node b and some leaf of T_b. By using a greedy search to select this leaf, we can enforce **P4**. It is not difficult to enforce **P5** by (possibly) modifying that part of the tree decomposition corresponding to T_b: Any violating vertex v is removed from X_P, so that v then appears only in bags indexed by descendants of the modified path. Since the trunks are chosen in a top-down manner, no node of T_b has yet been assigned to a trunk in $\mathcal{P}$; so the necessary modifications can be made without disturbing the previously-chosen trunks. We will claim that P satisfies two additional properties as well; these properties can also be enforced before choosing any other trunk from T_b.

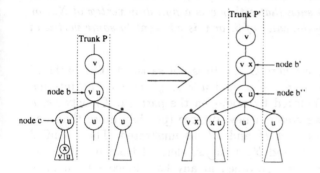

Fig. 2. Enforcing Claim 15 ("$*$" indicates there may be multiple similar subtrees)

Claim 15. *Let b be a node of $P \in \mathcal{P}$; let u be the drop vertex of X_b; and let v be a non-drop vertex of X_b. Suppose that $c \in V(T \backslash P)$ is the only child of b whose bag contains v. We claim there is no cut-vertex $x \in X_{T_c} - X_b$ of the subgraph underlying T_c that separates u from v.*

The left-hand side of Figure 2 illustrates the situation where Claim 15 is not satisfied: By Definition 9, the drop vertex u of X_b belongs to the bag indexed by each child of b. The figure shows how P is converted into a new trunk P' of a perturbed simple tree decomposition, such that $x \in X_{P'}$: This is done by splitting b into two nodes b', b'', and inserting x into the bag indexed by each of them. This operation can be applied repeatedly until Claim 15 is satisfied.

Property 14 and Claim 15 are needed to enable a CMS formula to determine the vertex set of each trunk-graph $R \in \mathcal{R}$. The next claim allows the CMS formula to deduce the structure of some pyramid in R, so that a fixed path decomposition can be encoded for R.

Claim 16. *Let R be the trunk-graph of some $P \in \mathcal{P}$; and suppose $b \in V(P)$ has an ancestor $a \in V(P)$ such that there are two vertex-disjoint paths $H, H' \sqsubseteq R$*

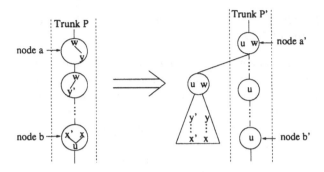

Fig. 3. Enforcing Claim 16

between the drop vertex of X_b and some vertex of X_a. We claim that there is an internal vertex v (of H or H') such that either v is a non-drop vertex of X_a, or v belongs to the bag indexed by the child of b, or v is adjacent to some vertex in $V(R) - V(H) - V(H')$.

Figure 3 illustrates the situation where Claim 16 is not satisfied: The paths H and H' have vertex sequences $u, x \ldots y, w$ and $u, x' \ldots y', w$; and the internal vertices are confined to bags indexed by nodes on the path (in T) between a and b. To enforce the claim, we create a new subtree (possibly more than one of them) for the internal vertices of H and H', as illustrated. The modified trunk P' contains nodes (say, a' and b') that are derived from a and b; no internal vertex of H or H' is now contained in any bag indexed by a node of P'. This operation does not create any new violations of Claim 15; so both claims are simultaneously satisfied after the operation can no longer be applied. Furthermore, neither operation (sketched in Figs. 2 and 3) may cause a violation of Definition 9 or Property 14.

Theorem 17. *Any 2-connected partial 3-tree admits a width-3 simple tree decomposition $(T, \mathcal{X})$ with a trunk hierarchy $\mathcal{R}$, in which each trunk-graph satisfies Property 14; and Claims 15 and 16 are satisfied by the trunk $P \sqsubseteq T$ of each trunk-graph.*

6 Encoding A Canonical Tree Decomposition

In this section, we develop a CMS formula to encode the structure of a fixed tree decomposition of a 2-connected partial 3-tree G. To do this, we use the trunk hierarchy $\mathcal{R}$ constructed in Section 5. We describe how a CMS formula can deduce the vertex set and encode a fixed path decomposition for each trunk-graph in $\mathcal{R}$. The collection of path decompositions can then be assembled into a *canonical* tree decomposition $(T, \mathcal{X})$ of G: A *witness* $v \in V(G)$ is associated with each node of T; and a CMS predicate can define "edges" among pairs of witnesses to describe a tree on $V(G)$ that is isomorphic to T. Each witness shall be a drop vertex of its corresponding bag.

Definition 18. A rooted tree decomposition $(T, \mathcal{X})$ of a graph G is called *canonical* if the following two predicates can be encoded within a CMS formula:

- **Bag**$(v, X) \equiv$ "$v \in V(G)$ is the unique witness for $X \in \mathcal{X}$"
- **Parent**$(p, c) \equiv$ **Bag**$(c, X) \wedge$ **Bag**$(p, X') \wedge$ "X' is indexed by the parent of the node indexing X"

Throughout this section, $G = (V, E)$ is a 2-connected partial 3-tree with either two or three terminals; and $\mathcal{R}$ is a trunk hierarchy of a simple width-3 tree decomposition of G, as constructed in Section 5. We will describe how to obtain a CMS formula in which the **Bag** and **Parent** predicates can be defined. Note that each vertex in $V(G) - V_{\text{term}}(G)$ is a non-terminal vertex of exactly one trunk-graph in $\mathcal{R}$.

Definition 19. If v is a non-terminal vertex of G, then $R(v)$ denotes the unique trunk-graph in $\mathcal{R}$ such that v is a non-terminal vertex of $R(v)$.

Using Lemma 6, each vertex v can encode any constant amount of information pertaining to its role in the trunk-graph $R(v)$. This allows the CMS formula to encode which of the non-terminal vertices belong to each axis, and which of them is the apex, of some fixed pyramid (Def. 12). For a designated apex v_1, we will inductively identify each other vertex v_i of $R(v_1)$ with the help of a unique edge incident to an already-identified non-terminal vertex v_j (where $j \leq i - 1$ in the vertex order given by Property 14). Such an edge can also encode information concerning the role of v_i in the trunk-graph. A (non-proper) vertex coloring is used for the inductive identification of vertices:

Proposition 20. *The vertex set V can be partitioned into thirteen color classes such that the non-terminal vertices of each $R \in \mathcal{R}$ belong to a common color class C; and if $t \in V_{\text{term}}(R)$, then C does not contain any vertex of $R(t)$.*

In other words, C is distinct from the color class of each terminal t of R, and C is distinct from the color class of each terminal of $R(t)$. To prove Proposition 20, we color the trunk-graphs in a top-down manner; so at most twelve color classes are unsuitable for the vertices of any trunk-graph. The next lemma (illustrated by Figure 4) follows from Claim 15:

Lemma 21. *Suppose a trunk-graph $R \in \mathcal{R}$ has a child $R' \in \mathcal{R}$; and let $v_i \in V_{\text{term}}(R')$, $v_j \in V_{\text{term}}(R') - V_{\text{term}}(R)$. Let U be the union of the non-terminal vertices over R' and all descendants of R'; and let G' be the subgraph of G induced by $U \cup \{v_i\}$. If $u, v \in U$ such that $\{v_j, v\} \in E(G)$ and u belongs to the same color class as v_i, then $G' \backslash u$ contains a path between v and v_i.*

We will use Lemma 21 to identify the (say r) vertices of $R(v_1)$, for each designated apex v_1. Let $v_1, v_2, \ldots, v_r$ be the these vertices, ordered to respect Property 14. Using the vertex coloring (of Proposition 20) a CMS formula can identify the vertices of R inductively. Property 14 gives three different conditions

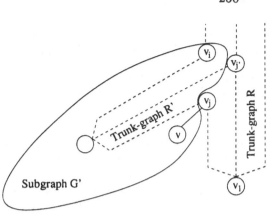

Fig. 4. G' is not cut by any other vertex with the color of v_i

by which a vertex v_i may be identified: in each case, there is a designated non-terminal vertex v_j (where $j \leq i - 1$) that interacts with v_i in some fixed manner:

Condition C1: $\{v_j, v_i\} \in E$. The edge of G can encode the fact that its endpoints belong to the same trunk-graph. If appropriate, v_i can also be identified as a terminal in this way.

Condition C2: $\{v_i, v_j\} = V_{\text{term}}(R')$, where R' is a child of R. It follows that v_j is adjacent to some non-terminal vertex (say v) of either R' or a descendant of R'. The edge $\{v_j, v\}$ can encode (by Lemma 6) the fact that the trunk-graph containing v_j also contains the "correctly-colored" cut-vertex of $G \backslash v_j$ for which v is separated into a minimal-sized component. It follows from Lemma 21 that v_i is this minimizing cut-vertex.

Condition C3: $\{v_i, v_j, v_{j'}\} = V_{\text{term}}(R')$, where $j' \leq i - 1$, and R' is a child of R. This case is not much different from **C2**. An edge of G can encode the fact that the next vertex v_i is identified as the minimizing cut-vertex of $G \backslash \{v_j, v_{j'}\}$, over all possible choices of $v_{j'}$.

The next lemma follows from the above observations:

Lemma 22. *CMS logic can encode a predicate* **trunk**(v_1, V') *which holds iff* V' *is the vertex set of* $R(v_1)$. *CMS predicates* **term**$_j(v_1, t)$, *for* $1 \leq j \leq 3$, *can also be defined to uniquely identify each terminal* t *of* $R(v_1)$. *(If* $R(v_1)$ *has only two terminals, then say* **term**$_3(v_1, t)$ *is never satisfied.)*

In order to obtain a canonical path decomposition of $R(v_1)$, we will encode the three vertex sequences (*i.e.* the axes) of a fixed pyramid by directing some of the edges of R. Lemma 5 says that a direction can be encoded for each edge in $E(R) \cap E(G)$. The endpoints of any other edge $e \in E(R) - E(G)$ are terminals of a common child of R. By Lemma 22, these terminals can be identified with a unique apex; so e can be represented in the CMS formula by that apex. A particular apex may represent no more than $\binom{3}{2} = 3$ edges; and by Lemma 6, a direction can be encoded for each of them.

If $(A, \rightarrow)$ is an axis of a pyramid, then the subgraph of R induced by A consists of a collection of paths. If G happens to be 3-connected, then this is just a single path between the apex and some terminal. Otherwise, an axis may contain "breaks" between its maximal paths. Since G is 2-connected, no more than one of the axes may be broken at any given node of a path decomposition. After directing all of the axial edges (as explained above), the CMS formula can identify each maximal path of the axes. By directing some of the cross edges as well, the CMS formula can encode how these maximal paths are placed into the correct sequence. To obtain this encoding, we make use of Claim 16; details can be found in [14].

Lemma 23. *The CMS formula can define the vertex orders "$\rightarrow_i$" $(1 \leq i \leq 3)$ of the axes of a fixed pyramid in each trunk-graph of $\mathcal{R}$.*

By Lemma 4, then, predicates can also be defined for the transitive closure "$\rightarrow_i^+$" of "$\rightarrow_i$".

Each axis gives part of an elimination order for the vertices in the corresponding trunk-graph. By interleaving these orders in a fixed manner, we can obtain a canonical path decomposition $(P, \mathcal{X})$. The leaf bag contains the apex and the first vertex of each axis. Each other bag contains, inductively, the k maximal vertices of the preceding (child) bag, as well as the immediate successor of one of them. Property **D5** (Def. 12) guarantees that at least one of the maximal vertices has no cross edges extending beyond the others; so we can "advance" along the corresponding axis. These bags are well-defined if we impose an order on the axes, and adopt the convention that we inductively advance along 1st axis whenever possible, otherwise the 2nd axis if possible, and otherwise the 3rd.

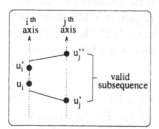 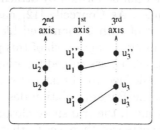

Fig. 5. Choosing a bag of the canonical path decomposition

Lemma 24. *The CMS formula can encode a fixed path decomposition for each $R \in \mathcal{R}$.*

Proof sketch. Each bag (except for the leaf bag) corresponds to a unique pair of consecutive vertices $u_i, u_i' \in A_i$ $(1 \leq i \leq 3)$; say $u_i \rightarrow_i u_i'$. The other two vertices $u_j \in A_j$ $(1 \leq j \leq 3;\ j \neq i)$ must be contained in the *valid subsequence* between u_j' and u_j'' (see Figure 5), where u_j' is the last vertex that is adjacent to u_i (or any vertex preceding u_i), and u_j'' is the first vertex that is adjacent to u_i' (or any vertex following u_i'). If no vertex of the valid subsequence is adjacent to any vertex of A_i, then *any* vertex in that subsequence could be chosen to be u_j. To effect the precedence convention, we choose $u_j = u_j''$ if $j < i$, and choose $u_j = u_j'$ if $j > i$. In the general case, we choose u_j as close as possible to u_j'' if $j < i$; and choose u_j as close as possible to u_j' if $j > i$. The exact choice is determined by the cross edges (which must obey property **D5** of Definition 12). The right-hand panel of Figure 5 illustrates an example for $i = 2$. □

Now, to obtain a canonical tree decomposition, we need only encode parents for the roots of all but one of these path decompositions from Lemma 24. By Lemma 22, the CMS formula can determine each terminal set $V_{\text{term}}(R)$. If $V_{\text{term}}(R) = V_{\text{term}}(G)$, then R is the root of the trunk hierarchy. Otherwise, the parent of R is the unique trunk-graph $R' \in \mathcal{R}$ for which $V_{\text{term}}(R) \subseteq V(R')$ and $V_{\text{term}}(R) \not\subseteq V_{\text{term}}(R')$. CMS can encode this, and then **Parent**(p, c) can be defined where c is the witness of the root of the path decomposition of R, and p is the witness of the closest node to the root of the path decomposition of R' such that $V_{\text{term}}(R)$ is a subset of the corresponding bag.

Theorem 25. *There is a canonical tree decomposition for any 2-connected partial 3-tree.*

7 Conclusion

We have shown that the predicates **Bag** and **Parent** (Def. 18) can be defined to describe a canonical tree decomposition $(T, \mathcal{X})$ of a partial 3-tree G. We assumed that G was 2-connected, but it is easy to remove this restriction because a graph that is not 2-connected can be formed by attaching its 2-connected *blocks* together in a tree-like manner. As Courcelle [12] has noted, a CMS formula can determine the vertex set of each 2-connected block; each of these can be decomposed independently using the approach of this paper; then the collection of these tree decompositions can be assembled together.

A CMS predicate can be defined to obtain the set $V' \subseteq V(G)$ of witnesses representing nodes of T. So the **Parent** relation describes a tree on V' that is isomorphic to T. Using Lemma 4, the transitive closure of the **Parent** relation can be defined; CMS statements can then verify that a tree is described. It is not difficult to also encode the statements defining a tree decomposition (Def. 1) in CMS logic. The formula developed in Section 6 can be augmented with these additional statements to ensure its sufficiency for describing a canonical tree decomposition $(T, \mathcal{X})$. It now follows from Courcelle [11, Proposition 5.4] that a CMS formula can encode whether or not a tree automaton accepts a binary tree decomposition that is derived by adding "dummy nodes" to $(T, \mathcal{X})$. See [14] for details of the construction.

Theorem 26. *Definability equals recognizability of partial 3-trees.*

The generalization to k-connected partial k-trees is quite straightforward: A trunk hierarchy is constructed in the same way as described. Property 14 and Claim 15 generalize easily for this trunk hierarchy; and Claim 16 becomes trivial. The results of Section 6 also generalize easily; in fact, the proofs become much simpler in the case of k-connected partial k-trees.

Theorem 27. *Definability equals recognizability of k-connected partial k-trees.*

References

1. S. Arnborg. Efficient algorithms for combinatorial problems on graphs with bounded decomposability. *BIT*, 25:2–33, 1985.
2. S. Arnborg, D.G. Corneil, and A. Proskurowski. Complexity of finding embeddings in a k-tree. *SIAM J. Alg. Disc. Meth.*, 8:277–284, 1987.
3. S. Arnborg, J. Lagergren, and D. Seese. Easy problems for tree decomposable graphs. *J. Algorithms*, 12:308–340, 1991.
4. S. Arnborg and A. Proskurowski. Linear time algorithms for NP-hard problems restricted to partial k-trees. *Disc. Appl. Math.*, 23:11–24, 1989.
5. S. Arnborg, A. Proskurowski, and D.G. Corneil. Forbidden minors characterization of partial 3-trees. *Disc. Math.*, 80:1–19, 1990.
6. M.W. Bern, E.L. Lawler, and A.L. Wong. Linear-time computation of optimal subgraphs of decomposable graphs. *J. Algorithms*, 8:216–235, 1987.
7. H.L. Bodlaender. Dynamic programming on graphs with bounded treewidth. In *Lecture Notes in Computer Science (Proc. 15th ICALP)*, volume 317, pages 105–119. Springer-Verlag, 1988.
8. H.L. Bodlaender. A linear time algorithm for finding tree-decompositions of small treewidth. In *Proc. 25th STOC*, pages 226–234, 1993.
9. B. Bollobás. *Extremal Graph Theory*. Academic Press, London, 1978.
10. R.B. Borie, R.G. Parker, and C.A. Tovey. Automatic generation of linear-time algorithms from predicate calculus descriptions of problems on recursively constructed graph families. *Algorithmica*, 7:555–581, 1992.
11. B. Courcelle. The monadic second-order logic of graphs. I. Recognizable sets of finite graphs. *Information and Computation*, 85:12–75, 1990.
12. B. Courcelle. The monadic second-order logic of graphs. V. On closing the gap between definability and recognizability. *Theoret. Comput. Sci.*, 80:153–202, 1991.
13. F. Gécseg and M. Steinby. *Tree Automata*. Akadémiai Kiadó, Budapest, 1984.
14. D. Kaller. *Monadic Second-Order Logic and Linear-Time Algorithms for Graphs of Bounded Treewidth*. PhD thesis, Simon Fraser University, Burnaby, B.C., Canada, 1996.
15. S. Mahajan and J.G. Peters. Regularity and locality in k-terminal graphs. *Disc. Appl. Math.*, 54:229–250, 1994.
16. N. Robertson and P.D. Seymour. Graph minors. XX. Wagner's conjecture. To appear.
17. N. Robertson and P.D. Seymour. Graph minors. II. Algorithmic aspects of treewidth. *J. Algorithms*, 7:309–322, 1986.
18. K. Takamizawa, T. Nishizeki, and N. Saito. Linear-time computability of combinatorial problems on series-parallel graphs. *JACM*, 29:623–641, 1982.

One, Two, Three, Many, or: Complexity Aspects of Dynamic Network Flows with Dedicated Arcs *

Bettina Klinz   Gerhard J. Woeginger

Institut für Mathematik B, TU Graz
Steyrergasse 30, A-8010 Graz, Austria
{klinz,gwoegi}@opt.math.tu-graz.ac.at

Abstract. A dynamic network consists of a directed graph with a source s, a sink t and capacities and integral transit times on the arcs. We investigate the computational complexity of dynamic network flow problems where sending flow along an arc blocks this arc as long as the transmission continues. Such arcs are called *dedicated* arcs. We are mainly interested in questions of the type "Given an integral time bound T and an integral flow value v, is it possible to transmit v flow units from the source s to the sink t within T time units?". The complexity of this question strongly depends on whether the values T and v are encoded in binary, in unary, or are constant. We provide a complete classification of all variants of this problem from the computational complexity point of view. Our results establish a sharp borderline between easy and difficult cases of this dynamic network flow problem.

We prove that in the dedicated arc model it is NP-hard to find the maximum flow value v that can be transmitted within $T = 3$ time units, whereas the maximum flow value v for $T = 2$ can be computed in polynomial time. Moreover, we prove that it is NP-hard to find the minimum time T during which $v = 2$ units of flow can be transmitted, whereas the corresponding question for $v = 1$ can be answered in polynomial time. Finally, we prove that the variant where T is encoded in unary and where v is not part of the input can be solved in polynomial time.

1 Introduction

A dynamic network is defined by a directed graph $G = (N, A)$ with a source $s \in N$ and a sink $t \in N$, together with nonnegative capacities u_a and nonnegative integral transit times τ_a for every arc $a \in A$. In a feasible dynamic flow, at most u_a units of flow can enter the arc a within each integral time step θ. More precisely, these flow units leave the tail of arc a at time θ, then traverse arc a and finally reach the head of a at time $\theta + \tau_a$. The corresponding time interval $[\theta, \theta + \tau_a]$ is called the *transmission period* of these flow units.

---

* This research has been supported by the Spezialforschungsbereich F 003 "Optimierung und Kontrolle", Projektbereich Diskrete Optimierung.

In the classical dynamic flow model as introduced by Ford and Fulkerson [3], it is assumed that flow can be sent along an arc a at *each* integer time step $\theta \geq 0$. In this paper, we investigate the variant of this model where sending flow along an arc a *blocks* this arc as long as the flow traverses the arc. In other words, during the transmission period of the flow units sent along an arc a, no further flow can be sent along a; further flow may enter the arc a only when the current flow has reached the head of a.

Due to this property the arcs will be referred to as *dedicated* arcs and the corresponding dynamic flow model as *dedicated arc model*. The dedicated arc model has applications e.g. in a class of telecommunication problems with distributed control where transmission along datalines is insecure and thus has to be monitored and acknowledged by the sending and by the receiving nodes.

We will mainly investigate the following formulation of the problem which we denote by DA-DF (*dynamic flow problem with dedicated arcs*): Given a dynamic network, an integer time bound T and an integer value v, does there exist a dynamic flow in the dedicated arc model that transmits v units of flow from the source s to the sink t within T time units? We will prove that

- when T is encoded in binary, then problem DA-DF is NP-hard for every constant $v \geq 2$ and it is solvable in polynomial time for $v = 1$.
- when v is encoded in unary, then problem DA-DF is NP-hard for every constant $T \geq 3$ and it is solvable in polynomial time for $T \leq 2$.
- when v is constant and T is encoded in unary, then problem DA-DF is solvable in polynomial time.

All these complexity results are summarized in Table 1. Note that in this table, the problems become more difficult when going horizontally to the right or when moving vertically downwards. With this, it is easy to verify that the five entries marked by a superscript of the form [·] already imply all the information contained in this table. The two polynomial time results [a] and [e] are straightforward and are stated as Observations 3.1 and 4.1. The NP-hardness result [c] is proven in Theorem 3.2, and the NP-hardness result [d] is derived in Theorem 4.2. Finally, the polynomial time algorithm for [b] is described in Theorem 5.1. All in all, our results establish a sharp and precise borderline between easy and difficult cases of problem DA-DF.

Related work. In the classical dynamic flow model (i.e. without dedicated arcs), all problems investigated in this paper (and even more general versions of them) can be solved in polynomial time (see Ford and Fulkerson [3] and Burkard, Dlaska and Klinz [2] for the case of a single source and a single sink and Hoppe and Tardos [5, 6] for multiterminal generalizations).

Papadimitriou, Serafini and Yannakakis [8] study a closely related dynamic network flow problem with dedicated arcs, though they do not use the language of dynamic flows to describe their problem. The main difference between their and our model is that (i) their model does not require that flow is sent along an arc only at integral time units and that (ii) their model allows only one unit of flow to be sent along an arc (i.e. in our terminology, all capacities u_a

		Time T			
		=1 or 2	const ≥ 3	unary	binary
Flow v	=1	P	P	P	P[a]
	const ≥ 2	P	P	P[b]	NP[c]
	unary	P	NP[d]	NP	NP
	binary	P[e]	NP	NP	NP

Table 1. The computational complexity of problem DA-DF. Entries P mean that the corresponding variant is solvable in polynomial time, entries NP mean that the corresponding variant is NP-hard.

are set to one). Translated into our language, one of the results in [8] implies that the problem DA-DF is strongly NP-hard already for unit capacities (the corresponding construction in [8] sends flow only at integral times). Note that this yields another argument for the NP-hardness of problem DA-DF when v and T are encoded in unary (cf. Table 1). Moreover, [8] considers a variant with an infinite time bound where the goal is to maximize the long-run average throughput through the network. It is shown that the optimum throughput rate can be computed in polynomial time. Hence, the results in [8] once more demonstrate that dynamic flow problems with a finite time bound are often harder than their asymptotic counterparts with an infinite time bound (see also Orlin [9] vs. Klinz and Woeginger [7]).

Applications. Dynamic flow problems arise in many applications, e.g. in production-distribution systems, communications systems, truck and railway scheduling and building evacuation problems (see the surveys by Aronson [1] and Powell, Jaillet and Odoni [10] for further details). The basic assumption made in the classical dynamic flow model is that flow units which are still in transit along an arc a do not prevent further flow units to be sent along this arc. However, there exist many applications where this property does not hold.

The following example from telecommunication is taken from Papadimitriou, Serafini and Yannakakis [8]: Suppose, we wish to transmit messages through a communication network with insecure data lines from a source s to a destination t. To monitor the data transmission, a communication protocol is used which requires that an acknowledgement is received by the sender before the communication can proceed further. As a consequence, never more than one packet of data can be sent along a data line at the same time, and a packet which arrives at the head node of a data line a cannot continue its way to the destination before the tail node of a receives an acknowledgement. (So, in this application the transit times τ_a are taken to be twice the time it takes sending a message along the arc a.)

Another possible area of application of dynamic network flows with dedicated arcs are problems where sending flow along an arc models using a "resource" (e.g. a certain tool or machine in manufacturing problems) that is available only once and cannot be used by more than one "process" at the same time.

Organization of the paper. Section 2 gives a formal description of the problem DA-DF. Section 3 presents the complexity results for the case where the flow value v is constant and the time bound T is encoded in binary. Section 4 investigates the complexity of problem DA-DF for a constant time bound T, and Section 5 deals with the case of a constant flow value v and a time bound T that is encoded in unary. Finally, Section 6 finishes the paper with a short discussion.

2 Definitions and Preliminaries

Let $G = (N, A)$ be a directed (multi) graph with node set N and arc set A where each arc $a \in A$ is characterized by its *tail* $t(a)$ and its *head* $h(a)$; a is directed from $t(a)$ to $h(a)$. For each node i we denote the set of arcs a with $t(a) = i$ by $A^+(i)$ and the set of arcs a with $h(a) = i$ by $A^-(i)$. If there is a single arc a with $t(a) = i$ and $h(a) = j$, this arc will also be referred to by (i, j).

A *path* P in G from node j to node k is an alternating sequence of nodes and arcs such that $P = (i_0, a_1, i_1, a_2 \ldots a_p, i_p)$, $i_0 = j$, $i_p = k$ and for each $r = 1, \ldots, p$ either arc a_r has head i_r and tail i_{r-1} or else it has head i_{r-1} and tail i_r. In the former case the arc is called a *forward arc* of the path; in the latter case it is called a *backward arc*. A path is called *directed* if every arc is a forward arc and *simple* if no node is repeated. A *cycle* is a path for which the initial node i_0 coincides with the final node i_p.

A *dynamic network* $\mathcal{N} = (G, u, \tau, s, t)$ consists of a directed (multi) graph G with *source* s and *sink* t, and two numbers attached to each arc $a \in A$, namely a nonnegative integer *capacity* u_a and a nonnegative integer *transit time* τ_a. Given a directed path $P = (i_0, a_1, i_1, a_2 \ldots a_p, i_p)$, we let $\tau(P) = \sum_{r=1}^{p} \tau_{a_r}$ denote its transit time. For simplicity we assume that no arc enters the source s and no arc leaves the sink t, and that G contains no directed cycle Q with transit time $\tau(Q) = 0$.

Let $f_a(\theta)$ denote the flow which leaves the tail node $t(a)$ at time θ along the arc a. This flow arrives at the head node $h(a)$ at time $\theta + \tau_a$. In order to allow the flow to arrive in an inner node $i \neq s, t$ at time θ_1, then wait there for some time and leave node i again at time $\theta_2 > \theta_1$, we introduce so-called *hold-over arcs* modelled by loops a with $h(a) = t(a) = i$. These loops get a transit time of one and infinite capacity.

For notational convenience let A_0 denote the union of the original arc set A and the set of loops introduced for modelling hold-overs. Following Ford and Fulkerson [3] the mapping $f : A_0 \times \{0, 1, \ldots, T\} \to \mathbb{N}_0^+$ is said to be a *feasible dynamic flow* (or just dynamic flow) if the following two groups of constraints are satisfied:

$$\sum_{\substack{a \in A_0 \\ t(a)=i}} f_a(\theta) - \sum_{\substack{a \in A_0 \\ h(a)=i}} f_a(\theta - \tau_a) = 0 \quad \text{for all } i \in N \setminus \{s,t\}, \theta \in \{0,\dots,T\} \quad (1)$$

$$0 \le f_a(\theta) \le u_a \qquad\qquad\qquad \text{for all } a \in A, \theta \in \{0,\dots,T\} \quad (2)$$

where for notational convenience we assume throughout that $f_a(\theta) = 0$ for $\theta < 0$. The equations (1) require that for any time $\theta \in \{0,\dots,T\}$ and any node $i \ne s,t$, the amount of flow which enters node i at time θ equals the amount of flow which leaves node i at time θ. The inequalities (2) require that the capacity constraints are fulfilled for each arc $a \in A$ and each time $\theta \in \{0,\dots,T\}$.

In the dedicated arc model, a feasible dynamic flow f additionally has to fulfill the following constraint for all arcs $a \in A$.

$$f_a(\theta^*) > 0 \quad \Longrightarrow \quad f_a(\theta) = 0 \quad \text{for all } \theta \in \{\theta^* + 1, \dots, \theta^* + \tau_a - 1\}. \quad (3)$$

Note that the flow conservation constraints (1) imply that the net amount which leaves the source equals the net amount which enters the sink, i.e.

$$\sum_{\theta=0}^{T} \sum_{a \in A^+(s)} f_a(\theta) = \sum_{\theta=0}^{T} \sum_{a \in A^-(t)} f_a(\theta - \tau_a) =: |f|. \quad (4)$$

$|f|$ is called the *value* of the dynamic flow f.

In the remaining part of the paper we will investigate the complexity of the following basic dynamic network flow problem in the dedicated arc model:

Given a dynamic network $\mathcal{N} = (G, u, \tau, s, t)$, a nonnegative integral time bound T and a nonnegative integral flow value v, the *dynamic flow problem with dedicated arcs*, DA-DF, asks for a feasible dynamic flow f with value $|f| = v$ and time bound T which obeys the constraints (1)—(3).

If there were no dedicated arcs, problem DA-DF could be solved in polynomial time by applying the ingenious maximum dynamic flow algorithm of Ford and Fulkerson [3]. In this paper we will show that the introduction of the additional constraints (3) makes the problem much harder.

One of the essential differences between the classical model for dynamic flow and the dedicated arc model is that in the former model, simple dynamic flow problems like the maximum dynamic flow problem always have an optimal solution which does not make use of hold-over arcs (see [3]). This is not any longer the case in the dedicated arc model as is illustrated by the following example.

Example: Consider the graph G with three nodes s, 1 and t and two dedicated arcs $a = (s,1)$ and $a' = (1,t)$. Let $u_a = 2$, $\tau_a = 2$, $u_{a'} = 4$ and $\tau_{a'} = 5$. The maximum amount of flow that can be sent from s to t within $T = 9$ units of time is equal to 4 if waiting in the inner node 1 is allowed, while it reduces to 2 if this is not allowed.

It turns out, however, that the possibility of waiting in inner nodes does not influence the complexity status of problem DA-DF in general. Our NP-hardness constructions in Sections 3 and 4 have the property that there does not exist a feasible dynamic flow which uses hold-over arcs.

3   One and Two: Constant Flow Value and Binary Time Bound

In this section we study the computational complexity of problem DA-DF in the case where the flow value v is constant and not part of the input, and where the time bound T is encoded in binary. We show that for this case the borderline between NP-hardness and polynomial time solvability lies between $v = 1$ and $v = 2$.

Observation 3.1 *The problem DA-DF is solvable in polynomial time for $v = 1$ and T encoded in binary.*

Proof. This case is trivial, since the solution consists of sending one unit of flow along a single path from s to t. Hence, DA-DF turns into the problem of computing a path from s to t with minimum total transit time. This shortest path problem can be solved in polynomial time by standard methods.   □

Next, we give an NP-hardness proof for the case $v = 2$. The proof is done by a reduction from the NP-hard EVEN-ODD PARTITION problem which can be stated as follows (cf. Garey and Johnson [4]):

EVEN-ODD PARTITION
Instance. A set $K = \{1, \dots, 2d\}$ of $2d$ objects and a positive integer size β_k for each object k, $k = 1, \dots, 2d$, such that $\sum_{k=1}^{d} \beta_k = 2B$.
Question. Does there exist a subset $K' \subseteq K$ such that $\sum_{k \in K'} \beta_k = \sum_{k \in K \setminus K'} \beta_k$ holds and such that for all $k = 1, \dots, d$, $|K' \cap \{2k-1, 2k\}| = 1$ holds?

Here it is important that the numbers β_i are encoded in binary, since EVEN-ODD PARTITION becomes solvable in polynomial time if these numbers are encoded in unary (cf. Garey and Johnson [4]).

Let an instance of EVEN-ODD PARTITION be given. We construct an instance of DA-DF in the following way (for an illustration see Figure 1): The graph $G = (N, A)$ consists of $d + 1$ nodes and $2d$ arcs as follows. Let $N = \{i_1, i_2, \dots, i_d, i_{d+1}\}$ with $s = i_1$ and $t = i_{d+1}$. We introduce for every k, $k = 1, \dots, d$, two arcs a_{2k-1} and a_{2k}, both with tail node i_k and head node i_{k+1}. The transit time of the arc a_k, $k = 1, \dots, 2d$, is set to $2B + \beta_k$ and its capacity is set to one.

Claim 1 *The given instance of EVEN-ODD PARTITION has a solution if and only if the constructed instance of DA-DF allows a flow with value $v = 2$ within time $T = (2d + 1)B$.*

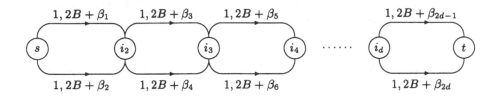

Fig. 1. Network for the proof of Claim 1. Every arc has as first label its capacity and as second label its transit time.

Proof. In case the EVEN-ODD PARTITION instance has a solution, the corresponding solution sets K' and $K \setminus K'$ correspond in a natural way to two edge-disjoint paths of length $(2d+1)B$ going from s to t. Sending a flow of value one along each of these two paths proves the (Only if)-part.

The proof of the (If)-part is based on the following observation: If a unit of flow leaves the sink at time zero or later and arrives at the sink at time $(2d+1)B$ or earlier, then for every $k = 2,\ldots,d$, it has to visit the node i_k somewhere in the time interval $[2(k-1)B+1,\ 2kB-1]$. Indeed, every arc has a transition time of at least $2B+1$ and so the distance to the source s is at least $2(k-1)B+1$, which shows the lower bound. Similarly, the distance to the sink t is at least $2(d+1-k)B+1$ and in order to arrive in t before time $T = (2d+1)B$ the upper bound must be obeyed.

Now suppose that a feasible flow with $v = 2$ exists and that some arc a from node i_k to i_{k+1} is used *twice*: By the above argument, both time points at which this arc is entered must lie in the interval $[2(k-1)B+1,\ 2kB-1]$. But this is impossible, since the transmission of the first flow unit blocks the arc for at least $2B$ time units. Hence, every arc has to be used exactly once and the flow has to move along two edge-disjoint paths. It is straightforward to verify that both paths have a total transition time of exactly $(2d+1)B$ and induce a solution for the EVEN-ODD PARTITION instance. □

Summarizing, we have proved the following theorem.

Theorem 3.2 *For a time bound T that is encoded in binary and for every constant flow value $v \geq 2$, the problem DA-DF is NP-hard.* □

4 Two and Three: Constant Time Bound

In this section we study the computational complexity of problem DA-DF in the case where the time bound T is constant and not part of the input. We show that for this case the borderline between NP-hardness and polynomial time solvability lies between $T = 2$ and $T = 3$.

Observation 4.1 *The problem* DA-DF *is solvable in polynomial time if the time bound T is a constant ≤ 2 and if the flow value v is encoded in binary.*

Proof. Since $T \leq 2$ holds, arcs with transit time greater than two cannot be used by a feasible dynamic flow. For arcs with transit time zero or one, condition (3) does not impose any additional restrictions on the flow. Furthermore, it is easy to see that we cannot send any flow along arcs with transit time two later than at time zero: The earliest other possibility for the flow to enter such an arc would be time one, and then the flow could not reach the source before time three. Hence, also for such arcs the condition (3) does not impose any additional restrictions on the flow.

Summarizing, for $T = 2$ problem DA-DF turns into a standard dynamic network flow problem and therefore, it can be solved in polynomial time by applying standard methods pioneered by Ford and Fulkerson [3]. □

Next, we give an NP-hardness proof for the case when $T = 3$ and v is encoded in unary. The proof is done by a reduction from the following variant of the NP-hard EXACT COVER BY 3-SETS problem:

EXACT COVER BY 3-SETS

Instance. A set $X = \{x_1, \ldots, x_{3q}\}$ of $3q$ elements and a collection $C = \{c_1, \ldots, c_{2q}\}$ of 3-element subsets of X, $|C| = 2q$. Every element of X is contained in at least one member of C.

Question. Does C contain a subcollection $C^* \subseteq C$ such that every element of X occurs in exactly one member of C^*? (Note that in this case, $|C^*| = q$ must hold).

The formulation of **EXACT COVER BY 3-SETS** in Garey and Johnson [4] does not impose the restriction $|C| = 2q$ on the input (which makes the ratio $|C|/|X|$ equal $\frac{2}{3}$). However, by repeatedly duplicating a triple in C (which increases the ratio $|C|/|X|$) or by repeatedly introducing a new triple with three new elements (which decreases this ratio), every unrestricted instance of EXACT COVER BY 3-SETS may be transformed into an equivalent instance with $|C|/|X| = \frac{2}{3}$. Hence, our formulation of EXACT COVER BY 3-SETS is also NP-hard.

Now let an instance of **EXACT COVER BY 3-SETS** be given. From this, we construct an instance of DA-DF in the following way (for an illustration see Figure 2): The graph $G = (N, A)$ consists of $7q + 4$ nodes and $17q + 2$ arcs. The nodes in N are partitioned into six layers $N_1, N_2, \ldots, N_6$ and the arcs in A will either connect nodes in consecutive layers N_k and N_{k+1} or they will go from N_4 to N_6. We define $N_1 = \{s\}$, $N_2 = \{y, z\}$ and $N_6 = \{t\}$. For every triple c_i in C, the layer N_3 contains a corresponding node c_i', and layer N_4 contains another corresponding node c_i''. The fifth layer N_5 contains for every element x_j in X a corresponding node x_j'.

The arcs are defined as follows. There are two arcs (s, y) and (s, z) that leave the source. Both have transit time 0, arc (s, y) has capacity q and arc (s, z) has capacity $2q$. For every i, $i = 1, \ldots, 2q$, we introduce the four arcs (y, c_i'), (z, c_i'),

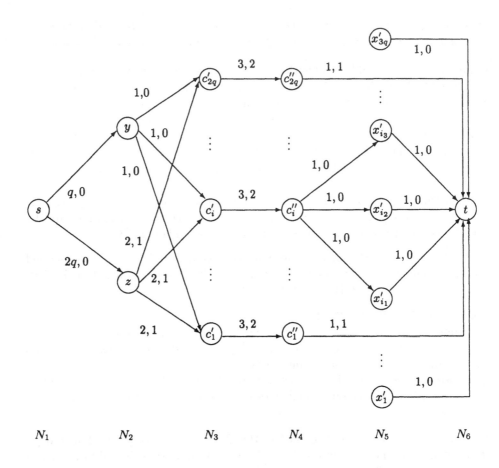

Fig. 2. Network for the proof of Claims 2 and 3. Every arc has as first label its capacity and as second label its transit time. For the sake of clarity not all arcs between layers N_4 and N_5 resp. N_4 and N_6 are depicted.

(c_i', c_i'') and (c_i'', t). Arc (y, c_i') has capacity 1 and transit time 0, arc (z, c_i') has capacity 2 and transit time 1, arc (c_i', c_i'') has capacity 3 and transit time 2, and finally arc (c_i'', t) has capacity 1 and transit time 1. Moreover, we introduce for every triple $c_i = \{x_{i_1}, x_{i_2}, x_{i_3}\}$ in C the three arcs (c_i'', x_{i_1}'), (c_i'', x_{i_2}') and (c_i'', x_{i_3}'). All these $6q$ arcs have capacity 1 and transit time 0. Finally, every node x_j' is connected to the sink t by another arc with capacity 1 and transit time 0. This completes the description of the graph $G = (N, A)$.

Claim 2 *If the given instance of* EXACT COVER BY 3-SETS *has a solution, then the constructed instance of DA-DF allows a feasible dynamic flow with value $4q$ within time $T = 3$.*

Proof. Consider the set C^* that solves the EXACT COVER BY 3-SETS instance. We send two separate amounts of flow, an amount of q and an amount of $3q$. The first amount will use the q arcs (c_i', c_i'') that correspond to triples that are *not* contained in C^* and the second one will use the q arcs (c_i', c_i'') that correspond to triples that are contained in C^*. This takes care of the critical arcs with transit time two. All other arcs will be used by the two flows at distinct times. Clearly, this is sufficient to establish the claim.

The first q units of flow are scheduled as follows. They start at time 0 in the source s and are sent along the arc (s, y) to the node y, from where they are immediately passed on to the q nodes c_i' that correspond to triples not in C^*. Then they move along the arcs (c_i', c_i'') and arrive at time 2 in layer N_4, where they occupy q pairwise distinct nodes. From layer N_4, they finally take the arcs (c_i'', t) and arrive at time $T = 3$ in the sink t.

The remaining $3q$ units of flow are scheduled as follows. At time 0, we send $2q$ units of flow along (s, z). From z these units then move on to the q nodes c_i' that correspond to triples in C^*. Clearly, in every such c_i' there arrive 2 flow units at time 1 where they are merged with other flow units that were routed via node y: At time 1 we send q units of flow along (s, y) and immediately pass them also on to the q nodes c_i' that correspond to triples in C^*. All in all, this yields that at time 1 in every node c_i' corresponding to a triple in C^* there are exactly 3 units of flow available. From each such node c_i' its three units of flow are passed on to the corresponding node c_i'' via the arcs (c_i', c_i''). At the nodes c_i'' a splitting of flow takes place. Since every element x_j is contained in exactly one triple in C^*, the $3q$ units of flow available at time 2 in nodes c_i'' corresponding to triples in C^* can be passed on to the layer N_5 simultaneously. Consequently, at time 2 exactly one unit of flow arrives at each of the nodes x_j'. Hence, by using the unit capacity arcs (x_j', t), all $3q$ units of flow will finally arrive at time $T = 3$ in the sink t. $\square$

Claim 3 *If the constructed instance of* DA-DF *allows a feasible dynamic flow with value* $v = 4q$ *within time* $T = 3$, *then the given instance of* EXACT COVER BY 3-SETS *has a solution.*

Proof. The proof of this claim is structured into a sequence of five simple observations (i)–(v) on a flow of value $v = 4q$:

(i) Node y is at (transit time) distance 2 from the sink, and hence only flow that enters y at time 0 or 1 can reach the sink at time $T = 3$. Similarly, node z is at (transit time) distance 3 from the sink and only flow that enters z at time 0 can reach the sink at time $T = 3$. Because of the capacity constraints on the two arcs (s, y) and (s, z), a flow of value $4q$ can only be achieved if q units of flow are sent along (s, y) at times 0 and 1, and $2q$ units of flow are sent along the arc (s, z) at time 0.

(ii) We claim that at time 1 at most $3q$ units of flow may have their current position somewhere in $N_1 \cup N_2 \cup N_3$: Otherwise, these $> 3q$ units could not arrive in N_4 before time $T = 3$ because the direct "shortcut arcs" from N_4 to N_6 are too slow, whereas the fast arcs via N_5 can only transport $3q$ flow units in time (recall that there are only $3q$ edges of capacity 1 that connect N_5 to N_6).

(iii) Hence, the q units of flow that are transported along the arc (s,y) at time 0 must have already left the nodes in the set $N_1 \cup N_2 \cup N_3$ at time 1. Since they were transported to q pairwise distinct nodes c_i', they are traversing and blocking exactly q of the arcs (c_i', c_i'') at time 1. The other $3q$ units of flow that arrive in N_3 at time 1 must use the remaining q arcs (c_i', c_i'') in order to go on to layer N_4 without delay.

(iv) Consequently, $3q$ units arrive (divided in groups of 3 units) at q distinct nodes c_i'' at time 3. Note that every c_i'' has four outgoing arcs: A slow "shortcut arc" that connects it directly to the sink (and cannot be used since it is too slow), and three fast arcs with transit time 0 that connect it to the three nodes x_{i_1}', x_{i_2}' and x_{i_3}' (here we assume that $c_i = \{x_{i_1}, x_{i_2}, x_{i_3}\}$ holds). Since the fast arcs have capacity 1, exactly one unit of flow is sent along every fast edge emanating from the q distinct nodes c_i'' where 3 units of flow are located.

(v) Every node x_j' in N_5 has only one outgoing arc, and this arc has capacity 1. Hence, the flow visits all $3q$ nodes in N_5. In other words, the q triples that correspond to the arcs (c_i', c_i'') traversed during the time interval $[1,3]$ cover all $3q$ elements in X. Clearly, the resulting set $C^* \subseteq C$ constitutes a solution for the EXACT COVER BY 3-SETS instance. This completes the proof of the claim.   □

Theorem 4.2 *For a flow value v that is encoded in unary and for every constant time bound $T \geq 3$, the problem DA-DF is NP-hard.*

Proof. All numbers involved in the above construction are either small constants or they depend linearly on q. Since the encoding of the EXACT COVER BY 3-SETS instance has length $\Omega(q)$, our reduction is strongly polynomial.   □

Note that we used only the three numbers 0, 1 and 2 as transit times in the NP-hardness construction above. Hence, already the introduction of arcs with transit time 2 makes problem DA-DF NP-hard.

5 Many: Constant Flow Value and Unary Time Bound

In this section, we study the computational complexity of problem DA-DF in the case where the flow value v is constant (and not part of the input) and where the time bound T is encoded in unary. We show that this case is solvable in polynomial time. The key idea is to observe that there is just a reasonable number of interesting time points and just a constant number of flow units, which makes it possible to trace and tabulate the way of every single flow unit through the network.

Theorem 5.1 *For a time bound T that is encoded in unary and for every constant flow value v, the problem DA-DF is solvable in polynomial time.*

Proof. Let $\mathcal{N} = (G, u, \tau, s, t)$ be a dynamic network with underlying graph $G = (N, A)$, and let v and T be a given flow value and a given time bound. We proceed as follows.

First, we remove all arcs with transit times greater than T from G. Then we subdivide and replace every arc $a \in A$ with transit time $\tau_a \geq 2$ by a chain of $\tau_a - 1$ new nodes and τ_a new arcs. These $\tau_a - 1$ new nodes form the set CORR(a) of *pseudo-nodes* corresponding to a. All τ_a pseudo-arcs in this new chain receive transit time 1 and the same capacity as the original arc a. This results in a new graph $G^* = (N^*, A^*)$ whose nodes are either pseudo-nodes or original nodes in N, and whose arcs are either pseudo-arcs with transit time 1 or original arcs with transit time 0 or 1. For notational convenience, we set CORR(a) = $\emptyset$ for all arcs a with transit time 0 or 1.

Next, we introduce for every integer timepoint θ, $\theta = 0, \ldots, T$, and for every v-tuple $(p_1, \ldots, p_v)$ of (not necessarily pairwise distinct) nodes over N^*, a corresponding state STA$[\theta; p_1, \ldots, p_v]$. The meaning of state STA$[\theta; p_1, \ldots, p_v]$ will be that "at time θ, the flow unit number i is located at node p_i for $i = 1, \ldots, v$". Such a state is called a *feasible* state, if

(a) for all arcs $a \in A$ and for all indices $1 \leq i < j \leq v$, $p_i \in$ CORR(a) and $p_j \in$ CORR(a) implies $p_i = p_j$, and
(b) for all arcs $a \in A$, the cardinality of $\{i \mid p_i \in$ CORR(a)$\}$ does not exceed the capacity u_a.

Intuitively speaking, condition (a) prevents further flow units from entering an arc that is already transmitting another flow unit, and condition (b) guarantees that the flow obeys the capacity constraints on arcs a with $\tau_a \geq 2$.

We say that a feasible state STA$_1$ = STA$[\theta; p_1, \ldots, p_v]$ *leads* to another feasible state STA$_2$ = STA$[\theta + 1; p_1', \ldots, p_v']$, if and only if there exists a feasible dynamic flow in the network G^* that sends v units of flow from the (artificial) sources $p_1, \ldots, p_v$ to the (artificial) sinks $p_1', \ldots, p_v'$ within a single time unit.

It is easily verified that there exists a dynamic flow with dedicated arcs in the original network G that sends v units of flow from source to sink within time T, if and only if there exists a sequence of feasible states STA$_0, \ldots,$ STA$_T$ such that

- STA$_0$ = $[0; s, s, \ldots, s]$ and STA$_T$ = $[T; t, t, \ldots, t]$, and
- for $0 \leq i \leq T - 1$, state STA$_i$ leads to STA$_{i+1}$.

Thus, the dynamic flow problem reduces to a simple reachability problem which can be solved by standard techniques.

To analyse the time and space complexity of the above construction, note that the number of nodes in N^* is $O(|N| + T|A|) = O(T|A|)$ and the number of arcs in A^* is $O(|A| + T|A|) = O(T|A|)$. The number of feasible states is bounded by $O(T^{v+1}|A|^v)$. Since T is encoded in unary and v is a constant, the size of the new graph G^* and the number of states is polynomial in the input length of the problem. Similarly as in the proof of Observation 4.1, we may argue that it can be decided in polynomial time whether some state STA$_1$ leads to some other state STA_2 by applying standard techniques for classical dynamic network flow problems (i.e. without dedicated arcs). Hence, the whole construction can be performed in polynomial time. □

6 Conclusion

In this paper we investigated the computational complexity of a basic dynamic network flow problem with dedicated arcs. We established a sharp borderline between sixteen easy respectively difficult cases of this problem (cf. Table 1), depending on whether the flow and time parameters are small constants, arbitrary constants, encoded in unary or encoded in binary. The borderline has essentially been drawn between the flow bounds $v = 1$ and $v = 2$ and between the time bounds $T = 2$ and $T = 3$.

Finally, we suggest as an open problem to investigate polynomial time approximation algorithms for the dynamic network flow problem with dedicated arcs. Note, however, that it follows from Theorem 4.2 that the maximum amount of flow that can be sent from the source s to the sink t within time T with respect to the dedicated arc model cannot be approximated with a performance guarantee $\rho > 2/3$ unless P=NP. Similarly, it follows from Theorem 3.2 that the minimum time which is needed for sending v units of flow from s to t in the dedicated arc model cannot be approximated with a performance guarantee $\rho < 4/3$ unless P=NP.

References

1. J.E. Aronson, A survey on dynamic network flows, *Annals of Operations Research* **20**, 1989, 1–66.
2. R.E. Burkard, K. Dlaska and B. Klinz, The quickest flow problem, *ZOR Methods and Models of Operations Research* **37**, 1993, 31–58.
3. L.R. Ford and D.R. Fulkerson, Constructing maximal dynamic flows from static flows, *Operations Research* **6**, 1958, 419–433.
4. M.R. Garey and D.S. Johnson, *Computers and Intractability, A Guide to the Theory of NP-Completeness*, Freeman, San Francisco, 1979.
5. B. Hoppe and É. Tardos, Polynomial time algorithms for some evacuation problems, in: *Proceedings of the 5-th Annual ACM-SIAM Symposium on Discrete Algorithms*, 1994, pp. 433–441.
6. B. Hoppe and É. Tardos, The quickest transshipment problem, in: *Proceedings of the 6-th Annual ACM-SIAM Symposium on Discrete Algorithms*, 1995, pp. 512–521.
7. B. Klinz and G.J. Woeginger, Minimum cost dynamic flows: The series-parallel case, in: *Proceedings of the 4th International IPCO Conference*, Springer Lecture Notes in Computer Science **920**, (E. Balas and J. Clausen, eds.), Springer-Verlag, Berlin, Heidelberg, 1995, pp. 329–343.
8. C.H. Papadimitriou, P. Serafini and M. Yannakakis, Computing the throughput of a network with dedicated lines, *Discrete Applied Mathematics* **42**, 1993, 271–278.

9. J.B. Orlin, Minimum convex cost dynamic network flows, *Mathematics of Operations Research* **9**, 1984, 190–207.

10. W.B. Powell, P. Jaillet and A. Odoni, Stochastic and dynamic networks and routing, in: *Handbooks in Operations Research and Management Science: Network Routing*, Vol. 8, (M.O. Ball, T.L. Magnanti, C.L. Monma and G.L. Nemhauser, eds.), Elsevier Science Publishers B.V., North Holland, Amsterdam, 1995, pp. 141–295.

Approximate Maxima Finding of Continuous Functions Under Restricted Budget (Extended Abstract)[*]

Evangelos Kranakis[14] Danny Krizanc[14] Andrzej Pelc[24] David Peleg[3]

[1] Carleton University, School of Computer Science, Ottawa, ON, K1A 5B6, Canada. E-mail: {kranakis,krizanc}@scs.carleton.ca
[2] Département d'Informatique, Université du Québec à Hull, Hull, Québec J8X 3X7, Canada. E-mail: pelc@uqah.uquebec.ca
[3] Department of Applied Mathematics and Computer Science, The Weizmann Institute of Science, Rehovot 76100, Israel. E-mail: peleg@wisdom.weizmann.ac.il.
[4] Research supported in part by NSERC (Natural Sciences and Engineering Research Council of Canada) grants.

Abstract. A function is distributed among nodes of a graph in a "continuous" way, i.e., such that the difference between values stored at adjacent nodes is small. The goal is to find a node of maximum value by probing some nodes under a restricted budget. Every node has an associated cost which has to be paid for probing it and a probe reveals the value of the node. If the total budget is too small to allow probing every node, it is impossible to find the maximum value in the worst case. Hence we seek an *Approximate Maxima Finding (AMF)* algorithm that offers the best worst-case guarantee g, i.e., for any continuous distribution of values it finds a node whose value differs from the maximum value by at most g.

Approximate Maxima Finding in graphs is related to a generalization of the multicenter problem and we get new results for this problem as well. For example, we give a polynomial algorithm to find a minimum cost solution for the multicenter problem on a tree, with arbitrary node costs.

1 Introduction

Suppose that a certain function is distributed among nodes of a graph G in a "continuous" way, namely, the difference between the function values stored at adjacent nodes is small. A *searcher* situated in one of the graph nodes seeks to find a node of maximum value by *probing* some of the nodes. What limits the scope of the search is the fact that probes are expensive, and the searcher operates under a restricted budget. More specifically, every node has an associated *cost* which has to be paid for probing it, and a probe reveals the value of the node. If the total budget is too small to allow probing every node, it is impossible

---

[*] Part of this research was performed while the fourth author was visiting Carleton University as a COGNOS scholar.

to find the maximum value in the worst case. Hence we seek an *Approximate Maxima Finding (AMF)* algorithm that offers the best *guarantee g*, i.e., for any continuous distribution of values it finds a node whose value differs from the maximum value by at most g.

There are a number of possible schemes of pricing probes, and in what follows we look at the problem of Approximate Maxima Finding under three such schemes. The simplest is the *unit cost* model, where accessing any node costs one unit. The second and most general pricing scheme is based on the assumption that probing different nodes costs different amounts. We refer to this scheme as the *arbitrary cost* model. Finally, in some cases it is plausible that the price is higher for remote nodes (namely, ones far away from the location of the searcher). This is captured by the *distance cost* model, in which the cost of probing a node is proportional to its distance in the graph from the searcher's location h_0.

Example 1 - Least loaded server: Consider a distributed network of servers. The network employs a distributed local load balancing scheme, by which each server periodically compares its load with that of its neighbors, and passes some of the jobs in its queue to certain neighbors, or takes some load off certain other neighbors, according to their relative loads. As a result of this scheme, neighboring servers are nearly balanced, so the load function is more or less continuous on the entire network. However, looking at a three-dimensional "topological map" representing the loads at the various servers, there may still be "hills" at certain regions of the network, representing areas of highly loaded servers, or "valleys" representing areas of lightly loaded servers. Note that as this load function constantly changes, one usually knows the topology of the network but not the current load distribution at any given moment.

Thus a client in need of a free server for an urgent job may still profit from seeking a relatively unloaded server, even if that server is in a remote location on the network. However, the search itself incurs a delay on the job, so it makes sense to limit the search by imposing a budget restriction, and settle for the best server found within the restricted search. In this example, it is probably appropriate to use either the unit cost or the distance cost models, according to the specific characteristics of the underlying communication network at hand. □

Example 2 - WWW access: Our original interest in the above problem was motivated by the following example. Consider the issue of searching the World Wide Web for a particular data. The expanding use of the Internet for commercial purposes makes it increasingly plausible that users will be required to pay for navigating on the Web in the future. This is already the case to a certain extent, as some sites charge for access to particular Web pages, containing, e.g., on-line magazines or encyclopedias. On the other hand, telecommunication companies that are owners of cables linking sites of the Internet may start charging per use rather than a flat access fee.

Consider an undirected graph whose nodes are Web pages. Two nodes are adjacent if there is a hypertext link between them in at least one direction. Consider a user at a given home page h_0 and consider its connected component

G in the above graph. The extremely large number of existing hypertext links joining various pages of the Web guarantees that for most home pages this component G is a large portion of the entire Web. The user searches for data on a particular subject. From the point of view of this task, every Web page of G can be assigned an integer (positive or negative) value representing how closely its content matches the data sought by the user. The home page of the user has value 0. It is reasonable to assume that (in most cases) values of adjacent nodes do not differ much. Indeed, pages represented by those nodes are joined by a hypertext link in at least one direction. Such pages are unlikely to have large difference in values from the point of view of a given data searching task. Suppose, for example, that the user seeks data on *fault-tolerant routing in graphs* and assigns the value v to the Web page *algorithms on graphs*. This page can have hypertext links to *routing* or *fault-tolerant algorithms on graphs*, that will be of slightly larger value for our user, as they approach the investigated subject more closely, or a link to *planarity testing* that is of slightly lesser value, as it deviates from the subject, but is unlikely to have a hypertext link to *Babylonian mythology* that would be clearly of much smaller value than v, although very interesting by itself.

With every Web page we associate a non-negative cost of examining its content. There are several natural ways of pricing such data acquisition. The *unit cost* model can apply in certain cases, and the accesses can be either charged by the telecommunication company for the transmission of data or by the owner of the page. In case of accessing data banks, it may be more natural to assume that different sites will charge differently for their various pages, which justifies using the *arbitrary cost* model. In the case where the price is charged by the telecommunication company it is also plausible that the price will be higher for remote Web pages, hence using the *distance cost* model may be appropriate. □

It turns out that the AMF problem with restricted budget is similar to a generalization of the multicenter problem [3], where a set of k nodes of a graph is sought that minimizes the maximum distance from every node of the graph. Our results imply solutions for some variants of this problem that are of independent interest. For example, we give a polynomial algorithm to find a minimum cost solution for the multicenter problem on a tree, with arbitrary node costs.

1.1 Results of the paper

Most of our results concern the case when the graph G is a chain or, more generally, a tree. For general graphs we show that basic hardness and approximability results can be readily derived by relating the problem to a variant of the *multicenter problem* [3, 4].

For the case of the chain we give formulas indicating which nodes should be probed for the unit and distance cost to get optimal guarantee. For the arbitrary cost we construct a simple optimal algorithm based on shortest paths and give an even faster, linear time approximation, if the costs of all nodes have limited range. In case of arbitrary trees, an optimal polynomial time algorithm for unit cost follows from a result in [3], while we construct an optimal polynomial time algorithm for arbitrary costs. For the distance cost a faster greedy algorithm can be applied. We also remark that AMF in general graphs is an NP-hard problem for all considered ways of pricing.

The paper is organized as follows. In section 2 we formally describe the problem. In section 3 we discuss adaptive versus oblivious algorithms for AMF. We also show the connection of AMF under unit cost to the multicenter problem. Sections 4 and 5 contain AMF algorithms for chains and trees, respectively. Finally, section 6 contains conclusions and open problems.

2 Statement of the Problem

Let $G = (V, E)$ be an undirected graph. For arbitrary nodes x and y, $dist(x, y)$ denotes the length of the shortest path in G joining x and y. Let h_0 be a fixed node. Let $F : V \to Z$ be an integer valued function representing values of nodes. We assume that $F(h_0) = 0$ and that F is *continuous* (in a "discrete" sense), namely, $|F(x) - F(y)| \in \{0, 1\}$, whenever nodes x and y are adjacent. Call a function F satisfying the above properties a *value function* and let $\mathcal{F}$ denote the set of all value functions. For any value function $F \in \mathcal{F}$, $F_{max} = \max(F(v) : v \in V)$. Let $\mathcal{C} : V \to R^+$ be a function representing the cost of probing nodes of the graph. For a set of nodes $P = \{v_1, ..., v_k\}$, we denote the total cost of P by

$$\mathcal{C}(P) = \sum_{i=1}^{k} \mathcal{C}(v_i).$$

We refer to the constant cost function $\mathcal{C} \equiv 1$ as the *unit cost*, and to the cost function satisfying $\mathcal{C}(x) = dist(x, h_0)$ for all nodes x, as the *distance cost*.

Fix a graph G, a node h_0 and a cost function $\mathcal{C}$. The input of an Approximate Maxima Finding (AMF) algorithm is a positive real B representing the budget allowed for the AMF task and a value function $F \in \mathcal{F}$. The algorithm can *probe* any set of nodes P, whose total cost is within the budget, i.e., such that $\mathcal{C}(P) \leq B$. In the beginning, only the value $F(h_0) = 0$ is known. In the course of the algorithm execution only the values of nodes that have already been probed become known. Decision on which node to probe next can be made dynamically, on the basis of the values discovered so far. For a given value function $F \in \mathcal{F}$ and a given budget B, the output of the algorithm $\mathcal{A}$ is the set $\mathcal{P}(\mathcal{A}, F, B)$ of all nodes that have been probed, together with the node h_0. Among them the node of highest value

$$\mathcal{A}_{max}(F, B) = \max(F(v) : v \in \mathcal{P}(\mathcal{A}, F, B))$$

can be chosen.

For a given AMF algorithm $\mathcal{A}$ and a given input budget B, the *guarantee* of $\mathcal{A}$ for B is defined as

$$\mathcal{G}(\mathcal{A}, B) = \max_{F \in \mathcal{F}}(F_{max} - \mathcal{A}_{max}(F, B)).$$

An AMF algorithm $\mathcal{A}^*$ is *optimal* if, for any budget B and any AMF algorithm $\mathcal{A}$, $\mathcal{G}(\mathcal{A}^*, B) \leq \mathcal{G}(\mathcal{A}, B)$. Let

$$\text{Rad}(\mathcal{A}, F, B) = \max_{v \in V} \min_{u \in \mathcal{P}(\mathcal{A}, F, B)} (dist(u, v)),$$

and let

$$\text{Rad}(\mathcal{A}, B) = \max_{F \in \mathcal{F}} (\text{Rad}(\mathcal{A}, F, B)).$$

Note that these two "radius parameters" are dependent on $\mathcal{A}$ and F, in that the choice of the probe set $\mathcal{P}(\mathcal{A}, F, B)$ can be made by $\mathcal{A}$ dynamically on the basis of the F values discovered along the way. In contrast, one can define a *static* radius parameter, based only on the graph G and the budget B, as follows: For a set of nodes S, let

$$\text{Rad}(S) = \max_{v \in V} \min_{u \in S} (dist(u, v)).$$

Let $\mathcal{S}$ be the collection of all sets $S \subseteq V$ whose total cost is within the budget, i.e., such that $\mathcal{C}(S) \leq B$. Now let

$$\text{Rad}(B) = \min_{S \in \mathcal{S}} (\text{Rad}(S)).$$

In the sequel we rely on the following central observation, based on the continuity of the value functions.

Lemma 1. *For every algorithm $\mathcal{A}$ and budget B, $\mathcal{G}(\mathcal{A}, B) = \text{Rad}(\mathcal{A}, B)$.* □

As a consequence, an optimal AMF algorithm $\mathcal{A}^*$ can be thought of as an algorithm minimizing $\text{Rad}(\mathcal{A}, B)$ for any budget B.

3 Adaptive vs. Oblivious Algorithms

Since values of nodes are not known *a priori*, there are two natural classes of AMF algorithms. An *oblivious* algorithm is required to decide on the entire set of probed nodes at once, while an *adaptive* algorithm can probe nodes one by one, computing the next node to probe on the basis of values of nodes already probed. At first glance, adaptive algorithms are more flexible and thus should be able to provide better guarantees. While this is likely to be true on average, we have the following simple result showing that in the *worst case*, adaptivity does not help in Approximate Maxima Finding.

Proposition 2. *For any given graph G, node h_0 and cost function $\mathcal{C}$, there exists an oblivious AMF algorithm $\mathcal{A}^{ob}$ which is optimal (among all AMF algorithms, including adaptive ones).*

PROOF Let $\mathcal{A}^*$ be an optimal AMF algorithm. Fix an input budget B and let $P = \{h_0, v_1, ..., v_k\}$ be the output of $\mathcal{A}^*$ for the value function $F \equiv 0$. We claim that the oblivious algorithm $\mathcal{A}^{ob}$ that outputs the set P for input B (regardless of the value function) is also optimal. We have to show that, for any B and any $\mathcal{A}$,

$$\max_{F \in \mathcal{F}}(F_{max} - \mathcal{A}^{ob}_{max}(F, B)) \leq \max_{F \in \mathcal{F}}(F_{max} - \mathcal{A}_{max}(F, B)).$$

Fix B. Let $S = \mathcal{A}^{ob}(F, B)$ and $r = \text{Rad}(\mathcal{A}^{ob}, F, B)$ (regardless of $F \in \mathcal{F}$). Take any function $F \in \mathcal{F}$. We have

$$F_{max} - \mathcal{A}^{ob}_{max}(F, B) \leq \text{Rad}(\mathcal{A}^{ob}, F, B) = r.$$

Hence, for any AMF algorithm $\mathcal{A}$ it suffices to show a value function $F \in \mathcal{F}$, such that $r \leq F_{max} - \mathcal{A}_{max}(F, B)$. Let Φ be a value function such that $\Phi(v) = 0$, for all $v \in S$ and $\Phi(v) = r$, for some $v \in V$. Such a function exists by definition of r. By definition of S, $\mathcal{A}^*_{max}(\Phi, B) = 0$. Since $\mathcal{A}^*$ was optimal, $\mathcal{A}(\Phi, B) \leq 0$, for any $\mathcal{A}$ and hence $\Phi_{max} - \mathcal{A}_{max}(\Phi, B) \geq r$. □

The idea behind the above proof further reveals the following.

Proposition 3. *For any given graph G, node h_0 and cost function C, the guarantee given by the (oblivious) optimal AMF algorithm $\mathcal{A}^*$ satisfies $\mathcal{G}(\mathcal{A}^*, B) = \text{Rad}(B)$.* □

In view of the above proposition, in the rest of the paper we will restrict attention to non-adaptive algorithms. It is now easy to notice the similarity of unit cost Approximate Maxima Finding to the *multicenter problem* [3]. In the latter problem, the input is a graph G and a positive integer k. The goal is to find a set S of k nodes that minimizes $\max_{v \in V} \min_{u \in S}(dist(u, v))$. Hence finding the optimal AMF algorithm for a graph G, node h_0, unit cost function C and budget k is equivalent to solvingr the multicenter problem for G and $k + 1$ with the additional restriction that h_0 be in the multicenter S. Both problems are polynomially equivalent.

4 Algorithms for Chains

In this section we present optimal AMF algorithms for the $(n + 1)$-node chain with nodes $0, 1, ..., n$.

4.1 Unit cost function

For the unit cost function, optimal AMF for the chain is given by the following proposition. We omit the easy proof.

Proposition 4. *Let G be the $(n + 1)$-node chain and h_0 its end-point 0. Then the optimal AMF algorithm with budget k has guarantee $g = \lceil \frac{n}{2k+1} \rceil$ and consists of probing nodes $2g, 4g, ... 2kg$.* □

4.2 Distance cost function

For the distance cost function we also obtain a formula indicating which nodes should be probed by an optimal algorithm.

Proposition 5. *Let G be the $(n + 1)$-node chain and h_0 its end-point 0. Let $P_k = n\frac{k(k+1)}{2k+1}$. Suppose that the budget B satisfies $P_{k-1} < B \leq P_k$, for some $k = 1, 2, ..., n-1$. Then the optimal AMF algorithm with budget B has guarantee $g = \lceil \frac{n}{k} - \frac{B}{k^2} \rceil$ and consists of probing nodes $n - g$, $n - 3g$,..., $n - (2k - 1)g$.*

PROOF First note that since $B > P_{k-1} = n\frac{k(k-1)}{2k-1}$, we have $(2k - 1)g < n$ and hence all proposed probes are feasible (belong to the set $\{1, ..., n\}$). On the other hand,

$$\sum_{i=1}^{k}(n - (2i - 1)g) \leq B,$$

hence the proposed probes are within the budget. Since distances between consecutive probes are $2g$ and the probe $n - g$ is at distance g from the node n, in order to show that our algorithm has guarantee g it is enough to show that the distance between $h_0 = 0$ and the closest probe is at most $2g$. We have

$$g \geq \frac{n}{k} - \frac{B}{k^2} \geq \frac{n}{k} - \frac{P_k}{k^2} = \frac{n}{2k + 1},$$

which implies $n - (2k - 1)g \leq 2g$, as needed.

Suppose that some other algorithm gives a better guarantee $g' < g$ for the same budget B. Thus the largest probed node must be larger than $n - g$, the second largest must be larger than $n - 3g$, etc., the smallest must be larger than $n - (2k - 1)g$. Contradiction follows, as the total cost must be larger than

$$\sum_{i=1}^{k}\left(n - (2i - 1)\left(\frac{n}{k} - \frac{B}{k^2}\right)\right) = B. \quad \square$$

4.3 Arbitrary cost function

Let B be the given budget and g the desired guarantee. Assume that the cost of probing node i is given by $c(i), i \leq n$. For a given budget we consider the problem of moving from node 0 to node n such that the total cost of visited nodes does not exceed the given budget B.

Theorem 6. *There is an $O(n^2 \log n)$ algorithm which for a given chain with nodes $0, 1, ..., n$ and budget B determines a set of probes having the smallest possible guarantee g and not exceeding the given budget.*

PROOF We transform the problem into a shortest path problem as follows. Fix g. For each node i add the $m = \min\{g, n - i\}$ directed edges

$$(i, i + 1), (i, i + 2), \ldots, (i, i + m).$$

Each of these edges is given weight 0. Now replace each node i with a directed edge of weight $c(i)$. The above problem is now transformed into the problem of finding a path from 0 to n which has total weight at most B. This problem is easily solved using the modified Dijkstra's algorithm (e.g. see [1]) whose running time is $O(n^2)$.

For the given budget B we execute binary search on the guarantee g, $(g \leq n)$. The resulting algorithm has running time $O(n^2 \log n)$. □

Remark: In cases where the complexity of the above algorithm is of concern, one may be willing to settle for *weaker* guarantees, dependent on the range of possible costs, and get a faster algorithm for determining the probe set. In particular, if it is known that the costs of the nodes are taken from a given range $[1, X]$ for some sufficiently small X, then it is possible to set

$$k = \lfloor B/X \rfloor,$$

and employ the rule of Subsect. 4.1. For this rule we can show the following.

Theorem 7. *For a given chain with nodes 0, 1,..., n, for costs taken from the range $[1, X]$ and budget B, the above explicit rule determines a set of probes not exceeding the given budget, and having a guarantee at most Xg, where g is the smallest possible guarantee.* □

5 Algorithms for Trees

In this section we study the problem assuming the graph G at hand is a tree. For trees, if the model is based on a unit cost function, then by the observation made at the end of Section 3, the problem can be solved in polynomial time by employing a simple variant of the algorithm of [3] for the multicenter problem. Hence in the remainder of this section we consider the distance cost function and the case of arbitrary costs.

5.1 Distance cost function

In this subsection we give an algorithm for the case where C is the distance cost function and the graph is a tree.

We are given a tree T and a budget B. For a candidate value g, Algorithm $\mathcal{A}_g$ returns a (minimum cost) set $\mathcal{P}$ of probes with radius $\text{Rad}(\mathcal{P}) = g$. It follows from Prop. 3 that the lowest value of g for which the cost of the solution produced by the algorithm fits the given budget B is the optimal guarantee.

Algorithm $\mathcal{A}_g$:

1. $\mathcal{P} \leftarrow \{h_0\}$;
2. mark the root h_0 and every node v such that $\text{Depth}(v) \leq g$ as **ok**.
3. **while** T still contains unmarked nodes **do**
   (a) Let l be a deepest unmarked node;

(b) Let c be the g-th ancestor of l;

(c) $\mathcal{P} \leftarrow \mathcal{P} \cup \{c\}$;

(d) Mark every node v of distance at most g from c as **ok**;

Let $\mathcal{P} = \{c_1, c_2, \ldots, c_k\}$ be the set of probes and $L = \{l_1, l_2, \ldots, l_k\}$ the set of *witnesses* selected in step 3(a). We prove the following lemmas.

Lemma 8. *For $i < j$, the nodes l_i and l_j are at distance at least $2g$.*

PROOF The node l_i is at distance g from c_i. Since $i < j$, l_j cannot be marked **ok** during the execution of the algorithm at node c_i. Therefore it must be the case that l_j is marked **ok** at the execution of the algorithm at c_j. Since the topology is a tree it is clear that l_i and l_j are at distance at least $2g$. $\square$

Lemma 9. *For any set $\mathcal{P}'$ of probes such that $\mathrm{Rad}(\mathcal{P}') = g$,*

$$\mathcal{C}(\mathcal{P}) \leq \mathcal{C}(\mathcal{P}').$$

PROOF Let $\mathcal{P}'$ be a set of probes with radius $\mathrm{Rad}(\mathcal{P}') = g$. For every i, let $c_i' \in \mathcal{P}'$ be the probe within distance at most g of l_i. By Lemma 8 the probes c_i' are pairwise distinct. By the choice of the probes c_i and c_i',

$$\mathrm{Depth}(c_i') \geq \mathrm{Depth}(l_i) - g = \mathrm{Depth}(c_i).$$

It follows that

$$\mathcal{C}(\mathcal{P}') \geq \sum_{i=1}^{k} \mathrm{Depth}(c_i') \geq \sum_{i=1}^{k} \mathrm{Depth}(c_i) = \mathcal{C}(\mathcal{P}).$$

The lemma follows. $\square$

We can now prove the following theorem.

Theorem 10. *There is an $O(n \log d)$ algorithm which for a given tree of size n and depth d, and budget B determines a set of probes having the smallest possible guarantee g and not exceeding the given budget.*

PROOF Given a tree T and a budget B we execute binary search on the possible values of g in order to find the smallest g and a set $\mathcal{P}$ of probes such that $\mathcal{C}(\mathcal{P}) \leq B$, and $\mathrm{Rad}(\mathcal{P}) = g$. To complete the proof we argue that $\mathrm{Rad}(B) = g$. The theorem then follows by Prop. 3.

To see why $\mathrm{Rad}(B) = g$, note that had there existed a smaller value $g' < g$ and a set of probes $\mathcal{P}'$ such that $\mathcal{C}(\mathcal{P}') \leq B$ and $\mathrm{Rad}(\mathcal{P}') = g'$, then when invoking algorithm $\mathcal{A}$ with parameter g', the resulting minimum cost solution $\mathcal{P}''$ should have cost no more than B, and our search should have ended up with g' instead of g; contradiction. $\square$

5.2 Arbitrary costs

In this section we give an optimal algorithm for the case where C is an arbitrary cost function and the graph is a tree rooted at the node h_0.

Definition 11. For any tree T, rooted at a node r, a set of probes, $v_1, \ldots, v_k$ *achieves guarantee g with surplus s* $(-g \leq s \leq g)$ if for all $v \in T$ such that $dist(v, r) \geq -s$, there exists an i such that $dist(v_i, v) \leq g$ and if $s \geq 0$ there exists an i such that $dist(v_i, r) \leq g - s$.

Let $X_g^s(r)$ be the cost of the minimum cost set of probes that achieves guarantee g with surplus s for a tree with root r. Then we have the following lemma:

Lemma 12. *Let T be a tree with root r where r has children $r_1, \ldots, r_d$. If $d > 0$ then*

$$
X_g^s(r) = \begin{cases}
C(r) + \sum_{j=1}^d X_g^{-g}(r_j), & \text{if } s = g \\
\min\{X_g^{s+1}, \quad \min_i\{X_g^{s+1}(r_i) + \sum_{j\neq i} X_g^{-s}(r_j)\}\}, & \text{if } 0 \leq s < g \\
\min\{X_g^{s+1}, \sum_{j=1}^d X_g^{s+1}(r_j)\}, & \text{if } -g \leq s < 0.
\end{cases}
$$

If $d = 0$ then

$$
X_g^s(r) = \begin{cases}
C(r), & \text{if } 0 \leq s \leq g \\
0, & \text{if } g \leq s < 0.
\end{cases}
$$

PROOF (Sketch.) Consider the case $s = g$. For a set of probes to achieve guarantee g with surplus g the root must be chosen. Once the root has been chosen the subtrees only require surplus $-g$. Using the fact that $X_g^s(r) \leq X_g^{s+1}(r)$ for $-g \leq s < g$ (since a set of probes with surplus $s + 1$ is also a set of nodes with surplus s), the formula follows. The other cases are similar. □

Theorem 13. *For any n node tree T (with root r) and cost function C there exists an optimal (oblivious) AMF algorithm with running time $O(n^2 \log n)$.*

PROOF Using Lemma 12 one can devise a dynamic programming algorithm to compute $X_g^g(r)$ starting at the leaves and moving up the tree to the root. By definition $X_g^g(r)$ is equal to the minimum budget B required to achieve a guarantee g on the given tree with $h_0 = r$. Using binary search on g, one can compute the maximum g that can be achieved for a certain budget B. It is straightforward to adapt the algorithm to compute the required probe set. The running time of the dynamic programming algorithm is $O(\sum_{v \in T} g \cdot deg(v)) = O(n^2)$ and it is applied $O(\log n)$ times. □

Note that the dynamic programming algorithm alluded to above can easily be adapted to solve the problem of finding a minimum cost solution to the multicenter problem on a tree, where the nodes are assigned arbitrary weights.

6 Conclusion

6.1 Arbitrary graphs

For arbitrary graphs, the problem of optimizing the guarantee under a given budget becomes NP-hard. This has been shown for the unit cost multicenter problem in [3], and a simple variation gives the same result for AMF under the unit cost model, hence it is clearly applicable also for the arbitrary cost model. It is also straightforward to establish hardness for the distance cost model, say, via a reduction from Exact 3-Cover (X3C). In summary we have:

Proposition 14. *The AMF problem for arbitrary graphs is NP-hard under the unit cost, distance cost and arbitrary cost models.* □

As for approximations, straightforward variants of the approximation algorithms of [4] for the multicenter problem yield also approximation algorithms for the unit cost model and the arbitrary cost model. The latter naturally encompasses also the distance cost model, with the same performance guarantees. Hence we have:

Proposition 15. *There exists an approximation algorithm for the AMF problem on arbitrary graphs*

1. *with approximation ratio 2 under the unit cost model, and*
2. *with approximation ratio 3 under the distance or arbitrary cost models.* □

6.2 Directions for future research

All our results apply to the worst-case behavior of the problem, and are restricted to deterministic AMF algorithms. It may be both of theoretical interest and practical significance to study *average case* behavior of Approximate Maxima Finding (assuming some random distribution of the value and/or cost functions), as well as to develop randomized algorithms for the problem. In both cases, it is anticipated that the equivalence between adaptive and oblivious strategies will disappear, and adaptive algorithms will perform better.

Another interesting and potentially realistic variant of the problem is based on assuming that the cost of probing a node is somehow related to its quality, hence also to the probability of obtaining a good solution by probing it.

References

1. T. H. Cormen, C. E. Leiserson, and R. L. Rivest, "Introduction to Algorithms", Electrical Engineering and Computer Science Series, M.I.T. Press, 1990.
2. S. L. Hakimi, E. F. Schmeichel, and M. Labbé, "On Locating Path- or Tree-Shaped Facilities on Networks", Networks, Vol. 23 (1993) 543 - 555.
3. O. Kariv and S. L. Hakimi, "An Algorithmic Approach to Network Location Problems. I: the *p*-Centers", SIAM J. Appl. Math., Vol. 37, No 3, Dec. 1979.
4. D. S. Hochbaum and D. B. Shmoys, "Powers of Graphs: A Powerful Approximation Technique for Bottleneck Problems", in Proceedings of STOC 1984, pp. 324 - 333.

The Optimal Cost Chromatic Partition Problem for Trees and Interval Graphs

Leo G. Kroon
Erasmus University Rotterdam
P.O. Box 1738, NL-3000 DR Rotterdam
The Netherlands

Arunabha Sen
Department of Computer Science and Engineering
Arizona State University
Tempe, AZ 85287
USA

Haiyong Deng
Computer Aided Design Group
Intel Corporation
Santa Clara, CA 95052
USA

Asim Roy
Department of Decision and Information Sciences
Arizona State University
Tempe, AZ 85287
USA

Abstract. In this paper we study the Optimal Cost Chromatic Partition (OCCP) problem for trees and interval graphs. The OCCP problem is the problem of coloring the nodes of a graph in such a way that adjacent nodes obtain different colors and that the total coloring costs are minimum.

In this paper we first give a linear time algorithm for the OCCP problem for trees. The OCCP problem for interval graphs is equivalent to the Fixed Interval Scheduling Problem with machine-dependent processing costs. We show that the OCCP problem for interval graphs can be solved in polynomial time if there are only two different values for the coloring costs. However, if there are at least four different values for the coloring costs, then the OCCP problem for interval graphs is shown to be NP-hard. We also give a formulation of the latter problem as an integer linear program, and prove that the corresponding coefficient matrix is perfect if and only if the associated intersection graph does not contain an odd hole of size 7 or more as a node-induced subgraph. Thereby we prove that the Strong Perfect Graph Conjecture holds for graphs of the form $K \times G$, where K is a clique and G is an interval graph.

1 Introduction

In this paper we study the Optimal Cost Chromatic Partition (OCCP) problem for trees and interval graphs. The general OCCP problem can be described as follows: given a graph $G = (V, E)$ with n nodes and a sequence of coloring costs $(k_1, \ldots, k_n)$, find a proper coloring $C(v) \in \{1, \ldots, n\}$ of each node $v \in V$ such that the total coloring costs $\sum_{v=1}^{n} k_{C(v)}$ are minimum. Here a coloring is proper if adjacent nodes have different colors. Thus an alternative formulation of the OCCP problem is the following: Given a graph $G = (V, E)$ with n nodes and a sequence of coloring costs $(k_1, \ldots, k_n)$, partition the vertex set V into independent sets $V_1, \ldots, V_s$ such that $\sum_{c=1}^{s} k_c |V_c|$ is minimum. Here all nodes in the independent set V_c are colored by the same color c. Without loss of generality we assume $k_c \leq k_{c'}$ whenever $c < c'$ throughout this paper.

The OCCP problem was introduced by Supowit [22] in the VLSI context. He referred to the problem as the *Weighted Coloring Problem* of a graph. However, the term *Weighted Coloring Problem* is used by Grötschel et al. [12] to describe a different problem. In order to avoid further confusion, we refer to the problem as the *Optimal Cost Chromatic Partition* problem. This problem is a generalization of the *Chromatic Sum* problem introduced by Kubicka & Schenk [17].

It is not difficult to show that the OCCP problem is NP-hard for arbitrary graphs. Supowit [22] deals with the OCCP problem for circle graphs. It follows from the results of Garey et al. [10] that this problem is NP-hard. Furthermore, Sen et al. [21] consider the OCCP problem for permutation graphs.

In this paper we study the OCCP problem for trees and interval graphs. The OCCP problem for interval graphs is equivalent to the Fixed Interval Scheduling Problem (FISP) with machine-dependent processing costs. In this scheduling problem each job j to be carried out requires processing during a fixed time interval (s_j, f_j). A sufficient number of machines is available, and all jobs may be carried out by all machines. However, the processing costs are machine-dependent. That is, if job j is carried out by machine m, then the associated processing costs are k_m. The objective is to find a feasible non-preemptive schedule for all jobs against minimum total processing costs.

The outline of this paper is as follows: In Section 2 we describe an integer linear program that can be used to solve the general OCCP problem. Next, in Section 3 we show that the OCCP problem for trees can be solved in linear time. In Section 4 we study the OCCP problem for interval graphs. We show that this problem can be solved in polynomial time if there are only two different values for the coloring costs. However, if there are at least four different values for the coloring costs, then the problem is shown to be NP-hard. We improve the given integer linear program by describing all clique inequalities, and we prove that the corresponding coefficient matrix is perfect if and only if the associated intersection graph does not contain an odd hole of size 7 or more as a node-induced subgraph. Thereby we show that the Strong Perfect Graph Conjecture holds for graphs of the form $K \times G$, where K is a clique and G is an interval graph.

2 Model formulation

In this section we give a formulation of the OCCP problem as an integer linear program. To that end, suppose we have an instance I of the OCCP problem containing a graph $G = (V, E)$ with n nodes and a sequence of coloring costs $(k_1, \ldots, k_n)$. Then the integer program to solve this instance of the OCCP problem uses the binary decision variables $x_{v,c}$ indicating whether or not node v is colored by color c $(v, c = 1, \ldots, n)$. The objective and the constraints of the model can be described as follows:

$$\min \sum_{v=1}^{n} \sum_{c=1}^{n} k_c x_{v,c} \tag{0}$$

subject to

$$\sum_{c=1}^{n} x_{v,c} = 1 \qquad\qquad v = 1, \ldots, n \tag{1}$$

$$x_{v,c} + x_{v',c} \leq 1 \qquad\qquad (v, v') \in E;\ c = 1, \ldots, n \tag{2}$$

$$x_{v,c} \in \{0,1\} \qquad\qquad v = 1, \ldots, n;\ c = 1, \ldots, n \tag{3}$$

The objective function (0) specifies that we are interested in minimizing the total coloring costs. Constraints (1) require each node to be colored exactly once. Constraints (2) guarantee that two nodes v and v' that are connected by an edge $(v, v') \in E$ are colored by different colors. Note that these constraints could be tightened by replacing them by the corresponding *clique* constraints. Unfortunately, for arbitrary graphs the number of cliques and clique constraints may be extremely large, and it may require quite some time to find them. However, for trees and interval graphs the number of cliques is linear in the number of nodes. In fact, the only cliques of a tree are the edges. Furthermore, in Section 4.2 we will show how for an interval graph the constraints (2) can be replaced by the corresponding clique constraints. Finally, the constraints (3) declare the variables $x_{v,c}$ as binary variables.

The coefficient matrix corresponding to the restrictions (1) to (3) associated with an instance I of the OCCP problem is called $M(I)$. Note that $M(I)$ is a zero/one matrix. A zero/one matrix M is said to be *perfect* if the polyhedron

$$P(M) = \{X \mid M.X \leq \mathbf{1} \text{ and } X \geq 0\}$$

has only integral extremal points. Here the vector $\mathbf{1}$ is a vector containing all 1's. It follows that the problem (0) to (3) can be solved by applying a linear programming algorithm if the matrix $M(I)$ is perfect. Padberg [18] gives a complete characterisation of perfect matrices in terms of forbidden submatrices.

3 The OCCP problem for Trees

Sen et al. [21] show that the coefficient matrix $M(I)$ associated with the integer linear program (1) to (3) is perfect if the graph G is a tree. Thus the OCCP problem for trees can be solved in polynomial time by any polynomial linear programming algorithm (Grötschel et al. [12]).

However, in this section we describe a linear time algorithm for solving the OCCP problem for trees. To that end, let T be a tree with n nodes, and let $(k_1, \ldots, k_n)$ be the corresponding sequence of coloring costs. In this section we denote the degree of node v, which is the number of different neighbours of node v, by D_v. The maximum node degree over all nodes is denoted by D.

Next, we will show that for any graph G the number of colors used by a proper coloring of minimum costs may be bounded by $D + 1$. This upper bound is sharp, as is demonstrated by the complete graphs (cliques).

Lemma 1. *For a graph G there exists a proper coloring C of minimum costs such that $C(v) \leq D_v + 1$ for all nodes v.*

Proof. Let C be a proper coloring of minimum costs, and suppose there is a node v with $C(v) = c > D_v + 1$. Then there exists a color c' with $c' \leq D_v + 1$ such that $c' \neq C(v')$ for all neighbour v' of v. Now we define the following alternative coloring C' of the graph G.

$$C'(v) = \begin{cases} c' & \text{if } v = n, \\ C(v) & \text{if } v \neq n. \end{cases}$$

Clearly, C' is a proper coloring of G. Furthermore, $k_{c'} \leq k_c$, since $c' < c$. Thus C' is also a coloring of minimum costs. By repeating this argument as often as necessary, we obtain a proper coloring of the graph of minimum costs satisfying the condition of the lemma. □

The algorithm for solving the OCCP problem for trees uses the result of Lemma 1 Recall that the chromatic number of a tree is 2 (or 1). On the other hand, it is not difficult to create an instance of the OCCP problem for trees for which the number of colors used in an optimal solution is arbitrarily large.

The algorithm for solving the OCCP problem for trees is based on the idea of dynamic programming. To that end, we first add a direction to the edges of T by choosing an arbitrary node as the root r, and by directing an edge $(v, v') \in E$ from v to v' if the unique path in T from root r to node v does not visit node v'. In the obtained directed tree each node v is the root of the subtree rooted at node v.

The algorithm to solve the OCCP problem for trees keeps track of the following information for each node v:

(i) $K_1(v)$, representing the minimum costs of coloring the subtree rooted at node v (this is called the *primary* coloring of the subtree rooted at node v),

(ii) $C(v)$, representing the color used for node v in the primary coloring of the subtree rooted at node v,

(iii) $K_2(v)$, representing the minimum costs of coloring the subtree rooted at node v with node v colored differently from $C(v)$ (this is called the *secondary coloring* of the subtree rooted at node v).

Obviously, if v is a leaf node, then $K_1(v) = c_1$, $C(v) = 1$, and $K_2(v) = c_2$.

If v is a non-leaf node with d children, $v_1, \ldots, v_d$, then first the values $K_1(v_i)$, $C(v_i)$, and $K_2(v_i)$ are determined recursively for $i = 1, \ldots, d$. Next, the coloring costs $K(v, c)$ are determined for $c = 1, \ldots, d+1$ by equation (4) below. The coloring costs $K(v, c)$ denote the costs of coloring the subtree rooted at node v, when node v is colored with color c and all subtrees rooted at the children of node v are colored as cheap as possible. Thus,

$$K(v, c) := k_c + \sum_{i:C(v_i) \neq c} K_1(v_i) + \sum_{i:C(v_i) = c} K_2(v_i) \quad \text{for } c = 1, \ldots, d+1 \quad (4)$$

Note that (4) can be calculated in an amount of time that is linear in the number of subtrees d by the following steps:

$$K := \sum_{i=1}^{d} K_1(v_i)$$

For $c := 1$ **to** $d+1$ **do** $\Delta(c) := K$

For $i := 1$ **to** d **do**

        If $C(v_i) \leq d+1$ **then** $\Delta(C(v_i)) := \Delta(C(v_i)) + K_2(v_i) - K_1(v_i)$

For $c := 1$ **to** $d+1$ **do** $K(v, c) := k_c + \Delta(c)$

Given the coloring costs $K(v, c)$ for $c = 1, \ldots, d+1$, the values $K_1(v)$, $C(v)$, and $K_2(v)$ are computed using the following formulas.

$$K_1(v) := \min \{ K(v, c) \mid c = 1, \ldots, d+1 \}, \tag{5}$$

$$C(v) := \operatorname{argmin} \{ K(v, c) \mid c = 1, \ldots, d+1 \}, \tag{6}$$

$$K_2(v) := \min \{ K(v, c) \mid c = 1, \ldots, d+1; \; c \neq C(v) \}. \tag{7}$$

In (5) the minimum costs of coloring the subtree rooted at node v are determined, and in (6) the corresponding color is determined. Next, in (7) the costs of a secondary coloring of the subtree rooted at node v are determined.

Obviously, the complexity of the above algorithm for coloring node v is $\mathcal{O}(d)$, where d is the number of children of node v. Therefore the overall complexity of the algorithm for computing the minimum costs of coloring the entire tree is proportional to the number of edges, which, in a tree, gives rise to an $\mathcal{O}(n)$ algorithm.

4 The OCCP problem for interval graphs

An interval graph $G = (V, E)$ is a graph where each node corresponds to a time interval (s_j, f_j), and where two nodes are connected by an edge if and only if the corresponding intervals are overlapping. Thus the OCCP problem for an interval graph G can be considered as the problem of coloring the corresponding intervals (s_j, f_j) in such a way that overlapping intervals obtain different colors, and such that the total coloring costs are minimum. The costs of coloring an interval with a certain color only depend on the color to be used.

As was mentioned already in the introduction of this paper, the OCCP problem for interval graphs is equivalent to the Fixed Interval Scheduling Problem (FISP) with machine-dependent processing costs. In this scheduling problem each job j to be carried out requires processing during a fixed time interval (s_j, f_j). A sufficient number of machines is available, and all jobs may be carried out by all machines. However, the processing costs are machine-dependent. That is, if job j is carried out by machine m, then the associated processing costs are k_m. The objective is to find a feasible non-preemptive schedule for all jobs against minimum total processing costs. It follows that the intervals and the colors of the OCCP problem for interval graphs correspond with the jobs and the machines of FISP with machine-dependent processing costs, respectively.

Several other variants of FISP have been considered in the literature (cf. Arkin & Silverberg [1], Dondeti & Emmons [4, 5], Fischetti, Martello & Toth [6, 7, 8], and Kolen & Kroon [13, 14, 15, 16]). The computational complexity of these variants of FISP has been studied extensively.

4.1 Computational complexity

In this section we prove that the OCCP problem for interval graphs is NP-hard if there are at least four different values for the coloring costs. This result is somewhat surprising, because almost every other combinatorial problem related to interval graphs, such as the computation of the chromatic number, the maximum independent set, the maximum clique, and the dominating set can be accomplished in polynomial time.

On the other hand, if the first s coloring costs are equal and the last $n - s$ coloring costs are equal as well, i.e., $k_1 = k_2 = \ldots = k_s = A_1$, and $k_{s+1} = k_{s+2} = \ldots = k_n = A_2$ (where it may be assumed, without loss of generality, that $A_1 < A_2$), then the optimal solution is obtained if the largest s-colorable subgraph is colored with the colors $1, \ldots, s$, and the other nodes are colored with the remaining colors. For interval graphs, the problem of finding the largest s-colorable subgraph can be solved in polynomial time by the greedy algorithm of Yanakakis & Gavril [25]. Note that, if there are only two different values for the coloring costs, then the OCCP problem for any graph is equivalent to the problem of finding the largest s-colorable subgraph of the graph.

Next we show that the OCCP problem for interval graphs is NP-hard if there are at least four different values for the coloring costs. The proof uses a reduction from the problem Numerical Three Dimensional Matching (N3DM), which is defined in following way:

Instance of N3DM:
- A positive integer t and $3t$ rational numbers a_i, b_i and c_i satisfying $\sum_{i=1}^{t}(a_i + b_i + c_i) = t$ and $0 < a_i, b_i, c_i < 1$ for $i = 1, \ldots, t$.

Question:
- Do there exist permutations ρ and σ of $\{1, \ldots, t\}$ such that $a_i + b_{\rho(i)} + c_{\sigma(i)} = 1$ for $i = 1, \ldots, t$?

It is well-known that N3DM is NP-complete in the strong sense (Garey and Johnson [9]). Therefore any problem in NP that is more general than N3DM is NP-complete as well. The proof of Theorem 2 is illustrated with the following instance of N3DM with $t = 3$: $(a_1, a_2, a_3) = (1/8, 1/4, 3/8)$, $(b_1, b_2, b_3) = (1/8, 1/4, 1/2)$ and $(c_1, c_2, c_3) = (1/4, 3/8, 3/4)$. This instance of N3DM is a yes-instance, since $a_1 + b_1 + c_3 = a_2 + b_3 + c_1 = a_3 + b_2 + c_2 = 1$.

Theorem 2. *The OCCP problem for interval graphs is NP-hard if there are at least four different values for the coloring costs.*

The idea behind the proof is that as many as possible intervals should be colored by the cheaper colors and as few as possible should be colored by the more expensive ones.

Proof. The theorem is proved by a reduction from N3DM. Hence let I_1 be an instance of N3DM containing the integer t and the rational numbers a_i, b_i and c_i for $i = 1, \ldots, t$. Next, for $i, j = 1, \ldots, t$ the rational numbers A_i, B_j and $X_{i,j}$ are chosen in such a way that all these numbers are different and that $4 < A_i < 5 < B_j < 6$ and $7 < X_{i,j} < 9$ for $i, j = 1, \ldots, t$.
  Based on these data, an instance I_2 of the OCCP problem for interval graphs is constructed. The intervals that have to be colored in I_2 are the following.

t times $(0, 1)$,	$(11 - c_k, 13)$	for $k = 1, \ldots, t$,
t times $(1, 2)$,	$t - 1$ times $(0, A_i)$	for $i = 1, \ldots, t$,
t times $(13, 14)$,	$t - 1$ times $(3, B_j)$	for $j = 1, \ldots, t$,
$t^2 - t$ times $(0, 3)$,	$(A_i, X_{i,j})$	for $i, j = 1, \ldots, t$,
$t^2 - t$ times $(12, 14)$,	$(B_j, X_{i,j})$	for $i, j = 1, \ldots, t$,
$(0, B_j)$ for $j = 1, \ldots, t$,	$(X_{i,j}, 14)$	for $i, j = 1, \ldots, t$,
$(2, A_i)$ for $i = 1, \ldots, t$,	$(X_{i,j}, 10 + a_i + b_j)$	for $i, j = 1, \ldots, t$.

Furthermore, for coloring these intervals there are t different colors with costs 0, there are $t^2 - t$ different colors with costs 1, there are t^2 different colors with costs 2, and all other colors have costs $12t^2$. Note that the construction of I_2 can be carried out in a polynomial amount of time. The instance I_2 constructed from an instance I_1 of N3DM defined above is shown in Figure 1.
  Now we will prove the following statement: I_1 is a yes-instance of N3DM if and only if the minimum total coloring costs for I_2 don't exceed $11t^2 - 5t$.

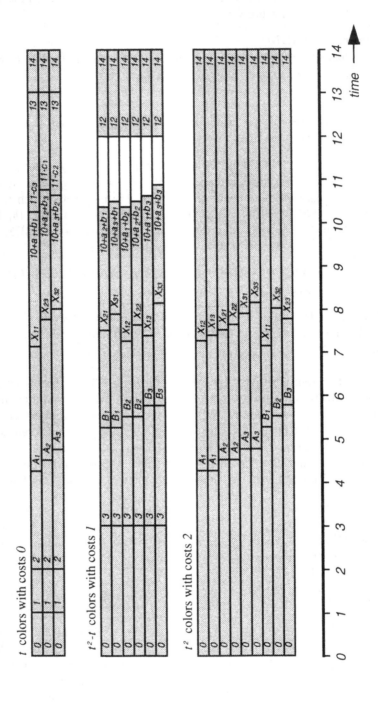

Figure 1. An instance of the OCCP problem for interval graphs.

Suppose the minimum total costs of coloring all intervals of I_2 don't exceed $11t^2 - 5t$. Then only the $2t^2$ cheapest colors are used, because otherwise the total costs would be at least $12t^2$. Furthermore, the overlap of the intervals during the time interval $(0,10)$ and during the time interval $(12,14)$ equals $2t^2$, as can be verified easily. Therefore at least $2t^2$ colors are required to color all intervals. It can be concluded that all intervals are colored by t colors with costs 0, by $t^2 - t$ colors with costs 1, and by t^2 colors with costs 2.

Since all intervals $(A_i, X_{i,j})$ and $(B_j, X_{i,j})$ are overlapping, and since the total number of these intervals $(= 2t^2)$ equals the total number of used colors, each color is used for coloring exactly one of these intervals. Furthermore, the overlap of the intervals in the time interval $(0,10)$ and in the time interval $(12,14)$ equals the total number of used colors. Therefore, if an interval finishes either in the time interval $(0,10)$ or in the interval $(12,14)$, then the next interval that is colored by the same color fits seamlessly to its predecessor.

Thus an interval $(A_i, X_{i,j})$ will be colored by the same color as a sequence of intervals of the form $(0, 1)$, $(1, 2)$, $(2, A_i)$, or as one of the intervals $(0, A_i)$. An interval $(B_j, X_{i,j})$ will be colored by the same color as a sequence of intervals of the form $(0, 3)$, $(3, B_j)$, or as an interval $(0, B_j)$. Furthermore, an interval $(A_i, X_{i,j})$ or $(B_j, X_{i,j})$ will be colored by the same color as a sequence of intervals of the form $(X_{i,j}, 10 + a_i + b_j)$, $(11 - c_k, 13)$, $(13, 14)$, as a sequence of intervals of the form $(X_{i,j}, 10 + a_i + b_j)$, $(12, 14)$, or as one of the intervals $(X_{i,j}, 14)$.

From this list of sequences, it can be concluded that each color colors at least 3 intervals and at most 7. If each color with costs 0 colors 7 intervals, each color with costs 1 colors 5 intervals, and each color with costs 2 colors 3 intervals, then the total costs equal $t \times 7 \times 0 + (t^2 - t) \times 5 \times 1 + t^2 \times 3 \times 2 = 11t^2 - 5t$. Also, any other assignment of colors to sequences of intervals will lead to higher total coloring costs. Thus if the total costs don't exceed $11t^2 - 5t$, then each color with costs 0 colors a sequence of intervals of the form $(0, 1)$, $(1, 2)$, $(2, A_i)$, $(A_i, X_{i,j})$, $(X_{i,j}, 10 + a_i + b_j)$, $(11 - c_k, 13)$, $(13, 14)$, where each i and each k occur exactly once. Further, each color with costs 1 colors a sequence of intervals of the form $(0, 3)$, $(3, B_j)$, $(B_j, X_{i,j})$, $(X_{i,j}, 10 + a_i + b_j)$, $(12, 14)$, where each j occurs exactly $t - 1$ times. Hence among the intervals $(X_{i,j}, 10 + a_i + b_j)$ that are colored by the colors with costs 0 each j also occurs exactly once.

The fact $\sum_{i=1}^{t}(a_i + b_i + c_i) = t$ implies that, if two consecutive intervals are colored by the same color with costs 0, then they fit seamlessly after another. It can be concluded that, if an interval $(X_{i,j}, 10 + a_i + b_j)$ is colored by the same color with costs 0 as an interval $(11 - c_k, 13)$, then $10 + a_i + b_j = 11 - c_k$. This means that $a_i + b_j + c_k = 1$. So if we define $\rho(i) = j$ and $\sigma(i) = k$ whenever interval $(X_{i,j}, 10 + a_i + b_j)$ is colored by the same color with costs 0 as interval $(11 - c_k, 13)$, then ρ and σ are the required permutations for I_1. It can be concluded that I_1 is a yes-instance of N3DM.

Conversely, given a feasible solution for I_1, the construction can be reversed to find a feasible coloring for all intervals of I_2 with total costs $11t^2 - 5t$. As N3DM is NP-complete, the OCCP problem for interval graphs is NP-hard. $\square$

In the above proof it is assumed that the four different values for the coloring costs are 0, 1, 2, or $12t^2$. However, the statement "I_1 is a yes-instance of N3DM if and only if the minimum total costs of all intervals of I_2 don't exceed $11t^2 - 5t$" also holds if the coloring costs are 0, 1, 2, and 3. However, in that case it takes somewhat more effort to see that in a coloring with minimum total costs no intervals are colored by colors with costs 3.

As was noted already, if there are only two different values for the coloring costs, then the OCCP problem for interval graphs can be solved in polynomial time. Hence an open question still to be answered asks for the computational complexity of the OCCP problem for interval graphs, if there are exactly three different values for the coloring costs. We conjecture this problem to be NP-hard as well. This is a subject for further research.

4.2 Improved model formulation

In this section we improve the integer linear program (0) to (3) by replacing constraints (2) by the corresponding *clique* constraints. To that end, let I be an instance of the OCCP problem for interval graphs containing n intervals (s_j, f_j) and a sequence of coloring costs $(k_1, k_2, \ldots, k_n)$. Furthermore, suppose the set $\{t_r | r = 0, \ldots, R\}$ contains the start times of the intervals in chronological order. That is, $\{t_r | r = 0, \ldots, R\} = \{s_j | j = 1, \ldots, n\}$ and $t_{r-1} < t_r$ for $r = 1, \ldots, R$. Then constraints (2) may be replaced by the following constraints (2'):

$$\sum_{\{j | s_j \le t_r < f_j\}} x_{j,c} \le 1 \qquad c = 1, \ldots, n; \ r = 0, \ldots, R. \tag{2'}$$

Constraints (2') state that at most one of the intervals overlapping the start time t_r of a certain interval can be colored with color c. Note that, if two intervals are overlapping, then either the first interval overlaps the start time of the second interval, or vice versa. Thus constraints (2') guarantee that overlapping intervals are colored by different colors. Note that the number of constraints (2') does not exceed the number of intervals n.

In the following section necessary and sufficient conditions for the coefficient matrix $M'(I)$ of the integer program (0), (1), (2'), (3) to be perfect are expressed in terms of the associated *intersection* graph $G(I)$. The intersection graph of a zero/one matrix M is a graph G containing exactly one node for each column of the matrix M. Two nodes x and y of G are connected if and only if $M_{r,x}.M_{r,y} = 1$ for some row r of the matrix M.

Note that the nodeset of $G(I)$ is the set $\{v_{j,c} | j, c = 1, \ldots, n\}$. Two different nodes $v_{j,c}$ and $v_{j',c'}$ are connected if $j = j'$ and $c \ne c'$, or if $c = c'$ and $(s_j, f_j) \cap (s_{j'}, f_{j'}) \ne \emptyset$. An edge connecting two nodes $v_{j,c}$ and $v_{j',c'}$ with $j = j'$ and $c \ne c'$ is called a *bridge edge*. An edge connecting the nodes $v_{j,c}$ and $v_{j',c'}$ with $c = c'$ and $(s_j, f_j) \cap (s_{j'}, f_{j'}) \ne \emptyset$ is called an *overlap edge*.

Note that for $c = 1, \ldots, n$ the subgraph $G_c(I)$ induced by the set of nodes $\{v_{j,c} | j = 1, \ldots, n\}$ is equivalent to the interval graph G corresponding to the intervals (s_j, f_j). The subgraphs $G_c(I)$ are coupled by the bridge edges. It follows that $G(I)$ has the form $K \times G$, where K is a clique and G is an interval graph.

Lemma 3. *If I is an instance of the OCCP problem for interval graphs, then $M'(I)$ is a clique matrix of $G(I)$.*

Proof. It follows from the definition of $G(I)$ that each clique of $G(I)$ contains either bridge edges or overlap edges, but not edges of both types.

If K is a maximum clique of $G(I)$ containing only bridge edges, then there exists an interval j such that the nodeset of K is the set $\{v_{j,c}|c = 1, \ldots, n\}$. Thus K is represented in $M'(I)$ by the row corresponding to the constraint $\sum_{c=1}^{n} x_{j,c} = 1$.

Next, let K be a maximum clique of $G(I)$ containing only overlap edges. Suppose K is a subgraph of $G_c(I)$ corresponding to color c, and let the nodeset of K be denoted by $\{v_{j,c}|j \in S\}$. Then the intervals p and q are defined as follows: $s_p = \max\{s_j|j \in S\}$ and $f_q = \min\{f_j|j \in S\}$. Now, by definition, $(s_p, f_p) \cap (s_q, f_q) \neq \emptyset$, which implies $s_p < f_q$. Note that this is trivially true if $p = q$. The definition of p and q implies that the time interval (s_p, f_q) is a subset of all intervals in the set S. It follows that $S = \{j|s_j \leq s_p < f_j\}$. Thus K is represented in $M'(I)$ by the row corresponding to the constraint $\sum_{\{j|s_j \leq s_p < f_j\}} x_{j,c} \leq 1$ in $(2')$. This completes the proof of Lemma 3. $\qquad\square$

4.3 Strong Perfect Graph Conjecture

In this section necessary and sufficient conditions for the matrix $M'(I)$ to be perfect are expressed in terms of the associated intersection graph $G(I)$. To that end, an undirected graph $G = (V, E)$ is said to be *perfect* if for each subset $V' \subset V$ the node-induced subgraph $G' = (V', E')$ has the property

$$\alpha(G') \times \omega(G') \geq |V'|, \tag{8}$$

where $\alpha(G')$ denotes the size of a maximum independent set in G' and where $\omega(G')$ denotes the size of a maximum clique in G'. If a graph G is not perfect, then it contains a node-induced *p-critical* subgraph G'. A p-critical graph is not perfect, but all of its node-induced subgraphs are perfect.

It is well known that interval graphs are perfect. Typical examples of non-perfect graphs are the *odd holes* and the *odd antiholes* of size 5 or more. A hole is a graph with nodes $1, \ldots, n$ and edges $(v, v + 1)$ for $v = 1, \ldots, n - 1$ as well as the edge $(n, 1)$. An antihole is the complement of a hole.

If G is an odd hole of size 5 or more, then $\alpha(G) = (|V| - 1)/2$ and $\omega(G) = 2$. Similarly, if G is an odd antihole of size 5 or more, then $\alpha(G) = 2$ and $\omega(G) = (|V| - 1)/2$. Thus if G is an odd hole or an odd antihole of size 5 or more, then inequality (8) is not satisfied for G itself. It follows that any graph that contains an odd hole or an odd antihole of size 5 or more as a node-induced subgraph is not perfect. The converse of this statement is called the *Strong Perfect Graph Conjecture* (Berge [2]).

Strong Perfect Graph Conjecture. *If a graph G is not perfect, then it contains an odd hole or an odd antihole of size 5 or more as a node-induced subgraph.*

Another formulation of the Strong Perfect Graph Conjecture states that the odd holes and the odd antiholes of size 5 or more are the only p-critical graphs.

For several classes of graphs the Strong Perfect Graph Conjecture has been validated. For example, it holds for *toroidal* graphs (Grinstead [11]), for *claw-free* graphs (Parthasarathy & Ravindra [19]), for *planar* graphs (Tucker [23]), and for K_4-*free* graphs (Tucker [24]). However, a general proof of the Strong Perfect Graph Conjecture is still awaited at this moment.

In Theorem 4 we will show that the Strong Perfect Graph Conjecture also holds for graphs $G(I)$ where I is an instance of the OCCP problem for interval graphs. In other words, the Strong Perfect Graph Conjecture holds for graphs of the form $K \times G$, where K is a clique and G is an interval graph.

Theorem 4. *If I is an instance of the OCCP problem for interval graphs, then $G(I)$ is perfect if and only if $G(I)$ does not contain an odd hole of size 5 or more as a node-induced subgraph.*

Proof. Let I be an instance of the OCCP problem for interval graphs, and let $G(I)$ be the associated intersection graph. Suppose $G(I)$ is not perfect. Then $G(I)$ contains a node-induced p-critical subgraph G'. It is well known that each node of a p-critical subgraph G' is contained in exactly $\omega(G')$ maximal cliques of G', if $\omega(G')$ denotes the size of a maximum clique of G' (Padberg [18]).

As for $c = 1, \ldots, n$ the interval graph $G_c(I)$ is perfect, G' contains at least one bridge edge of $G(I)$, say between $G_1(I)$ and $G_2(I)$. Thus the subgraphs $G' \cap G_1(I)$ and $G' \cap G_2(I)$ are not empty. Now we consider the interval p defined by $f_p = \min\{f_j | n_{j,1} \in G'\}$. According to the definition of interval p, all intervals j with $v_{j,1} \in G'$ that are overlapping interval p are also overlapping each other. It follows that all nodes in the set $\{v_{j,1} \in G' | (s_j, f_j) \cap (s_p, f_p) \neq \emptyset\}$ make up a maximum clique of G'. This clique contains only overlap edges. Note that $v_{p,1}$ is an element of this clique. Node $v_{p,1}$ is contained in at most 1 other maximum clique, namely a clique containing only bridge edges. Thus node $v_{p,1}$ is contained in at most 2 maximum cliques. As G' is p-critical, each node is contained in at least 2 maximum cliques. It follows that $v_{p,1}$ is contained in exactly 2 maximum cliques. As a consequence, $\omega(G') = 2$. The latter implies that G' is an odd hole of size 5 or more. □

The result of Theorem 4 can be improved slightly. Indeed, if the node-induced subgraph G' of $G(I)$ would be an odd hole of size 5, then G' would contain exactly 2 bridge edges and exactly 3 overlap edges. (Indeed, recall that G' can not be a subgraph of one of the graphs $G_c(I)$, since the latter are perfect). Let the corresponding intervals be called 1, 2 and 3, and let the corresponding colors be called 1 and 2. Then the nodes of G' are indicated by $v_{1,1}, v_{2,1}, v_{1,2}, v_{2,2}$ and $v_{3,2}$. Since there is an edge between the nodes $v_{1,1}$ and $v_{2,1}$, the intervals 1 and 2 are overlapping. But then there is also an edge between the nodes $v_{1,2}$ and $v_{2,2}$. This implies that G' contains a chord and, as a consequence, it is not an odd hole. Thus the following corollaries have been obtained.

Corollary 5. *If I is an instance of the OCCP problem for interval graphs, then $G(I)$ is perfect if and only if it does not contain an odd hole of size 7 or more as a node-induced subgraph.*

Corollary 6. *If I is an instance of the OCCP problem for interval graphs, then $M'(I)$ is perfect if and only if $G(I)$ does not contain an odd hole of size 7 or more as a node-induced subgraph.*

Proof. This result is a consequence of the following well-known result: if the zero/one matrix M is a clique matrix of its associated intersection graph G, then M is perfect if and only if G is perfect (Chvatal [3]). □

5 Final Remarks

In this paper we have studied the Optimal Cost Chromatic Partition problem for trees and interval graphs. First, we have given an algorithm based on the idea of dynamic programming for solving the OCCP problem for trees which runs in linear time.

Furthermore, we have shown that the OCCP problem for chordal graphs can be solved in polynomial time if there are only two different values for the coloring costs. This result is also valid for interval graphs, since each interval graph is a chordal graph. However, if there are at least four different values for the coloring costs, then the OCCP problem for interval graphs is NP-hard.

We have presented a formulation of the problem as an integer program with coefficient matrix $M'(I)$. The matrix $M'(I)$ is shown to be a clique matrix of its associated intersection graph $G(I)$. This implies that perfectness of the matrix $M'(I)$ is equivalent to perfectness of the graph $G(I)$. Thus, if $G(I)$ is perfect, then the problem can be solved by linear programming. We have shown that $G(I)$ is perfect if and only if it does not contain an odd hole of size 7 or more as a node induced subgraph. Thereby we have proved that the Strong Perfect Graph Conjecture is valid for graphs of the form $K \times G$, where K is a clique and G is an interval graph.

References

1. E.M. Arkin, and E.L. Silverberg. Scheduling jobs with fixed start and finish times. *Discrete Applied Mathematics*, 18 (1987), 1–8.
2. C. Berge. Färbung von Graphen deren sämtliche bzw. deren ungerade Kreise starr sind. *Wiss. Z. Martin-Luther Univ. Halle-Wittenberg*. Math.-Natur. Reihe, 114 (1961).
3. V. Chvatal. On certain polytopes associated with graphs. *J. Comb. Theory*, B-18 (1975) 138–154.
4. V.R. Dondeti, and H. Emmons. Job scheduling with processors of two types. *Operations Research*, 40 (1992) S76–S85.
5. V.R. Dondeti, and H. Emmons. Algorithms for preemptive scheduling of different classes of processors to do jobs with fixed times. *European Journal of Operational Research*, 70 (1993) 316–326.

6. M. Fischetti, S. Martello, and P. Toth. The Fixed Job Schedule Problem with spread time constraints. *Operations Research*, 6 (1987) 849–858.

7. M. Fischetti, S. Martello, and P. Toth. The Fixed Job Schedule Problem with working time constraints. *Operations Research*, 3 (1989) 395–403.

8. M. Fischetti, S. Martello and P. Toth. Approximation algorithms for Fixed Job Schedule Problems. *Operations Research*, 40 (1992) S96–S108.

9. M.R. Garey, and D.S. Johnson. *Computers and intractability: A guide to the theory of NP-Completeness*. Freeman, San Fransisco, 1979.

10. M.R. Garey, D.S. Johnson, G.L. Miller and C.H. Papadimitriou. The complexity of coloring circular arcs and chords. *SIAM Journal of Alg. Disc. Meth*, 1 (1980) 216–227.

11. C. Grinstead. The perfect graph conjecture for toroidal graphs. *Annals of Discrete Mathematics*, 21 (1984) 97–101.

12. M. Grötschel, L. Lovasz, and A. Schrijver. *Geometric algorithms and combinatorial optimization*, (1988). Springer Verlag.

13. A.W.J. Kolen, and L.G. Kroon. On the computational complexity of (maximum) class scheduling. *European Journal of Operational Research*, 54 (1991), 23–38.

14. A.W.J. Kolen, and L.G. Kroon. Licence Class Design: complexity and algorithms. *European Journal of Operational Research*, 63 (1992), 432–444.

15. A.W.J. Kolen, and L.G. Kroon. On the computational complexity of (maximum) shift class scheduling. *European Journal of Operational Research*, 64 (1993), 138–151.

16. A.W.J. Kolen, and L.G. Kroon. Analysis of shift class design problems. *European Journal of Operational Research*, 79 (1994), 417–430.

17. E. Kubicka, and A. Schwenk. Introduction to chromatic sums. *Proc. ACM Computer Science Conference*, (1989).

18. M.W. Padberg. Perfect Zero-One matrices. *Mathematical Programming*, 6 (1974) 180–196.

19. K.R. Parthasarathy, and G. Ravindra. The Strong Perfect Graph Conjecture is true for $K_{1,3}$-free graphs. *J. Comb. Theory*, B-21 (1976) 212–223.

20. A. Sen, H. Deng, and S. Guha. On a graph partition problem with an application to VLSI layout. *Information Processing Letters*, 43 (1991) 87–94.

21. A. Sen, H. Deng, and A. Roy. On a graph partition problem with application to multiprocessor scheduling. TR-92-020, Department of Computer Science and Engineering, Arizona State University.

22. K.J. Supowit. Finding a maximum planar subset of a set of nets in a channel. *IEEE Trans. on Computer Aided Design*, CAD 6, 1 (1987) 93–94.

23. A. Tucker. The Strong Perfect Graph Conjecture for planar graphs. *Canad. J. Math*, 25 (1973) 103–114.

24. A. Tucker. The validity of the perfect graph conjecture for K_4-free graphs. *Annals of Discrete Mathematics*, 21 (1984) 149–157.

25. M. Yanakakis, and F. Gavril. The maximum k-colorable subgraph problem for chordal graphs. *Information Processing Letters*, 24 (1987), 133–137.

Modifying Networks to Obtain Low Cost Trees

S. O. Krumke[1] and H. Noltemeier[1] and M. V. Marathe[2] and S. S. Ravi[3] and
K. U. Drangmeister[1]

[1] University of Würzburg, Am Hubland, 97074 Würzburg, Germany.
Email: {krumke,noltemei,drangmei}@informatik.uni-wuerzburg.de.
[2] Los Alamos Nat. Lab. P.O. Box 1663, MS M986, Los Alamos, NM 87545, USA.
Research supported by Department of Energy under contract W-7405-ENG-36.
Email: madhav@c3.lanl.gov.
[3] University at Albany - SUNY, Albany, NY 12222, USA.
Email: ravi@cs.albany.edu.

Abstract. We consider the problem of reducing the edge lengths of a given network so that the modified network has a spanning tree of small total length. It is assumed that each edge e of the given network has an associated function c_e that specifies the cost of shortening the edge by a given amount and that there is a budget B on the total reduction cost. The goal is to develop a reduction strategy satisfying the budget constraint so that the total length of a minimum spanning tree in the modified network is the smallest possible over all reduction strategies that obey the budget constraint.

We show that in general the problem of computing an optimal reduction strategy for modifying the network as above is NP-hard and present the first polynomial time approximation algorithms for the problem, where the cost functions c_e are allowed to be taken from a broad class of functions. We also present improved approximation algorithms for the class of treewidth-bounded graphs when the cost functions are linear. Our results can be extended to obtain approximation algorithms for more general network design problems such as those considered in [9, 10].

Keywords: Location Theory, Approximation Algorithms, Parametric Search, Computational Complexity, NP-hardness.

1 Introduction

The problem of computing a minimum spanning tree for a network is a well studied problem in computer science. In this paper, we consider a variant of the problem where the goal is to shorten the edge lengths of a given network so that the length of a minimum spanning tree in resulting network is as small as possible. The problem is considered in a context where there is a cost associated with shortening a link and there is a budget constraint on the total cost of shortening the edges. Such a problem models situations wherein an organization desires to take advantage of technological advances to upgrade the communications network interconnecting its branches and has allocated a fixed budget to do so. The goal is to devise a strategy that upgrades the links of the network so that the total upgrading cost is within the chosen budget, and the length of a minimum spanning tree in the upgraded network is the smallest among all strategies that obey the budget constraint. A precise definition of the problem is given in the next section. Such problems arise in diverse areas such as design of high speed communication networks [13], video on demand [15], teleconferencing [14], VLSI design [2, 3, 22], database retrieval [20], etc.

Most of the network improvement problems considered in this paper are NP-hard. Given these hardness results, we aim at finding efficient approximation algorithms for these problems. Define an (α, β)-approximation algorithm as a polynomial-time algorithm that produces a solution within α times the optimal function value, violating the budget constraint by a factor of at most β.

The main contribution of this paper is to develop a framework for formulating such network improvement problems. We provide the first polynomial time (α, β)-approximation algorithms for several versions of the problem. In the next section we formally define the problem considered in this paper and summarize our results.

2 Definitions and Summary of Results

Let $G = (V, E)$ be an undirected graph. Associated with each edge $e \in E$, there are two nonnegative values as follows: $\ell(e)$ denotes the *length* or the *weight* of the edge e and $\ell_{\min}(e)$ denotes the *minimum length* to which the edge e can be reduced. Consequently, we assume throughout the presentation that $\ell_{\min}(e) \le \ell(e)$. The nonnegative *cost function* c_e indicates how expensive it is to reduce the length of e by a certain amount. We assume without loss of generality that $c_e(0) = 0$ for all edges $e \in E$.[4]

A *reduction strategy* (or simply *reduction*) on the edges of G specifies how to reduce the ℓ-length of each edge e to a value in the range $[\ell_{\min}(e), \ell(e)]$. Given a budget B, we define a *feasible reduction* to be a nonnegative function r defined on E with the following properties: For all edges $e \in E$, $\ell(e) - r(e) \ge \ell_{\min}(e)$ and $\sum_{e \in E} c_e(r(e)) \le B$. If r is a (feasible) reduction, in G we can consider the graph G with edge weights given by the "reduced lengths", namely $(\ell - r)(e) := \ell(e) - r(e)$ $(e \in E)$.

Let T be a spanning tree of G. The *total length* of T under the weight function ℓ, denoted by $\ell(T)$, is defined to be the sum of the lengths of the edges that are in T. We denote the total weight of a minimum total length spanning tree with respect to the weight function ℓ by $\mathrm{MST}_G(\ell)$. Similarly, if r is a (feasible) reduction in G then $\mathrm{MST}_G(\ell - r)$ denotes the weight of a MST with respect to the reduced lengths $\ell(e) - r(e)$ $(e \in E)$. We omit the graph G in the subscript whenever such an omission does not cause any ambiguity.

We are now ready to state the general problem studied in this paper. The goal is to shorten the edges in the graph such that the resulting graph has a spanning tree of small weight. A formal statement of the problem is given in the next definition:

---

[4] Any reduction will incur a minimum cost of $\sum_{e \in E} c_e(0)$ and we can subtract this sum from the budget B in advance.

Definition 1. The *Budget Minimum Total Cost Spanning Tree Problem* BMST is to find a feasible reduction r such that $\text{MST}_G(\ell - r)$ has the least possible value.

Definition 2. Let $\alpha, \beta \geq 1$ be constants. We say that an algorithm is an (α, β)-*approximation algorithm* for BMST, if for each instance, the algorithm returns a reduction r of cost at most βB such that

$$\frac{\text{MST}_G(\ell - r)}{\text{MST}_G(\ell - r^*)} \leq \alpha, \tag{1}$$

where r^* denotes an optimal edge-reduction on G of cost at most B.

Example: Consider the graphs given in Figure 1. Figure 1(a) shows a graph G where each edge e is associated with the three values $(\ell(e), \ell_{\min}(e), c_e)$. The third parameter c_e represents the cost of reducing the length of the edge by a unit amount, i.e., the cost function on each edge in this simple example is linear and is given by $c_e(t) = c_e \cdot t$. The result of a modification of G is shown in Figure 1(b). The edges belonging to the minimum spanning tree are drawn as dashed lines. The modification corresponding to Figure 1(b) involves a cost of 24 and the weight of the resulting tree is 7. Figure 1(c) shows the graph with edge lengths resulting from a reduction that is optimal among all reductions of cost no more than 22. There, the weight of the spanning tree resulting from the reduction is 4. Thus, the reduction of Figure 1(b) is a $(7/4, 24/22)$-approximation to an optimal solution with budget 22.

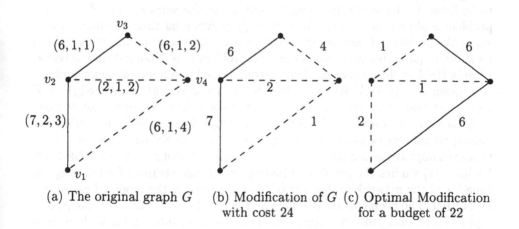

(a) The original graph G (b) Modification of G (c) Optimal Modification

with cost 24   for a budget of 22

Fig. 1. An example of a graph modification via edge reductions.

The results obtained in this paper are summarized in Table 1. The approximation algorithm for BMST can be extended significantly. For example, using

Trees	Treewidth bounded graphs	General graphs
"Easy" Solvable by a Greedy algorithm for linear reduction costs	NP-hard $(1 + \varepsilon, 1 + \xi)$-approximable for any fixed $\varepsilon, \xi > 0$	NP-hard $(1 + 1/\gamma, 1 + \gamma)$-approximable[5] for any fixed $\gamma > 0$

Table 1. Approximation and Hardness Results for BMST.

our ideas in conjunction with the results of Goemans et. al. [10], we can obtain similar approximation results for finding budget constrained minimum-cost generalized Steiner trees, minimum-cost k-edge connected subgraphs and other network design problems specified by weakly supermodular functions introduced in that paper.

The remainder of the paper is organized as follows. In Section 3 we briefly discuss the structure of an optimal solution for linear reduction costs. Section 4 presents our approximation algorithm for general graphs, while in Section 5 we provide an improved version for the class of treewidth-bounded graphs. Section 6 contains the hardness results for the problems tackled in the paper.

As far as we know, the problems considered in this paper have not been previously studied. Recently in an independent effort, Frederickson and Solis-Oba [6] considered the problem of increasing the weight of the minimum spanning tree in a graph subject to a budget constraint where the cost functions are assumed to be linear in the weight increase. In contrast to the work presented here, this problem is shown to be solvable in strongly polynomial time. Berman [1] considers the problem of shortening edges in a given tree to minimize the weight of its shortest path tree and proves that the problem is polynomial time solvable. Plesnik [18] has shown that the budget-constrained minimum diameter problem (i.e., given a graph $G(V, E)$ with a length $\ell(e)$ and cost $c(e)$ for each edge $e \in E$ and a cost budget B, select a subset E' of edges so that the total cost of edges in E' is at most B and the diameter of the graph formed by E' is a minimum among all subsets satisfying the budget constraint) is NP-hard. He also shows that even approximating the solution to within a factor of less than 2 is NP-hard. Phillips [17] studies the problem of finding an optimal strategy for reducing the capacity of the network so that the residual capacity in the modified network is minimized. The problems studied here and in [1, 17] can be broadly classified as types of bicriteria problems. Recently, there has been substantial work on finding efficient approximation algorithms for a variety of bicriteria problems (see [11, 12, 16, 19, 20, 21, 22] and the references therein).

---

[5] The length of the spanning tree produced is at most $(1 + 1/\gamma)$ times that of an optimal tree plus an additive constant of ε that can be made arbitrarily close to zero.

3 Structure of an Optimal Solution

In this section we comment on the structure of optimal solutions to the BMST-problem for linear reduction costs on the edges, i.e. $c_e(t) = c_e \cdot t$ for all $e \in E$ and constants c_e. We also look at special cases of the problem that can be solved in polynomial time.

First, suppose that the given budget B is zero. Then BMST reduces to the well known minimum spanning tree problem (with length function ℓ), and is known to be optimally solvable by classical algorithms (e.g. Prim's algorithm [4]). Similarly, if $B = +\infty$ (i.e., there is no bound on the cost of upgrading the network), the BMST problem again reduces to the minimum spanning tree problem but this time with edge-lengths given by $\ell_{\min}$.

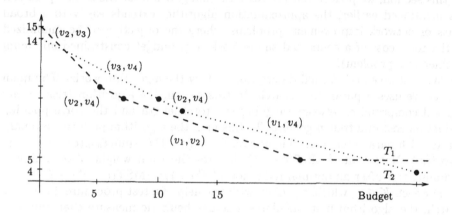

Fig. 2. Remaining weight of the trees T_1 and T_2 as a function of the budget.

Optimal solutions to BMST also exhibit some structure in the general case (i.e., $B \notin \{0, +\infty\}$). Any (feasible) reduction r induces a tree in a natural way, namely a minimum spanning tree T_r in the modified graph. Observe that the quality of the solution produced via the reduction r depends solely on the weight of T_r, so all the cost incurred in upgrading edges not in T_r is wasted. Moreover, for any *fixed* tree T in G, the Greedy strategy that successively reduces a cheapest available edge is an optimal reduction strategy. Thus, if we already knew a minimum spanning tree T_{r^*} corresponding to an optimal reduction r^*, we could solve BMST quite easily.

This observation also suggests a very simple exponential time algorithm for solving BMST: Enumerate all spanning trees in G, apply the Greedy strategy to each of them and then select the best solution. Unfortunately, a graph G with n nodes can have n^{n-2} different spanning trees.

We now discuss the *sensitivity* of optimal reduction strategies to changes in the given budget B. If we fix a spanning tree and plot the weight of that tree as a function of the money spent on it in a Greedy manner, we see that each piece corresponds to a budget range where one particular edge e is shortened. Thus it is easy to see that the piece has slope $-1/c_e$.

Figure 2 shows the plots corresponding to the tree T_1 consisting of the edges (v_2, v_3), (v_2, v_4), (v_1, v_2) and the tree T_2 consisting of the edges (v_3, v_4), (v_2, v_4) and (v_1, v_4) taken from the example graph of Figure 1. As can be seen from Figure 2, the plots for different trees can cross each other multiple times. If we plot the weights of all spanning trees on the same set of axes, the lower envelope gives the optimal remaining weight per budget. It is easy to see that the lower envelope can have an exponential number of linear pieces.

4 Approximation Algorithms on General Graphs

In this section, we present our approximation algorithm for the BMST problem. As mentioned earlier, the approximation algorithm extends easily to a broad class of network improvement problems where the objective to be minimized is the total cost of a connected subnetwork (e.g. budget constrained minimum Steiner tree problem).

We first give an informal description of how the algorithm works. The main procedure uses a parametric search. In this search, the algorithm tries to find a good compromise between weighing the total length and the corresponding reduction cost of a tree in general. To this end, the algorithm performs a binary search with parameter K on the interval $\mathcal{I} := [(n-1)/\gamma \cdot \min \ell_{\min}(e), (n-1)/\gamma \cdot \max \ell(e)]$. Note that if $\mathrm{MST}(\ell - r^*)$ denotes the total weight of a minimum spanning tree after an optimal reduction r^* then $1/\gamma \cdot \mathrm{MST}(\ell - r^*) \in \mathcal{I}$.

For each $K \in \mathcal{I}$, which is probed with the help of a test procedure during the search, the algorithm first calculates a coarse heuristic measure that indicates how important it is to shorten an edge. Then, for each edge e in the graph, the blend of its length and the reduction cost is refined by using the information in the cost function c_e. After calculating such *compound costs* for the edges, we compute a MST with respect to these costs. The algorithm stops when a good blend has been found, meaning in this context that there exists a tree of total compound cost that is small compared to the current parameter K.

For large values of K the reduction costs on the edges are weighted more than their lengths and the algorithm will tend to reduce the edge lengths only by a small amount, resulting in low overall reduction costs and more or less heavy trees. Also, since K is large, the test on the compound cost of the minimum spanning tree computed will succeed. The algorithm now tries to reduce K as much as possible and find a minimum $K \in \mathcal{I}$ such that it can successfully compute a light compound cost spanning tree. Figures 3 and 4 show our approximation algorithm and the test procedure respectively.

4.1 Correctness and Performance Guarantee.

The performance guarantee provided by the algorithm Heuristic-BMST is summarized in the following theorem.

Procedure Heuristic-BMST(γ, ε)

1 Perform a binary search on the interval $\mathcal{I} := [\dfrac{(n-1)\min\limits_{e \in E}\ell_{\min}(e)}{\gamma}, \dfrac{(n-1)\max\limits_{e \in E}\ell(e)}{\gamma}]$ with a spacing of ε to find the minimum value K' such that Test-Blend(K') returns "Yes".

2 Let T' be the tree generated by Test-Blend(K') and let t_e $(e \in T')$ be the corresponding "fine tuned" blend parameters.
   Define the reduction r by $r(e) := 0$ if e is not included in T and by $r(e) := t_e$ otherwise.

4 **return** r and T.

Fig. 3. Main Procedure for the approximation of BMST.

Procedure Test-Blend(K)

1 **Comment:** This procedure tries to estimate whether in the current blend of lengths and reduction costs, the costs are weighted strongly enough (i.e., K large enough) resulting in a low cost reduction. For this purpose, it uses the heuristic measure computed in Step 2.

2 **for** each edge e let $h_K(e) = \min\limits_{t \in [0, \ell(e) - \ell_{\min}(e)]} \left(\ell(e) - t + \frac{K}{B} c_e(t) \right)$.
   Also, let t_e be the value of t which achieves the value $h_K(e)$.

3 Compute a minimum spanning tree T in G using the weight $h_K(e)$ for each $e \in E$. Let $h_K(T)$ denote the cost of this spanning tree.

4 **if** $h_K(T) \leq (1 + \gamma)K$ **then return** "Yes" **else return** "No".

Fig. 4. Test procedure used for the approximation of BMST.

Theorem 3. *For any fixed $\gamma, \varepsilon > 0$, Heuristic-BMST is an approximation algorithm for BMST that finds a solution whose length is at most $(1 + \frac{1}{\gamma})$ times that of a minimum length spanning tree plus an additive constant of at most ε, and the total cost of the improvement is at most $(1 + \gamma)$ times the budget B.*

The proof of Theorem 3 relies mainly on the following lemma, which ensures that the binary search in the main procedure works correctly. In stating this lemma, we use the notation introduced in the two procedures (Heuristic-BMST and Test-Blend) described above.

Lemma 4. *Define F on $\mathbb{R}_{>0}$ by $F(K) := \frac{\mathrm{MST}_G(h_K)}{K}$. Then F is monotonically nonincreasing on $\mathbb{R}_{>0}$.*

Proof. Let $K^{(1)}$ and $K^{(2)}$ be two positive numbers such that $K^{(1)} < K^{(2)}$. For $i = 1, 2$, let $T^{(i)}$ be a minimum spanning tree in G under the cost function $h_{K^{(i)}}$. Then

$$h_{K^{(i)}}(T^{(i)}) = \sum_{e \in T^{(i)}} h_{K^{(i)}}(e) =$$

$$= \underbrace{\sum_{e \in T^{(i)}} (\ell(e) - t_e^{(i)})}_{=: \, L^{(i)}} + \frac{K^{(i)}}{B} \underbrace{\sum_{e \in T^{(i)}} c_e(t_e^{(i)})}_{=: \, C^{(i)}} =: L^{(i)} + \frac{K^{(i)}}{B} C^{(i)}.$$

Here $t_e^{(i)}$ are the values chosen in Step 2 of **Test-Blend** which minimize $\ell(e) - t + \frac{K^{(i)}}{B} c_e(t)$ on the interval $[0, \ell(e) - \ell_{\min}(e)]$. By dividing the last equation by $K^{(i)}$ we obtain for $i = 1, 2$

$$F(K^{(i)}) = \frac{L^{(i)}}{K^{(i)}} + \frac{C^{(i)}}{B}.$$

In the next step we find an upper bound for $F(K^{(2)})$. To this end, we estimate the weight of each edge in $T^{(1)}$ under the cost function $h_{K^{(2)}}$. Let $e \in T^{(1)}$ be an arbitrary edge. Then by the choice of $t_e^{(2)}$ in Step 2 of **Test-Blend** we have that

$$h_{K^{(2)}}(e) = \ell(e) - t_e^{(2)} + \frac{K^{(2)}}{B} c_e(t_e^{(2)}) = \min_{t \in [0, \ell(e) - \ell_{\min}(e)]} \left(\ell(e) - t + \frac{K^{(2)}}{B} c_e(t) \right)$$

$$\leq \ell(e) - t_e^{(1)} + \frac{K^{(2)}}{B} c_e(t_e^{(1)}). \tag{2}$$

Summing up the inequalities in (2) over all $e \in T^{(1)}$, we obtain:

$$h_{K^{(2)}}(T^{(1)}) \leq L^{(1)} + \frac{K^{(2)}}{B} C^{(1)}. \tag{3}$$

Dividing (3) by $K^{(2)}$ and using $h_{K^{(2)}}(T^{(2)}) \leq h_{K^{(2)}}(T^{(1)})$ this results in

$$F(K^{(2)}) \leq \frac{L^{(1)}}{K^{(2)}} + \frac{C^{(1)}}{B} < \frac{L^{(1)}}{K^{(1)}} + \frac{C^{(1)}}{B} = F(K^{(1)}). \tag{4}$$

The strict inequality in the chain above stems from the fact that $K^{(1)} < K^{(2)}$. This completes the proof of the lemma. $\qquad\square$

Corollary 5. *If the procedure* **Test-Blend** *returns "Yes" for some $K' > 0$ then it also returns "Yes" for all $K \geq K'$. Thus, the binary search in* **Heuristic-BMST** *works correctly.*

Proof. Let T' be a minimum spanning tree with respect to $h_{K'}$. Then, since the test procedure **Test-Blend**(K') returns "Yes" we have that $h_{K'}(T') \leq (1 + \gamma)K'$; i.e., $F(K') \leq (1 + \gamma)$. Thus it follows by Lemma 4 that $F(K) \leq (1 + \gamma)$ for all $K \geq K'$. Since $F(K) = \frac{\mathrm{MST}_G(h_K)}{K}$, this is equivalent to saying that $\mathrm{MST}_G(h_K) \leq (1 + \gamma)K$ for all $K > K'$. $\qquad\square$

Proof of Theorem 3: Let r^* be an optimal feasible reduction and let T^* be a minimum spanning tree in G with respect to the weight function $\ell - r^*$. For the sake of shorter notation let $L^* := (\ell - r^*)(T^*)$ be its total weight in the graph with the edge lengths resulting from the optimal reduction r^*.

We now show **Test-Blend** would return "Yes" if called with the value $\tilde{K}$ which is the smallest value in the ε-spacing of the interval $\mathcal{I}$ satisfying $\tilde{K} \geq L^*/\gamma$. Thus, $\tilde{K}$ is some rational number satisfying

$$\tilde{K} = L^*/\gamma + \varepsilon' , \text{ where } 0 \leq \varepsilon' < \varepsilon. \tag{5}$$

For each edge $e \in T^*$ we can estimate the weight $h_{K^*}(e)$ similar to inequality (2) in the proof of Lemma 4. This way, we see that the weight of T^* under $h_{\tilde{K}}$ is no more than $L^* + \frac{\tilde{K}}{B}B$. Consequently, the minimum spanning tree with respect to $h_{\tilde{K}}$ that would be found by the procedure during the call has $h_{\tilde{K}}$-weight at most

$$L^* + \tilde{K} \overset{(5)}{=} \gamma(\tilde{K} - \varepsilon') + \tilde{K} \leq (1 + \gamma)\tilde{K}.$$

Hence, the test in Step 4 of **Test-Blend** would be successful and the procedure would return "Yes". Since we know by Corollary 5 that the binary search correctly locates a minimum value K', it follows that $K' \leq \tilde{K} = L^*/\gamma + \varepsilon'$. Let T' be the minimum spanning tree found by Test-Blend(K'). Since $K', B \geq 0$ and $c_e(t) \geq 0$ for all t, we have:

$$h_{K'}(T') = \sum_{e \in T'} (\ell(e) - t'_e) + \frac{K'}{B} \sum_{e \in T'} c_e(t'_e) \geq \sum_{e \in T'} (\ell(e) - t'_e). \tag{6}$$

Here again the numbers t'_e are the values of t chosen in Step 2 of the test procedure. For the reduction r which is calculated in Step 2 of **Heuristic-BMST**, it now follows from (6) that

$$\text{MST}_G(\ell - r) \leq (\ell - r)(T') \leq h_{K'}(T'). \tag{7}$$

Moreover,

$$h_{K'}(T_{K'}) \leq h_{K'}(T^*) \leq L^* + \frac{K'}{B}B \leq L^* + \tilde{K} \overset{(5)}{=} L^* + \frac{L^*}{\gamma} + \varepsilon'$$

$$\leq (1 + \frac{1}{\gamma})L^* + \varepsilon = (1 + \frac{1}{\gamma})\text{MST}_G(\ell - r^*) + \varepsilon. \tag{8}$$

Using this result in (7), we get $\text{MST}_G(\ell - r) \leq (1 + \frac{1}{\gamma})\text{MST}_G(\ell - r^*) + \varepsilon$, which proves the claimed performance of the algorithm with respect to the weight of an MST in the graph after applying the reduction r.

We now estimate the cost of the reduction r found by our heuristic. Note that the cost of r is exactly $\sum_{e \in T'} c_e(t_e)$. We have

$$\frac{K'}{B} \sum_{e \in T'} c_e(t'_e) \leq \sum_{e \in T'} (\ell(e) - t'_e + \frac{K'}{B}c_e(t'_e)) = h_{K'}(T') \leq (1 + \gamma)K'.$$

Dividing the last chain of inequalities by $\frac{K'}{B}$ yields that the budget B is violated by a factor of at most $1 + \gamma$ as claimed in the theorem. $\qquad \square$

4.2 Running Time

We now show that the algorithm can be implemented to run in polynomial time for a broad class of reduction cost functions c_e on the edges of the graph. Let $L_{\max} = \max_{e \in E} \ell(e)$. Then the total number of calls to Procedure **Test-Blend** is in $\mathcal{O}(\log(\frac{nL_{\max}}{\gamma\varepsilon}))$. Since γ and ε are fixed, the test procedure is called only a polynomial number of times. Thus, to prove that the overall running time of the algorithm is polynomial, it suffices to show that each execution of **Test-Blend** can be completed in polynomial time. Here, the only fact to show is that we can minimize the function $f_e(t) := \ell(e) - t + \frac{K}{B} c_e(t)$ on the compact interval $\mathcal{I}' := [0, \ell(e) - \ell_{\min}(e)]$ in Step 2 of the procedure in polynomial time. The rest of the procedure consists of computing a minimum spanning tree which can be done in $\mathcal{O}(n + m \log \beta(m, n))$ time using the algorithm of Gabow et. al. [7], where $\beta(m, n) = \min\{i \mid \log^{(i)} n \leq m/n\}$.

Consider the execution of **Test-Blend** for a given value of K. Observe that in Step 2 the number $\ell(e)$ is an additive constant and $\frac{K}{B}$ is a constant factor. Thus, the constrained minimization of f_e can be done easily for the following sample classes of functions c_e:

1. Linear functions, i.e. $c_e(t) = c_e \cdot t$ for a constant c_e: Then f_e is a linear function in t and the minimum is attained at one of the endpoints of $\mathcal{I}'$. Minimizing f_e can be done in constant time. Thus, the total running time of the heuristic is $\mathcal{O}(\log(\frac{nL_{\max}}{\gamma \varepsilon})(n + m \log \beta(m, n)))$.

2. Concave functions: Let $\Delta(e) := \ell(e) - \ell_{\min}(e)$. Then, for any $0 < \lambda < 1$ we have by the concavity of c_e (which implies the concavity of f_e):

$$f_e(\lambda \cdot 0 + (1 - \lambda)\Delta(e)) \geq \lambda f_e(0) + (1 - \lambda)f_e(\Delta(e)) \geq \min\{f_e(0), f_e(\Delta(e))\}.$$

   Thus, the minimum of f_e is again attained either at 0 or at $\ell(e) - \ell_{\min}(e)$.

3. Differentiable convex functions where we can find a root of the equation $c'_e(t) = \frac{B}{K}$ explicitly.

4. Functions that are piecewise of one of the types described above. Observe that the number of pieces is polynomial in the input size.

For the first three classes of functions mentioned above, the total effort of our algorithm consists essentially of $\mathcal{O}(\log(\frac{nL_{\max}}{\gamma \varepsilon}))$ minimum spanning tree computations. This not only results in an overall polynomial time but also in a complexity that is feasible in practice.

4.3 Notes on the Algorithm

It should be noted here that **Heuristic-BMST** can be modified easily to handle the case when the reduction is required to be either integer valued or to satisfy $r(e) \in \{0, \ell(e) - \ell_{\min}(e)\}$ for all $e \in E$. In this case, Step 2 of **Test-Blend** is modified in such a way that the minimization is carried out only over the integers in $[0, \ell(e) - \ell_{\min}(e)]$ or over the two element set $\{0, \ell(e) - \ell_{\min}(e)\}$ respectively. Due to lack of space we omit the details.

Integer valued reductions are helpful to model discrete steps of improvement, e.g. the addition of a number of communication links parallel to already existing ones in the network. Reductions that take values only from $\{0, \ell(e) - \ell_{\min}(e)\}$ can be used to model the insertion of alternative edges to the graph G, with the reduction of the edge e corresponding to the construction of a new edge e' parallel to e with length $\ell_{\min}(e)$.

So far, we have assumed that the function $f_e(t) = \ell(e) - t + \frac{K}{B} c_e(t)$ can be minimized *exactly*. This indeed is not necessary to obtain an approximation algorithm with a constant factor approximation for BMST. In fact, one can show that if in Step 2 of procedure Test-Blend we find a value t' satisfying

$$f_e(t') \leq \alpha \cdot \min_{t \in [0, \ell(e) - \ell_{\min}(e)]} f_e(t)$$

for some $\alpha \geq 1$ and modify Step 4 to check whether the compound weight of the tree is at most $\alpha^2(1 + \gamma)K$, this will lead to a polynomial time algorithm which produces a reduction of cost at most $\alpha^2(1 + \gamma)B$ and a corresponding MST of total length at most $\alpha(1 + 1/\gamma)$ times that of an optimal tree plus an additive constant of ε.

5 Improved Approximation Ratios for Treewidth Bounded Graphs and Linear Reduction Costs

In this section we present our improved approximation results for the class of treewidth bounded graphs under the additional assumption that the reduction costs on the edges are *linear*. The basic idea behind the algorithm in this section is to reduce the problem of improving the tree to some appropriately chosen bicriteria problem. To this end we recall the following result from [16]:

Theorem 6. *1. There is a polynomial-time algorithm that, given an undirected graph G on n nodes with two nonnegative integral costs E and F on its edges, a bound $\mathcal{E}$, and a fixed $\gamma > 0$, constructs a spanning tree of G of total E-cost at most $(1 + \gamma)\mathcal{E}$ and of total F-cost at most $(1 + 1/\gamma)$ times that of the minimum-F-cost of any spanning tree with total E-cost at most $\mathcal{E}$.*
2. For the class of treewidth-bounded graphs, there is a polynomial time algorithm that returns a spanning tree of total E-cost at most $\mathcal{E}$ and of total F-cost at most $(1 + \varepsilon)$ times that of any spanning tree with total E-cost at most $\mathcal{E}$. □

We will use the second part of the theorem to obtain an improved approximation. We note here that using first part of Theorem 6 instead, we could also construct a $((1 + \frac{1}{\gamma}), (1 + \xi)(1 + \gamma))$ approximation algorithm for BMST on general graphs for any fixed $\xi > 0$, if we restrict ourselves to linear reduction costs. Since our algorithm from the last section already gives us a $(1 + \frac{1}{\gamma}, 1 + \gamma)$ performance for far more general classes of cost functions, this is not as interesting. Also, our approximation algorithm from Section 4 does not need any additional space, while the construction presented below transforms the original graph into a graph having more nodes and edges.

In the sequel we sketch how our improved algorithm Heuristic-TW-BMST works for treewidth-bounded graphs with linear reduction costs. First we transform the original graph into another graph that can be fed into the algorithm from Theorem 6. This is done in the following manner. Fix $\xi > 0$. For each edge $e = (u, v)$ in the graph, let the integer b_e be chosen so that $(1 + \xi)^{b_e} \leq \ell(e) - \ell_{\min}(e) \leq (1 + \xi)^{b_e + 1}$. Add $b_e + 2$ new vertices r_k, $k = -1, 0, \ldots, b_e$, which are joined together in a simple cycle. For all $k, -1 \le k \le b_e$, join r_k to both u and v. For $k \ge 0$, the edge (u, r_k) has E-cost $E(u, r_k) := \ell(e) - (1 + \xi)^k$ and F-cost $(1 + \xi)^k c_e$, while the edge (u, r_{-1}) has E-cost $\ell(e)$ and F-cost 0. All the edges (r_k, v) and (r_k, r_{k+1}) have their E-cost and F-cost set to zero. Let G be the original graph and G' be the graph obtained as a result of the transformation. Also, let tw(G) and tw(G') denote the treewidths of G and G' respectively. We have the following observation.

Lemma 7. *Whenever* tw$(G) \ge 3$, tw$(G) = $ tw(G'). $\qquad\qquad\square$

The approximation algorithm for the class of treewidth bounded graphs now works as follows. After transforming G into G' with two edge weight functions E and F, we use a binary search to find the smallest integer L' in the interval $[(n-1) \min \ell_{\min}(e), (n-1) \max \ell(e)]$ such that the algorithm referred to in Part 2 of Theorem 6 called with the parameters L' for the E-cost bound $\mathcal{E}$ and $\varepsilon > 0$ returns a tree T' of F-cost at most $(1 + \xi)B$. For each edge $e = (u, v)$, we then define a reduction r on e by $r(e) := 0$ if (u, r_{-1}) is included in T' and otherwise by $r(e) := (1 + \xi)^k$, where $0 \le k \le b_e$ is the minimum value such that (u, r_k) is included in T'. The quality of the approximate solution produced is indicated in the following theorem.

Theorem 8. *For the class of treewidth bounded graphs with linear reduction costs the following statement holds: For all fixed $\varepsilon, \xi > 0$,* Heuristic-TW-BMST *is a polynomial time $((1 + \varepsilon), (1 + \xi))$-approximation algorithm for* BMST. $\qquad\square$

6 Hardness Results

In this section we will prove the BMST problem to be hard even for very restricted class of graphs and the most simple reduction cost functions.

Theorem 9. BMST *is NP-hard, even when restricted to series-parallel graphs and even when all the reduction cost functions c_e are linear, i.e., $c_e(t) = c_e \cdot t$ for all $e \in E$.*

Proof. We use a reduction from Continuous Multiple Choice Knapsack which is known to be an NP-complete problem (c.f. [8, MP11]). An instance of CMC-Knapsack is given by a finite set U of n items, a size $s(u)$ and value $v(u)$ for each item, a partition $U_1 \cup \cdots \cup U_k$ of U into disjoint sets and two integers S and K. The question is, whether there is a choice of a unique element $u_i \in U_i$, for each $1 \le i \le k$, and an assignment of rational numbers $r_i, 0 \le r_i \le 1$ to these elements such that $\sum_{i=1}^{k} r_i s(u_i) \le S$ and $\sum_{i=1}^{k} r_i v(u_i) \ge K$.

Given an instance of CMC-Knapsack we construct a graph $G = (V, E)$ in the following way: We let $V = U \cup \{X, T, T_1, \ldots, T_k\}$, $E := E_1 \cup E_2 \cup E_3$ with $E_1 := \{(X, u) : u \in U\}$, $E_2 := \{(u, T_i) : u \in U_i, i = 1, \ldots, k\}$ and $E_3 := \{(T_i, T) : i = 1, \ldots, k\}$. The graph constructed this way is obviously series-parallel with terminals X and T.

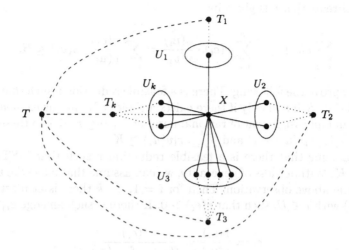

Fig. 5. Graph used in the reduction from Continuous Multiple Choice Knapsack.

Define $D := \max\{v(u) : u \in U\}$. For each edge $(x, u) \in E_1$, let $\ell(x, u) := D, \ell_{\min}(x, u) := D - v(u), c(x, u) := s(u)/v(u)$. For all edges $e \in E_2$ we let $\ell(e) := \ell_{\min}(e) := c_e := 0$, and for all edges $e \in E_3$ we define $\ell(e) := \ell_{\min}(e) := 3D$ and $c_e := 0$. Set the bound B on the total cost to be S.

The graph is shown in Figure 5. The dotted edges are of weight 0 while the dashed ones have weight $3D$. Any MST in G has weight $kD + 3D$.

By the construction, any feasible reduction can only reduce the length of the edges in E_1. Assume that r is a feasible reduction. Observe that the MST in G with edge lengths given by $(\ell - r)$ will always include *all* edges from E_2 (which are of weight 0) and *exactly one* edge from E_3, regardless of which edges from E_1 are affected by the reduction. Observe also that for any fixed $i \in \{1, \ldots, k\}$, any MST in the modified graph will contain *exactly one* of the edges of the form (X, u'), where $u' \in U_i$. Consequently, reducing the length of *more than one* edge (X, u') with $u' \in U_i$ will *not* improve the quality of the solution, but cost money from the budget B. We thus have:

Observation: *If r is a feasible reduction for the instance of BMST defined above and the weight of an MST in the modified graph is Y, then there is always a feasible reduction r', which for each $i \in \{1, \ldots, k\}$ reduces at most one of the edges (X, u), $u \in U_i$ and the weight of an MST with respect to $(\ell - r')$ is also equal to Y.* □

Let r be any reduction as defined in the above observation and for $i = 1, \ldots, k$ let $e_i = (X, u_i)$ be the unique edge from x to U_i affected by the reduction. The weight of an MST with respect to $(\ell - r)$ is then given by

$$3D + \sum_{i=1}^{k}(\ell(e_i) - r(e_i)) = 3D + k \cdot D - \sum_{i=1}^{k} r(e_i). \tag{9}$$

The cost of reduction r is given by

$$\sum_{i=1}^{k} r(e_i)c_{e_i} = \sum_{i=1}^{k} r(e_i) \cdot \frac{s(u_i)}{v(u_i)} = \sum_{i=1}^{k} \frac{r(e_i)}{v(u_i)} \cdot s(u_i) \leq B. \tag{10}$$

We now prove the following: There is a feasible reduction r such that $MST(\ell - r) \leq (3+k)D - K$ if and only if there exists a choice of a unique element $u_i \in U_i$, $1 \leq i \leq k$ and an assignment of rational numbers $r_i, 0 \leq r_i \leq 1$ to these elements such that $\sum_{i=1}^{k} r_i s(u_i) \leq B$ and $\sum_{i=1}^{k} r_i v(u_i) \geq K$.

First, assume that there is a feasible reduction r such that $MST(\ell - r) \leq (3+k)D - K$. Without loss of generality, we can assume that r has the properties stated in the above observation. Then for $i = 1, \ldots, k$ there is at most one edge $e_i = (X, u)$ with $u \in U_i$ such that $r(e_i) > 0$. If there is such an edge e_i, we define

$$r_i := \frac{r(e_i)}{v(u_i)} = \frac{r(e_i)}{\ell(e_i) - \ell_{\min}(e_i)} \tag{11}$$

and let $u_i := u$. If for all edges (X, u) with $u \in U_i$ we have $r(e_i) = 0$, we simply let $r_i := 0$ and choose $u_i \in U$ arbitrarily. It follows readily from the definition and the feasibility of the reduction r that $r_i \in [0, 1]$. Moreover, using equation (10) we see that $\sum_{i=1}^{n} r_i s(u_i) \leq B \leq B_1$. Using equation (9) and the fact that the weight $MST_G(\ell - r)$ is no more than $(3 + k)D - K$ we obtain

$$\sum_{i=1}^{n} r_i v(u_i) = \sum_{i=1}^{n} \frac{r(e_i)}{v(u_i)} \cdot v(u_i) = \sum_{i=1}^{n} r(e_i) \geq K.$$

Conversely, if we can pick unique elements u_i from the sets U_i and find rational numbers $r_i \in [0, 1]$ such that $\sum_{i=1}^{n} r_i s(u_i) \leq B$ and $\sum_{i=1}^{n} r_i v(u_i) \geq K$, we can define a reduction r by $r(X, u_i) := r_i v(u_i) = r_i(\ell(x, u_i) - \ell_{\min}(x, u_i))$ for $i = 1, \ldots, k$ and $r(e) := 0$ for all other edges. It follows that r is indeed feasible, and using equation (9) we see that the MST in the modified graph is no heavier than $(3 + k)D - K$. $\square$

References

1. O. Berman, "Improving The Location of Minisum Facilities Through Network Modification," *Annals of Operations Research*, 40(1992), pp. 1–16.
2. J. Cong, A. B. Kahng, G. Robins, M. Sarafzadeh and C. K. Wong, "Provably Good Performance Driven Global Routing," *IEEE Transactions on Computer Aided Design*, 11(6), 1992, pp. 739–752.
3. J. P. Cohoon and L. J. Randall, "Critical Net Routing," *IEEE Intern. Conf. on Computer Design*, 1991, pp. 174–177.
4. T. H. Cormen, C. E. Leiserson and R. L. Rivest, *Introduction to Algorithms*, McGraw-Hill Book Co., Cambridge, MA, 1990.
5. W. Cunningham, "Optimal Attack and Reinforcement of a Network," *J. ACM*, 32(3), 1985, pp. 549–561.

6. G.N. Frederickson and R. Solis-Oba, "Increasing the Weight of Minimum Spanning Trees", *Proceedings of the Sixth Annual ACM-SIAM SODA '96*, Jan. 1996, pp. 539–546.

7. H. N. Gabow, Z. Galil, T. H. Spencer and R. E. Tarjan, "Efficient Algorithms for Finding Minimum Spanning Trees in Undirected and Directed Graphs," *Combinatorica*, 6 (1986), pp. 109–122.

8. M. R. Garey and D. S. Johnson, *Computers and Intractability: A Guide to the Theory of NP-Completeness*, W. H. Freeman and Co., San Francisco, CA, 1979.

9. M. X. Goemans, and D. P. Williamson, "A General Approximation Technique for Constrained Forest Problems", *SIAM J. Computing*, 24 (1995), pp. 296–317.

10. M. X. Goemans, A. V. Goldberg, S. Plotkin, D. B. Shmoys, E. Tardos and D. P. Williamson, "Improved Approximation Algorithms for Network Design Problems," *Proceedings of the Fifth Annual ACM-SIAM SODA '94*, Jan. 1994, pp. 223–232.

11. R. Hassin, "Approximation Schemes for the Restricted Shortest Path Problem," *Math. of OR*, 17(1), 1992, pp. 36–42.

12. D. Karger and S. Plotkin, "Adding Multiple Cost Constraints to Combinatorial Optimization Problems, with Applications to Multicommodity Flows," *Proc. 27th Annual ACM Symp. on Theory of Computing* (STOC'95), May 1995, pp. 18–25.

13. B. Kadaba and J. Jaffe, "Routing to Multiple Destinations in Computer Networks," *IEEE Trans. on Communication*, Vol. COM-31, Mar. 1983, pp. 343–351.

14. V. P. Kompella, J. C. Pasquale and G. C. Polyzos, "Two Distributed Algorithms for the Constrained Steiner Tree Problem," Technical Report CAL-1005-92, Computer Systems Laboratory, University of California, San Diego, Oct. 1992.

15. V. P. Kompella, J. C. Pasquale and G. C. Polyzos, "Multicast Routing for Multimedia Communication," *IEEE/ACM Transactions on Networking*, 1993, pp. 286–292.

16. M. V. Marathe, R. Ravi, S. Sundaram, S. S. Ravi, D. J. Rosenkrantz, and H. B. Hunt III, "Bicriteria Network Design Problems", *Proc. ICALP'95*, July 1995, pp. 487–498.

17. C. Phillips, "The Network Inhibition Problem," *Proc. 25th Annual ACM STOC'93*, May 1993, pp. 288–293.

18. J. Plesnik, "The Complexity of Designing a Network with Minimum Diameter," *Networks*, 11 (1981), pp. 77–85.

19. R. Ravi, M. V. Marathe, S. S. Ravi, D. J. Rosenkrantz and H. B. Hunt III, "Many Birds with one Stone: Multi–objective Approximation Algorithms," *Proc. 25th Annual ACM STOC'93*, May 1993, pp. 438–447.

20. R. Ravi, "Rapid Rumor Ramification: Approximating the Minimum Broadcast Time," *Proceedings of the 35th Annual FOCS'94*, Nov. 1994, pp. 202–213.

21. A. Warburton, "Approximation of Pareto optima in Multiple–Objective, Shortest Path Problems," *Oper. Res.*, 35 (1987), pp. 70–79.

22. Q. Zhu, M. Parsa and W. Dai, "An Iterative Approach for Delay Bounded Minimum Steiner Tree Construction," Technical Report UCSC-CRL-94-39, University of California, Santa Cruz, Oct 1994.

On the Hardness of Allocating Frequencies for Hybrid Networks

Ewa Malesińska[1]* and Alessandro Panconesi[2]**

[1] TU Berlin, MA 6-1, Str. des 17 Juni 136, 10623 Berlin, Germany
Email: malesin@math.tu-berlin.de
[2] FU Berlin, Informatik, Takustr. 9, 14195 Berlin, Germany
Email: ale@inf.fu-berlin.de

Abstract. This paper studies the channel stability number, a combinatorial function that has been introduced for evaluating frequency allocation plans for hybrid cellular networks. We present several results concerning the approximability of this function in the case of complete graphs and analyze how different constraints influence its computational complexity.

1   Introduction

In this paper we study a combinatorial problem arising in the context of frequency allocation strategies for cellular networks. Before defining the problem and listing our results, it is perhaps best to give the relevant context so that the genesis and the relevance of the problem can be better appreciated.

In the frequency allocation problem for mobile telephone networks we are given a so-called interference graph and a frequency spectrum consisting of a set of carrier frequencies. Vertices in the graph represent stations, each requiring a certain number of frequencies. Two nodes in the graph are adjacent if the signals transmitted by the corresponding stations can interfere. For technical reasons, two frequencies f and g assigned to the same station must satisfy $|f - g| > \delta$, where typically $\delta = 2$. Frequencies assigned to interfering stations must satisfy this inequality with $\delta = 0$ or 1. The optimization goal is to assign frequencies so that the expected traffic can be supported keeping at the same time the interferences low. This is formalized in several ways according to the context. The allocation strategies currently in use in all large, *i.e.* non local, networks follow a *static* approach: frequencies are assigned once for all. Reallocation happens rarely, usually every two or three months or when the network undertakes restructuring (stations are added, replaced or deleted). This static approach has several shortcomings [ZE'93]. In particular, it does not make a good use of the limited number of frequencies that are reserved for one cellular telephone network, a number usually in the order of 50. Consider for instance

---

* Supported by the graduate school "Algorithmic Discrete Mathematics". The graduate school is supported by the Deutsche Forschungsgemeinschaft, grant WE 1265/2-1.
** Supported by a research fellowship of the Alexander von Humboldt Foundation

a typical metropolis like Berlin or New York City; during the day, phone call traffic tends to be very heavy in the downtown area and light in the suburbs. In the evening the pattern is just reversed. Ideally, a network should be able to adapt to such changes and allocate frequencies dynamically as needed. In the above example the traffic pattern is predictable, but in reality this will not be the case; the network should be able to reconfigure dynamically according to (unpredictable) contingencies. No doubt this will be a common scenario in the next future and several dynamic allocation strategies have already appeared in the literature (see for instance [DJLS'94, DV'93, EB'91]). Between the current situation and the appearance of dynamic networks however, there will be a transition phase where networks will be *hybrid, i.e.* partly dynamic and partly static. It is this type of networks that we are concerned with in this paper. A central problem arising when considering hybrid networks is the following: allocating frequencies to the static part of the network constraints the available frequency spectrum for vertices of the dynamic part. Such an assignment might be too restrictive and it is important to be able to detect whether this is the case. A related and rather important problem is to devise computationally efficient methods allowing to compare different frequency assignments to the static part of the network, *i.e.* to decide which of two assignments is the best with respect to the dynamic part.

Therefore, in this paper we define and study the following combinatorial problem. An input instance I consists of an interference graph $G = (V, E)$ modeling the dynamic part of the network and, for each vertex u, a *list of available frequencies* $L(u)$ and two requirements $\max(u)$ and $\min(u)$ denoting the minimum and maximum number of frequencies required by the station corresponding to u. The $L(u)$'s are subsets of some linear order $\mathcal{F}$– the frequency spectrum– which is also part of the input G. The optimization goal is to select subsets $S(u) \subseteq L(u)$ in order to maximize

$$\sum_{u \in V} |S(u)|$$

subject to some or all of the following constraints: (a) for all u, $\max(u) \geq |S(u)| \geq \min(u)$; (b) for all $(u, v) \in E$, if $f \in S(u)$ and $g \in S(v)$ then $|f - g| > 0$ (*i.e.* $S(u) \cap S(v) = \emptyset$); (c) [co-site constraint] for all vertices u, if $f \in S(u)$ and $g \in S(u)$, $f \neq g$, then $|f - g| > 1$. We call the maximum value of $\sum_{u \in V} |S(u)|$ the *channel stability number* of I and denote it with $ch(I)$.

The input subsets $L(u)$ model the fact that frequencies allocated to the static neighbors of u are not available for u any more. Constraint (a) models the fact that for dynamic stations the actual number of frequencies needed can vary unpredictably within two values. Although a station, strictly speaking, functions as long as there is one available channel, in practice each station, in order to service its area satisfactorily, requires a minimum number of channels. This parameter, denoted as $\min(u)$, is an estimate done by the network managers. Constraint (b) ensures that no two neighboring stations use the same frequency. In real systems, a stronger version of constraint (c) is required, namely $|f - g| > \delta > 0$. Usually,

$\delta = 2$. The simpler condition $|f - g| > 1$ does not affect the validity of our results except that approximation factors should be scaled down by a δ factor. The function $ch(I)$ measures how much "flexibility" is left to the dynamic part of the network.

This problem simultaneously generalizes other well-known combinatorial problems related to vertex coloring: the T-coloring problem, list coloring, set coloring and the k-th stability number, also known as the partial k-coloring problem, where one seeks a vertex induced k-colorable subgraph of maximum size [B'89, H'80, JT'95, T'89].

In this paper, we study the channel stability number problem for the special case of cliques. The importance of this special case stems from several facts. First, graphs coming from real systems have the clique number bounded by a rather small constant, in the order of 15 or so. Such a condition is simply enforced by the managers of real networks who, in order to avoid severe signal interference, make sure that large cliques are never created. When a clique is too large it can be broken in several ways for instance, by decreasing the emission power so that stations sufficiently far apart cease to interfere. If, as a result, some regions are left underserviced new stations are introduced as appropriate. The net result of such operations is replacement of a large clique by two or more smaller ones. In this fashion the maximum vertex degree is kept under control and usually is in the order of 20. Some old stations might have higher degrees–in the order of 50– but they are being gradually modified and their degree is expected to decrease. It should also be pointed out that the topology of these graphs, although unknown, is by no means arbitrary. In particular, it resembles intersection graphs of roughly disk shaped regions. Such graphs have several nice properties. In particular, their chromatic number is bounded by a constant times the clique number [MBHRR'95, GSW'94].

These considerations are corroborated by the experimental evidence available. When allocating frequencies telephone companies routinely compute all maximal cliques of the whole graph using backtracking algorithms. Such computations take the order of minutes even for the largest existing networks which have as many as 5,000 nodes [P'95]. Given that all maximal cliques are available studying complete graphs is all the more relevant. On the one hand, a given fixed allocation plan should pass the test of performing well on each clique. On the other, it is plausible that good upper estimates of $ch(\cdot)$ for the whole network can be computed from the cliques. For instance a (hopefully good) upper bound can be computed by constructing an appropriate clique cover of the whole graph G and by summing up the clique bounds. For this to work, it is necessary to deal with cliques efficiently.

In a previous paper, Malesińska showed that when constraint (c) is dropped the problem for cliques is solvable in polynomial time by reducing it to bipartite matching, and that when (c) is introduced the problem becomes non approximable in a rather strong sense (unless P=NP): to find any feasible solution satisfying (a) through (c) is NP-hard [M'95].

In this paper, we continue this line of research. After some preliminary definitions in Section 2, we study the approximation complexity of $ch(I)$ in Section 3. Given that under constraints (a) through (c) even finding a feasible solution is NP-hard, we study the problem under relaxed constraints. First, we show that, interestingly, if any feasible solution is available then a 1/3-approximation can be computed in polynomial time. Then, we show that when constraint (a) is relaxed by dropping the set of conditions $|S(u)| \geq \min(u)$, for all u, the problem becomes 1/2-approximable. It cannot, however, be approximated to any degree of accuracy because, as we show, it is MAX SNP-hard– a fact which rules out the existence of polynomial-time approximation schemes for the problem, unless $P = NP$. In Section 4 we show that the approximation ratio of 1/2 can be improved if the input instances satisfy a certain "sparsity" condition of the lists $L(u)$ of admissible frequencies. Finally, in Section 5 we give some "density" conditions for these lists that, when satisfied, make the co-site constraints have no influence on the maximum value of $ch(I)$.

2 Preliminaries

We recall some well-known notions and definitions from the theory of approximation algorithms (see e.g. [CLR'90, PY'91]). An NP maximization (minimization) problem A is α-*approximable* if there exists a polynomial time algorithm $\mathcal{A}$ which, for all inputs I, produces a solution $\mathcal{A}(I)$ whose value is at least (at most) α times the optimal value. We say then that the algorithm $\mathcal{A}$ has the performance guarantee α. Alternatively, the quality of an approximation algorithm can be measured by its relative error, *i.e.*

$$\max_I \frac{|opt_A(I) - \mathcal{A}(I)|}{opt_A(I)}.$$

Note that if the relative error of an approximation algorithm $\mathcal{A}$ is bounded by ϵ then $\mathcal{A}$ has the performance guarantee $1 - \epsilon$ if A is maximization problem, and $1 + \epsilon$ if A is a minimization problem.

A problem A *L-reduces* to another problem B if we can find two constants α and β and a pair (f, g) of polynomially computable functions such that: (a) for all instances I of A, $f(I)$ is an instance of B such that $opt_B(f(I)) \leq \alpha\, opt_A(I)$; and (b) given a feasible solution b of $f(I)$ whose value is $c(b)$ then, $a = g(b, f(I))$ is a feasible solution of I of value $c(a)$ such that $|opt_A(I) - c(a)| \leq \beta\,|opt_B(f(I)) - c(b)|$. Taken together, these two conditions imply that if B is approximable with worst case relative error ϵ then A is approximable with the relative error $\alpha\beta\epsilon$ or, conversely, that if there exist a limit δ_0 such that A cannot be $(1 - \delta_0)$-approximated (or $(1 + \delta_0)$-approximated if A is a minimization problem) then, B cannot be approximated within $1 - \delta_0\alpha^{-1}\beta^{-1}$ (or $1 + \delta_0\alpha^{-1}\beta^{-1}$).

In a seminal paper, Papadimitriou and Yannakakis introduced a class of combinatorial problems called MAX SNP which has many natural complete problems with respect to L-reductions [PY'91]. Among these is MAX 3SAT-B whose input is a boolean formula $F(x_1, \dots, x_n)$ in conjunctive normal form such that each clause has at most three literals and each variable appears at most

B times. The optimization goal of MAX 3SAT-B is to find a truth assignment satisfying the maximum number of clauses. Recent breakthroughs in the theory of approximation algorithms show that if a problem A is MAX SNP– hard w.r.t. L-reductions then there exist some δ_A such that A cannot be δ_A-approximated provided that P$\neq$NP (see [A'94, BGS'93, H'94] among others).

As defined in the introduction $ch(I)$ denotes the *channel stability number* of an instance $I = (G, L, \min, \max)$ consisting of the interference graph G, lists of available frequencies $L(u)$ and minimum and maximum channel requirements. If no minimum frequency requirements have to be observed– or, equivalently, if $\min(u) = 0$, for all u– than the corresponding channel stability number is denoted by $\widetilde{ch}(I)$. Analogously, if only conditions (a) and (b) are considered and the co-site constraints are ignored then the channel stability number is denoted by $ch^*(I)$.

As customary, K_n denotes the complete graph on n vertices. In this paper, $ch(I), \widetilde{ch}(I)$ and $ch^*(I)$ are studied when the underlying graph topology is a complete graph. Moreover, note that in this paper the notions of frequency and channel are used interchangeably as logical rather than technical terms and that they are often represented as colors.

3 Approximability of the channel stability number

In this section we show some results on the approximability of the channel stability number. It is an obvious necessary condition for the existence of a polynomial time approximation algorithm for any optimization problem that at least one feasible solution can be computed in polynomial time. However, the following theorem is proven in [M'95]:

Theorem 1 *Consider the class of instances of* $ch(I)$ *defined on a complete graph and satisfying* $\max(v) = \min(v) = r(v)$. *Then, it is NP-complete to decide if there is a feasible solution, i.e. a collection of* $S(u) \subseteq L(u)$ *satisfying constraints (a) through (c).*

In view of this, it makes sense to study the approximability of the channel stability number only if we have some information about the feasibility of the instance. The next result shows that if at least one feasible solution exists then $ch(I)$ is $\frac{1}{3}$-approximable.

Theorem 2 *Consider an instance I of the channel stability problem defined on a complete graph K_n and a frequency spectrum C. Let $M = \max_{u \in V(K_n)}\{\max(u)\}$. If the instance I is feasible then $ch(G)$ can be $\frac{1}{3}$-approximated in time $O(Mn|C|\sqrt{Mn + |C|})$. If additionally one feasible solution S is known for I then the algorithm yields not only the approximated value of $ch(G)$ but also a new set of feasible sublists reaching this value.*

Proof.
We claim that if there is at least one feasible solution for the instance I satisfying all requirements (a) through (c) then $\frac{1}{3}ch^*(I) \leq ch(I) \leq ch^*(I)$, where

$ch^*(\cdot)$ denotes the number of channels that can be assigned when only conditions (a) and (b) are observed. This yields a polynomial time approximation algorithm with the performance guarantee $\frac{1}{3}$ as $ch^*(I)$ can be computed in time $O(Mn|C|\sqrt{Mn+|C|})$ [M'95]. Note that in this approach no feasible solution for the instance I needs to be known explicitly. The knowledge of its existence alone is enough to assure the performance guarantee.

Now, in order to complete the proof we construct a solution P by combining any feasible solution S satisfying all constraints with a solution S' that selects $ch^*(I)$ channels, but does not necessary observe the co-site constraints. The new assignment P satisfies all the constraints and $\sum_{v \in V} |P(v)| \geq \frac{1}{3}ch^*(I)$. It is constructed in two stages. First, $\min(v)$ colors from the solution S are assigned to each vertex $v \in V$. Let us consider these colors according to their linear order. When a channel c is added to $P(v)$ then the set $S'(v)$ is reduced to $S'(v) \setminus \{c - 1, c, c+1\}$. If $c \in S'(w)$ for another vertex $w \in V$ then c is removed from the list $S'(w)$. Hence, in the first stage $\sum_{v \in V} \min(v)$ channels are introduced to the solution P and at most three times as much elements are removed from the solution S'.

Let us denote the remaining elements of the solution S' by $S'_r(v)$, $v \in V$. If now $S'_r(v)$ satisfies the condition $\lceil |S'_r(v)|/2 \rceil \leq \max(v) - \min(v)$ for all vertices $v \in V$ then in the second stage every second element of the lists $S'_r(v)$ can be added to the solution P and the proof is completed since the overall value of the solution $\sum_{v \in V} |P(v)|$ is at least $\frac{1}{3}ch^*(I)$. Otherwise, let us assume that there is a vertex $w \in V$ such that $\lceil |S'_r(w)|/2 \rceil = \max(w) - \min(w) + x$, for some $0 < x \leq \min(w)$. It means that $\min(w)$ - the number of channels added to the list $P(w)$ from the list $S(w)$ - exceeded at least by x the number of channels removed during the first stage from the list $S'(w)$. That is there were at least x channels in the list $S(w)$ added to $P(w)$ such that neither they were in $S'(w)$ nor any of the neighboring channels had to be removed from $S'(w)$ when they were added to $P(w)$. Even when some of these x channels were removed from $S(v)$, for one vertex $v \neq w$ per channel, we can remove x additional elements from the list $S'_r(w)$ and the ratio between the number of channels introduced to the solution P and the number of channels deleted from the solution S' won't drop below $\frac{1}{3}$. Afterwards, $\lceil |S'_r(v)|/2 \rceil \leq \max(v) - \min(v)$. The same method can be applied to all vertices $v \in V$ for which $\lceil |S'_r(v)|/2 \rceil > \max(v) - \min(v)$ and then every second remaining element of $S'_r(v)$ is added to the solution $P(v)$. At the end $\sum_{v \in V} |P(v)| \geq \frac{1}{3}ch^*(I)$. $\square$

The next result shows that the $\min(u)$ requirements have a great impact on the computational complexity of the problem. Recall that the channel stability number for the special instances which do not have minimum requirements– i.e. $\min(u) = 0$ for all u– is denoted by $\widetilde{ch}$.

Proposition 1 *If I is defined on a complete graph and $\max(v) \leq M$ for all v then $\widetilde{ch}(I)$ can be 0.5-approximated in time $O(Mn|C|\sqrt{Mn+|C|})$, where $|C|$ denotes the number of frequencies in the spectrum.*

Proof. It can be easily seen that $\frac{ch^*(I)}{2} \leq \tilde{ch}(I) \leq ch^*(I)$. Namely, when there are no minimum channel requirements then one can construct a feasible solution having at least value $ch^*(I)/2$ by taking every second element from an optimal solution that may violate the co-site constraints. Hence, the algorithm for $ch^*(I)$ can be used to approximate $\tilde{ch}(I)$. □

Note that this reasoning remains valid when the condition $\min(u) = 0$, for all u, is replaced by $\min(v) \leq 1$, for all v.

Now the question can be asked if the channel stability number of a complete graph is $(1-\epsilon)$–approximable for every rational $0 < \epsilon < 1$, when all minimum frequency requirements equal zero. Unfortunately, it follows from the next theorem that even in that case there is a constant $c > 0$ such that there is no deterministic polynomial time algorithm for $\tilde{ch}(I)$ with the performance guarantee c, unless $P = NP$.

Theorem 3 *Computing $\tilde{ch}(I)$ is MAX SNP-hard.*

The proof of Theorem 3 requires several steps. The first is to establish the MAX SNP-hardness of a special class of MAX 3SAT instances. In [PY'91] it is proven that MAX 3SAT-B is MAX SNP-hard for any $B \geq 8$. In order to prove Theorem 3 we need to show that MAX 3SAT-B remains MAX SNP-hard when the instances are restricted to $B = 6$ and moreover have a special clause structure.

Definition 1 (MAX 3SAT*) *Let I be a set of clauses of length 3 or 2 such that each variable occurs in exactly 4, 5 or 6 clauses. Moreover, if a variable x_i occurs in 4 or 6 clauses then these are respectively 2 or 3 pairs of symmetric clauses of the form $x_i \vee \bar{x}_j$ and $\bar{x}_i \vee x_j$. If a variable x_i appears in 5 clauses then these are two pairs of symmetric clauses and one arbitrary clause of length 3. MAX 3SAT* problem is to find a maximum number of simultaneously satisfiable clauses in I.*

A slight modification of the original MAX SNP-hardness proof of Papadimitriou and Yannakakis establishes the MAX SNP-hardness of MAX 3SAT*. Here we only explain the idea of the modification. For the complete proof we refer to the full version of the paper [MP'96].

Lemma 1 *MAX 3SAT* is MAX SNP-hard.*

Proof: In Theorem 2(b) in [PY'91] Papadimitriou and Yannakakis define an L-reduction from the MAX 3SAT to the MAX 3SAT-B problem. The instances of MAX 3SAT-B obtained in the proof have already a special structure. The only difference between them and the MAX 3SAT* instances is that in the first case there can be some variables occuring in four pairs of symmetric clauses. The reason for this is the degree of graphs F_m's that are used to define a transformation from MAX 3SAT to MAX 3SAT-B. F_m's are composed of binary trees and a cubic c-expander ([A'87]) defined on the leaves of the trees. Hence, the maximum vertex degree in F_m is 4. However, it can be reduced to 3 and then the resulting boolean formula has the 3SAT* structure. Namely, instead of directly connecting leaves of the trees we can first hang a binary tree with four vertices on each leaf. In that way we obtain three copies of each original leaf that are now connected by the expander. It can be shown that the modified graphs F_m's with maximum vertex degree 3 retain all the required properties and that the transormation that they define is an L-reduction. □

We now exhibit an L-reduction from MAX 3SAT* to $\widetilde{ch}(\cdot)$ when the underlying graph topology is that of complete graphs. We define a transformation f from the MAX 3SAT* problem into the channel stability problem. If an instance I of MAX 3SAT* has m clauses and n variables, then $I' = f(I)$ is defined on a complete graph with $3n + m$ vertices and the set of channels $C = \{x_i^1, \overline{x}_i^1, x_i^2, \overline{x}_i^2, x_i^3, \overline{x}_i^3, a_i, b_i, c_i, d_i \mid i=1,\ldots,n\}$. The first six channels represent the occurrences of a variable x_i, $i = 1,\ldots,n$. Positive literals are represented by *positive* channels and negative literals by *negative* ones. The occurence in a clause of length 3 is always represented by the channel with upper index 1. Two literals from a pair of a symmetric clauses are represented by two channels with the same upper index. For each variable x_i we introduce 3 nodes t_i, p_i and r_i with lists of channels: $L(t_i) = \{b_i, \overline{x}_i^1, a_i, \overline{x}_i^2, x_i^3, \overline{x}_i^3, x_i^2, c_i, x_i^1, d_i\}$, $L(p_i) = \{b_i, d_i\}$ and $L(r_i) = \{a_i, c_i\}$. The sequence $L(t_i)$ defines a linear order of *consecutive* channels, *i.e.* nearby channels are incompatible. We define $\max(t_i) = 5$, $\max(p_i) = \max(r_i) = 1$. Intuitively, the task of group t_i, r_i, p_i is to ensure that only two set assignments for $S(t_i)$, $S(p_i)$ and $S(r_i)$ reach the maximum requirements. For $S(t_i)$ it is either $\{b_i, a_i, x_i^3, x_i^2, x_i^1\}$ or $\{\overline{x}_i^1, \overline{x}_i^2, \overline{x}_i^3, c_i, d_i\}$. For each clause c_j we introduce one node k_j, $j = 1,\ldots,m$ and associate with it the set of channels representing the literals of the respective clause. We set $\max(k_j) = 1$. If a variable i occurs in 5 clauses than either the channel x_i^1 or $\overline{x}_i^1$ is not admissible for any *clause*-node. If a variable i occurs 4 times than both of these channels are absent in the lists of admissible channels for *clause*-nodes.

Let us illustrate this transformation by means of an example. Given a boolean formula I of the form

$$(x_1 \vee \overline{x}_2 \vee x_3) \wedge (x_1 \vee \overline{x}_2) \wedge (\overline{x}_1 \vee x_2) \wedge (x_2 \vee \overline{x}_3) \wedge (\overline{x}_2 \vee x_3) \wedge (x_1 \vee \overline{x}_3) \wedge (\overline{x}_1 \vee x_3)$$

we obtain an instance I' of the channel stability problem defined on a complete graph K_{16} with the following lists of admissible channels:

$$L(t_i) = \{b_i, \overline{x}_i^1, a_i, \overline{x}_i^2, x_i^3, \overline{x}_i^3, x_i^2, c_i, x_i^1, d_i\}, i = 1, 2, 3,$$
$$L(p_i) = \{b_i, d_i\}, \qquad i = 1, 2, 3,$$
$$L(r_i) = \{a_i, c_i\}, \qquad i = 1, 2, 3,$$
$$L(c_1) = \{x_1^1, \overline{x}_2^1, x_3^1\},$$
$$L(c_2) = \{x_1^2, \overline{x}_2^2\}, \qquad L(c_3) = \{\overline{x}_1^2, x_2^2\},$$
$$L(c_4) = \{x_2^3, \overline{x}_3^2\}, \qquad L(c_5) = \{\overline{x}_2^3, x_3^2\},$$
$$L(c_6) = \{x_1^3, \overline{x}_3^3\}, \qquad L(c_7) = \{\overline{x}_1^3, x_3^3\}.$$

A solution S of I' is defined to be *regular* if and only if, for every $i \in \{1,\ldots,n\}$, (a) each $S(t_i)$ is either $\{b_i, a_i, x_i^3, x_i^2, x_i^1\}$ or $\{\overline{x}_i^1, \overline{x}_i^2, \overline{x}_i^3, c_i, d_i\}$; (b) no channel is assigned both positive and negated to some *clause*-nodes, *i.e.* for no x_i and indices j, l, c, d, $x_i^j \in S(k_c)$ and $\overline{x}_i^l \in S(k_d)$.

Any satisfying assignment induces, in the obvious way, a regular solution. Conversely, a regular solution defines a proper truth assignment for the original 3SAT formula. The difficulty of the proof is to show that any given feasible solution can be transformed into a regular one whose value is no worse than the original.

Lemma 2 *If I is an instance of the MAX 3SAT* problem and $I' = f(I)$ has a solution S_1 of value c, then I' has also a solution S_2 of value at least c satisfying the additional property:*

$$\neg(\exists i, \exists j \ s.t. \ \{x_i^j, \overline{x}_i^j\} \subseteq S(t_i)).$$

Proof. Such pairs of channels can be iteratively removed from the solution S_1 for $i = 1,\ldots,n$ and $j = 1, 2, 3$ without decreasing the value of the solution:

- (j=1) If $\{x_i^1, \overline{x}_i^1\} \subseteq S(t_i))$ then $|S(t_i)| \leq 4$. Moreover, $b_i \notin S(t_i)$ and $d_i \notin S(t_i)$. Since $S(p_i)$ can contain only one of the colors b_i and d_i, the other one can replace x_i^1 or $\overline{x}_i^1$ in $S(t_i)$. The value of the modified solution does not change.
- (j=2) If $\{x_i^2, \overline{x}_i^2\} \subseteq S(t_i))$ then $|S(t_i)| \leq 4$ and none of the channels: a_i, x_i^3, $\overline{x}_i^3$, c_i is in $S(t_i)$. Moreover, we can now assume that only one of the channels x_i^1 and $\overline{x}_i^1$ belongs to $S(t_i)$. If $x_i^1 \notin S(t_i)$ then x_i^2 can be replaced in $S(t_i)$ by c_i and $S(r_i)$ can be set to $\{a_i\}$. Otherwise, if $\overline{x}_i^1 \notin S(t_i)$ then $\overline{x}_i^2$ is replaced in $S(t_i)$ by a_i and $S(r_i)$ is set to $\{x_i^c\}$. The value of the modified solution does not decrease.
- (j=3) Since x_i^3 and $\overline{x}_i^3$ are consecutive they could not have been together chosen for $S(t_i)$. □

Lemma 3 *If I is an instance of the MAX 3SAT* problem and $I' = f(I)$ has a solution S_1 of value c, then I' has also a solution S_2 of value at least c satisfying the additional property:*

$$\neg(\exists i, \exists j \ and \ two \ clause\text{-}nodes \ k_k \ and \ k_l \ s.t. \ x_i^j \in S(k_k) \ and \ \overline{x}_i^j \in S(k_l)).$$

Proof. Note that if $x_i^j \in S(k_k)$ and $\overline{x}_i^j \in S(k_l)$ for some $i \in \{1,\ldots,n\}$, $j \in \{1, 2, 3\}$ then k_k and k_l must correspond to a pair of symmetric clauses of the form $x_i^j \vee \overline{x}_s^u$ and $\overline{x}_i^j \vee x_s^u$. By Lemma 2 we can assume that either $\overline{x}_s^u$ or x_s^u does not belong to $S(t_s)$. If $x_s^u \notin S(t_s)$, then we can set $S(k_l) = \{x_s^u\}$ instead of $\{\overline{x}_i^j\}$. Similarly, if $\overline{x}_s^u \notin S(t_s)$, then we can set $S(k_k) = \{\overline{x}_s^u\}$. The value of the modified solution S_2 equals the value of the original solution S_1. □

Lemma 4 *If I is an instance of the MAX 3SAT* problem and $I' = f(I)$ has a solution S such that $|S(t_i) \cup S(p_i) \cup S(r_i)| = 7$, for any $i = 1, \ldots, n$, then either only positive or only negative channels corresponding to the variable x_i are chosen to the sets $S(k_l)$, where k_l represents a clause, $l = 1, \ldots, m$.*

Proof. If $|S(t_i) \cup S(p_i) \cup S(r_i)| = 7$ then $|S(t_i)| = 5$, $|S(p_i)| = 1$ and $|S(r_i)| = 1$. Then, the only two possibilities for $S(t_i)$ are $\{\overline{x}_i^1, \overline{x}_i^2, \overline{x}_i^3, c_i, d_i\}$ and $\{b_i, a_i, x_i^3, x_i^2, x_i^1\}$. Hence, all the channels representing the variable x_i that are chosen for some *clause*-nodes are either positive or negative. $\square$

Lemma 5 *If I is an instance of the MAX 3SAT* problem and $I' = f(I)$ has a solution S_1 of value c, then I' has also a regular solution S_2 of value not less than c.*

Proof. By Lemma 3 we can assume that for any $i = 1, \ldots, n$ maximally three channels corresponding to the variable x_i are chosen to the sets $S(k_l)$, where k_l represents a clause and $l = 1, \ldots, m$. Hence, the requirements that either only positive channels or only negative channels are chosen for the *clause*-nodes can be violated at most by one channel. W.l.o.g. assume that one negative and two positive channels corresponding to the variable x_i are selected for some *clause*-nodes. In that case, by Lemma 4, $|S(t_i) \cup S(p_i) \cup S(r_i)| < 7$. Hence, instead of assigning the *negative* channel to the *clause*-node, we can add it to $S(t_i)$. More precisely, $S(t_i)$ can be set to $\{\overline{x}_i^1, \overline{x}_i^2, \overline{x}_i^3, c_i, d_i\}$, $S(p_i) = \{b_i\}$ and $S(r_i) = \{a_i\}$. By this modification the value of the solution does not decrease. $\square$

Proof of Theorem 3.
We claim that the function f defined above is an L-reduction of the MAX 3SAT* problem into the channel stability problem in complete graphs. Clearly, for any instance I and $I' = f(I)$, we have $OPT(I') \geq OPT(I) + 7n$. On the other hand, by Lemma 5 each solution of I' of value $7n + k$ can be made regular. Then, defining the truth assignment in I according to the channels selected for the *clause*-nodes yields a solution for I of value k. Therefore, $OPT(I') \leq OPT(I) + 7n$. For any instance I of MAX 3SAT* we have $4n \leq 3m$ and $OPT(I) \geq \frac{m}{2}$. It follows that $OPT(I') \leq \frac{23}{2} OPT(I)$ and $|k - OPT(I)| = |7n + k - OPT(I')|$. Hence, f is an L-reduction, where the constant α can be set to $\frac{23}{2}$ and $\beta = 1$. $\square$

4 Sparse lists of admissible channels

The channel stability number of complete graphs can be easily computed when the co-site constraints can be ignored, but the problem becomes NP-hard when the constraints have to be observed. It would be interesting to know if this difference disappear when the lists of admissible channels satisfy certain additional conditions. As a partial answer to this question we first consider instances with relatively short lists of channels and show how the approximation results can be improved when $|L(v)| < 2 \max(v)$ for every $v \in V$. In Section 5 some conditions for dense lists of admissible channels are defined, under which the co-site constraints have no influence on the value of an optimal solution.

The approximation ratio can be improved using the notion of *capacity* of lists of admissible channels for some vertices $A \subseteq V$.

Definition 2 *Given a set of linearly ordered channels $C = \{1, \ldots, |C|\}$ and a complete graph K_n with lists of admissible channels $L(v)$ for every vertex $v \in V(K_n)$, the capacity(A), $A \subseteq V(K_n)$, is defined as the maximum value of the sum $\sum_{v \in A} |S(v)|$ subject to the condition that $S(v) \subseteq L(v)$ and $S(v) \cap S(u) = \emptyset$ for all $v, u \in A$ as well as to the co-site constraints.*

Note that in this definition there is no bound on the cardinality of $S(v)$. The usefulness of the capacity notion stems from the fact that it can be efficiently computed for any set of vertices $A \subseteq V(K_n)$ and then used to approximate the value of $\widetilde{ch}(I)$. For the computation of *capacity* A we need an auxiliary digraph $G = (W, F)$. It has one vertex v_c for every vertex $v \in A$ and every channel c such that $c \in L(v)$ and an arc (v_c, w_{c+1}), for all $w \neq v$. The interpretation of this arc is that if channel c is assigned to the vertex v then $c + 1$ can be assigned to w. Let $\mathcal{P}$ be a family of vertex-disjoint paths in G having the following two properties. For every channel c there is at most one vertex v_c contained in any of the paths $\mathcal{P}$ and for any channel $c + 1$ such that a path P ends at level c there is no path containing any vertex from the level $c + 1$. By the definition of the digraph G the family $\mathcal{P}$ corresponds to an assignment of channels to vertices fulfilling the requirements from Definition 2. Conversely, any feasible channel assignment uniquely defines such a path family. Therefore,

$$capacity(A) = \max_{\mathcal{P}} \{ \sum_{P \in \mathcal{P}} |P| \}.$$

It is easy to see that the family $\mathcal{P}$ maximizing the above expression can be found in time $O(|W| + |F|)$. This can be done applying BFS on the digraph G. For details see the full version of the paper [MP'96]. Using the notion of capacity the next theorem gives a better performance ratio than Proposition 1 for instances with short lists of admissible channels.

Theorem 4 *Assume that we are given an instance I with a set of linearly ordered channels $C = \{1, \ldots, |C|\}$ and a complete graph K_n with lists of admissible channels $L(v)$ and maximum channel requirements $\max(v)$ for every vertex $v \in V(K_n)$. If there is a constant α such that for every vertex $v \in V(K_n)$*

$$\frac{capacity(v)}{\max(v)} \leq \alpha$$

then $\frac{capacity(V)}{\alpha} \leq \widetilde{ch}(I) \leq capacity(V)$.

Proof:
Consider a family of disjoint paths $\mathcal{P}$ defined as above and covering $capacity(V)$ nodes. Such a family induces an assignment $S(v)$ of channels to vertices: $S(v) :=$ $\{c : \exists P \in \mathcal{P} \text{ s.t. } v_c \in P\}$. Note that $|S(v)| \leq capacity(v)$, for every vertex $v \in V(K_n)$. In a proper assignment no selected sublist $S(v)$ should be longer than $\max(v)$ and therefore, all longer lists $S(v)$ have to be reduced. However, for each v, at most $capacity(v) - \max(v)$ channels are removed which, in the worst case, constitutes a $(1 - \frac{1}{\alpha})$ fraction of the number of channels assigned to v. The remaining channels form a proper assignment at least of value $capacity(V)/\alpha$.
□

5 Dense lists of admissible channels

In this section we go back to the study of $ch(\cdot)$, namely we consider again constraints (a) through (c). We examine certain "density" conditions on the input lists $L(v)$'s, which guarantee that the value of optimal solutions does not depend on the co-site constraints.

Theorem 5 *Assume that we are given an instance I with a set of linearly ordered channels $C = \{1, \ldots, |C|\}$, a complete graph K_n with lists of admissible channels $L(v)$ and minimum and maximum channel requirements $\min(v)$ and $\max(v)$ for every vertex $v \in V(K_n)$. Moreover, assume that there is a constant $k < n - 1$ such that the instance has the following two properties:*

(1) $\forall A \subseteq V(K_n) \ (|A| \leq k + 1 \Rightarrow |\bigcup_{v \in A} L(v)| > 4 \sum_{v \in A} \max(v))$
(2) $\forall A \subseteq V(K_n) \ (|A| = k \Rightarrow \bigcup_{v \in A} L(v) = C)$.

Then the value of an optimal solution of the channel stability problem is the same when the co-site constraints are observed and when they are ignored. Moreover, a feasible solution exists iff $\sum_{v \in V} \min(v) \leq |C|$ and then

$$ch(I) = \min\{|C|, \sum_{v \in V} \max(v)\}.$$

Proof. Given any solution S of an instance I of the channel stability problem with minimum requirements we call a channel c to be *free* for a vertex v iff none of the channels $\{c - 1, c, c + 1\}$ has been chosen to $S(v)$. A channel c is *used* by a vertex v iff $c \in S(v)$.

Let us point out some consequences of the properties (1) and (2). It follows from the first property that for any set of vertices A, $|A| \leq k + 1$, the colors from the set $\bigcup_{v \in A} L(v)$ can be partitioned into disjoint groups of $4 \max(v)$ colors per vertex, so that only colors from the list $L(v)$ are in the group associated with any vertex v, $v \in A$. For one vertex one more color can be selected. In a proper assignment S satisfying all requirements, the selection of a channel c for a vertex v may cause that maximally three admissible channels: $c - 1$, c and $c + 1$ are not free for v. Hence, given any admissible assignment S and a set of vertices A, $|A| \leq k+1$, the number of *free* channels in A exceeds at least by one the number of *used* channels.

The property (2) guarantees that for every channel c and an arbitrary set of vertices A of cardinality $k+2$ there are at least three vertices $u \in A$ such that $c \in L(u)$. Hence, for any assignment S violating only the co-site constraints, if color c and one of the colors $c-1$, $c+1$ have been chosen to $S(v)$ for a vertex $v \in A$ then c is free for at least one vertex $w \in A$.

Clearly, if $\sum_{v \in V} \min(v) \leq |C|$ then there is no feasible solution. Otherwise, it follows from the Hall property (see e.g. [B'85]) that $\min\{|C|, \sum_{v \in V} \max(v)\}$ channels can be assigned to the vertices when the co-site constraints are ignored. Moreover, we claim that all violations of the co-site constraints can be removed from such an optimal assignment S. Namely, assume that there is a vertex v that has been assigned two consecutive channels c and $c+1$. Then S can be modified to a solution S' that has the same value, assigns no new pairs of consecutive channels and such that $c \notin S(v)$. This is achieved with the help of an auxiliary directed tree T with labeled edges that is iteratively augmented until the new solution S' is found. The vertex v is defined to be the root of T. Then, new nodes are added to T according to the following principle. Let w be the vertex just added to T. If (a) the channel c is *free* for w or, (b) there is a channel d that is free for w and is not used by any other vertex then, the unique path from v to w defines a possible transformation from S to S'. Namely, in the first case the channel c is assigned to w instead of the channel that labeled the last arc of the path and in the second case d is allocated to w. Moreover, for every arc (v_1, v_2) in the path labeled with c^*, the channel c^* is removed from the list of channels assigned to v_2 and is now allocated to v_1. By the second property when $|V(T)| = k+2$ then c is free for at least one vertex in $V(T)$ and hence, the transformation from S to S' is found. Otherwise, if $|V(T)| < k+2$ and none of the two cases applies then by the first property there is a vertex $u \in V(T)$ and a channel f such that f is free for u and is not used by any of the vertices in T. Then, there is a vertex $z \notin V(T)$, such that $f \in S(z)$. The tree T is now extended by the arc (u, z) labeled f. Since in each such step a new vertex is added to $V(T)$ and $k \leq n-2$ the sought transformation is found at most after $k+1$ steps. Moreover, the same method can be iteratively applied to all pairs of consecutive channels that are assigned to one vertex and at the end we obtain a solution satisfying all constraints and assigning $\min\{|C|, \sum_{v \in V} \max(v)\}$ channels. $\square$

The requirements in Theorem 5 are quite restrictive and therefore, it would be good to know if there are some lighter conditions on the lists of admissible channels under which the co-site constraints have no impact on the channel stability number $ch(\cdot)$. However, the following two examples show that none of the properties 1 and 2 alone can be sufficient.

Example 1. Consider a complete graph K_n with vertices $\{v_1, \ldots, v_n\}$. For the vertices v_i, $i = 1, \ldots n-1$ the lists of admissible channels are defined as $L(v_i) =$

$\{1, 2, \ldots, n-1\}$ and maximum channel requirements are $\max(v) = 1$. For the last vertex v_n, $L(v_n) = \{1, 2, \ldots, n, n+1\}$ and $\max(v_n) = 2$. The minimum channel requirements of all vertices equal zero. Then $ch(I) = n$ but an optimal solution that does not observe the co-site constraints achieves value $n + 1$. On the other hand, this instance satisfies the first property from Theorem 5 for any constant $k < \lfloor \frac{n}{4} \rfloor - 1$.

Example 2. In this example we also consider $n + 1$ channels and a complete graph K_n. For the first $n - 1$ vertices the lists of admissible channels are defined as $L(v_i) = \{1, 2, \ldots, n+1\}$ and $\max(v_i) = 1$. For the last vertex v_n we set $L(v_n) = \{1, 2\}$ and $\max(v_n) = 2$. Then, as in the previous example, $ch(I) = n$ and an optimal solution that does not observe the co-site constraints achieves value $n + 1$. However, the second property from Theorem 5 is satisfied for every subset of vertices A, $|A| > 2$.

6 Conclusions

In this work we have studied the channel stability function $ch(I)$ of an instance I defined on a graph G. This function can be used to select a channel assignment plan for the fixed part of a mixed cellular network leaving enough freedom in the dynamic part. In particular we have studied the computational complexity and approximability of $ch(I)$ when G is a complete graph and the restriction of $ch(I)$ to $\widetilde{ch}(I)$ when there are no minimum channel requirements. It has been shown that $\widetilde{ch}(I)$ can be approximated up to a constant factor but unfortunately the problem is still MAX SNP-hard. Moreover, the influence of the co-site constraints on the complexity of $\widetilde{ch}(I)$ and $ch(I)$ has been examined.

The future work should concentrate on the design of algorithms for the general topology of the dynamic part of the network. The results for complete graphs can be used in the computation of upper bounds on $ch(I)$. One possibility is to find an appropriate disjoint clique cover of G and then sum up known bounds for the cliques. This idea is now being tested by running experiments with real life graphs; the outcome is so far encouraging. Another approach to the evaluation of lists of admissible channels in the dynamic part of the network could be based on the comparison of the channel stability number for different maximal cliques of G. In that case it remains to be examined how the values for single cliques should be combined together into a function characterizing the whole graph.

It would be also interesting to study the complexity of the channel stability number when the frequency spectrum is independent of the input.

References

[A'87]   M.Ajtai. Recursive construction for 3-regular expanders. *Proc. 28th Annual IEEE Symp. on Foundations of Computer Science* (1987), 295-304.

[A'94]   S. Arora. Probabilistic checking of proofs and the hardness of approximation problems. PhD Thesis, U.C. Berkeley, 1994. Available via anonymous ftp as Princeton TR94-476.

[BGS'93]   M. Bellare, O. Goldreich, and M. Sudan. Free bits, PCPs and non-approximability – towards tight results. Technical Report ECCC TR95-24, Revised version, September 1995. Extended abstract in *Proc. 25th ACM Symp. on Theory of Computing* (1993), 113–131, 1993.

[B'85]   C. Berge. Graphs. *North-Holland Math. Library, Vol. 6, Part 1, Elsevier Science Publishers* (1985).

[B'89]   C. Berge. Minimax relations for the partial q-colorings of a graph. *Disc. Mathematics 74* (1989), 3-14.

[CLR'90]   T.H.Cormen, C.E.Leiserson and R.L.Rivest. Introduction to Algorithms. The MIT Press, Cambridge, McGraw Hill, 1990.

[DJLS'94]   G. Dahl, K. Jörnsten, G. Løvnes, S. Svaet. Graph optimization problems in connection with the management of mobile communication systems. *Telecommunications Systems*, Vol. 3 (1994), 319-340.

[DV'93]   D. Dimitrijević, J. Vučetić. Design and Performance Analysis of the Algorithms for Channel Allocation in Cellular Networks. *IEEE Transactions on Vehicular Technology*, Vol. 42 (1993), 526-534.

[GSW'94]   A. Gräf, M. Stumpf, G. Weißenfels. On coloring unit disk graphs. Johannes Gutenberg-Universität Mainz (1994).

[EB'91]   H. Eriksson, R. Bownds. Performance of Dynamic Channel Allocation in the DECT System. *41st IEEE Vehicular Technology Conference* (1991), 693-698.

[H'80]   W.K. Hale. Frequency Assignment: Theory and Applications. *Proc. of the IEEE, Vol. 68* (1980), 1497-1514.

[H'94]   J. Håstad. Recent results in hardness of approximation. *Proc. of 3rd Scandinavian Workshop on Algorithm Theory* (1994), Springer–Verlag LNCS 824, pp. 231–239.

[JT'95]   T. Jensen, B. Toft. *Graph Coloring Problems*. John Wiley & Sons, Inc., 1995.

[M'95]   E. Malesińska. An Optimization Method for the Channel Assignment in Mixed Environments. *Proc. of the 1st ACM Int. Conf. on Mobile Computing and Networking* (1995), 210-217.

[MP'96]   E. Malesińska, A. Panconesi. On the Hardness of Allocating Frequencies for Hybrid Networks. Preprint No. 498/1996, Fachb. Mathematik, TU Berlin.

[MBHRR'95]   M.V. Marathe, H. Breu, H.B. Hunt III, S.S. Ravi, D.J. Rosenkrantz. Simple Heuristics for Unit Disk Graphs. *Networks, Vol.25* (1995), 59-68.

[PY'91]   Ch.H. Papadimitriou, M. Yannakakis. Optimization, Approximation, and Complexity Classes. *Journal of Computer and System Sciences 43* (1991), 425-440.

[P'95]   J. Plehn. Private communication.

[T'89]   B.A. Tesman. T-colorings, list T-colorings, and set T-colorings of graphs. *RUTCOR Res. Rept. RRR 57-89* Rutgers University, New Brunswick, NJ (1989).

[ZE'93]   J. Zander, H. Eriksson. Asymptotic Bounds on the Performance of a Class of Dynamic Channel Assignment Algorithms. *Wireless communications: future directions*, ed. by J.M. Holtzman, D.J. Goodman, Kluwer Ac. Publ. (1993), 259-274.

Homogeneous Sets and Domination Problems[*]

Falk Nicolai[1] and Thomas Szymczak[2]

[1] Gerhard–Mercator–Universität — GH Duisburg, FB Mathematik, FG Informatik
D–47048 Duisburg, Germany
e–mail : nicolai@informatik.uni-duisburg.de
[2] Universität Rostock, FB Informatik, Lehrstuhl für Theoretische Informatik,
Albert–Einstein–Str. 21, D–18051 Rostock, Germany
e–mail : szymczak@informatik.uni-rostock.de

Abstract. In this paper we consider the relation of homogeneous sets (sometimes called modules) to the domination problems r–dominating clique and connected r–dominating set by investigating homogeneous extensions of graphs.

At first, we show that homogeneous extensions of a hereditary graph class $\mathcal{G}$ can be recognized nearly as efficiently as the graphs of $\mathcal{G}$, itself. The algorithm is based on modular decomposition.

In the main part of this work we show, that efficient algorithms solving the r–dominating clique and the connected r–dominating set problem (and thus the Steiner tree problem) on a hereditary graph class $\mathcal{G}$ lead to efficient algorithms on their homogeneous extensions.

Applying these results to homogeneous extensions of trees we get efficient algorithms solving these problems in linear sequential and polylogarithmic parallel time using a linear number of processors.

1 Introduction

Location problems play an important role in network design. Let us given a network structure G and two kinds of processors (suppliers and receivers) which have to be assigned to the vertices of G. To each receiver a value is associated indicating the radius within which it can receive information from suppliers. Assuming that the production effort for suppliers is much higher than for receivers we have to minimize the number of suppliers which are necessary to provide all receivers with information. If all receivers are of the same type, i.e. their radius value is identical, then this is exactly the well-known k–domination problem on graphs (cf. [5]) : Compute a minimum cardinality set D such that for each vertex v outside D there is at least one vertex inside D of distance at most k. Assigning a supplier to each vertex of D then minimizes their number. If we allow different types of receivers, i.e. the radii do not coincide, then we have the more general r–domination problem : Given a graph G and a radius function $r : V(G) \to \mathbb{N}$, compute a minimum cardinality set D such that for each vertex v outside D there is at least one vertex inside D of distance at most $r(v)$. Note that $r(v) = 0$

[*] This work is supported by the German Research Community DFG.

means that this vertex must belong to D. So we can extend a given substructure of the network by assigning radius zero to already installed suppliers.

In this paper we consider r–domination problems for homogeneous extensions of trees, i.e. graphs obtained from trees by substituting the vertices v of the tree by graphs G_v such that all vertices of G_v have the same neighbours outside G_v as v had in G. We can interpret these graphs as a generalization of the tree interconnection network where each processor is again a network of a certain type, or as a model to describe the granularity of the processors in a tree network.

The paper is organized as follows. In the first part we consider homogeneous extensions of graphs. We start by presenting the definition and some basic properties. Next, we show that homogeneous extensions of a hereditary graph class $\mathcal{G}$ can be recognized nearly as efficiently as the graphs of $\mathcal{G}$ itself. These algorithms are based on modular decomposition.

For a connected graph G of at least two vertices the *homogeneous extension* $G' := \mathsf{HExt}(G, v, H)$ of G via a graph H in v is the graph obtained by substituting v by H such that the vertices of H have the same neighbours outside H as v had in G. Let r, r' be radius functions of G, G', respectively, such that r, r' coincide on $V(G') \setminus V(H) = V(G) \setminus \{v\}$ and $r(v) = \min\{r'(h) : h \in V(H)\}$. We show in detail, how to compute a minimum r'–dominating clique (resp. minimum connected r'–dominating set) of G' from a minimum r–dominating clique (resp. minimum connected r–dominating set) of G. In [9] a similar result is given for the Steiner tree problem : Given a graph $G = (V, E)$ and a set $T \subset V$ one has to determine a minimal (with respect to inclusion) set S such that $T \subseteq S$ and $G(S)$ is connected. Observe, that the Steiner tree problem is a special case of the connected r–dominating set problem. Indeed, it is equivalent to compute a connected r–dominating set where $r(v) := 0$ for each $v \in T$ and $r(v) := |V(G)|$ for all other vertices. Note, that the Steiner tree problem can be interpreted as finding an optimal query in a relational database scheme.

In the second part we apply the results of the first one to homogeneous extensions of trees. At first we summarize the results contained in [15] for computing a minimum r–dominating set, a minimum r–dominating clique and a minimum connected r–dominating set on trees. This will form the basis of the corresponding algorithms for their homogeneous extensions. Then the above approach and an efficient algorithm computing all r–dominating vertices of a tree lead to efficient sequential and parallel algorithms for the problems r–dominating clique and connected r–dominating set for homogeneous extensions of trees. Finally we outline the idea of a linear time sequential algorithm solving the r–dominating set problem on homogeneous extensions of trees (cf. [15]).

Note that most of the classical graph problems like computing a maximum clique, a maximum independent set, a minimum clique cover, the chromatic number and index, the Hamiltonian circuit problem are $\mathbb{NP}$–complete for homogeneous extensions of trees. This can be easily seen, since for an arbitrary graph $G = (V, E)$ and a vertex $v \notin V$ the graph G' obtained from G by joining v to each vertex of G, i.e. $N(v) = V(G)$, is the homogeneous extension of an edge.

2 Homogeneous Extensions of Graphs

2.1 Preliminaries

Let $N_k(v)$ denote the *k-th neighbourhood* of v, i.e. the set of vertices at distance k to v. For convenience we write $N(v)$ instead of $N_1(v)$.

A set $H \subseteq V$ is called *homogeneous* iff any pair of vertices of H has the same neighbourhood outside H :

$$N(u) \cap (V \setminus H) = N(v) \cap (V \setminus H) \qquad \text{for all } u, v \in H.$$

A homogeneous set H is *proper* iff $|H| < |V|$.

Let H be a proper homogeneous set of G containing at least two vertices and let $v_H \in H$. Then the graph $\mathsf{HRed}(G, H, v_H)$ obtained from G by deleting $H \setminus \{v_H\}$, i.e. contracting H to a representing vertex v_H, will be called the *homogeneous reduction* of G (via H).

Conversely, for a connected graph G of at least two vertices the *homogeneous extension* $\mathsf{HExt}(G, v, H)$ of G via a graph H in v is the graph obtained by substituting v by H such that the vertices of H have the same neighbours outside H as v had in G.

Note that it is necessary for structural reasons to have at least two vertices in G since otherwise each graph can be represented as the homogeneous extension of a single vertex by itself.

As usual we denote by $\mathsf{HExt}^*(\mathcal{G})$ the transitive closure of the graph class $\mathcal{G}$ with respect to homogeneous extensions.

Let G be a homogeneous extension of a graph G'. Then we can define an *extension sequence* $\sigma = ((v_1, H_1), \ldots, (v_k, H_k))$ such that :

1. $G_0 := G'$.
2. For each $i = 1, \ldots, k$ vertex v_i belongs to G_{i-1}.
3. For each $i = 1, \ldots, k$ we define $G_i := \mathsf{HExt}(G_{i-1}, v_i, H_i)$.
4. $G = G_k$.

We will write $G = \mathsf{HExt}(G', \sigma)$. Moreover, let $\sigma_i := ((v_1, H_1), \ldots, (v_i, H_i))$.

Lemma 1 (Extension sequence in normal form). *Let G be a homogeneous extension of a graph G' and let $\sigma = ((v_1, H_1), \ldots, (v_k, H_k))$ be an extension sequence. Then there is an extension sequence $\tau = ((u_1, F_1), \ldots, (u_l, F_l))$ such that $\{u_1, \ldots, u_l\} \subseteq V(G')$ and $\mathsf{HExt}(G', \sigma) = \mathsf{HExt}(G', \tau)$.*

Proof. Let i be the smallest index in σ such that $v_i \notin V(G')$. Then v_i must be in H_j for some $j < i$. Define $H := \mathsf{HExt}(H_j, v_i, H_i)$ and

$$\tau_{i-1} := ((v_1, H_1), \ldots, (v_{j-1}, H_{j-1}), (v_j, H), (v_{j+1}, H_{j+1}), \ldots, (v_{i-1}, H_{i-1})).$$

It is easy to verify that $\mathsf{HExt}(G', \sigma_i) = \mathsf{HExt}(G', \tau_{i-1})$. The assertion follows by induction. $\qquad \square$

An extension sequence τ according to the above lemma is called to be in *normal form*.

2.2 Recognition of Homogeneous Extensions

Let $\mathcal{G}$ be a hereditary graph class, i.e. if $G \in \mathcal{G}$ then each induced subgraph F of G is in $\mathcal{G}$, too. To recognize homogeneous extensions of graphs of $\mathcal{G}$ we use modular decomposition. Two homogeneous sets H_1, H_2 *overlap* iff their intersection and their mutual differences are nonempty. A homogeneous set H is *overlap-free* iff there is no other homogeneous set overlapping H. Since for any overlap-free homogeneous set $H \subset V$ there is exactly one minimal overlap-free homogeneous set H' containing H properly we obtain a parent function by $\mathsf{parent}(H) := H'$. Thus, using V as root, this gives a tree of homogeneous sets called the *module tree* $T_M(G)$. This tree can be computed in linear sequential time (cf. [14], [6]) and in parallel time $O(\log^2 n)$ using $O(n + m)$ processors on a CRCW–PRAM as shown in [7].

Let $\mathcal{H}(G)$ denote the set of all maximal proper homogeneous sets of a graph $G = (V, E)$. Since each singleton of V is a homogeneous set, $\mathcal{H}(G)$ covers V for graphs with at least two vertices. Thus, either $\mathcal{H}(G)$ is a partition of V or there are at least two homogeneous sets in $\mathcal{H}(G)$ which intersect. Before presenting the recognition algorithm we give the relationship between the maximal proper and maximal overlap-free homogeneous sets.

Lemma 2. *If $\mathcal{H}(G)$ is a partition of V then $\mathcal{H}(G)$ is exactly the set of maximal proper overlap-free homogeneous sets of G.*

Proof. Obviously it is sufficient to show that each homogeneous set of $\mathcal{H}(G)$ is overlap-free. By assuming the contrary let H_1 be a homogeneous set of $\mathcal{H}(G)$ overlaped by a homogeneous set H_2. Since H_2 is homogeneous it must be contained in some homogeneous set $H_3 \neq H_1$ of $\mathcal{H}(G)$. But then $H_1 \cap H_3 \neq \emptyset$ is a contradiction. $\square$

Lemma 3 [9]. *If $\mathcal{H}(G)$ is not a partition of V then $\mathcal{H}(G) = \{V \setminus V_i : i = 1, \ldots, l\}$ where $V_1, \ldots, V_l$ are the connected components of the complement $\overline{G}$ of G, and $l \geq 3$.*

Two disjoint subsets X and Y of the vertex set V of a graph G form a *join*, denoted by $X \bowtie Y$, iff each vertex of X is adjacent to every vertex of Y.

Lemma 4. *If $\mathcal{H}(G)$ is not a partition of V then the maximal proper overlap-free homogeneous sets of G are exactly the connected components $V_1, \ldots, V_l$ of $\overline{G}$.*

Proof. At first we show that each V_i is overlap-free. Assume the contrary and let H be a homogeneous set that overlaps V_i. Since $V \setminus V_i$ is homogeneous by Lemma 3 we conclude $(H \cap V_i) \bowtie (H \cap (V \setminus V_i))$. Hence $(H \cap V_i) \bowtie (V_i \setminus H)$ which is a contradiction to Lemma 3.

Now assume that there is some overlap-free homogeneous set H properly containing V_i. Again we consider $V \setminus V_i$ which is homogeneous by Lemma 3. From $V_i \subset H$ we conclude $H \cap (V \setminus V_i) \neq \emptyset$ and $H \setminus (V \setminus V_i) \neq \emptyset$. Since H is overlap-free we must have $(V \setminus V_i) \setminus H = \emptyset$ implying $H = V$. Thus V_i is maximal overlap-free. $\square$

Our recognition algorithm works as follows : At first we compute the module tree $T_M(G)$. If the root of $T_M(G)$ has only two children then the input graph is a homogeneous extension of an edge. Otherwise we check whether the children of the root of $T_M(G)$ are maximal homogeneous sets. If so then we shrink all these modules to representing vertices and check whether the obtained graph is in $\mathcal{G}$. If not then the input graph is a homogeneous extension of an edge.

Input : A connected graph $G = (V, E)$, $|V| \geq 2$.
Output : A graph $G' \in \mathcal{G}$ and an extension sequence σ in normal form
   such that $G = \mathsf{HExt}(G', \sigma)$ if $G \in \mathsf{HExt}^*(\mathcal{G})$ or 'NO'.

begin
(1)   Let $K_2 := (\{x, y\}, \{xy\})$.
(2)   Compute the module tree $T_M(G)$.
(3)   Let $M_1, \ldots, M_l$ be the neighbours of the root of $T_M(G)$.
(4)   **if** $l \leq 2$ **then**
(5)   **if** $K_2 \in \mathcal{G}$ **then return**$(K_2, \sigma = ((x, G_{M_1}), (y, G_{M_2})))$
(6)   **else return**('NO')
(7)   **else** $H_1 := M_1$; $H_2 := \bigcup_{i=2}^{l} M_i$
(8)   **if** $H_1 \bowtie H_2$ **then**
(9)   **if** $K_2 \in \mathcal{G}$ **then return**$(K_2, \sigma = ((x, G_{H_1}), (y, G_{H_2})))$
(10)   **else return**('NO')
(11)   **else for all** $i \in \{1, \ldots, l\}$ shrink M_i to v_i
(12)   Let G' be the resulting graph.
(13)   **if** $G' \in \mathcal{G}$ **then return**$(G', \sigma = ((v_1, G_{M_1}), \ldots, (v_l, G_{M_l})))$
(14)   **else return**('NO')
end.

Theorem 5. *Let $\mathcal{G}$ be a hereditary class of graphs. Then, homogeneous extensions of graphs of $\mathcal{G}$ can be recognized*

- *sequentially as efficiently as graphs of $\mathcal{G}$ can be recognized,*
- *in $O(t + \log^2 |V|)$ parallel time with $O(p + |V| + |E|)$ processors where (t, p) is a (time, proc)–complexity for the parallel recognition of the graph class $\mathcal{G}$.*

Proof. By the definition of the module tree we have only to consider the neighbours $M_1, \ldots, M_l$ of its root. If the number l of these neighbours is two then by Lemma 3 $\mathcal{H}(G)$ is a partition of $V(G)$. So steps $(4) - (6)$ are correct. If $l \geq 3$ then Lemma 3 implies that $\mathcal{H}(G)$ is a partition of $V(G)$ if and only if $H_2 := \bigcup_{i=2}^{l} M_i$ is not homogeneous in G. This is checked in step (8) (recall that G is connected). Therefore the algorithm is correct.

To verify the time and processor bounds we note that the module tree can be computed in linear sequential time (cf. [14], [6]), and in $O(\log^2 n)$ parallel time using $O(n + m)$ processors as shown in [7]. The test in step (8) can easily be done within this bounds. $\qquad\square$

2.3 The r–Dominating Clique Problem

Let G, G' be graphs such that $G := \mathsf{HExt}(G', v_H, H)$. Further let r, r' be radius functions of G, G', respectively, such that r, r' coincide on $V(G) \setminus H$ and $r'(v_H) = \min\{r(h) : h \in H\}$.[3] In the next lemma we investigate how to compute a minimum r–dominating clique of G from a minimum r'–dominating clique of G'. With H^0 we denote the set of vertices v of H with $r(v) = 0$.

Lemma 6. *Let $G = \mathsf{HExt}(G', v_H, H)$ and C' be a minimum r'-dominating clique in G'. If H^0 is a clique then we obtain a minimum r-dominating clique C of G by*

$$C := \begin{cases} C' & : \quad v_H \notin C' \\ (C' \setminus \{v_H\}) \cup S & : \quad v_H \in C' \end{cases}$$

where S is for $H^0 \neq \emptyset$:

$$S := \begin{cases} H^0 & : C' = \{v_H\} \text{ and } H^0 \text{ } r\text{-dominates } H, \text{ or} \\ & |C'| \geq 2 \\ H^0 \cup \{w\} & : C' = \{v_H\} \text{ and } H^0 \text{ does not } r\text{-dominate } H, \, w \in N(H) \setminus H \end{cases}$$

and for $H^0 = \emptyset$:

$$S := \begin{cases} \{v\} & : C' = \{v_H\} \text{ and } v \in H \text{ } r\text{-dominates } H, \text{ or} \\ & |C'| \geq 2, \, v \in H \\ \{w\} & : C' = \{v_H\} \text{ and } w \in G \setminus H \text{ } r\text{-dominates } G \\ \{v, w\} & : \text{otherwise, where } v \in H, \, w \in N(H) \setminus H. \end{cases}$$

Proof. We distinguish between four cases :

Case 1. $v_H \notin C'$.
Since C' r'–dominates G' there must be a vertex u in C' such that $d(u, v_H) \leq r'(v_H)$. Since $r'(v_H) = \min\{r(h) : h \in H\}$ and H is homogeneous C' is a r–dominating clique in G, too.

Case 2. $v_H \in C'$ and $H^0 \neq \emptyset$.
This case is obvious.

Case 3. $C' = \{v_H\}$ and $H^0 = \emptyset$.
If there is any vertex in $G \setminus H$ r–dominating G then the correctness follows from Case 1. Otherwise, v_H is the only vertex which r'–dominates G'. If there is a vertex v within H r–dominating H then this vertex r–dominates G, too, since $r'(v_H) = \min\{r(h) : h \in H\}$. Finally, if there is neither a vertex in $G \setminus H$ nor a vertex in H r–dominating G then the cardinality of a minimum r–dominating clique must be at least 2. Obviously, any pair v, w of vertices $v \in H$ and $w \in N(H) \setminus H$ r–dominates G.

Case 4. $v_H \in C'$, $|C'| \geq 2$ and $H^0 = \emptyset$.
In this case there is nothing to show. □

[3] For convenience we identify H with the vertex set $V(H)$.

Theorem 7. *Let $\mathcal{G}$ be a hereditary class of graphs on which the r–dominating clique problem can be solved in polynomial time. Then this can be done on the graph class $\mathsf{HExt}^*(\mathcal{G})$, too.*

Proof. Let $G \in \mathsf{HExt}^*(\mathcal{G})$ and r be a radius function on G. Using the recognition-algorithm of section 2.2 we compute in polynomial time a graph G' with vertices $v_1, \ldots, v_n$ and graphs $H_1, \ldots, H_n$ such that $G = \mathsf{HExt}(G', \sigma)$, $\sigma = ((v_1, H_1), \ldots, (v_n, H_n))$. (Note, that since $G \in \mathsf{HExt}^*(\mathcal{G})$ we can skip the tests $K_2 \in \mathcal{G}$, $G' \in \mathcal{G}$ performed in the recognition–algorithm.) To compute a minimum r–dominating clique in G we proceed in three steps.

In the first one we compute, for each $i \in \{1, \ldots, n\}$, the value $r'(v_i) := \min\{r(h) : h \in H_i\}$ and the set $(H_i)^0$. Further we test if $(H_i)^0$ is complete. If this is not so for any i then we stop, G has no r–dominating clique.

After terminating the first step we compute in the second step a minimum r'–dominating clique in G'. If the size of this clique equals one then we compute all r'–dominating vertices in G'. This can be done in polynomial time using the distance matrix.

In the third step we process σ in natural order and construct a minimum r–dominating clique in G from a minimum r'–dominating clique in G' according to Lemma 6. This can be done in polynomial time by taking into account the following remarks :

- A proper homogeneous set H is r–dominated by a subset $U \subseteq H$ if and only if each vertex $v \in H$ with $r(v) = 0$ belongs to U and each vertex $v \in H \setminus U$ with $r(v) = 1$ has a neighbour in U. This can be tested in time $\sum\limits_{v \in H \setminus U} \deg(v)$.

- After each extension step $G_{i+1} = \mathsf{HExt}(G_i, v_i, H_i)$ we have to compute the set D_{i+1} of all r–dominating vertices in G_{i+1} from the set D_i of all r–dominating vertices in G_i. If $v_i \notin D_i$ then clearly $D_{i+1} = D_i$; if $v_i \in D_i$ then we compute the set D' of all vertices of H_i that r–dominate H_i in G_{i+1}. Then $D_{i+1} = (D_i \setminus \{v_i\}) \cup D'$.

Consequently, the algorithm works in polynomial time. $\qquad\qquad\square$

Now we give a short idea how to parallelize the above algorithm.

1. Compute a graph $G' \in \mathcal{G}$ with $G = \mathsf{HExt}(G', \sigma)$, σ an extension sequence in normal form.
2. For each $v \in V(G')$ define $r'(v) := \min\{r(u) : u \in H_v\}$ where H_v is the homogeneous set substituted into v (possibly $H_v = \{v\}$).
3. Compute a minimum r'–dominating clique C' in G'.
4. If $|C'| = 1$ then compute in parallel all r'–dominating vertices of G' by using the distance matrix. Let S be this set. If there is no vertex in G with r–value zero then we check in parallel for each vertex $v \in S$ whether H_v is r–dominated by a single vertex within H_v — then this vertex r–dominates G, too — or not, in this case we need an edge to r–dominate G.

If there are vertices in G with r–values zero then these vertices must belong to some H_v. Hence $S = \{v\}$. Now check whether H_v^0 is complete and r–dominates H_v. If the latter is not true then we must add a vertex $w \in N(H_v)$ to H_v^0.

5. If $|C'| \geq 2$ then we proceed according to Lemma 6.

Summarizing the above results we obtain

Theorem 8. *Let $\mathcal{G}$ be a hereditary class of graphs on which the r–dominating clique problem can be solved in polylogarithmic parallel time with a polynomial number of processors. Then this can be done on the graph class $\mathsf{HExt}^*(\mathcal{G})$, too.*

2.4 The Connected r–Dominating Set Problem

Now we want to consider connected r–dominating sets in G. We will use the above algorithm up to the handling for the sets H^0.

Lemma 9. *Let $G = \mathsf{HExt}(G', v_H, H)$ and let D' be a minimum connected r'–dominating set in G'. We obtain a minimum connected r–dominating set D of G by*

$$D := \begin{cases} D' & : \quad v_H \notin D' \\ (D' \setminus \{v_H\}) \cup S & : \quad v_H \in D' \end{cases}$$

where S is for $H^0 \neq \emptyset$:

$$S := \begin{cases} H^0 & : D' = \{v_H\} \text{ and } H^0 \text{ is connected and } r\text{-dominates } H \\ & \quad \text{or } |D'| \geq 2 \\ H^0 \cup \{w\} & : \text{otherwise, where } w \in N(H) \setminus H \end{cases}$$

and for $H^0 = \emptyset$:

$$S := \begin{cases} \{v\} & : D' = \{v_H\} \text{ and } v \in H \ r\text{-dominates } H, \quad \text{or} \\ & \quad |D'| \geq 2, v \in H \\ \{w\} & : D' = \{v_H\} \text{ and } w \in G \setminus H \ r\text{-dominates } G \\ \{v, w\} & : \text{otherwise, where } v \in H, w \in N(H) \setminus H. \end{cases}$$

Proof. Analogously to the proof of Lemma 6. $\qquad\qquad\qquad\qquad\qquad\square$

Instead of testing H^0 to be complete we must here test for connectedness. This can be done in linear sequential and polylogarithmic parallel time with respect to H^0. Therefore, we immediately obtain

Theorem 10. *Let $\mathcal{G}$ be a hereditary class of graphs.*

- *If the connected r–dominating set problem on $\mathcal{G}$ can be solved in polynomial time, so this can be done for graphs of $\mathsf{HExt}^*(\mathcal{G})$, too.*
- *Provided the connected r–dominating set problem can be solved in polylogarithmic parallel time using a polynomial number of processors, then this can be done for graphs of the class $\mathsf{HExt}^*(\mathcal{G})$, too.*

3 Homogeneous Extensions of Trees

In this section we apply the results of the previous one to homogeneous extensions of trees. First note that Theorem 5 immediately gives

Corollary 11. *Homogeneous extensions of trees can be recognized in linear sequential time and in $O(\log^2 n)$ parallel time using a linear number of processors.*

3.1 Efficient Algorithms on Trees

An usual method for solving problems on trees is the algebraic tree computation (ATC, cf. [12]) : Let T be a rooted regular binary tree – i.e. a rooted tree such that every inner vertex of T has exactly two children. Furthermore, let S be a set, $B \subseteq \{f : f : S \times S \to S\}$ be a subset of binary functions on S and $U \subseteq \{g : g : S \to S\}$ be a subset of unary functions on S. Then there are two kinds of algebraic tree computations :

B–ATC. *Bottom–up algebraic tree computation* :
The leaves are labeled by elements of S, the inner vertices by elements of B and the edges by elements of U. Recursively we can now evaluate the labeled tree in the following sense. If v is a vertex labeled by f with children u_1, u_2 being leaves labeled by $L(u_1)$, $L(u_2)$, then delete u_1, u_2 in the tree and update the label of v by $f(g(L(u_1)), h(L(u_2)))$ where g and h denote the labels of the edges vu_1 and vu_2, respectively. The B–ATC problem is to compute the final label of the root.
In [1] it is shown that if the B–ATC problem is decomposable then it can be solved on a EREW-PRAM in time $O(\log n)$ using $O(n/\log n)$ processors via the update rule illustrated in Figure 1.

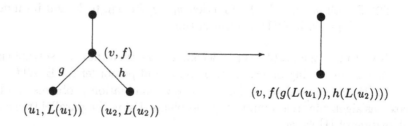

Fig. 1. The update rule for the B–ATC problem.

Hereby a B–ATC problem is *decomposable* iff the following two conditions are fulfilled :
(B1) The sets B and U are indexed and their elements can be computed in $O(1)$ sequential time.

(B2) For all $g_i, g_j, g_k \in U$, $f_l \in B$ and $a \in S$ the functions $x \mapsto g_i(f_l(g_j(x), g_k(a)))$ and $x \mapsto g_i(f_l(g_k(a), g_j(x)))$ belong to U and their indicies can be computed in $O(1)$ sequential time.

Moreover in [1] it is shown that the final labels for all internal vertices can be computed within the same time and processor bound.

T–ATC. *Top–down algebraic tree computation* :

The root is labeled by an element of S, the inner vertices and the edges are labeled by elements from U. The T–ATC problem is to compute the final labels for all vertices according to the rule $L(v) := h(g(L(u)))$ where u is the father of v, the edge vu is labeled by g and v is labeled by h.

In [1] it is shown that if the T–ATC problem is decomposable it can be solved on a EREW-PRAM in time $O(\log n)$ using $O(n/\log n)$ processors via the update rule illustrated in Figure 2.

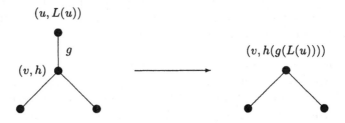

Fig. 2. The update rule for the T–ATC problem.

Similar to the B–ATC problem a T–ATC problem is *decomposable* iff the following two conditions are fulfilled :

(T1) The set U is indexed and its elements can be computed in $O(1)$ sequential time.

(T2) For all $g_i, g_j \in U$ the function $g_i \circ g_j$ belong to U and its index can be computed in $O(1)$ sequential time.

Note that if the problems are decomposable then they are solvable in sequential linear time using preorder for T–ATC and postorder for B–ATC.

In [15] we show that for trees several r–domination problems can be formulated as algebraic tree computation problems. Using the parallel tree contraction algorithm of [1] we get

Theorem 12 [15]. *The minimum r–dominating clique problem and the minimum connected r–dominating set problem for trees can be solved in $O(\log |V|)$ time using $O(|V|)$ processors on a CREW-PRAM.*

The r–dominating set problem is more difficult to solve in parallel because the functions involved in the algebraic tree computation are more complex. In sequential a minimum r–dominating set can be computed in linear time (cf. [3]).

Theorem 13 [15]. *On a CREW-PRAM the minimum r–dominating set problem for trees can be solved in $O(\log n \log \log n)$ time using $O(n)$ processors.*

The proof of Theorem 13 generalizes the one of [13] where the authors only consider the case $r = \text{const.}$.

In what follows we present an optimal algorithm computing all r–dominating vertices of a tree. This algorithm avoids the use of the distance matrix for solving the domination problems on homogeneous extension of trees (see Theorem 7).

We define the following parameters for every vertex v of a tree T :

$$l_{in}(v) := \min_{u \in T_v} \{r(u) - d(u, v)\}, \qquad l_{out}(v) := \min_{u \in T \setminus T_v} \{r(u) - d(u, v)\}.$$

Note that any vertex v with l_{in}–value at least zero r–dominates the whole subtree T_v. Moreover, a vertex v r–dominates T if and only if $l_{in}(v) \geq 0$ and $l_{out}(v) \geq 0$.

In order to compute all r–dominating vertices of a tree we compute both the l_{in}– and l_{out}–values.

The algorithm works in two steps. In the first one we compute, from the leaves to the root, the l_{in}–values. In the second step the l_{out}–values are computed from the root to the leaves and the marking of the r–dominating vertices is done.

In the sequel let w be the root of T. Further, let $e(x)$ denote the *eccentricity* of x, i.e. the maximum over all distances $d(x, y)$, $y \in V$.

Algorithm ALLRDV.

Step 1 For $N_{e(w)}(w)$ to $N_0(w)$ do levelwise :
For each vertex v of the current level with children $u_1, \ldots, u_k$ compute

$$l_{in}(v) := \min\{r(v), \min_{i=1,\ldots,k} \{l_{in}(u_i) - 1\}\}.$$

Step 2 Define $l_{out}(w) := l_{in}(w)$.
If $l_{out}(w) \geq 0$ then mark w.
For $N_0(w)$ to $N_{e(w)-1}(w)$ do :
For each vertex v of the current level with children $u_1, \ldots, u_k$ compute

$$l_{out}(u_j) := \min\{r(v) - 1, l_{out}(v) - 1, \min_{i \neq j}\{l_{in}(u_i) - 2\}\}.$$

If $\min\{l_{in}(u_j), l_{out}(u_j)\} \geq 0$ then mark u_j.

Lemma 14. *The above algorithm* **ALLRDV** *computes all r–dominating vertices of a tree and can be implemented to run in linear sequential and logarithmic parallel time using a linear number of processors.*

Proof. The verification of the correctness is straightforward. To check the sequential time bound we have only to consider the computation of the l_{out}–values since there we need some informations of the siblings. So let v be a vertex with children $u_1, \ldots, u_k$. Before computing $l_{out}(u_i)$, $i = 1, \ldots, k$, we determine a vertex u_{i_0} such that $l_{in}(u_{i_0}) = \min\{l_{in}(u_j) : j = 1, \ldots, k\}$.

This can be done in time $O(\deg(v))$. Now we are able to compute $l_{out}(u_j)$ for $j = 1, \ldots, i_0 - 1, i_0 + 1, \ldots, k$ in constant time. To get $l_{out}(u_{i_0})$ compute $\min\{l_{in}(u_j) : j = 1, \ldots, i_0 - 1, i_0 + 1, \ldots, k\}$. Again, this can be done in time $O(\deg(v))$. Thus computing the l_{out}–values of all children of v takes $O(\deg(v))$ steps.

A parallel variant of the above algorithm via algebraic tree computations and the proof of its complexity is contained in [15]. □

Thus, by the proofs of the Theorems 7, 8 and 10 we get the following

Corollary 15. *For homogeneous extensions of trees a minimum r–dominating clique and a minimum connected r–dominating set can be computed*

- *in linear sequential time,*
- *in $O(\log^2 |V|)$ parallel time using $O(|V| + |E|)$ processors.*

3.2   The r–Dominating Set Problem

In this section we want to outline the idea of the linear time algorithm solving the r–dominating set problem on homogeneous extensions of tree — for details we refer to [15].

For every homogeneous set H of G we denote by H^0 the set of vertices of H with r–value zero. By definition H^0 must be included in any r–dominating set of G. Thus, if $H^0 \neq \emptyset$ we may reduce H^0 to a single vertex (which is adjacent to all neighbours of H^0), and at the end of the algorithm we replace this single vertex by the whole set H^0. So we may assume $|H^0| \leq 1$. But now each proper homogeneous set H of G can be r–dominated by at most two vertices (recall $|V(G)| \geq 2$). Hence we transform G into a graph consisting of vertices of the following two types :

1. If there is a vertex v of H which r–dominates H then H is r–dominated by a single vertex. In this case v_H is a usual vertex. We define $r(v_H) := \min\{r(v) : v \in H\}$.
2. In all other cases v_H is a meta–vertex consisting of two nonadjacent inner vertices v_1, v_2 such that :
   If $H^0 \neq \emptyset$ then $r(v_1) := 0$ and $r(v_2) := 1$, otherwise $r(v_1) := r(v_2) := 1$.

So, for homogeneous extensions of trees, we obtained a tree of vertices and meta–vertices for which the r–dominating set problem can be solved in linear time. The idea of the algorithm is similar to the one of [3] for computing r–dominating sets in dually chordal graphs. The algorithm dismantles the leaves of a tree T and updates certain parameters of the fathers of these leaves. The basic rule is to choose vertices for domination closest to the root, i.e. as long as possible we remove leaves without adding these ones to the current dominating set R. The r–value of a leaf x of the current tree T (i.e. the tree consisting of all unprocessed vertices) represents the distance within x and all still undominated vertices of T_x must be dominated in T. So $r(x) = \infty$ says that all processed vertices of the subtree T_x and x itself are r–dominated by the current set R. Additionally we use a function $c : V(G) \to \mathbb{N}$ indicating the minimum distance of a current vertex x to a member of the current r–dominating set R in T_x. So if $c(x) \le r(x)$ then x is already r–dominated by R. If $c(x) > r(x)$ then x is r–dominated by some vertex z in the current tree if its father y is r'–dominated by z where $r'(y) := \min\{r(y), r(x) - 1\}$. This works well for usual vertices as proved in [3].

For meta–vertices we cannot use the above technique. Indeed, it is impossible to decide during the removal of a leaf with a meta–vertex as father whether this leaf has to be added to the r–dominating set or not. To illustrate this problem consider the example in Figure 3. We draw meta–vertices as rectangles and use pairs (x, i) to indicate that vertex x has r–value i. In the left graph $\{c, d\}$ is the unique minimum r–dominating set. On the other hand, in the right one there is no minimum r–dominating set containing c.

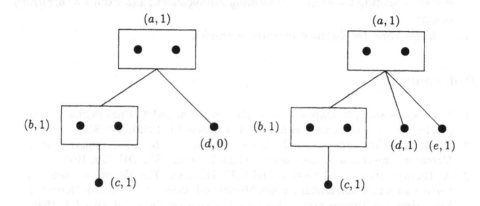

Fig. 3. The problem with meta–vertices.

To avoid this decision we introduce four parameters A_v, B_v, C_v, D_v for meta–vertices v describing all possibilities for local r–domination. Hereby, A_v and B_v contain vertices such that T_v is completely r–dominated by $R \cup A_v$ resp. $R \cup B_v$. The difference between these two sets is that A_v must contain at least one (inner) vertex of v whereas B_v must not contain any vertex of v. On the other hand, the sets C_v and D_v contain vertices such that $T_v \setminus \{v\}$ is r–dominated but not v itself. Note that these sets contain only already processed vertices, i.e. vertices from T_v.

In [15] we describe the update rules for the set parameters of meta–vertices and show that this can be done in linear time. Hence we get

Theorem 16. *The r–dominating set problem can be solved in linear time for homogeneous extensions of trees.*

4 Conclusions

In [16] we extend the technique of meta–vertices to a general reduction theorem for homogeneous sets yielding a polynomial time algorithm solving the r–dominating set problem on graphs which can be recursively generated from a single vertex by adding leaves and performing homogeneous extensions of arbitrary graphs. For distance–hereditary graphs, a proper subclass of these graphs, we even get a linear time algorithm.

So, as a summary we have

Theorem 17 [15, 16]. *The r–dominating set problem can be solved*

1. *in linear time for homogeneous extensions of trees,*
2. *in time $O(|V||E|)$ for graphs which can be recursively generated from a single vertex by adding leaves and performing homogeneous extensions of arbitrary graphs,*
3. *in linear time for distance–hereditary graphs.*

References

1. K. ABRAHAMSON, N. DADOUN, D.G. KIRKPATRICK and T. PRZYTYCKA, A simple parallel tree contraction algorithm, *J. Algorithms* 10 (1989), 287–302.
2. A. BRANDSTÄDT, Special graph classes — a survey, *Technical Report* Gerhard–Mercator–Universität — Gesamthochschule Duisburg SM–DU–199, 1991.
3. A. BRANDSTÄDT, V.D. CHEPOI and F.F. DRAGAN, The algorithmic use of hypertree structure and maximum neighbourhood orderings, *Technical Report* Gerhard–Mercator–Universität — Gesamthochschule Duisburg SM–DU–244, 1994.
4. A. BRANDSTÄDT, F.F. DRAGAN, V.D. CHEPOI and V.I. VOLOSHIN, Dually chordal graphs, *Technical Report* Gerhard–Mercator–Universität — Gesamthochschule Duisburg SM–DU–225, 1993; extended abstract in *Proceedings* of WG'93, *Springer, Lecture Notes in Computer Science* 790, 237–251.
5. G.J. CHANG and G.L. NEMHAUSER, The k–domination and k–stability problems on sun–free chordal graphs, *SIAM J. Algebraic and Discrete Methods*, 5 (1984), 332–345.
6. A. COURNIER and M. HABIB, A simple linear algorithm to build modular decomposition trees, *Technical Report* LIRMM 94-063, Montpellier 1994.
7. E. DAHLHAUS, Efficient parallel modular decomposition, *Proceedings* of WG'95, *Springer, Lecture Notes in Computer Science* 1017, 290–302.
8. A. D'ATRI and M. MOSCARINI, Distance–hereditary graphs, Steiner trees and connected domination, *SIAM J. Comp.* 17 (1988), 521–538.
9. A. D'ATRI, M. MOSCARINI and A. SASSANO, The Steiner tree problem and homogeneous sets, *Proceedings* of MFCS'88, *Springer, Lecture Notes in Computer Science* 324, 249–261.
10. F.F. DRAGAN and F. NICOLAI, r–dominating problems in homogeneously orderable graphs, *Technical Report* Gerhard–Mercator–Universität — Gesamthochschule Duisburg SM–DU–275, 1995, extended abstract in *Proceedings* of FCT'95, *Springer, Lecture Notes in Computer Science* 965, 201–210.

11. A. GIBBONS and W. RYTTER, Efficient parallel algorithms, *Cambridge University Press*, 1988.
12. X. HE, Efficient parallel algorithms for solving some tree problems, *in '24th Allerton Conference on Communication, Control and Computing'*, 1986, 777–786.
13. Y. HE and Y. YESHA, Efficient parallel algorithms for r–dominating set and p–center problems on trees, *Algorithmica* (1990), 129–145.
14. R. MCCONNELL and J. SPINRAD, Linear–Time Modular Decomposition and Efficient Transitive Orientation of Comparability Graphs, *Fifth Annual ACM–SIAM Symposium of Discrete Algorithms* (1994), 536–545.
15. F. NICOLAI and T. SZYMCZAK, r–Domination problems on trees and their homogeneous extensions, *Technical Report* Gerhard–Mercator–Universität — Gesamthochschule Duisburg SM–DU–309, 1995.
16. F. NICOLAI and T. SZYMCZAK, Homogeneous Sets and Domination — A Linear Time Algorithm for Distance–Hereditary Graphs, *Manuscript* 1996.

Independent Spanning Trees of Product Graphs

Koji Obokata, Yukihiro Iwasaki, Feng Bao, and Yoshihide Igarashi

Department of Computer Science
Gunma University, Kiryu, 376 Japan

Abstract. A graph G is called an n-channel graph at vertex r if there are n independent spanning trees rooted at r. A graph G is called an n-channel graph if for every vertex u, G is an n-channel graph at u. Independent spanning trees of a graph play an important role in fault-tolerant broadcasting in the graph. In this paper we show that if G_1 is an n_1-channel graph and G_2 is an n_2-channel graph, then $G_1 \times G_2$ is an $(n_1 + n_2)$-channel graph. We prove this fact by a construction of $n_1 + n_2$ independent spanning trees of $G_1 \times G_2$ from n_1 independent spanning trees of G_1 and n_2 independent spanning trees of G_2.

1 Introduction

For a pair of graphs $G_1 = (V_1, E_1)$ and $G_2 = (V_2, E_2)$, the product of G_1 and G_2, denoted by $G_1 \times G_2$, is a graph with the vertex set $V_1 \times V_2 = \{(x, y) \mid x \in V_1, y \in V_2\}$ and the edge set such that two vertices (u_1, u_2) and (v_1, v_2) are adjacent in $G_1 \times G_2$ if and only if either $u_1 = v_1$ and $u_2 v_2 \in E_2$, or $u_2 = v_2$ and $u_1 v_1 \in E_1$. The definition of the product of two graphs can be generalized to the product of n graphs in the natural way. $G_1 \times G_2 \times G_3$ is $(G_1 \times G_2) \times G_3$ or $G_1 \times (G_2 \times G_3)$. Note that $(G_1 \times G_2) \times G_3$ and $G_1 \times (G_2 \times G_3)$ are isomorphic. The product of n graphs $G_1 \times G_2 \times \cdots \times G_n$ is $(G_1 \times \cdots \times G_k) \times (G_{k+1} \times \cdots \times G_n)$ for some k $(1 \leq k \leq n - 1)$, where each G_i $(1 \leq i \leq n)$ is called a component of $G_1 \times G_2 \times \cdots \times G_n$. To avoid undesirable cases, throughout this paper we assume that each component of a product graph has at least two vertices.

Some of popular interconnection networks are product graphs. For example, the n-dimensional hypercube Q_n is $Q_{n-1} \times K_2 = Q_{n-2} \times K_2 \times K_2 = \cdots = K_2 \times K_2 \times \cdots \times K_2$, and an n-dimensional generalized hypercube Q_n^t is $Q_{n-1}^t \times K_t = Q_{n-2}^t \times K_t \times K_t = \cdots = K_t \times K_t \times \cdots \times K_t$, where K_t is the complete graph of order t. The $(m_1 \times \cdots \times m_n)$-mesh is $L_{m_1} \times \cdots \times L_{m_n}$, and the $(m_1 \times \cdots \times m_n)$-torus is $R_{m_1} \times \cdots \times R_{m_n}$, where L_i and R_i are a linearly linked graph of order i and a ring of order i, respectively. The hyper de Bruijn graph $HD(m, n)$ is $Q_m \times D_n$, and the hyper Petersen graph HP_n is $Q_{n-3} \times P$, where D_n and P are the binary de Bruijn graph of order 2^n and the Petersen graph, respectively. Denote the vertex connectivity of a graph G by $\kappa(G)$. Youssef [6] showed that for a pair of graphs G_1 and G_2, $\kappa(G_1 \times G_2) = \kappa(G_1) + \kappa(G_2)$.

A set of paths connecting a pair of vertices in a graph are said to be internally disjoint if and only if any pair of paths of the set have no common vertices and no common edges except for their extreme vertices. Two spanning trees of a graph $G = (V, E)$ are said to be independent if they are rooted at the same vertex, say r, and for each vertex v in V, the two paths from r to v, one path in each tree, are internally disjoint. A set of spanning trees of G are said to be independent if they are pairwise independent. A graph G is called an n-channel graph at vertex r, if there are n independent spanning trees rooted at r of G. If G is an n-channel graph at every vertex, G is called an n-channel graph. For example, $R_3 \times R_3$ is a 4-channel graph, and 4 independent spanning trees rooted at vertex r are shown in Figure 1.

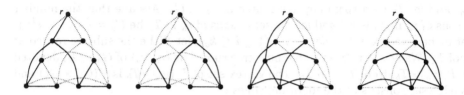

Fig. 1. 4 Independent spanning trees of $R_3 \times R_3$.

Itai and Rodeh [4] gave a linear time algorithm for finding two independent spanning trees in a biconnected graph. Cheriyan and Maheshwari [3] showed how to find three independent spanning trees of $G = (V, E)$ in $O(|V||E|)$ time. Zehavi and Itai [7] also showed that for any 3-connected graph G and any vertex r there are three independent spanning trees rooted at r. They conjectured [5][7] that any κ-vertex connected graph has κ independent spanning trees rooted at an arbitrary vertex r. This conjecture is still open for any $\kappa > 3$.

It has been shown that broadcasting through independent spanning trees are efficient and reliable [1][2][4]. In fact, if G is an n-channel graph and the source vertex is not faulty, then there exists a broadcasting scheme that tolerates up to $n - 1$ faults of the crash type and up to $\lfloor (n - 1)/2 \rfloor$ faults of the Byzantine type even in the worst case. All transmissions by such a broadcasting scheme contribute to the majority voting to obtain the correct message, and its communication complexity is optimal to tolerate up to $\lfloor (n - 1)/2 \rfloor$ faults of the Byzantine type [1][2].

In general it is very hard to construct n independent spanning trees rooted at the same vertex of a given n-connected graph. In this paper we focus attention on the construction of independent spanning trees of a given product graph. We show that if G_1 is an n_1-channel graph and G_2 is an n_2-channel graph, then $G_1 \times G_2$ is an $(n_1 + n_2)$-channel graph. The proof of this fact is by a construction of $n_1 + n_2$ independent spanning trees of $G_1 \times G_2$ from n_1 independent spanning trees of G_1 and n_2 independent spanning trees of G_2. This construction is not straightforward. We use some sophisticated modifications of the independent spanning trees of the component graphs. From our construction we can say that if for each component graph G_i $(1 \le i \le n)$, the vertex connectivity of G_i coincides with the number of independent spanning trees rooted at the same vertex of G_i, then the vertex connectivity and the number of independent spanning trees rooted at the same vertex of $G_1 \times \cdots \times G_n$ coincide.

2 Spanning Trees of Product Graphs

We first define an operation "$*$" on spanning trees. The set of vertices and the set of edges of a graph G are denoted by $V(G)$ and $E(G)$, respectively. The cardinality of a set α is denoted by $|\alpha|$. Let G_a and G_b be two graphs, r_a be a vertex of G_a, and r_b be a vertex of G_b. Let T_a be a spanning tree rooted at r_a of G_a, and let T_b be a spanning tree rooted at r_b of G_b. Assume that the number of sons of r_a in T_a is k_a, and let the set of sons of r_a in T_a be $C_a = \{s_a^1, \cdots, s_a^{k_a}\}$. Let v_a be a vertex in C_a. For each i ($1 \le i \le k_a$), let S_a^i be the subtree rooted at s_a^i of T_a. We now construct a spanning tree rooted at (r_a, r_b) of $G_a \times G_b$, denoted by $T_a(v_a) * T_b$, from T_a and T_b. The vertex set of $T_a(v_a) * T_b$ is $V(G_a \times G_b)$ and its edge set consists of the following edges :

(1) For each u in C_a, connect (r_a, r_b) with (u, r_b).
(2) For each $y_1 y_2 \in E(T_b)$, if $u \in C_a$ then connect (u, y_1) with (u, y_2).
(3) For each i ($1 \le i \le k_a$), if $x_1 x_2 \in E(S_a^i)$ and $y \in V(G_b)$ then connect (x_1, y) with (x_2, y).
(4) For each $y \in V(G_b) - \{r_b\}$, connect (r_a, y) with (v_a, y).

Assume that the number of sons of r_b in T_b is k_b, and let the set of sons of r_b in T_b be $C_b = \{s_b^1, \cdots, s_b^{k_b}\}$. Let v_b be a vertex in C_b. For each i ($1 \le i \le k_b$), let S_b^i be the subtree rooted at s_b^i of T_b. Symmetrically we can construct $T_a * T_b(v_b)$. The vertex set of $T_a * T_b(v_b)$ is $V(G_a \times G_b)$ and its edge set consists of the following edges :

(1) For each u in C_b, connect (r_a, r_b) with (r_a, u).
(2) For each $x_1 x_2 \in E(T_a)$, if $u \in C_b$ then connect (x_1, u) with (x_2, u).
(3) For each i ($1 \le i \le k_b$), if $y_1 y_2 \in E(S_b^i)$ and $x \in V(G_a)$ then connect (x, y_1) with (x, y_2).
(4) For each $x \in V(G_a) - \{r_a\}$, connect (x, r_b) with (x, v_b).

Examples of $T_a(v_a) * T_b$ and $T_a * T_b(v_b)$ are shown in Figure 2.

To specify a path in $G_a \times G_b$ we use the following notations. If $x_1 x_2$ is an edge of a subgraph T of G_a, then path with length 1 from (x_1, y) to (x_2, y) is denoted by $(x_1, y) \xrightarrow{T} (x_2, y)$. The reflexive and transitive closure of $\xrightarrow{T}$ is denoted by $\xRightarrow{}$. Similarly, if $y_1 y_2$ is an edge of a subgraph T' of G_b, then path with length 1 from (x, y_1) to (x, y_2) is denoted by $(x, y_1) \xrightarrow{T'} (x, y_2)$. The reflexive and transitive closure of $\xrightarrow{T'}$ is denoted by $\xRightarrow{T'}$.

Lemma 1. *Let T_a and T_b be a spanning tree rooted at r_a of G_a and a spanning tree rooted at r_b of G_b, respectively, and let v_a and v_b be a son of r_a in T_a and a son of r_b in T_b, respectively. Then each of $T_a(v_a) * T_b$ and $T_a * T_b(v_b)$ is a spanning tree rooted at (r_a, r_b) of $G_a \times G_b$.*

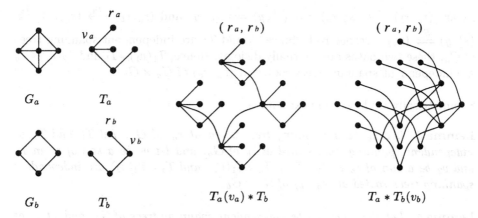

Fig. 2. Examples of $T_a(v_a) * T_b$ and $T_a * T_b(v_b)$

Proof. We consider $T_a(v_a) * T_b$. The sets of edges constructed by rule (1), rule (2), rule (3), and rule (4) have no common edges, and the numbers of edges constructed by these rules are $|C_a|$, $|C_a||E(T_b)|$, $(|E(T_a)| - |C_a|)|V(G_b)|$, and $|V(G_b)| - 1$, respectively. Hence, the number of edges of $T_a(v_a) * T_b$ is $|C_a|(1 + |E(T_b)| - |V(G_b)|) + (|E(T_a)| + 1)|V(G_b)| - 1 = |V(G_a)||V(G_b)| - 1$. Let (x, y) be a vertex of $G_a \times G_b$. If $y = r_b$, the path from (r_a, r_b) to (x, y) is $(r_a, r_b) \xrightarrow{T_a} (x, r_b)$. If $x = r_a$ and $y \neq r_b$, it is $(r_a, r_b) \xrightarrow{T_a} (v_a, r_b) \xRightarrow{T_b} (v_a, y) \xrightarrow{T_a} (r_a, y)$. Suppose that $x \neq r_a$, $y \neq r_b$ and $x \in V(S_a^i)$. Then it is $(r_a, r_b) \xrightarrow{T_a} (s_a^i, r_b) \xRightarrow{T_b} (s_a^i, y) \xRightarrow{S_a^i} (x, y)$. Hence, for an arbitrary (x, y) in $V(G_a) \times V(G_b)$, there is a path from (r_a, r_b) to (x, y) in $T_a(v_a) * T_b$. Therefore, $T_a(v_a) * T_b$ is a spanning tree of $G_a \times G_b$. Symmetrically we can prove that $T_a * T_b(v_b)$ is also a spanning tree of $G_a \times G_b$. □

Lemma 2. *Let T_1 and T_2 be independent spanning trees rooted at r_a of G_a, and let v_1 be a son of r_a in T_1 and v_2 be a son of r_a in T_2. Let T_b be a spanning tree rooted at r_b of G_b. Then $T_1(v_1) * T_b$ and $T_2(v_2) * T_b$ are independent spanning trees rooted at (r_a, r_b) of $G_a \times G_b$.*

Proof. Let us consider the paths from (r_a, r_b) to (x, y) in $T_1(v_1) * T_b$ and $T_2(v_2) * T_b$. If $x = r_a$ and $y \neq r_b$, these two paths are $(r_a, r_b) \xrightarrow{T_1} (v_1, r_b) \xRightarrow{T_b} (v_1, y) \xrightarrow{T_1} (r_a, y)$ and $(r_a, r_b) \xrightarrow{T_2} (v_2, r_b) \xRightarrow{T_b} (v_2, y) \xrightarrow{T_2} (r_a, y)$. Since T_1 and T_2 are independent spanning trees of G_a, $v_1 \neq v_2$. Hence, these two paths are internally disjoint. If $x \neq r_a$ and $y = r_b$, these two paths are $(r_a, r_b) \xRightarrow{T_1} (x, r_b)$ and $(r_a, r_b) \xRightarrow{T_2} (x, r_b)$. Since T_1 and T_2 are independent spanning trees of G_a, these two paths are internally disjoint. Let $x \neq r_a$ and $y \neq r_b$. Suppose that x is in the subtree S_a^i rooted at s_a^i, a son of r_a in T_1, and in the subtree S_a^j rooted s_a^j, a son of r_a in T_2. Then paths from (r_a, r_b) to (x, y) in $T_1(v_1) * T_b$ and $T_2(v_2) * T_b$ are $(r_a, r_b) \xrightarrow{T_1} (s_a^i, r_b) \xrightarrow{T_b} (s_a^i, y) \xrightarrow{S_a^i} (x, y)$ and $(r_a, r_b) \xrightarrow{T_2} (s_a^j, r_b) \xrightarrow{T_b} (s_a^j, y) \xrightarrow{S_a^j} (x, y)$, respectively. Since T_1 and T_2 are independent spanning trees of G_a, these two paths are internally disjoint. Hence, $T_1(v_1) * T_b$ and $T_2(v_2) * T_b$ are independent spanning trees rooted at (r_a, r_b) of $G_a \times G_b$. □

Symmetrically we have the next lemma.

Lemma 3. *Let T_a be a spanning tree rooted at r_a of G_a. Let T_1 and T_2 be independent spanning trees rooted at r_b of G_b, and let v_1 be a son of r_b in T_1 and v_2 be a son of r_b in T_2. Then $T_a * T_1(v_1)$ and $T_a * T_2(v_2)$ are independent spanning trees rooted at (r_a, r_b) of $G_a \times G_b$.*

Lemma 4. *Let $T_{a,1}$ and $T_{a,2}$ be independent spanning trees of G_a, and let v_a be a son of r_a in $T_{a,1}$. Let $T_{b,1}$ and $T_{b,2}$ be independent spanning trees of G_b, and let v_b be a son of r_b in $T_{b,1}$. Then $T_{a,1}(v_a) * T_{b,2}$ and $T_{a,2} * T_{b,1}(v_b)$ are independent spanning trees rooted at (r_a, r_b) of $G_a \times G_b$.*

Proof. Consider the paths from (r_a, r_b) to (x, y) in $T_{a,1}(v_a) * T_{b,2}$ and in $T_{a,2} * T_{b,1}(v_b)$. If $x = r_a$ and $y \neq r_b$, then these paths are $(r_a, r_b) \xrightarrow{T_{a,1}} (v_a, r_b) \xrightarrow{T_{b,2}} (v_a, y) \xrightarrow{T_{a,1}} (r_a, y)$ and $(r_a, r_b) \xrightarrow{T_{b,1}} (r_a, y)$, and they are internally disjoint. If $x \neq r_a$ and $y = r_b$, then these paths are $(r_a, r_b) \xrightarrow{T_{a,1}} (x, r_b)$ and $(r_a, r_b) \xrightarrow{T_{b,1}} (r_a, v_b) \xrightarrow{T_{a,2}} (x, v_b) \xrightarrow{T_{b,1}} (x, r_b)$, and they are internally disjoint. Let $x \neq r_a$ and $y \neq r_b$. Suppose that x is in the subtree rooted at s_a^i, a son of r_a in $T_{a,1}$ and y is in the subtree rooted at s_b^j, a son of r_b in $T_{b,1}$. Then the paths from (r_a, r_b) to (x, y) in $T_{a,1}(v_a) * T_{b,2}$ and in $T_{a,2} * T_{b,1}(v_b)$ are $(r_a, r_b) \xrightarrow{T_{a,1}} (s_a^i, r_b) \xrightarrow{T_{b,2}} (s_a^i, y) \xrightarrow{T_{a,1}} (x, y)$ and $(r_a, r_b) \xrightarrow{T_{b,1}} (r_a, s_b^j) \xrightarrow{T_{a,2}} (x, s_b^j) \xrightarrow{T_{b,1}} (x, y)$, respectively. Since $T_{a,1}$ and $T_{a,2}$ are independent spanning trees of G_a, and $T_{b,1}$ and $T_{b,2}$ are independent spanning trees of G_b, these two paths are internally disjoint. □

Suppose that G_a is an n_a-channel graph and G_b is an n_b-channel graph. Let $T_{a,1}, \cdots, T_{a,n_a}$ be n_a independent spanning trees rooted at r_a of G_a, and let $T_{b,1}, \cdots, T_{b,n_b}$ be n_b independent spanning trees rooted at r_b of G_b. For each i $(1 \leq i \leq n_a)$, let $v_{a,i}$ be a son of r_a in $T_{a,i}$, and for each j $(1 \leq j \leq n_b)$, let $v_{b,j}$ be a son of r_b in $T_{b,j}$. We now consider the following $n_a + n_b$ spanning trees:

$$T_{a,1}(v_{a,1}) * T_{b,1}, \; T_{a,2}(v_{a,2}) * T_{b,1}, \; \cdots, \; T_{a,n_a}(v_{a,n_a}) * T_{b,1},$$

$$T_{a,1} * T_{b,1}(v_{b,1}), \; T_{a,1} * T_{b,2}(v_{b,2}), \; \cdots, \; T_{a,1} * T_{b,n_b}(v_{b,n_b}).$$

From Lemma 2, Lemma 3 and Lemma 4, the $n_a + n_b - 2$ trees listed above excepting $T_{a,1}(v_{a,1}) * T_{b,1}$ and $T_{a,1} * T_{b,1}(v_{b,1})$ are independent spanning trees rooted at (r_a, r_b) of $G_a \times G_b$. Hence, the following proposition is immediate.

Proposition 5. *If G_a is an n_a-channel graph and G_b is an n_b-channel graph, then there are at least $n_a + n_b - 2$ independent spanning trees rooted at the same vertex of $G_a \times G_b$.*

The goal of this paper is to construct $n_a + n_b$ independent spanning trees rooted at the same vertex of the product graph of an n_a-channel graph and an n_b-channel graph. For this purpose we introduce another operation, denoted by "∘", on spanning trees. Let T_a be a spanning tree rooted at r_a of G_a, and let T_b be a spanning tree rooted at r_b of G_b. Let v_a be a son of r_a in T_a. The vertex set of $T_a(v_a) \circ T_b$ is $V(G_a \times G_b)$, and its edge set consists of the following edges:

(1) For each $x_1 x_2 \in E(T_a)$, connect (x_1, r_b) with (x_2, r_b).
(2) For each $y_1 y_2 \in E(T_b)$, if $x \in V(T_a) - \{r_a\}$ then connect (x, y_1) with (x, y_2).
(3) For each $y \in V(T_b) - \{r_b\}$, connect (r_a, y) with (v_a, y).

Let v_b be a son of r_b in T_b. The vertex set of $T_a \circ T_b(v_b)$ is $V(G_a \times G_b)$, and its edge set consists of the following edges:

(1) For each $y_1 y_2 \in E(T_b)$, connect (r_a, y_1) with (r_a, y_2).
(2) For each $x_1 x_2 \in E(T_a)$, if $y \in V(T_b) - \{r_b\}$ then connect (x_1, y) with (x_2, y).
(3) For each $x \in V(T_a) - \{r_a\}$, connect (x, r_b) with (x, v_b).

Examples of $T_a(v_a) \circ T_b$ and $T_a \circ T_b(v_b)$ are shown in Figure 3, where T_a, T_b, v_a, and v_b are given in Figure 2.

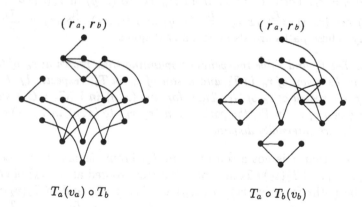

$$(r_a, r_b) \qquad\qquad (r_a, r_b)$$

$$T_a(v_a) \circ T_b \qquad\qquad T_a \circ T_b(v_b)$$

Fig. 3. Examples of $T_a(v_a) \circ T_b$ and $T_a \circ T_b(v_b)$

Lemma 6. *Let T_a and T_b be a spanning tree rooted at r_a of G_a and a spanning tree rooted at r_b of G_b, respectively, and let v_a and v_b be a son of r_a in T_a and a son of r_b in T_b, respectively. Then each of $T_a(v_a) \circ T_b$ and $T_a \circ T_b(v_b)$ is a spanning tree rooted at (r_a, r_b) of $G_a \times G_b$.*

Proof. We consider $T_a(v_a) \circ T_b$. The sets of edges constructed by rule (1), rule (2) and rule (3) have no common edges, and the numbers of edges constructed by these rules are $|E(T_a)|$, $(|V(G_a)|-1)|E(T_b)|$, and $|V(G_b)|-1$, respectively. Hence, the number of edges of $T_a(v_a) \circ T_b$ is $|E(T_a)|+(|V(G_a)|-1)|E(T_b)|+|V(G_b)|-1 = |V(G_a)||V(G_b)| - 1$. Let (x,y) be a vertex of $G_a \times G_b$. If $y = r_b$ then the path from (r_a, r_b) to (x,y) is $(r_a, r_b) \overset{T_a}{\Longrightarrow} (x, r_b)$. If $x = r_a$ and $y \neq r_b$, then it is $(r_a, r_b) \overset{T_a}{\longrightarrow} (v_a, r_b) \overset{T_b}{\Longrightarrow} (v_a, y) \overset{T_a}{\longrightarrow} (r_a, y)$. Suppose that $x \neq r_a$ and $y \neq r_b$. Then it is $(r_a, r_b) \overset{T_a}{\Longrightarrow} (x, r_b) \overset{T_b}{\Longrightarrow} (x, y)$. Hence, there is a path from (r_a, r_b) to any vertex (x,y) of $V(G_a \times G_b)$ in $T_a(v_a) \circ T_b$. Therefore, $T_a(v_a) \circ T_b$ is a spanning tree of $G_a \times G_b$. Symmetrically we can prove that $T_a \circ T_b(v_b)$ is also a spanning tree of $G_a \times G_b$. $\square$

Lemma 7. *Let T_a and T_b be a spanning tree rooted at r_a of G_a and a spanning tree rooted at r_b of G_b, respectively. Let v_a and v_b be a son of r_a in T_a and a son of r_b in T_b, respectively. Then $T_a(v_a) \circ T_b$ and $T_a \circ T_b(v_b)$ are independent spanning trees rooted at (r_a, r_b) of $G_a \times G_b$.*

Proof. From Lemma 6 both $T_a(v_a) \circ T_b$ and $T_a \circ T_b(v_b)$ are spanning trees rooted at (r_a, r_b) of $G_a \times G_b$. Consider the paths from (r_a, r_b) to (x,y) in $T_a(v_a) \circ T_b$ and in $T_a \circ T_b(v_b)$. If $x = r_a$ and $y \neq r_b$, then these paths are $(r_a, r_b) \overset{T_a}{\longrightarrow} (v_a, r_b) \overset{T_b}{\Longrightarrow} (v_a, y) \overset{T_a}{\longrightarrow} (r_a, y)$ and $(r_a, r_b) \overset{T_b}{\Longrightarrow} (r_a, y)$, and internally disjoint. Symmetrically, if $x \neq r_a$ and $y = r_b$, then these paths are $(r_a, r_b) \overset{T_a}{\Longrightarrow} (x, r_b)$ and $(r_a, r_b) \overset{T_b}{\longrightarrow} (r_a, v_b) \overset{T_a}{\longrightarrow} (x, v_b) \overset{T_b}{\longrightarrow} (x, r_b)$, and internally disjoint. Suppose that $x \neq r_a$ and $y \neq r_b$. Then the paths from (r_a, r_b) to (x,y) in $T_a(v_a) \circ T_b$ and in $T_a \circ T_b(v_b)$ are $(r_a, r_b) \overset{T_a}{\Longrightarrow} (x, r_b) \overset{T_b}{\Longrightarrow} (x, y)$ and $(r_a, r_b) \overset{T_b}{\Longrightarrow} (r_a, y) \overset{T_a}{\Longrightarrow} (x, y)$, respectively. These paths are also internally disjoint. $\square$

Lemma 8. *Let T_1 and T_2 be independent spanning trees rooted at r_a of G_a, and let v_1 and v_2 be a son of r_a in T_1 and a son of r_a in T_2, respectively. Let T_b be a spanning tree rooted at r_b of G_b. Then for any (x,y) in $V(G_a \times G_b)$ such that x is not any son of r_a in T_2, the paths from (r_a, r_b) to (x,y) in $T_1(v_1) \circ T_b$ and in $T_2(v_2) * T_b$ are internally disjoint.*

Proof. Assume that x is not a son of r_a in T_2. From Lemma 1 and Lemma 6 both $T_1(v_1) \circ T_b$ and $T_2(v_2) * T_b$ are spanning trees rooted at (r_a, r_b) of $G_a \times G_b$. Consider the paths from (r_a, r_b) to (x,y) in $T_1(v_1) \circ T_b$ and in $T_2(v_2) * T_b$. If $x = r_a$ and $y \neq r_b$, these paths are $(r_a, r_b) \overset{T_1}{\longrightarrow} (v_1, r_b) \overset{T_b}{\Longrightarrow} (v_1, y) \overset{T_1}{\longrightarrow} (r_a, y)$ and $(r_a, r_b) \overset{T_2}{\longrightarrow} (v_2, r_b) \overset{T_b}{\Longrightarrow} (v_2, y) \overset{T_2}{\longrightarrow} (r_a, y)$. Since T_1 and T_2 are independent spanning trees of G_a, $v_1 \neq v_2$ and these paths are internally disjoint. If $x \neq r_a$ and $y = r_b$, these paths are $(r_a, r_b) \overset{T_1}{\Longrightarrow} (x, r_b)$ and $(r_a, r_b) \overset{T_2}{\Longrightarrow} (x, r_b)$. Since T_1 and T_2 are independent spanning trees of G_a, they are also internally disjoint. Suppose that $x \neq r_a$, $y \neq r_b$, and $x \in V(S_2^i)$, where S_2^i is the subtree rooted at s_2^i, one of the sons of r_a in T_2. Then the paths from (r_a, r_b) to (x,y) in $T_1(v_1) \circ T_b$ and in $T_2(v_2) * T_b$ are $(r_a, r_b) \overset{T_1}{\Longrightarrow} (x, r_b) \overset{T_b}{\Longrightarrow} (x, y)$ and $(r_a, r_b) \overset{T_2}{\longrightarrow} (s_2^i, r_b) \overset{T_b}{\Longrightarrow}$

$(s_2^i, y) \xRightarrow{T_2} (x, y)$, respectively. Since T_1 and T_2 are independent spanning trees of G_a and x cannot be s_2^i from the assumption, they are internally disjoint. If $x \neq r_a$, $y \neq r_b$, and the condition for x in Lemma 8 is not satisfied, the paths from (r_a, r_b) to (x, y) in $T_1(v_1) \circ T_b$ and in $T_2(v_2) * T_b$ are not internally disjoint.

$\square$

Symmetrically we have the next lemma.

Lemma 9. *Let T_a be a spanning tree rooted at r_a of G_a. Let T_1 and T_2 be independent spanning trees rooted at r_b of G_b, and let v_1 and v_2 be a son of r_b in T_1 and a son of r_b in T_2, respectively. Then for any (x, y) in $V(G_a \times G_b)$ such that y is not any son of r_b in T_2, the paths from (r_a, r_b) to (x, y) in $T_a \circ T_1(v_1)$ and $T_a * T_2(v_2)$ are internally disjoint.*

Consider the following $n_a + n_b$ spanning trees rooted at (r_a, r_b) of $G_a \times G_b$:

$$T_{a,1}(v_{a,1}) \circ T_{b,1}, \ T_{a,2}(v_{a,2}) * T_{b,1}, \ \cdots, \ T_{a,n_a}(v_{a,n_a}) * T_{b,1},$$

$$T_{a,1} \circ T_{b,1}(v_{b,1}), \ T_{a,1} * T_{b,2}(v_{b,2}), \ \cdots, \ T_{a,1} * T_{b,n_b}(v_{b,n_b}).$$

These spanning trees are obtained from the $n_a + n_b$ spanning trees listed earlier in this section by replacing $T_{a,1}(v_{a,1}) * T_{b,1}$ and $T_{a,1} * T_{b,1}(v_{b,1})$ with $T_{a,1}(v_{a,1}) \circ T_{b,1}$ and $T_{a,1} \circ T_{b,1}(v_{b,1})$, respectively. In general, for any i $(2 \leq i \leq n_a)$, $T_{a,1}(v_{a,1}) \circ T_{b,1}$ and $T_{a,i}(v_{a,i}) * T_{b,1}$ are not independent. In general for any i $(2 \leq i \leq n_b)$, $T_{a,1} \circ T_{b,1}(v_{b,1})$ and $T_{a,1} * T_{b,i}(v_{b,i})$ are also not independent. Hence, in general the set of spanning trees listed above are not independent spanning trees of $G_a \times G_b$. To construct $n_a + n_b$ independent spanning trees, we need further modifications.

3 Construction of Independent Spanning Trees

Let $T_{a,1}, \cdots, T_{a,n_a}$ be n_a independent spanning trees rooted at r_a of G_a, and let T_b be a spanning tree rooted at r_b of G_b. For each i $(1 \leq i \leq n_a)$, let k_a^i be the number of sons of r_a in $T_{a,i}$, and let $C_{a,i} = \{s_{a,i}^1, \cdots, s_{a,i}^{k_a^i}\}$ be the set of sons of r_a in $T_{a,i}$. Let $C_a = C_{a,1} \cup C_{a,2} \cup \cdots \cup C_{a,n_a}$. For each $v \in V(T_{a,1}) - \{r_a\}$, let $p_{a,1}(v)$ be the parent of v in $T_{a,1}$. The variation of $T_{a,1}$, denoted by $var(T_{a,1})$, is a graph with vertex set $V(T_{a,1})$ and the edge set $E(T_{a,1}) \cup \{r_a x \mid x \in C_a - C_{a,1}\} - \{p_{a,1}(x)x \mid x \in C_a - C_{a,1}\}$. Note that $p_{a,1}(x)x$ is an edge connecting $p_{a,1}(x)$ and x, and that $r_a x$ is an edge connecting r_a and x. Apparently $var(T_{a,1})$ is also a spanning tree rooted at r_a of G_a. Similarly we can define $var(T_{b,1})$. Examples of $var(T_{a,1})$ and $var(T_{b,1})$ are shown in Figure 4.

We now modify $T_{a,1}(v_{a,1}) \circ T_b$, where $v_{a,1}$ is one of the sons of r_a in $T_{a,1}$. For each $x \in C_{a,i}$ $(2 \leq i \leq n_a)$ and each $y_1 y_2 \in E(T_b)$, edge $(x, y_1)(x, y_2)$ of $T_{a,1}(v_{a,1}) \circ T_b$ is also used in $T_{a,i}(v_{a,i}) * T_b$. For each $x \in C_a - C_{a,1}$ and each $y_1 y_2 \in E(T_b)$, we remove edge $(x, y_1)(x, y_2)$ from $T_{a,1}(v_{a,1}) \circ T_b$, and for each $x \in C_a - C_{a,1}$ and each $y \in V(G_b) - \{r_b\}$, we add edge $(p_{a,1}(x), y)(x, y)$. This modification is called the transformation. The graph obtained from $T_{a,1}(v_{a,1}) \circ T_b$ by the

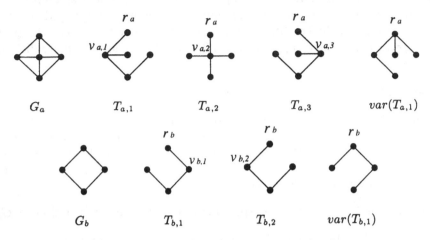

Fig. 4. Examples of $var(T_{a,1})$ and $var(T_{b,1})$

transformation is denoted by $tr(T_{a,1}(v_{a,1}) \circ T_b)$. More formally $tr(T_{a,1}(v_{a,1}) \circ T_b)$ is a graph with the vertex set $V(G_a \times G_b)$ and the following edge set:

$$E(T_{a,1}(v_{a,1}) \circ T_b) \cup \{(p_{a,1}(x), y)(x, y) \mid x \in C_a - C_{a,1}, \; y \in V(G_b) - \{r_b\}\}$$
$$- \{(x, y_1)(x, y_2) \mid x \in C_a - C_{a,1}, \; y_1 y_2 \in E(T_b)\}$$

For each $x \in C_a - C_{a,1}$ and each $y \in V(G_b) - \{r_b\}$, $(p_{a,1}(x), y)(x, y)$ is an edge of $G_a \times G_b$ since $p_{a,1}(x)x$ is an edge of $T_{a,1}$. From the construction of $tr(T_{a,1}(v_{a,1}) \circ T_b)$, it is also a spanning tree rooted at (r_a, r_b) of $G_a \times G_b$.

Let $T_{b,1}, \cdots, T_{b,n_b}$ be n_b independent spanning trees rooted at r_b of G_b, and let T_a be a spanning tree rooted at r_a of G_a. For each i ($1 \le i \le n_b$), let k_b^i be the number of sons of r_b in $T_{b,i}$, and let $C_{b,i} = \{s_{b,i}^1, \cdots s_{b,i}^{k_b^i}\}$ be the set of sons of r_b in $T_{b,i}$. Let $C_b = C_{b,1} \cup C_{b,2} \cup \cdots \cup C_{b,n_b}$. For each $v \in V(T_{b,1}) - \{r_b\}$, let $p_{b,1}(v)$ be the parent of v in $T_{b,1}$. Symmetrically we define $var(T_{b,1})$ and $tr(T_a \circ T_{b,1}(v_{b,1}))$, where $v_{b,1}$ is one of the sons of r_b in $T_{b,1}$. The variation of $T_{b,1}$, denoted by $var(T_{b,1})$, is a graph with the vertex set $V(T_{b,1})$ and the edge set $E(T_{b,1}) \cup \{r_b y \mid y \in C_b - C_{b,1}\} - \{p_{b,1}(y) \mid y \in C_b - C_{b,1}\}$. It is also a spanning tree rooted at r_b of G_b. The transformation of $T_a \circ T_{b,1}(v_{b,1})$, denoted by $tr(T_a \circ T_{b,1}(v_{b,1}))$, is a graph with the vertex set $V(G_a \times G_b)$ and the following edge set:

$$E(T_a \circ T_{b,1}(v_{b,1})) \cup \{(x, p_{b,1}(y))(x, y) \mid x \in V(G_a) - \{r_a\}, \; y \in C_b - C_{b,1}\}$$
$$- \{(x_1, y)(x_2, y) \mid y \in C_b - C_{b,1}, \; x_1 x_2 \in E(T_a)\}$$

The transformation of $T_a \circ T_{b,1}(v_{b,1})$ is also a spanning tree rooted at (r_a, r_b) of $G_a \times G_b$. Examples of $T_{a,1}(v_{a,1}) \circ var(T_{b,1})$ and $tr(T_{a,1}(v_{a,1}) \circ var(T_{b,1}))$ are shown in Figure 5, where $T_{a,1}, T_{b,1}$, and $v_{a,1}$ are given in Figure 4.

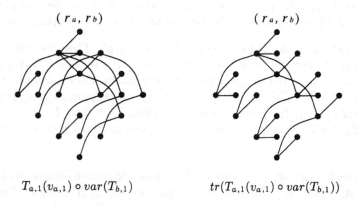

$$T_{a,1}(v_{a,1}) \circ var(T_{b,1}) \qquad\qquad tr(T_{a,1}(v_{a,1}) \circ var(T_{b,1}))$$

Fig. 5. Examples of $T_{a,1}(v_{a,1}) \circ var(T_{b,1})$ and $tr(T_{a,1}(v_{a,1}) \circ var(T_{b,1}))$

Assume that G_a is an n_a-channel graph and G_b is an n_b-channel graph. We now consider the following $n_a + n_b$ graphs:

$$tr(T_{a,1}(v_{a,1}) \circ var(T_{b,1})),\ T_{a,2}(v_{a,2}) * var(T_{b,1}),\ \cdots,\ T_{a,n_a}(v_{a,n_a}) * var(T_{b,1}),$$

$$tr(var(T_{a,1}) \circ T_{b,1}(v_{b,1})),\ var(T_{a,1}) * T_{b,2}(v_{b,2}),\ \cdots,\ var(T_{a,1}) * T_{b,n_b}(v_{b,n_b}),$$

where for each i ($1 \leq i \leq n_a$), $v_{a,i}$ is one of the sons of r_a in $T_{a,i}$, and for each j ($1 \leq j \leq n_b$), $v_{b,j}$ is one of the sons of r_b in $T_{b,j}$. From Lemma 1, Lemma 6 and the discussion above, these graphs are $n_a + n_b$ spanning trees rooted at (r_a, r_b) of $G_a \times G_b$. We show that these $n_a + n_b$ spanning trees are independent spanning trees rooted at (r_a, r_b) of $G_a \times G_b$.

Lemma 10. *For each i ($2 \leq i \leq n_a$) and each j ($2 \leq j \leq n_b$), $T_{a,i}(v_{a,i}) * var(T_{b,1})$ and $var(T_{a,1}) * T_{b,j}(v_{b,j})$ are independent spanning trees rooted at (r_a, r_b) of $G_a \times G_b$.*

Proof. From Lemma 4, $T_{a,i}(v_{a,i}) * T_{b,1}$ and $T_{a,1} * T_{b,j}(v_{b,j})$ are independent spanning trees rooted at (r_a, r_b) of $G_a \times G_b$. Consider the paths from (r_a, r_b) to (x, y) in $T_{a,i}(v_{a,i}) * T_{b,1}$ and in $T_{a,i}(v_{a,i}) * var(T_{b,1})$. If the former path does not contain any node (s, t) such that $t \in C_b - C_{b,1}$, then these two paths are identical. Suppose that the former path contains a node (s, t) such that $t \in C_b - C_{b,1}$. Assume that x is in $S_{a,i}^h$ (i.e., the subtree rooted at $s_{a,i}^h$ of $T_{a,i}$) or $x = r_a$. Then for the case where x is in $S_{a,i}^h$, the former path is

$$(r_a, r_b) \xrightarrow{T_{a,i}} (s_{a,i}^h, r_b) \xRightarrow{T_{b,1}} (s_{a,i}^h, t) \xRightarrow{T_{b,1}} (s_{a,i}^h, y) \xRightarrow{S_{a,i}^h} (x, y).$$ On the other hand, the latter path is $(r_a, r_b) \xrightarrow{T_{a,i}} (s_{a,i}^h, r_b) \xrightarrow{var(T_{b,1})} (s_{a,i}^h, t) \xrightarrow{var(T_{b,1})} (s_{a,i}^h, y) \xRightarrow{S_{a,i}^h} (x, y).$ Since $(s_{a,i}^h, t) \xrightarrow{var(T_{b,1})} (s_{a,i}^h, y)$ and $(s_{a,i}^h, t) \xRightarrow{T_{b,1}} (s_{a,i}^h, y)$ are identical paths, the latter path is obtained from the former path by cutting the subpath $(s_{a,i}^h, r_b) \xRightarrow{T_{b,1}}$

$(s_{a,i}^h, t)$ short to edge $(s_{a,i}^h, r_b)(s_{a,i}^h, t)$. Note that edge $(s_{a,i}^h, r_b)(s_{a,i}^h, t)$ is not used in $var(T_{a,1}) * T_{b,j}(v_{b,j})$. For the case where $x = r_a$, the two paths from (r_a, r_b) to (x, y) are similar to the two paths above if we replace the $s_{a,i}^h$'s by $v_{a,i}$'s and $(s_{a,i}^h, y) \overset{S_{a,i}^h}{\Longrightarrow} (x, y)$ by $(v_{a,i}, y) \overset{T_{a,i}}{\longrightarrow} (r_a, y)$. Then we can apply the same argument. For the paths from (r_a, r_b) to (x, y) in $T_{a,1} * T_{b,j}(v_{b,j})$ and in $var(T_{a,1}) * T_{b,j}(v_{b,j})$, we can apply the same argument. Hence, from Lemma 4 for each $(x, y) \in V(G_a \times G_b)$, the paths from (r_a, r_b) to (x, y) in $T_{a,1}(v_{a,1}) * var(T_{b,1})$ and in $var(T_{a,1}) * T_{b,1}(v_{b,1})$ are internally disjoint. Therefore, $T_{a,i}(v_{a,i}) * var(T_{b,1})$ and $var(T_{a,1}) * T_{b,j}(v_{b,j})$ are independent spanning trees rooted at (r_a, r_b) of $G_a \times G_b$. □

For clear description of the proofs of the following three lemmas we introduce functions $A_{a,1}$ and $A_{b,1}$. Remember that $p_{a,1}(v)$ and $p_{b,1}(v)$ denote the parents of v in $T_{a,1}$ and in $T_{b,1}$, respectively. For each $v \in V(T_{a,1}) - \{r_a\}$, let

$$A_{a,1}(v) = \begin{cases} A_{a,1}(p_{a,1}(v)) & \text{if } v \in C_a - C_{a,1} \\ v & \text{if } v \notin C_a - C_{a,1}. \end{cases}$$

Symmetrically, for each $v \in V(T_{b,1}) - \{r_b\}$, let

$$A_{b,1}(v) = \begin{cases} A_{b,1}(p_{b,1}(v)) & \text{if } v \in C_b - C_{b,1} \\ v & \text{if } v \notin C_b - C_{b,1}. \end{cases}$$

Lemma 11. *A pair of $tr(T_{a,1}(v_{a,1}) \circ var(T_{b,1}))$ and $tr(var(T_{a,1}) \circ T_{b,1}(v_{b,1}))$ are independent spanning trees rooted at (r_a, r_b) of $G_a \times G_b$.*

Proof. From Lemma 6 and a similar argument to the proof of Lemma 7, $T_{a,1}(v_{a,1}) \circ var(T_{b,1})$ and $var(T_{a,1}) \circ T_{b,1}(v_{b,1})$ are independent spanning trees rooted at (r_a, r_b) of $G_a \times G_b$.

Consider the paths from (r_a, r_b) to (x, y) in $tr(T_{a,1}(v_{a,1}) \circ var(T_{b,1}))$ and in $tr(var(T_{a,1}) \circ T_{b,1}(v_{b,1}))$. If $x \notin C_a - C_{a,1}$ and $y \notin C_b - C_{b,1}$, then the paths from (r_a, r_b) to (x, y) in $tr(T_{a,1}(v_{a,1}) \circ var(T_{b,1}))$ and in $T_{a,1}(v_{a,1}) \circ var(T_{b,1})$ are identical, and the paths from (r_a, r_b) to (x, y) in $tr(var(T_{a,1}) \circ T_{b,1}(v_{b,1}))$ and in $var(T_{a,1}) \circ T_{b,1}(v_{b,1})$ are identical. Hence, in this case the paths from (r_a, r_b) to (x, y) in $tr(T_{a,1}(v_{a,1}) \circ var(T_{b,1}))$ and in $tr(var(T_{a,1}) \circ T_{b,1}(v_{b,1}))$ are internally disjoint. If $x \notin C_a - C_{a,1}$ and $y \in C_b - C_{b,1}$, then the path from (r_a, r_b) to (x, y) in $tr(T_{a,1}(v_{a,1}) \circ var(T_{b,1}))$ is $(r_a, r_b) \overset{T_{a,1}}{\Longrightarrow} (x, r_b) \overset{var(T_{b,1})}{\longrightarrow} (x, y)$, and the path from (r_a, r_b) to (x, y) in $tr(var(T_{a,1}) \circ T_{b,1}(v_{b,1}))$ is $(r_a, r_b) \overset{T_{b,1}}{\Longrightarrow} (r_a, A_{b,1}(y)) \overset{var(T_{a,1})}{\Longrightarrow} (x, A_{b,1}(y)) \overset{T_{b,1}}{\Longrightarrow} (x, y)$. These two paths are internally disjoint. Symmetrically, for the case where $x \in C_a - C_{a,1}$ and $y \notin C_b - C_{b,1}$, the paths from (r_a, r_b) to (x, y) in $tr(T_{a,1}(v_{a,1}) \circ var(T_{b,1}))$ and in $tr(var(T_{a,1}) \circ T_{b,1}(v_{b,1}))$ are internally disjoint. If $x \in C_a - C_{a,1}$ and $y \in C_b - C_{b,1}$, then the path from (r_a, r_b) to (x, y) in $tr(T_{a,1}(v_{a,1}) \circ var(T_{b,1}))$ is $(r_a, r_b) \overset{T_{a,1}}{\Longrightarrow} (A_{a,1}(x), r_b) \overset{var(T_{b,1})}{\longrightarrow} (A_{a,1}(x), y) \overset{T_{a,1}}{\Longrightarrow} (x, y)$, and the path from (r_a, r_b) to (x, y) in $tr(var(T_{a,1}) \circ$

$T_{b,1}(v_{b,1}))$ is $(r_a, r_b) \overset{T_{b,1}}{\Longrightarrow} (r_a, A_{b,1}(y)) \overset{var(T_{a,1})}{\longrightarrow} (x, A_{b,1}(y)) \overset{T_{b,1}}{\Longrightarrow} (x, y)$. These paths are also internally disjoint. Therefore, $tr(T_{a,1}(v_{a,1}) \circ var(T_{b,1}))$ and $tr(var(T_{a,1}) \circ T_{b,1}(v_{b,1}))$ are independent spanning trees rooted at (r_a, r_b) of $G_a \times G_b$. □

Lemma 12. *For each* i $(2 \le i \le n_a)$, $tr(T_{a,1}(v_{a,1}) \circ var(T_{b,1}))$ *and* $T_{a,i}(v_{a,i}) * var(T_{b,1})$ *are independent spanning trees rooted at* (r_a, r_b) *of* $G_a \times G_b$.

Proof. Consider the paths from (r_a, r_b) to (x, y) in $tr(T_{a,1}(v_{a,1}) \circ var(T_{b,1}))$ and in $T_{a,i}(v_{a,i}) * var(T_{b,1})$. If $x \ne r_a$ and $y = r_b$, then these paths are $(r_a, r_b) \overset{T_{a,1}}{\Longrightarrow} (x, r_b)$ and $(r_a, r_b) \overset{T_{a,i}}{\Longrightarrow} (x, r_b)$. Since $T_{a,1}$ and $T_{a,i}$ are independent spanning trees of G_a, $v_{a,1} \ne v_{a,i}$ and these paths are internally disjoint. If $x = r_a$ and $y \ne r_b$, then these paths are $(r_a, r_b) \overset{T_{a,1}}{\longrightarrow} (v_{a,1}, r_b) \overset{var(T_{b,1})}{\Longrightarrow} (v_{a,1}, y) \overset{T_{a,1}}{\longrightarrow} (r_a, y)$ and $(r_a, r_b) \overset{T_{a,i}}{\longrightarrow} (v_{a,i}, r_b) \overset{var(T_{b,1})}{\Longrightarrow} (v_{a,i}, y) \overset{T_{a,i}}{\longrightarrow} (r_a, y)$. Since $T_{a,1}$ and $T_{a,i}$ are independent spanning trees of G_a, they are also internally disjoint. If $x \notin C_a \cup \{r_a\} - C_{a,1}$ and $y \ne r_b$ and if $s_{a,1}^j$ be in $C_{a,1}$ such that x is in the subtree rooted at $s_{a,1}^j$ in $T_{a,1}$, then the paths from (r_a, r_b) to (x, y) in $tr(T_{a,1}(v_{a,1}) \circ var(T_{b,1}))$ and in $T_{a,i}(v_{a,i}) * var(T_{b,1})$ are $(r_a, r_b) \overset{T_{a,1}}{\Longrightarrow} (s_{a,1}^j, r_b) \overset{var(T_{b,1})}{\Longrightarrow} (s_{a,1}^j, y) \overset{T_{a,1}}{\Longrightarrow} (x, y)$ and $(r_a, r_b) \overset{T_{a,i}}{\longrightarrow} (v_{a,i}, r_b) \overset{var(T_{b,1})}{\Longrightarrow} (v_{a,i}, y) \overset{T_{a,i}}{\Longrightarrow} (x, y)$. Since $T_{a,1}$ and $T_{a,i}$ are independent spanning trees of G_a, $s_{a,1}^j \ne v_{a,i}$ and they are internally disjoint. Suppose that $x \in C_a - C_{a,1}$, $y \ne r_b$ and x is in subtree $S_{a,i}^h$ of $T_{a,i}$. Then the former path from (r_a, r_b) to (x, y) is $(r_a, r_b) \overset{T_{a,1}}{\Longrightarrow} (A_{a,1}(x), r_b) \overset{var(T_{b,1})}{\Longrightarrow} (A_{a,1}(x), y) \overset{T_{a,1}}{\Longrightarrow} (x, y)$, and the latter path from (r_a, r_b) to (x, y) is $(r_a, r_b) \overset{T_{a,1}}{\longrightarrow} (s_{a,i}^h, r_b) \overset{var(T_{b,1})}{\Longrightarrow} (s_{a,i}^h, y) \overset{S_{a,i}^h}{\Longrightarrow} (x, y)$. Since $T_{a,1}$ and $T_{a,i}$ are independent spanning trees rooted at r_a of G_a, $(A_{a,1}(x), y) \overset{T_{a,1}}{\Longrightarrow} (x, y)$ in $tr(T_{a,1}(v_{a,1}) \circ var(T_{b,1}))$ and $(s_{a,i}^h, y) \overset{S_{a,i}^h}{\Longrightarrow} (x, y)$ have no common vertices excepting (x, y). Hence, in this case the paths from (r_a, r_b) to (x, y) in $tr(T_{a,1}(v_{a,1}) \circ var(T_{b,1}))$ and in $T_{a,i}(v_{a,i}) * var(T_{b,1})$ are internally disjoint. Therefore, a pair of these two trees are independent spanning trees rooted at (r_a, r_b) of $G_a \times G_b$. □

Symmetrically we have the next lemma.

Lemma 13. *For each* j $(2 \le j \le n_b)$, $tr(var(T_{a,1}) \circ T_{b,1}(v_{b,1}))$ *and* $var(T_{a,1}) * T_{b,j}(v_{b,j})$ *are independent spanning trees rooted at* (r_a, r_b) *of* $G_a \times G_b$.

Lemma 14. *For each* i $(2 \le i \le n_a)$, $tr(var(T_{a,1}) \circ T_{b,1}(v_{b,1}))$ *and* $T_{a,i}(v_{a,i}) * var(T_{b,1})$ *are independent spanning trees rooted at* (r_a, r_b) *of* $G_a \times G_b$.

Proof. Consider the paths from (r_a, r_b) to (x, y) in $tr(var(T_{a,1}) \circ T_{b,1}(v_{b,1}))$ and in $T_{a,i}(v_{a,i}) * var(T_{b,1})$. If $x \ne r_a$ and $y = r_b$, then these two paths are $(r_a, r_b) \overset{T_{b,1}}{\longrightarrow} (r_a, v_{b,1}) \overset{var(T_{a,1})}{\Longrightarrow} (x, v_{b,1}) \overset{T_{b,1}}{\longrightarrow} (x, r_b)$ and $(r_a, r_b) \overset{T_{a,i}}{\Longrightarrow} (x, r_b)$. These paths are internally disjoint. If $x = r_a$ and $y \neq r_b$, then these two paths are $(r_a, r_b) \overset{T_{b,1}}{\Longrightarrow} (r_a, y)$ and $(r_a, r_b) \overset{T_{a,i}}{\longrightarrow} (v_{a,i}, r_b) \overset{var(T_{b,1})}{\Longrightarrow} (v_{a,i}, y) \overset{T_{a,i}}{\longrightarrow} (r_a, y)$. These paths are also internally disjoint. Suppose that $x \neq r_a$, $y \notin C_b \cup \{r_b\} - C_{b,1}$, and x is in subtree $S_{a,i}^h$. Then these two paths are $(r_a, r_b) \overset{T_{b,1}}{\Longrightarrow} (r_a, y) \overset{var(T_{a,1})}{\Longrightarrow} (x, y)$ and $(r_a, r_b) \overset{T_{a,i}}{\longrightarrow} (s_{a,i}^h, r_b) \overset{var(T_{b,1})}{\Longrightarrow} (s_{a,i}^h, y) \overset{S_{a,i}^h}{\Longrightarrow} (x, y)$. Since $T_{a,1}$ and $T_{a,i}$ are independent spanning trees rooted at r_a of G_a, $(r_a, y) \overset{var(T_{a,1})}{\Longrightarrow} (x, y)$ and $(s_{a,i}^h, y) \overset{S_{a,i}^h}{\Longrightarrow} (x, y)$ have no common vertices excepting (x, y). Note that $s_{a,i}^h$ does not appear in $var(T_{a,1})$ or $s_{a,i}^h = x$. Hence, in this case the paths from (r_a, r_b) to (x, y) in $tr(var(T_{a,1}) \circ T_{b,1}(v_{b,1}))$ and in $T_{a,i}(v_{a,i}) * var(T_{b,1})$ are internally disjoint. Suppose that $x \neq r_a$, $y \in C_b - C_{b,1}$, and x is in subtree $S_{a,i}^h$, then the path from (r_a, r_b) to (x, y) in $tr(var(T_{a,1}) \circ T_{b,1}(v_{b,1}))$ is $(r_a, r_b) \overset{T_{b,1}}{\Longrightarrow} (r_a, A_{b,1}(y)) \overset{var(T_{a,1})}{\Longrightarrow} (x, A_{b,1}(y)) \overset{T_{b,1}}{\Longrightarrow} (x, y)$, and the path from (r_a, r_b) to (x, y) in $T_{a,i}(v_{a,i}) * var(T_{b,1})$ is $(r_a, r_b) \overset{T_{a,i}}{\longrightarrow} (s_{a,i}^h, r_b) \overset{var(T_{b,1})}{\longrightarrow} (s_{a,i}^h, y) \overset{S_{a,i}^h}{\Longrightarrow} (x, y)$. Note that $y \neq A_{b,1}(y)$. Hence, these two paths are also internally disjoint. Therefore, $tr(var(T_{a,1}) \circ T_{b,1}(v_{b,1}))$ and $T_{a,i}(v_{a,i}) * var(T_{b,1})$ are independent spanning trees rooted at (r_a, r_b) of $G_a \times G_b$. □

Symmetrically we have the next lemma.

Lemma 15. *For each j $(2 \leq j \leq n_b)$, $tr(T_{a,1}(v_{a,1}) \circ var(T_{b,1}))$ and $var(T_{a,1}) * T_{b,j}(v_{b,j})$ are independent spanning trees rooted at (r_a, r_b) of $G_a \times G_b$.*

Theorem 16. *If G_a is an n_a-channel graph and G_b is an n_b-channel graph, then $G_a \times G_b$ is an $(n_a + n_b)$-channel graph.*

Proof. It is immediate from Lemmas $2 - 4$ and Lemmas $10 - 15$. □

4 Concluding Remarks

We have shown a construction of $n_a + n_b$ independent spanning trees of the product graph of an n_a-channel graph G_a and n_b-channel graph G_b from n_a independent spanning trees of G_a and n_b independent spanning trees of G_b. Hence, we can construct $n_1 + n_2 + \cdots + n_m$ independent spanning trees of $G = G_1 \times G_2 \times \cdots \times G_m$ by successively applying the construction if for each i $(1 \leq i \leq m)$, n_i independent spanning trees of G_i are given. A broadcasting scheme sending multiple copies of the message from the source vertex along n independent spanning trees of a graph is efficient. In fact, its communication complexity is optimal for tolerating $\lfloor (n - 1)/2 \rfloor$ Byzantine faults and for tolerating $n - 1$ crash faults.

The following two problems arise from our approach if we consider general graphs.

(1) How can we construct independent spanning trees rooted at the same vertex of an arbitrary graph? This is a very hard problem. It is open whether

every n-connected graph is an n-channel graph. The problem has been solved only for $k \leq 3$ [3][4][7]. Furthermore, even if we know a method of constructing independent spanning trees of some families of graphs, the independent spanning trees obtained by the method may not have good properties. In practice, shallow spanning trees with regular structures are desirable.

(2) How can we design efficient broadcasting protocols, in particular for one-port broadcasting, based on message transmissions through independent spanning trees? Since such a broadcasting scheme consists of sub-broadcasts, each through one of the independent spanning trees, there are few hints how each vertex should use a strategy to achieve short broadcasting time.

These problems would be worthy for further investigation.

References

1. F. Bao, Y. Igarashi, K. Katano : Broadcasting in hypercubes with randomly distributed Byzantine faults. 9th International Workshop on Distributed Algorithms Le Mont-Saint-Michel LNCS **972** (1995) 215-229
2. F. Bao, Y. Igarashi, S. R. Öhring : Reliable broadcasting in product networks. IEICE Technical Report COMP **95-18** (1995) 57-66
3. J. Cheriyan, S. N. Maheshwari : Finding nonseparating induced cycles and independent spanning trees in 3-connected graphs. J. Algorithms **9** (1988) 507-537
4. A. Itai, M. Rodeh : The multi-tree approach to reliability in distributed networks. Information and Computation **79** (1988) 43-59
5. S. Khuller, B. Schieber : On independent spanning trees. Information Processing Letters **42** (1992) 321-323
6. A. Youssef : Cartesian product networks. the 1991 International Conference on Parallel Processing **I** (1991) 684-685
7. A. Zehavi, A. Itai : Three tree-paths. J. Graph Theory **13** (1989) 175-188

Designing Distrance-Preserving Fault-Tolerant Topologies *

Swamy K. Sitarama and Abdol-Hossein Esfahanian

Computer Science Department
Michigan State University
East Lansing, Michigan

Abstract. In this paper we introduce and study a new family of graphs called *distance preserving graphs*. A graph G is said to be *k-edge distance preserving* with respect to a spanning subgraph D, if there exist k edge-disjoint u-v paths in G of length at most $d_D(u,v)$ for every pair of nonadjacent vertices u, v of D. We study two design models each with a different optimality criterion. In the first model, minimizing the overall redundancy is considered. The second model considers regular fault-tolerant topologies with minimum regularity. The focus of this paper is on designs based on model 2. In particular, we construct *distance preserving graphs* with respect to *cycles*, *crowns*. We also prove that our constructions are optimal when $D = C_p$ and $k < \frac{p}{2}$. Further, given a graph G and a spanning subgraph D, we show that there exists a polynomial time algorithm to determine if G is k-edge distance preserving with respect to D. Finally we present a bound on the distance between adjacent vertices when k or fewer edges are removed.

1 Introduction

We study the following graph construction problem. Given a graph D and a nonnegative integer k, construct a graph $G \subseteq D$ in which between every pair of nonadjacent vertices u and v in D, there are at least k edge-disjoint paths, each having length at most $d_D(u,v)$, where $d_H(x,y)$ represents the *distance* (i.e., the number of edges in a shortest path) between vertex x and vertex y in a graph H. *Fault-tolerant* designs involving distance between every pair of nonadjacent nodes were first considered by Entringer *et al.* [7]. Preserving distance is particularly applicable in *communication networks* where the end-to-end packet delay is dependent on the distance traveled.

In topological design of communication networks (CN) whose underlying structure can be modeled by a graph, one fundamental consideration is *system-level fault-tolerance*. At this level, the type of faults to be tolerated are *processor* or *link failures*. A system is said to be *fault tolerant* if it can remain *functional* in presence of failures. Here, we consider a CN functional as long as the distance between every pair of nonadjacent nodes remains intact in the presence of failures.

---

* This work was supported in part by the NSF grant MIP-9204066.

The underlying topology of a CN is usually modeled by a *graph* $G(V, E)$ in which the *vertex* set V represents the set of processors and the *edge* set E represents the links. Graph theoretic terms not defined here can be found in [8]. If $u \in V(G)$ then $N_G(u)$ represents the set of all vertices adjacent to vertex u in G. The set $N_G(u)$ is called the *neighborhood* of u in G. The *degree* of a vertex $u \in V(G)$ is $deg_G(u) = |N_G(u)|$. The *distance* $d_G(u, v)$ between two distinct vertices u and v in G is defined as the length (in number of edges) of a shortest path joining these vertices. All graphs considered here are connected, undirected, without any loops or multiple edges.

Over the years researchers have considered different approaches in designing fault-tolerant CNs [5, 9, 10, 13, 14, 15]. The existing work, in general, have mainly studied preserving a given topology, connectivity, or diameter. A special class of graphs called *k-geodetically line connected* graphs were introduced. In these graphs, there are k edge-disjoint paths of minimum length between every pair of nonadjacent vertices. Some properties of k-geodetically line connected graphs are given in [7]. However, designing such graphs of minimum size is known only for a special case, namely, when diameter of the resulting graph is two [6]. The main subject of this paper is the corresponding graph construction problem.

The rest of the paper is organized as follows. Section 2 presents a theorem which characterizes the distance preserving graphs. In Section 3 we present two design models with different optimality criteria and their corresponding *costs*. Section 4 presents distance preserving graphs with respect to *cycles, crowns*. In Appendix A, we show that preserving the distance between nonadjacent vertices also sets a bound, in presence of edge removal, on the distance between vertices which were adjacent before any edge removal.

2 Preliminaries

Definition 1. Given a graph D and a nonnegative integer k, a graph G is said to be k-edge distance preserving with respect to the given graph D, if:

1. D is a spanning subgraph of G and,
2. every pair of nonadjacent vertices u and v in D are joined in G by at least k edge-disjoint paths, each having length at most $d_D(u, v)$.

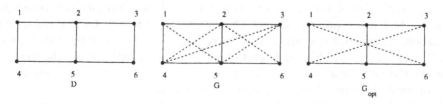

Fig. 1. $\{G, G_{opt}\} \subseteq \mathcal{F}_3(D)$.

Let $\mathcal{F}_k(D)$ denote the set of all graphs that are k-edge distance preserving with respect to a given graph D. Figure 1 shows three graphs D, G and G_{opt}; where $\{G, G_{opt}\} \subseteq \mathcal{F}_3(D)$. For any graph D on p vertices, it is easy to see that $\mathcal{F}_{p-1}(D) = \{K_p\}$. By definition, we have $\mathcal{F}_{k+1}(D) \subseteq \mathcal{F}_k(D)$ for any k, $1 \leq k < p - 1$; hence the set $\mathcal{F}_k(D)$ is not empty for any graph D and k, $1 \leq k < p$.

The results provided in the remaining part of this section can be used to check the validity of the constructions presented in this paper.

Lemma 2. *Let G be a k-edge distance preserving graph with respect to a given graph D. Then for every pair of vertices u and v with $d_D(u, v) = 2$, we have $|N_G(u) \cap N_G(v)| \geq k - 1$.*

Proof. By the definition of a k-edge distance preserving graph, G contains k edge-disjoint $u - v$ paths of length at most 2. Since G is a simple graph there can be at most one edge connecting the vertices u and v. There remain at least $k - 1$ paths and each path must have one intermediate vertex. Hence $|N_G(u) \cap N_G(v)| \geq k - 1$. Note that if it turns out that the edge $(u, v) \notin E(G)$, then we have $|N_G(u) \cup N_G(v)| \geq k$, since in this case all the k edge-disjoint $u - v$ paths in G are of length two.

Theorem 3. *Let D be a p-order graph and $k < p$ be a nonnegative integer. Further, let G be a graph having D as a spanning subgraph, and if u and v are nonadjacent vertices of D such that $d_D(u, v) \leq 3$, then there are k edge-disjoint paths between u and v in G, each having length $d_D(u, v)$ or less. Then $G \in \mathcal{F}_k(D)$.*

Proof. We will consider an arbitrary pair of vertices u and v in D and inductively construct a set of k edge-disjoint $u - v$ paths in G of length at most $d_D(u, v)$. The following notation is useful. For any two vertices x and y in a path P, we denote by $P_{x \sim y}$, the portion of P connecting the vertices x and y.

Basis : If $d_D(u, v) \leq 3$, then by the way G is defined there are k edge-disjoint $u - v$ paths in G of length at most $d_D(u, v)$.

Hypothesis : For every pair of nonadjacent vertices u and v in D, with $d_D(u, v) < n$ and $n \geq 4$, there are k edge-disjoint $u - v$ paths in G of length at most $d_D(u, v)$.

Induction Step : Consider a pair of nonadjacent vertices u and v in D such that $d_D(u, v) = n$; if no such pair exists we are done. Consider the vertex w on a shortest $u - v$ path in D such that $d_D(u, w) = n - 2$. By the induction hypothesis, there are k edge-disjoint $u - w$ paths in G of length at most $d_D(u, w)$. Let $\mathcal{P} = \{P_1, P_2, \ldots, P_k\}$ be such a set of k edge-disjoint paths. From the way we chose w, the vertices w and v are distance two apart in D and hence there are k edge-disjoint $w - v$ paths in G of length at most 2. Let $\mathcal{Q} = \{Q_1, Q_2, \ldots, Q_k\}$ be such a set of k edge disjoint paths. Also by Lemma 2 we know that at least $k - 1$ of the paths in $\mathcal{Q}$ are of length two. Without loss of generality, let $Q_i = w, q_i, v$,

$1 \le i < k$. Further let $q_k = w$ if $Q_k = w, v$; otherwise, $Q_k = w, q_k, v$. Using the paths in $\mathcal{P}$ and $\mathcal{Q}$ we construct, $\mathcal{R}$, a set of k edge-disjoint $u - v$ paths in G of length at most $d_D(u, v)$.

We call a path P in $\mathcal{P}$ as *used* if P is used to construct a $u - v$ path in $\mathcal{R}$. Initially all the paths in $\mathcal{P}$ are *unused*. Also, we mark a vertex q_i, $1 \le i \le k$, as *visited* if the edge $(q_i, v) \in E(G)$ is used in constructing a $u - v$ path in $\mathcal{R}$. Initially, all vertices q_i, $1 \le i \le k$, are marked *unvisited*.

Step 1: For each path $P \in \mathcal{P}$,

Case: $v \in V(P)$

If v is an intermediate vertex in P then $P_{u \sim v} \in \mathcal{R}$; mark P as *used*. The path $P_{u \sim v}$ is of length at most $d_D(u, v)$, because P is of length at most $d_D(u, w) < d_D(u, v)$. Note that any such path $P_{u \sim v}$ can have at most one edge (q_j, v), $1 \le j \le k$. If such a vertex q_j exists, mark q_j as *visited*.

Case: $v \notin V(P)$

In this case, the path P may contain zero or more *unvisited* vertices q_i, $1 \le i \le k$.

Subcase: P contains one or more unvisited vertices

Let q_j be the first *unvisited* vertex encountered when traversing the path P from u to w. Then we construct a new $u - v$ path in $\mathcal{R}$ by augmenting the path $P_{u \sim q_j}$ with edge (q_j, v), and then mark the vertex q_j as *visited*. It is easy to see that such a $u - v$ path is of length at most $d_D(u, v)$. Further, the constructed $u - v$ path is mutually edge-disjoint with all the existing paths in $\mathcal{R}$ as all paths in $\mathcal{P}$ are edge-disjoint and the vertex q_j is not *visited* before.

Subcase: P contains no unvisited vertex

In this case the path will remain *unused* and will be processed in Step 2.

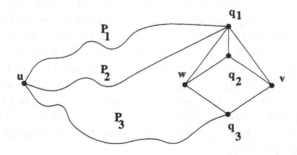

Fig. 2. Example: $k = 3$, $d_D(w, v) = 2$ and $d_G(u, w) \ge 3$

(Note: Each time we construct a path in $\mathcal{R}$ we mark at most one more vertex q_j, $1 \le j \le k$, as *visited*. Hence after the above construction the number of *unvisited* vertices in G is at least as many as the number of *unused* paths in $\mathcal{P}$.

If a vertex q_j, $1 \leq j \leq k$, is still *unvisited*, it implies that both the edges (w, q_j) and (q_j, v) do not exist in any *unused* path in $\mathcal{P}$.)

Step 2: For each *unused* path in $P \in \mathcal{P}$

Construct a new $u-v$ path by appending the path w, q_j, v to P, where q_j is some *unvisited* vertex. It is easy to see that each such $u - v$ path in G is mutually edge-disjoint with the paths in $\mathcal{R}$ and of length at most $d_D(u, w) + 2 \leq d_D(u, v)$. Hence, by using each of the k paths in $\mathcal{P}$ we construct a set of k edge-disjoint $u - v$ paths in G of length at most $d_D(u, v)$.

An example explaining the above construction is given in Figure 2. Figure 2 represents a portion of a graph $G \in \mathcal{F}_3(D)$, for some D. Let $\mathcal{P} = \{P_1, P_2, P_3\}$, where $P_1 = \{u, \ldots, q_1, v, q_2, w\}$, $P_2 = \{u, \ldots, q_1, w\}$ and $P_3 = \{u, \ldots, q_3, w\}$. The above construction first considers P_1. Since $v \in P_1$, $R_1 = P_{1_{u \sim v}}$ and vertex q_1 is marked *visited*. The path P_2 does not contain any *unvisited* vertex. Hence P_2 will still be marked *unused* and path P_3 is considered next. Since P_3 contains one unvisited vertex, q_3, the new path R_3 is $\{P_{3_{u \sim q_3}}, v\}$. The vertex q_3 is marked *visited*. Now in step 2 the remaining *unused* path, P_2 is considered. A new $u - v$ path is constructed by augmenting P_2 with path w, q_2, v.

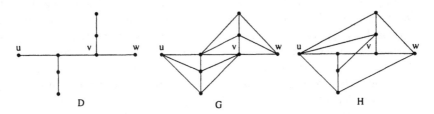

Fig. 3. Graphs G and H are not in $\mathcal{F}_3(D)$.

The above theorem implies that to prove a graph G belongs to $\mathcal{F}_k(D)$, it is enough to show that every pair of vertices u and v with $d_D(u, v) \leq 3$ are joined in G by k-edge disjoint paths of length at most $d_D(u, v)$. Figure 3 shows an example in which every pair of vertices having distance two in D are joined by 3 edge-disjoint paths of length at most two in G. However, for the vertices u and w, with $d_D(u, w) = 3$, there are only 2 edge-disjoint paths of length at most three in G. Similarly, all the vertices which are distance three apart in D are joined by 3 edge disjoint paths of length at most three in H; but the vertices u and v with $d_D(u, v) = 2$, are not joined by 3 edge disjoint paths of length at most two in H. Hence, the above two examples show that the sufficiency condition given in Theorem 3 is sharp.

It is well known that the problem of finding the maximum number of edge-disjoint paths of bounded length, say K, between a pair of vertices in a graph is *NP*-hard [12]. However, the problem is solvable in polynomial time when $K \leq 5$ [12]. Hence, given a graph G and a spanning subgraph D, Theorem 3 and the polynomial algorithm in [12] can be used to determine if G belongs to $\mathcal{F}_k(D)$ for some nonnegative integer k.

3 Design Models

In this section, we present two design models that tolerate edge failures, each with a different optimality criterion. In the first model, the optimality criterion is to minimize the number of edges. The second model considers regular fault-tolerant topologies with minimum regularity.

Model 1 *Given a graph D and a positive integer k, construct a graph, G, such that:*

(a) $G \in \mathcal{F}_k(D)$,
(b) for any $G' \in \mathcal{F}_k(D)$, we have $|E(G)| \leq |E(G')|$.

□

Since we consider only simple graphs it is assumed in the rest of the paper that $k < |V(D)|$. For any graph D on p vertices, if G belongs to $\mathcal{F}_k(D)$ then it is easy to see that $\delta(G) \geq k$. Hence, for any $G \in \mathcal{F}_k(D)$, we have $|E(G)| \geq \lceil \frac{kp}{2} \rceil$. Note that the example graph G in Figure 1 belongs to $\mathcal{F}_3(D)$; however, it is not of minimum size. The graph G_{opt} in Figure 1 is a minimum size graph. The graph G_{opt} in Figure 1 uses $\lceil \frac{kp}{2} \rceil$ edges. The following lemma gives an upper bound on the size of the augmentation.

Lemma 4. *Given any p-order graph D and a nonnegative integer $k < p$, there exists a graph $G \in \mathcal{F}_k(D)$ such that $|E(G)| - |E(D)| \leq \frac{k(2p-k-3)}{2}$.*

Proof. We will construct a new graph G from D by adding at most $\frac{k(2p-k-3)}{2}$ edges. Since D is a spanning subgraph of G we start with $V(G) = V(D)$ and $E(G) = E(D)$. Label the vertices of G as $v_1, v_2, \ldots, v_p$ such that $deg_D(v_i) \geq deg_D(v_{i+1})$, $1 \leq i < p$. Let $V_k = \{v_1, v_2, \ldots, v_k\} \subseteq V(G)$. If $(v_i, v_j) \notin E(G)$, for $1 \leq i \leq k$ and $1 \leq j \neq i \leq p$, then add the edge (v_i, v_j) to $E(G)$. Hence, the total number edges added is $\leq \sum_{i=1}^{k}(p-i-1) = \frac{k(2p-k-3)}{2}$. It is easy to see that such a graph G belongs to $\mathcal{F}_k(D)$. Note that if $k = p-1$, the above construction results in a complete graph.

The optimality criterion used in model 1 is to add a minimum number of edges. But, regular topologies are desired in designing modular systems [15]. Hence, we shift our optimality criterion from minimum number of edges to minimum regularity. In the rest of the paper, we study designing optimal graphs in this model.

Model 2 *Given a graph D and a positive integer k, construct a graph G such that:*

(a) $G \in \mathcal{F}_k(D)$,
(b) G is r-regular,
(c) for any r'-regular graph $G' \in \mathcal{F}_k(D)$, we have $r \leq r'$.

□

Let $p = |V(D)|$. Since $K_p \in \mathcal{F}_k(D)$, it is obvious that the upperbound for r is $p-1$. For the lower bound, since a graph $G \in \mathcal{F}_k(D)$ is at least k-edge connected, we have $r \geq k$. In the remaining part of this section, we prove that for certain values of k, there are no k-regular graphs in $\mathcal{F}_k(D)$, in which case the lower bound for r is $(k+1)$. But before proving this statement we give some useful properties of k-regular graphs in $\mathcal{F}_k(D)$.

Property 1 *For a given graph D, if there exists a k-regular graph $G \in \mathcal{F}_k(D)$ then for every pair of nonadjacent vertices u and v in D we have either $(u,v) \in E(G)$ or $N_G(u) = N_G(v)$.*

Proof. Let us consider an arbitrary pair of nonadjacent vertices u and v in D. We will prove the above property by induction on $d_D(u,v)$.
Basis: If $d_D(u,v) = 2$ then the validity of the property can be deduced from the proof of Lemma 2. If $d_D(u,v) = 3$, we will show that the edge (u,v) must belong to $E(G)$. Consider the vertex w on a shortest $u-v$ path in D such that $d_D(u,w) = 2$. Since the graph G is k-regular, the edge (w,v) must be used in one of the k edge-disjoint $u-w$ paths. Further, since any such $u-w$ path is of length at most two, the edge (u,v) must belong to $E(G)$. This implies $(u,v) \in E(G)$. This completes the verification of the basis.
Hypothesis: For every pair of nonadjacent vertices u and v in D with $2 \leq d_D(u,v) < n$, either $(u,v) \in E(G)$ or $N_G(u) = N_G(v)$.
Induction Step : Consider a pair of vertices u and v in D such that $d_D(u,v) = n$, and not satisfying the hypothesis; if no such pair exist we are done. Let x be the vertex on a shortest $u-v$ path in D such that $d_D(u,x) = n-2$. By induction hypothesis, we know that either $(u,x) \in E(G)$ or $N_G(u) = N_G(x)$. Note that from the way we choose x, the vertices x and v are distance two apart in D. Therefore, we have k edge-disjoint $x-v$ paths in G of length at most two. Now, we have two cases to consider.
Case 1 $(u,x) \in E(G)$: Since G is k-regular, one of the k edge-disjoint $x-v$ paths must use the edge (u,x). Further, any such path can be of length at most two. Hence the edge (u,v) must belong to $E(G)$.
Case 2 $N_G(u) = N_G(x)$: Let such a set be $\{y_1, y_2, \ldots, y_k\}$. Since G is k-regular, each of the k edge-disjoint $x-v$ paths in G must start with an edge (x, y_i), where $1 \leq i \leq k$. Since any such path can be of length at most two, the edge (y_i, v) must be in $E(G)$ for all i, $1 \leq i \leq k$. This implies $\{y_1, y_2, \ldots, y_k\} \subseteq N_G(v)$. Since $deg_G(v) = k$, we have $N_G(v) = \{y_1, y_2, \ldots, y_k\}$. Hence $N_G(u) = N_G(v) = \{y_1, y_2, \ldots, y_k\}$. This concludes the proof of the property.

Property 2 *Given a graph D, if there exists a k-regular graph $G \in \mathcal{F}_k(D)$ then the diameter of G is at most 2.* □

Proof. Immediate from property 1.

Theorem 5. *Given a graph D, if there exists a k-regular graph $G \in \mathcal{F}_k(D)$ then the order of the graph D (and G) is at most $2k$.*

Proof. Let $p = |V(D)| = |V(G)|$ and V_i be the set of all vertices that are distance $i > 0$ away from u in G. By property 2, we have $i \leq 2$. Since the graph G is k-regular $|V_1| = k$ and $|V_2| = p - k - 1$. By property 1, for every vertex $v \in V_2$ we have $N_G(u) = N_G(v)$; that is, every vertex in V_2 must be connected to all the k vertices in V_1. But each vertex in V_1 can be connected to at most $k - 1$ vertices in V_2, hence $p - k - 1 \leq k - 1$. This implies $p \leq 2k$.

Corollary 6. *For any p-order graph D and any nonnegative integer $k < \frac{p}{2}$, there does not exist a k-regular graph $G \in \mathcal{F}_k(D)$.* ◻

It is clear from Corollary 6 that for any $k < \frac{p}{2}$, the degree of regularity of a graph in $\mathcal{F}_k(D)$ is at least $(k + 1)$.

4 Special Topologies

In this paper, we design fault-tolerant topologies for *cycles, crowns*. These graphs are highly symmetric and attractive for communication networks due to their regularity, symmetry and small diameter. The connectivity and reliability of this class of graphs have been studied extensively by many researchers [1, 2, 3, 4, 11].

Definition 7. A chordal ring $C(p, h)$, where p is even and $h \leq \frac{p}{2}$ is odd, can be defined as a 3-regular graph with vertex set $\{v_o, v_1, \ldots, v_{p-1}\}$, and for each even vertex v_i in $V(C)$, $(v_i, v_{(i+1) \mod p}) \in E(C)$ and $(v_i, v_{(i+h) \mod p}) \in E(C)$.

A generalized chordal ring $C(p, < h_1, h_2, \ldots, h_r >)$ can be defined similarly with multiple odd chords $h_1 < h_2 < \ldots < h_r \leq \frac{p}{2}$.

Definition 8. [2] A r-regular crown, called r-crown, of order p is a bipartite graph with partitions $V_0 = \{v_0, v_2, \ldots, v_{p-2}\}$ and $V_1 = \{v_1, v_3, \ldots, v_{p-1}\}$; each v_i in V_0 is connected to $v_{(i+2j-1) \mod p}$, $0 \leq j < r$.

4.1 Cycle: C_p

In this section, we show that a class of $(k + 1)$-regular graphs is k edge distance-preserving with respect to C_p, a cycle of order p. For even values of k, the regularity of the distance-preserving graphs is odd. For a regular graph, both the degree and order cannot be odd. Hence to give a general construction, we assume that the order of the graph is even.

Theorem 9. *A p-order $(k + 1)$-crown belongs to $\mathcal{F}_k(C_p)$.*

---

[2] Since $|V_1| = \frac{p}{2}$, $p \geq 2r$.

Proof. Since crowns are Hamiltonian [2], C_p is a spanning subgraph of a p-order $(k+1)$-crown. Due to Theorem 3, it suffices to show that each pairs of vertices that are distance two apart in C_p are joined by k edge-disjoint paths of length at most two, and vertices that are distance three apart in C_p are joined by k edge-disjoint paths of length at most three. Note that in a C_p, only vertices of form v_i and v_{i+2} are distance two apart, and vertices of form v_i and v_{i+3} are distance three apart.

The following notation will be useful in the remaining part of the proof. If $\{v_0, v_1, \ldots, v_{p-1}\}$ is the vertex set of a graph then the vertex v_i is referred to as an *even* vertex if i is even and *odd* otherwise. In a k-crown, if v_i is an even vertex, we call the edge (v_i, v_{i+2j-1}), $0 \le j < k$, as the jth *forward chord* (FC) from vertex v_i and the jth *backward chord* (BC) from vertex v_{i+2j-1}. Specifically, if v_l is an odd vertex, the edge (v_l, v_{l-2j+1}) is the jth BC from v_l. Each even vertex in a k-crown has k FCs, and each odd vertex has k BCs.

First let us consider a pair of vertices, say v_i and v_{i+2}, that are distance two apart in C_p. Since crowns are known to be *point-symmetric graphs* [1], without loss of generality, we assume that v_i is an even vertex. By definition, in a $(k+1)$-crown we have $N(v_i) = \{v_{i+2j-1} \in V(G) \mid 0 \le j \le k\}$. Similarly, $N(v_{i+2}) = \{v_{i+2j+1} \in V(G) \mid 0 \le j \le k\}$. It is easy to see that $|N(v_i) \cap N(v_{i+2})| = k$.

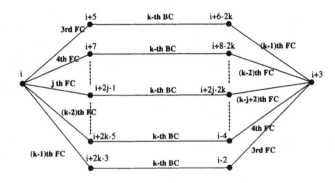

Fig. 4. A set of $k-3$ paths of length 3 in a $(k+1)$-crown.

Let us consider a pair of vertices, v_i and v_{i+3}, that are distance three apart in C_p. First we will construct a set of three paths between vertices v_i and v_{i+3} of length at most three. These paths are, $P_1 = \{v_i, v_{i+1}, v_{i+2}, v_{i+3}\}$, $P_2 = \{v_i, v_{i+3}\}$ and $P_3 = \{v_i, v_{i+2k-1}, v_{i+4}, v_{i+3}\}$. It is obvious that the above three paths between v_i and v_{i+3} are pair-wise edge (vertex) disjoint. The next $(k-3)$ paths are as follows. For each j, $3 \le j \le k-1$, we have, $P_j := \{v_i, v_{i+2j-1}, v_{i+2j-2k}, v_{i+3}\}$. Now we will show that the above $(k-3)$ paths are edge (vertex) disjoint.

Let $A_i = \{v_{i+5}, v_{i+7}, \ldots, v_{i+2j-1}, \ldots, v_{i+2k-3}\} = \{v_{i+2j-1} \in V(G) \mid 3 \le j \le k-1\} \subseteq N(v_i)$. Note that all the vertices in A_i are odd, distinct, and unused. Similarly, let $A_{i+3} = \{v_{i-2}, v_{i-4}, \ldots, v_{i+2j-2k}, \ldots, v_{i+6-2k}\} = \{v_{i+2j-2k} \in V(G) \mid 3 \le j \le k-1\} \subseteq N(v_{i+3})$. Note that all the vertices in A_{i+3} are even, distinct, and unused.

Now take the kth BC from each vertex in A_i. The set of vertices reached by using the kth BC from each vertex in A_i can be given as $\{v_{(i+2j-1)-(2k-1)} \mid 3 \leq j \leq k-1\} = \{v_{(i+2j-2k)} \mid 3 \leq j \leq k-1\} = A_{i+3}$. Hence from the $(k-3)$-vertices in A_i we can reach the $(k-3)$ vertices in A_{i+3}, using $k-3$ distinct edges. In a $(k+1)$-crown we have $N(v_i) \cap N(v_{i+3}) = \emptyset$. This implies that the constructed set of $(k-3)$ paths between v_i and v_{i+3} are pair-wise edge (in fact, vertex) disjoint. The set of $(k-3)$ edge disjoint paths between v_i and v_{i+3} is shown in Figure 4. This concludes the proof of the theorem.

Using Corollary 6, it is easy to see that the above construction is optimal. In the remaining part of this section we construct distance preserving graphs for crowns and chordal rings.

4.2 Crowns

Theorem 10. *A p-order $(r+k-1)$-crown belongs to $\mathcal{F}_k(r$-crown), for any non-negative integer $k \leq \frac{p-2r+2}{2}$.*

Proof. [3] By definition, a p-order r-crown is a spanning subgraph of a p-order $(r+k-1)$-crown. Due to Theorem 3, it suffices to show that each pairs of vertices that are distance two apart in an r-crown are joined by k edge-disjoint paths of length at most two, and vertices that are distance three apart in an r-crown are joined by k edge-disjoint paths of length at most three. Since crowns are known to be *point-symmetric graphs* [1], without loss of generality, we assume that our "source" vertex is v_0. First we will consider vertices that distance two apart in an r-crown. Note that in an r-crown all the pairs of vertices that are distance two apart are of the form v_0 and v_{2l}, $1 \leq l < r$. It is easy to see that in an $(r+k-1)$-crown, the k edge-disjoint paths joining v_0 to v_{2l}, $1 \leq l < r$, are v_0, v_{2j-1}, v_{2l}, for any $r-1 \leq j \leq r+k-2$.

Now we will consider all the vertices in an r-crown that are distance 3-away from v_0. In an r-crown, it is easy to verify that vertices v_{2r-2} and v_{p-2r+2} *span* all the vertices that are distance 3-away from v_0. In otherwords, a shortest path to all the distance 3-away vertices can be given by, (a) For each $1 \leq l < r$, $v_0, v_{2r-3}, v_{2r-2}, v_{2r+2l-3}$, and (b) For each $1 \leq l < r$, $v_0, v_{p-1}, v_{p-2r+2}, v_{p-2r+2l-1}$. Hence, if a vertex v_j is distance 3-away from v_0, we have either $2r-1 \leq j \leq 4r-5$ or, $p-2r+1 \leq j \leq p-3$. Given a vertex v_j which is distance 3-away from v_0 in an r-crown, the k edge-disjoint paths joining v_0 and v_j in an $(r+k-1)$-crown can be given as:

Case: $2r-1 \leq j \leq 4r-5$: We construct a set of k edge-disjoint paths by considering two sets of k vertices that are adjacent to v_0 and v_j. Since all the vertices adjacent to v_0 are odd[4] and all the vertices adjacent to v_j are even,

---

[3] Note that a p-order $(r+k-1)$-crown is only defined when $2(r+k-1) \leq p$.
[4] We call a vertex v_i odd (even), if i is odd (even).

by definition of an $(r + k - 1)$-crown, we have $N(v_0) \cap N(v_j) = \emptyset$. However, we have to show that all the k vertices used in $N(v_0)$ are distinct and all the k vertices used in $N(v_j)$ are distinct. The construction is divided into three subcases. A brief overview of each subcase is as follows. In subcase 1, we construct a set of $min(r, k)$ paths. In these paths, v_0 is adjacent to odd vertices between[5] v_0 and v_{2r-2} and v_j is adjacent to even vertices between v_0 and v_{2r-2}. In subcase 2, we construct an $r + 1^{th}$ path.

In subcase 3, we construct the rest of the $k - r - 1$ paths. In these paths v_0 is adjacent to $v_{2(r+k-2)-3}, v_{2(r+k-2)-5}, \ldots, v_{2(r+k-2)-2(k-r-1)-1}$. Notice that $2(r + k - 2) - 2(k - r - 1) - 1 > 2r - 3$, which implies that paths constructed in subcase 3 would not reuse the odd vertices used in subcase 1. Hence all the selected k vertices adjacent to v_0 are distinct. Similarly in the paths constructed in this subcase, v_j is adjacent to $v_{p-2}, v_{p-4}, \ldots, v_{p-2(k-r-1)}$. By definition of an $(r + k - 1)$-crown we have, $p \geq 2(r + k - 1)$ and since r is a positive integer, we have $p - 2(k - r - 1) > 2r - 2$, which guarantees that the even vertices used in subcase 1 are not reused. Now we are ready to construct a set of k edge-disjoint v_0-v_j paths of length at most 3 in an $(r + k - 1)$-crown.

Subcase 1: For each $0 \leq l < min(r, k)$ we have, $v_0, v_{(2r-2)-2l-1}, v_{(2r-2)-2l}, v_j$. By definition v_0 is adjacent to $v_{(2r-2)-2l-1}$. In an r-crown vertex v_j is adjacent to v_{2r-2}, which implies that in an $(r + k - 1)$-crown vertex v_j is connected to at least k even vertices before v_{2r-2}. In otherwords, v_j is connected to $v_{(2r-2)-2l}$. Note that when $k \geq r$ then the above given path is v_0, v_{-1}, v_0, v_j when $l = r - 1$. In thise case we should consider the edge v_0-v_j as a path.

Subcase 2: If $k = r + 1$ we have $v_0, v_{2(r+k-2)-1}, v_{j+1}, v_j$. We will show that v_{j+1} is joined to $v_{2(r+k-2)-1}$ in an $(r + k - 1)$-crown. By definition, in an $(r + k - 1)$-crown the vertex v_{j+1} is joined to all odd vertices in between $v_{(j+1)}$ and $v_{(j+1)+2(r+k-2)-1}$. Since j is positive, $2r - 1 \leq j \leq 4r - 5$, and $k = r + 1$, we can see that $v_{2(r+k-2)-1}$ is an odd vertex in between $v_{(j+1)}$ and $v_{(j+1)+2(r+k-2)-1}$. In otherwords, vertex v_{j+1} is joined to $v_{2(r+k-2)-1}$.

Subcase 3: If $k > r + 1$ we have $v_0, v_{(p-2l-2)+2(r+k-2)-1}, v_{p-2l-2}, v_j$, for each $0 \leq l < k-r-1$. Since $k > r+1$, vertex v_j is connected to $v_{p-2l-2i}$, $0 \leq l < k - r - 1$. Hence, it is enough to show that v_0 is adjacent to $v_{(p-2l-2)+2(r+k-2)-1}$, for each $0 \leq l < k - r - 1$. We know that vertex v_0 is adjacent to all odd vertices between v_1 and $v_{2(r+k-2)-1}$. Note that $p-2l-2+2(r+k-2)-1 \geq p$ and since vertex index operations are taken over modulo p, it remains to show that $1 \leq 2(r + k - 2) - 1 - 2l - 2 \leq 2(r + k - 2) - 1$. Since $k > r + 1 > 2$ and $0 \leq l < k - 3$, the above inequality holds.

Figure 5 shows a set of k edge-disjoint paths between v_0 and v_j when $k > r$.

---

[5] By "between v_i and v_j" we refer to the vertices on the clockwise v_i-v_j path in the spanning cycle of the given graph, as can be seen in Figure 5.

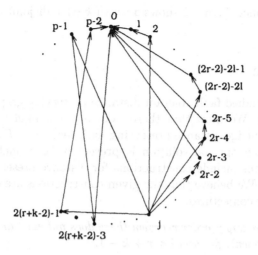

Fig. 5. k edge-disjoint paths between v_0 and v_j when $2r - 1 \leq j \leq 4r - 5$.

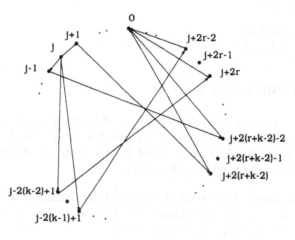

Fig. 6. k edge-disjoint paths between v_0 and v_j when $p - 2r + 1 \leq j \leq p - 3$.

Case: $p - 2r + 1 \leq j \leq p - 3$: In this case we consider vertices $v_{(j-2l+1)}$, $0 \leq l < k$, which are adjacent to v_j. From each of the k vertices, traverse through the $(r + k - 2)^{th}$ chord to reach a vertex adjacent to v_0. The k edge-disjoint paths for this case can be give as follows. For each $0 \leq l < k$, $v_0, v_{(j-2l+1)+2(r+k-2)-1}, v_{j-2l+1}, v_j$. We have to show that vertex v_0 is adjacent to vertex $v_{(j-2l+1)+2(r+k-2)-1}$. Since, $p - 2r + 1 \leq j \leq p - 3$ and vertex index operations are taken over modulo p, the vertex $v_{(j-2l+1)+2(r+k-2)-1}$ is in between $v_{2k-2l-3}$ and $v_{2(r+k-2)-(2l+3)}$. It is easy to see that for all values of l vertex v_0 is adjacent to odd vertices in between $v_{2k-2l-3}$ and $v_{2(r+k-2)-(2l+3)}$. Since we use the $(r + k - 2)^{th}$ chord from each of the k distinct vertices to reach a vertex adjacent to v_0, all the vertices used in

$N(v_0)$ are distinct. Figure 6 shows a set of k edge-disjoint paths between v_0 and v_j.

5 Conclusion

In this paper we studied fault-tolerant distance-preserving graphs with different optimality criteria. We show that there exist a polynomial time algorithm to determine if G is k-edge distance preserving with respect to D. When the given graph is a cycle, an optimal solution is presented for Model 2. Further, our preliminary solutions include constructions for distance-preserving graphs with respect to crowns. We believe that the given constructions are optimal and thus state the following conjectures.

Conjecture 1 *For any p-order r-crown there does not exist an i-regular p-order graph $G \in \mathcal{F}_k(r\text{-}crown)$, for any $i < r + k - 1$.*

Conjecture 2 *For any $C(p, h)$ there does not exist an i-regular graph $G \in \mathcal{F}_k(r\text{-}crown)$, for any*

1. *$i < 2k + 1$, if $h \geq 2k$.*
2. *$i < k + \lfloor \frac{h}{2} \rfloor + 1$, if $h < 2k$.*

6 On adjacent vertices in D

Corollary 11. [6] *Given a graph G and vertex set $\{u, x, v\} \subseteq V(G)$, if there exists a set of k edge disjoint u-x paths in G of length at most d and a set of k edge disjoint x-v paths in G of length at most 2, then there are k edge disjoint u-v paths in G, each having length $d + 2$ or less.*

To follow the proof of the next theorem a brief definition of *eccentricity* is necessary. Given a connected graph D and a vertex $u \in V(D)$ the $eccentricity_D(u) = max\{d_D(u, v) | v \in V(D)\}$.

Theorem 12. *Given a graph D and a graph $G \in \mathcal{F}_k(D)$ with $k < |V(D)|$. For any pair of adjacent vertices u and v in D and for any $F \subseteq E(G)$ with $|F| < k$, we have $d_{G-F}(u, v) \leq 5$.*

Proof. The proof is given by constructing a set of k edge disjoint u-v paths in G of length at most 5. Without loss of generality, assume that $eccentricity_D(u) \geq eccentricity_D(v)$. First consider the case when $eccentricity_D(u) \geq 3$. Let x be a vertex such that $d_D(u, x) = 3$. Let $P = u, l, m, x$ be a u-x path of length 3 in D. Since $(u, v) \in E(D)$, we have $d_D(v, x) \geq 2$; Otherwise, $d_D(u, x) < 3$. Similarly, $d_D(v, x) \leq 4$ because v and x are joined in G by a path of length 4 namely, v, u, l, m, x. In other words, we have $2 \leq d_D(v, x) \leq 4$.

---

[6] This Corollary is immediate from "Induction Step" of Proof for Theorem 3.

Case $d_D(v, x) = 2$: We have $d_D(u, x) = 3$ and $d_D(v, x) = 2$ and since $G \in \mathcal{F}_k(D)$, by Corollary 11, there are k edge disjoint u-v paths in G, each having length 5 or less.

Case $d_D(v, x) = 3$: Let us consider the vertex m in the path P. We know that $d_D(u, m) = 2$. Notice that $d_D(v, m) > 1$; Otherwise, $d_D(v, x) < 3$. Also, $d_D(v, m) \leq 3$ because v, u, l, m is a v-m path in D. In other words, we have $2 \leq d_D(v, m) \leq 3$. Since $G \in \mathcal{F}_k(D)$ and $d_D(u, m) = 2$, by Corollary 11 we have k edge disjoint u-v paths in G, each having length at most 5.

Case $d_D(v, x) = 4$: Consider the vertex m in the path P; we have $d_D(u, m) = 2$. Since we have a v-m path v, u, l, m in D, we have $d_D(v, m) \leq 3$. But note that if $d_D(v, m) < 3$, then $d_D(v, x) < 4$, a contradiction. Hence $d_D(v, m) = 3$. By Corollary 11, we have at least k edge disjoint u-v paths in G, each having length at most 5.

Now let us consider the case when $eccentricity_D(u) \leq 2$ and $eccentricity_D(v) \leq 2$. If $eccentricity_D(u) = 1$ or $eccentricity_D(v) = 1$, then all the vertices in D are adjacent to either u or v. Further, since $\delta(G) \geq k$ it is easy to see that $|N_G[u] \cap N_G[v]| \geq k$. This implies that there k edge disjoint u-v paths of length at most 2. Hence, without loss of generality, we assume that $eccentricity_D(u) = 2$ and $eccentricity_D(v) = 2$. If there exist a vertex w which is distance 2 away from both u and v in D then, by Corollary 11, we have k edge disjoint u-v paths in G, each having length at most 4 and we are done. If such a w does not exist then each vertex in $V(D)$ is either adjacent to u or adjacent to v. In such a case, we claim that there are k edge disjoint u-v paths in G of length at most 3. In the remaining part of the proof we enumerate a set of k edge disjoint u-v paths in G, each having length at most 3.

Since $(u, v) \in E(G)$ there is one u-v path of length 1 in G. Let $L = N_G(u) \cap N_G(v)$ and let $l = |L| \geq 0$. By using the vertices in L we can construct a set of l edge disjoint u-v paths of length 2. So far we have $l + 1$ number of edge disjoint paths of length at most 2. If $k \leq l + 1$ we are done. Otherwise, since $\delta(G) \geq k$ we have at least $k - l - 1$ *unused* vertices adjacent to u. Similarly, there are at least $k - l - 1$ *unused* vertices adjacent to v. Let us denote by M and N the set of unused vertices adjacent to u and v, respectively; It is easy to see that $M \cap N_G(v) = \emptyset$ and $N \cap N_G(u) = \emptyset$. Since $k > l + 1$, we have $M \neq \emptyset$, $N \neq \emptyset$. Notice that each vertex in M is distance 2 away from v in D, implying that there are at least k edge disjoint paths in G of length at most 2 from each vertex in M to v. Consider any vertex $u_i \in M$. Since $M \cap N_G(v) = \emptyset$, by lemma 2, we must have $N_G(u_i) \cap N_G(v) \geq k$. However, u_i can at most be connected to $l + 1$ vertices of $L \cup \{u\}$. The remaining $k - l - 1$ vertices in $N_G(u_i) \cap N_G(v)$ must belong to N. In other words, for each $u_i \in M$ we have $|N_G(u_i) \cap N| \geq k - l - 1$. Since $|M| \geq k - l - 1$ there exist a *matching* of size at least $k - l - 1$ from vertices in M to the vertices in N. Since all vertices in M are connected to u and all vertices in N are connected to v, there are $k - l - 1$ internally disjoint u-v paths, each having length 3 in G. To summarize this case, there is one u-v path of length 1, l u-v paths of length 2 and $k - l - 1$ u-v paths of length 3 in G.

366

Corollary 13. *Given a graph D and a graph $G \in \mathcal{F}_k(D)$, with $k < |V(D)|$. For any $F \subseteq E(G)$ with $|F| < k$, we have $diam(G - F) \leq max\{5, diam(D)\}$.*

References

1. F. Boesch and R. Tindell. Connectivity and symmetry in graphs. *Graphs and Applications, Proc. of 1st Colorado sym on Graph Theory*, pages 53–67, 1982.
2. F. Boesch and R. Tindell. Circulants and their connectivities. *Journal of Graph Theory*, 8:487–499, 1984.
3. F. Boesch and J. Wang. Reliable circulant networks with minimum transmission delay. *IEEE Tran on circuits and systems*, 32:1286–1291, December 1985.
4. F. Buckley and F. Harary. *Distance in Graphs*. Addison-Wesley Publishing Company, 1990.
5. F. R. K. Chung. Graphs with small diameter after edge deletion. *Discrete Applied Mathematics*, 37-38:73–84, 1992.
6. R. Entringer and D. E. Jackson. Minimum k-geodetically connected graphs. *Congressus Numerantium*, 36:303–309, 1982.
7. R. C. Entringer, D. Jackson, and P. J. Slater. Geodetic connectivity of graphs. *IEEE tran. on circuits and systems*, 24:460–463, August 1977.
8. F. Harary. *Graph Theory*. Addison-Wesley Publishing Company, 1968.
9. F. Harary and J. P. Hayes. Edge fault tolerance in graphs. *NETWORKS*, 23:135–142, 1993.
10. J. P. Hayes. A graph model for fault-tolerant computing systems. *IEEE tran. on Computers*, 25:875–884, September 1976.
11. X.D. Hu and F.K. Hwang. Reliabilities of chordal rings. *Networks*, 22:487–501, 1992.
12. A. Itai, Y. Perl, and Y. Shiloach. The complexity of finding maximum disjoint paths with length constraints. *NETWORKS*, 12:277–286, 1982.
13. K. Vijayan and U. S. R. Murty. On accessibility in graphs. *Sankhya, Series A*, 26:299–302, 1965.
14. W.W. Wong and C.K. Wong. Minimum k-hamiltonian graphs. *Journal of Graph Theory*, 8:155–165, 1984.
15. G. Zimmerman and A-H. Esfahanian. Chordal rings as fault-tolerant loops. *J. Desc. Applied Math.*, 99:563–573, December 1992.

Shortest Path Algorithms for Nearly Acyclic Directed Graphs

Tadao TAKAOKA

Department of Computer Science
Ibaraki University
Hitachi, Ibaraki 316, JAPAN
E-mail: takaoka@cis.ibaraki.ac.jp

Abstract. Abuaiadh and Kingston gave an efficient algorithm for the single source shortest path problem for a nearly acyclic graph with $O(m + n \log t)$ computing time, where m and n are the numbers of edges and vertices of the given directed graph and t is the number of delete-min operations in the priority queue manipulation. They use the Fibonacci heap for the priority queue. If the graph is acyclic, we have $t = 0$ and the time complexity becomes $O(m + n)$ which is linear and optimal. They claim that if the graph is nearly acyclic, t is expected to be small and the algorithm runs fast. In the present paper, we take another definition of acyclicity. The degree of cyclicity $cyc(G)$ of graph G is defined by the maximum cardinality of the strongly connected components of G. When $cyc(G) = k$ is small, we can say the graph is nearly acyclic and we give an algorithm for the single source problem with $O(m + n \log k)$ time complexity. Finally we give a hybrid algorithm that incorporates the merits of the above two algorithms.

1 Introduction

Since the classical algorithms for shortest paths were published by Dijkstra [3] for the single source problem and Floyd [4] for the all pairs problem, there have been many variations and improvements in either analysis or special classes of the given graphs. To name just a few, Moffat and Takaoka [7] gave an $O(n^2 \log n)$ expected time algorithm for the all pairs shortest path (APSP) problem, which is a considerable improvement from $O(n^3)$ of Floyd, where n is the number of vertices of the given graph. Fredman and Tarjan [6] gave an $O(m + n \log n)$ time algorithm for the single source problem using the data-structure, called a Fibonacci heap, where m is the number of edges of the given graph. This is an improvement of $O(n^2)$ of Dijkstra, when m is small. If the given graph is planar, Frederickson [5] gave an $O(n\sqrt{\log n})$ time algorithm for the single source problem and $O(n^2)$ time algorithm for the APSP problem. If the given graph is nearly tree, Chaudhuri and Zaroligas [2] gave an algorithm for the APSP problem with $O(t^4 n)$ time where t is the tree width of the graph.

Recently Abuaiadh and Kingston [1] gave an interesting result by restricting the given graph to being nearly acyclic. When they solve the single source problem, they distinguish between two kinds of vertices in $V - S$ where V is the set of vertices and S is the solution set, that is, the set of vertices to which the shortest distances from the source have been established by the algorithm. One is the set of vertices, "easy" ones, to which there are no edges from $V - S$, e.g., only edges from S. The other is the set of vertices, "difficult" ones, to which there are edges from $V - S$. To expand S, if there are easy vertices, those are included in S and distances to other vertices in $V - S$ are updated. If there are no easy vertices, the vertex with minimum tentative distance is chosen to be included in S. If the number of such delete-minimum operations is t, the authors show that the single source problem can be solved in $O(m + n \log t)$ time with use of a Fibonacci heap. If the graph is acyclic, $t = 0$ and we have $O(m + n)$ time. Since we have $O(m + n \log n)$ when $t = n$, the result is an improvement of Fredman and Tarjan with use of the new parameter t. The authors claim that if the given graph is nearly acyclic, t is expected to be small and thus we can have a speed up.

The definition of near acyclicity and the estimate of t under it is not clear, however. Take up an example graph $G = (V, E)$ such that $V = \{v_1, v_2, \cdots, v_n\}$ for even n and E is defined by

$$E = \{(v_i, v_{i+1}) \mid i = 1, 2, \cdots, n - 1\} \cup \{(v_i, v_{i-1}) \mid i = 2, 4, \cdots, n\}$$
$$\cup \{(v_1, v_i) \mid i = 3, 4, \cdots, n\}.$$

We give non-negative real numbers as edge costs to edges in such a way that $v_1, \cdots, v_n$ are included in S in this order. Then the graph is nearly acyclic in the sense given below, but $t = n/2$ and hence the complexity is $O(m + n \log n)$.

In the present paper, we give a new definition of near acyclicity by the maximum size k of strongly connected components of the given directed graph. Under this definition, we give an $O(m + k^2 n)$ time algorithm for the single source problem and an $O(mn + kn^2)$ time algorithm for the APSP problem in Section 2. Specifically for the above example graph our time is $O(m + n)$, that is linear, whereas the time by [1] is $O(m + n \log n)$. On the other hand, the efficiency of our algorithm worsens for a circular graph with $E = \{(v_i, v_{i+1}) \mid i = 1, \cdots, n - 1\} \cup \{(v_n, v_1)\}$ since k becomes n, whereas the algorithm in [1] performs well since $t = 0$. In Section 4, we improve the above results by establishing $O(m + n \log k)$ time for the single source problem and $(mn + n^2 \log k)$ time for the APSP problem. In Section 5, we give a hybrid algorithm that inherits merits from the algorithms in this paper and in [1].

2　Simple algorithms

Let $G = (V, E)$ be a directed graph where $V = \{v_1, \cdots, v_n\}$ and $E \subseteq V \times V$. The non-negative cost of edge (v_i, v_j) is denoted by $c(v_i, v_j)$. Let Tarjan's algorithm [8] compute strongly connected components (sc-components) $V_r, V_{r-1}, \cdots, V_1$ in this order. Let graph $\tilde{G}(\tilde{V}, \tilde{E})$ be defined by $\tilde{V} = \{V_1, \cdots, V_r\}$ and $\tilde{E}$, where there is an edge from V_i to V_j in $\tilde{E}$ if there is an edge (v, w) in E such that $v \in V_i$ and $w \in V_j$. That is, $\tilde{G}$ is the degenerated acyclic graph of G such that

V_i's are degenerated into single vertices and edges from V_i to V_j are degenerated into a single edge. The set $\tilde{V} = \{V_1, \cdots, V_r\}$ is topologically sorted in this order.

Let the graph $G_i = (V_i, E_i)$ be defined by $E_i = \{(v, w) \mid v \in V_i$ and $w \in V_i\}$. We solve the all pairs shortest path problem for each G_i. Let $D(v, w)$ be the shortest distance from v to w in G_i. Using this information, we solve the single source problem from source $v_0 \in V_1$ to all other vertices along the degenerated edges. We start with an algorithm for the simpler case of an acyclic graph.

Algorithm 1 $\{G = (V, E)$ is an acyclic graph.$\}$

1 Topologically sort V and assume without loss of generality $V = \{v_1, \cdots, v_n\}$
  where $(v_i, v_j) \in E \Leftrightarrow i < j$;
2 $d[v_1] := 0$; $\{v_1$ is the source$\}$
3 **for** $i := 2$ **to** n **do** $d[v_i] := \infty$;
4 **for** $i := 1$ **to** n **do**
5 **for** v_j such that $(v_i, v_j) \in E$ **do**
6   $d[v_j] := \min\{d[v_j], d[v_i] + c(v_i, v_j)\}$.

We expand the above algorithm to Algorithm 2 with line numbers expanded with dots.

Algorithm 2 $\{$Solve the single source problem for graph $G = (V, E)$ and source $v_0\}$

1.1 Compute sc-components $V_r, V_{r-1}, \cdots, V_1$
1.2 Solve the APSP problem for $G_1, G_2, \cdots, G_r$; $\{D$ computed$\}$
2.1 **for** $v \in V_1$ **do** $d[v] := \infty$;
2.2 $d[v_0] := 0$; $\{$For source v_0 let $v_0 \in V_1$ without loss of generality$\}$
3 **for** $i := 2$ **to** r **do for** $v \in V_i$ **do** $d[v] := \infty$;
4.1 **for** $i := 1$ **to** r **do begin**
4.2 **for** $v \in V_i$ **do for** $w \in V_i$ **do**
4.3 $d[w] := \min\{d[w], d[v] + D[v, w]\}$;
5   **for** V_j such that $(V_i, V_j) \in \tilde{E}$ **do**
6.1   **for** $v \in V_i$ and $w \in V_j$ such that $(v, w) \in E$ **do**
6.2 $d[w] := \min\{d[w], d[v] + c(v, w)\}$
7   **end**.

Lines 4.2 and 4.3 are to obtain shortest distances within sc-components whereas lines 6.1 and 6.2 are to update distances through edges between sc-components.

Lemma 1. *At the beginning of Line 5 in Algorithm 1, the shortest distances from v_1 to v_j $(j \leq i)$ are computed. Also at the beginning of line 5, distances computed in $d[v_j]$ $(j \geq i)$ are those of shortest paths that lie in $\{v_1, \cdots, v_{i-1}\}$ except for v_j.*

Proof. By induction. When $i = 1$, $d[v_1] = 0$ and the set of $v_1, \cdots, v_{i-1}$ is empty, and thus the lemma trivially holds. Assume the lemma is true for i. Since the paths from v_1 to v_i only go through the set $\{v_1, \cdots, v_{i-1}\}$, the shortest distance to v_i is already computed in $d[v_i]$. At lines 5 and 6, the shortest distances to v_j $(j > i)$ are updated through v_i. $\square$

Lemma 2. *At the beginning of line 5 in Algorithm 2, the shortest distances from v_0 to $v \in V_j$ $(j \leq i)$ are computed. Also at the beginning of line 5, the distances computed in $d[v]$ for $v \in V_j$ $(j \geq i)$ are those of shortest paths that lie in $V_i \cup \cdots \cup V_{i-1}$ except for vertices in V_j.*

Proof. Similar to the proof of Lemma 1. When $i = 1$, the shortest distances from v_0 to $v \in V_1$ are computed at lines 4.2 and 4.3. The second statement of the lemma is true since $V_1 \cup \cdots \cup V_{i-1}$ is empty.

Assume the lemma is true for i. Since the paths from v_0 to $v \in V_i$ only go through the set $V_1 \cup \cdots \cup V_{i-1} \cup V_i$ and the shortest distances to $v \in V_i$ through $V_1 \cup \cdots \cup V_{i-1}$ are computed, the shortest distances to $v \in V_i$ are computed at lines 4.2 and 4.3. At lines 5, 6.1 and 6.2, the shortest distances to $w \in V_j$ $(j > i)$ are updated through $v \in V_i$. $\qquad\square$

The all pairs version of Algorithm 1 is to compute shortest distances by changing the source from v_1 to v_{n-1}. The all pairs version of Algorithm 2 is similar. We change v_0 in V_1 and solve the single path problems. Then we take V_2 and choose all v_0 in V_2 and so on.

3 Analysis

To analyze the algorithm, we first establish the following lemma.

Lemma 3. *Let non-negative integer variables $x_1, x_2, \cdots, x_n$ satisfy the following conditions for constant integers k and x such that $0 \leq k \leq x$ and $x \leq kn$.*

$$(1) \quad x_1 + x_2 + \cdots + x_n = x$$
$$(2) \quad x_i \leq k \ (i = 1, \cdots, n).$$

Also let the maximum of the objective function $\sum_{i=1}^{n} f(x_i)$ be denoted by $\phi_n(x)$ where $f(x)$ is a convex function such that $f(0) = 0$. A solution $(x_1, \cdots, x_n)$ that gives $\phi_n(x)$ is given for $q = \lceil x/k \rceil$ in the following,

$$x_i = k \ (i = 1, \cdots, q - 1)$$
$$x_q = x - (q - 1)k$$
$$x_i = 0 \quad (i = q + 1, \cdots, n).$$

Note that $f(x)$ is convex means that $f(x_2) + f(x_3) \leq f(x_1) + f(x_4)$ if $x_1 \leq x_2 \leq x_3 \leq x_4$ and $x_1 + x_4 = x_2 + x_3$. The vector $(x_1, \cdots, x_n)$ that satisfies (1) and (2) is called a feasible solution.

Proof. The principle of optimality in dynamic programming is given by

$$\phi_n(x) = \max_{0 \leq x_n \leq k} \{f(x_n) + \phi_{n-1}(x - x_n)\}.$$

Let $(x_1, \cdots, x_n)$ be the solution for $\phi_n(x)$ and we assume $x_1 \geq x_2 \geq \cdots \geq x_n$ without loss of generality.

Case a. $x > kn - k$. In this case $(k, k, \cdots, k, x - nk + k)$ is the solution, since redistribution of positive δ by setting $x_i \leftarrow x_i - \delta$ and $x_n \leftarrow x_n + \delta$ may decrease the value of the objective function due to convexity of f.

Case b. $x \leq kn - k$. In this case, for any feasible solution $(x_1, \cdots, x_n)$ we can redistribute x_n to $(x_1, \cdots, x_{n-1})$, since $x \leq k(n - 1)$, and we may increase the value of the objective function because of convexity. Thus we have $\phi_n(x) = \phi_{n-1}(x)$. Repeating case b by setting $x_i \leftarrow 0$ ($i = n, n - 1, \cdots$), we eventually hit case a and we have the solution given in the theorem. $\qquad\square$

Now we analyze the algorithms. The computing time of Algorithm 1 is obviously $O(m + n)$. Its all pairs version takes $O(mn + n^2)$ time.

At line 1.2 of Algorithm 2, we use Floyd's algorithm. Then the time becomes $O(k_1^3 + \cdots + k_r^3)$ where k_i ($1 \leq i \leq r$) is the size of V_i. Let us assume that $k_i \leq k$ ($i = 1, \cdots, r$). From Lemma 3, we see that $O(k_1^3 + \cdots + k_r^3) \leq O((n/k)k^3) = O(k^2 n)$, since $k_1 + \cdots + k_r = n$ and x^3 is a convex function. The overall time for lines 4.2 and 4.3 is $O(k_1^2 + \cdots + k_r^2)$, which is $O(kn)$ from Lemma 3. The overall time for lines 5, 6.1 and 6.2 is $O(m)$. Hence the total time is $O(m + k^2 n)$.

When we apply Algorithm 2 to n sources, we note that we can perform lines 1.1 and 1.2 only once. Thus the total time becomes $O(n(m+kn)) = O(mn+kn^2)$, since $k^2 n \leq kn^2$. To summarize, we have the following definition and theorem.

Definition 4. The degree of cyclicity of graph G, denoted by $cyc(G)$, is defined to be the maximum cardinality of the strongly connected components of G.

Theorem 5. *Let $k = cyc(G)$. Then we can solve the single source problem and the APSP problem for G in $O(m+k^2 n)$ time and $O(mn+kn^2)$ time respectively.*

We can say that the given directed graph is nearly acyclic, if $cyc(G)$ is small.

4 More efficient algorithms

In this section we improve Algorithm 2 by not solving APSP problems for $G_1, G_2, \cdots, G_r$. We use a modified version of Fredman and Tarjan's algorithm [6] for the single source problem. Here we generalize the single source shortest path problem in the following way. We omit "shortest path" for simplicity.

Definition 6. The generalized single source (GSS) problem for a directed graph $G = (V, E)$ with the non-negative cost function $c(v, w)$ for edge (v, w) and the initial distances $d_0[v] \geq 0$ for $v \in V$ is to compute the shortest distances $d[w]$ for all $w \in V$. The shortest distance $d[w]$ is defined by

$$d]w] = \min_v \{d_0[v] + D(v, w)\},$$

where $D[v, w]$ is the shortest distance from v to w. The conventional single source problem has $d_0[v_1]=0$ and $d_0[v] = \infty$ for all other $v \in V$.

To solve the GSS problem, we have the following algorithm in which we do not actually compute D.

Algorithm 3

```
 1  for v ∈ V do d[v] := d₀[v];
 2  Organize V in a priority queue Q with d[v] as key;
 3  S := ∅;
 4  while S ≠ V do begin
 5      Find v from Q with minimum key and delete v from Q;
 6      S := S ∪ {v};
 7      for w ∈ V − S do begin
 8          d[w] := min{d[w], d[v] + c(v, w)};
 9          Reorganize Q with new d[w];
10      end
11  end.
```

The correctness of this algorithm follows from that of Dijkstra's algorithm if we attach a hypothetical source vertex v_0 and edges (v_0, v) with costs $d_0[v]$. If we use a Fibonacci heap, we can solve the GSS problem in $O(m + n \log n)$ time. Note that $O(n)$ time for make-heap is absorbed in the main complexity.

Now we use Algorithm 3 for lines 4.2 and 4.3 in Algorithm 2. Then the computing time becomes $O(m + \Sigma(m_i + k_i \log k_i))$. Since $x \log x$ is a convex function, we have the computing time given by $O(m + (n/k)k \log k) = O(m + n \log k)$ from Lemma 3. For the APSP problem we can use this new version n times. To summarize we have Theorem 2.

Theorem 7. *The single source and APSP problems for $G = (V, E)$ can be solved in $O(m + n \log k)$ time and $O(mn + n^2 \log k)$ time respectively where $k = cyc(G)$.*

5   Further improvement

Our algorithm in the previous section and that in [1] work well for different kinds of nearly acyclic graphs. They are, however, not incompatible. We use the following algorithm, Algorithm 4 in [1], in place of Algorithm 3 used at lines 4.2 and 4.3 in Algorithm 2. This way the two algorithms can compensate for each other. Let $out(v) = \{w \mid (v, w) \in E\}$.

Algorithm 4

```
 1  for v ∈ V do d[v] := d₀[v];
 2  Organize V in a priority queue Q with d[v] as key;
 3  S := ∅;
 4  while S ≠ V do begin
 5      if there is a vertex v in V − S with no incoming edge from V − S then
 6          Choose v
 7      else
```

8   Choose v from $V - S$ such that $d[v]$ is minimum;
9   Delete v from Q;
10   $S := S \cup \{v\}$;
11   **for** $w \in out(v) \cap (V - S)$ **do** $d[w] := min\{d[w], d[v] + c(v, w)\}$
12 **end**.

The priority queue Q is slightly modified in [1] from the Fibonacci heap in such a way that there is no delete-min operation, and no pointer to the minimum element in Q. Only delete operation is defined and the minimum is found when the trees of equal rank are linked. It is shown in [1] that a sequence of n delete, m decrease-key and t find-min operations is processed in $O(m + n \log t)$ time. The readers are referred to [1] for details. Now we give the following final algorithm.

Algorithm 5

1   Compute sc-components $V_r, V_{r-1}, \cdots, V_1$;
2.1 **for** $v \in V_1$ **do** $d[v] := \infty$;
2.2 $d[v_0] := 0$; {Let $v_0 \in V_1$ without loss of generality}
3   **for** $i := 2$ **to** r **do for** $v \in V_i$ **do** $d[v] := \infty$;
4.1 **for** $i := 1$ **to** r **do begin**
4.2   Use Algorithm 4 to solve the GSS for G_i;
5   **for** V_j such that $(V_i, V_j) \in \tilde{E}$ **do**
6.1   **for** $v \in V_i$ **do for** $w \in V_j$ **do**
6.2 $d[w] := min\{d[w], d[v] + c(v, w)\}$
7   **end**.

Suppose we use Algorithm 4 for the whole graph with the initial condition that $d_0[v_0] = 0$ and $d_0[v] = \infty$ for $v \neq v_0$ and the number of v's chosen at line 8 in (hypothetical) V_i is denoted by t_i. Denote also by t'_i the number of v's chosen at line 8 of Algorithm 4 used at line 4.2 in Algorithm 5. Let us call these vwetices "min-vertices." Although the orders in which vertices are included in S in the above two computations are different in general, we can show that $t'_i \leq t_i$ since if min-vertices $v, w \in V_i$ are included in S in this order, i.e., v first, in the latter computation, then they are included in S in the same relative order in the former computation. Let $s = max\{t'_i\}$ and $t = \sum t_i$. Then the time for Algorithm 5 is bounded by

$$O(m + \sum(m_i + k_i \log t'_i)) \leq O(m + n \log s).$$

Since $s \leq t$ and $s \leq k$, this algorithm is an improvement over those in the previous section and in [1]. Note that this algorithm runs in linear time for the two example graphs in Section 1.

6 Concluding Remarks

If we use a simple version of Dijkstra's algorithm where the priority queue is organized in a linear list, we have $O(m + kn)$ time for the single source problem where $k = cyc(G)$. When k is small, however, this version will be faster in practice.

When the degenerated acyclic graph $\tilde{G}$ is a tree, we can solve the single source problem in $O(k^2 n)$ time since $|\tilde{E}| = O(n)$ and hence $m = O(k^2 n)$. Although this is optimal in terms of n, we conjecture that the complexity is smaller. Whether it is $O(kn)$ is open.

Since we no longer follow Dijkstra's thesis "Compute shortest paths from shorter to longer," it will be hard to obtain a lower bound for the problem in this paper, based on the lower bound on sorting. We conjecture, however, our algorithm in Section 4 is optimal for the single source problem with $k = cyc(G)$.

References

1. Abuaiadh, D. and J.H. Kingston, Are Fibonacci heaps optimal? ISAAC'94, LNCS, pp. 442–450 (1994).
2. Chaudhuri, S. and C.D. Zaroliagis, Shortest path queries in digraphs of small treewidth, Proc. of 22nd Inter. Colloq., ICALP95, Szeged, Hungary, July 1995, in Lecture Notes on Computer Science Vol. 944, pp. 244–255, Springer-Verlag (1995).
3. Dijkstra, E.W., A note on two problems in connection with graphs, Numer. Math. Vol. 1, pp. 269–271 (1959).
4. Floyd, R.W., Algorithm 97: Shortest path, CACM, Vol. 5, No. , p. 345 (1962).
5. Frederickson, G.N., Fast algorithms for shortest paths in planar graphs, with applications, SIAM Jour. Comp., Vol. 16, No. 6, pp. 1004–1022 (1987).
6. Fredman, M.L. and R.E. Tarjan, Fibonacci heaps and their use in improved network optimization problems, JACM, Vol. 34, No. 3, pp. 596–615 (1987).
7. Moffat, A. and T. Takaoka, An all pairs shortest path algorithm with expected time $O(n^2 \log n)$, SIAM Jour. Comp., Vol. 16, No. 6, pp. 1023–1031 (1987).
8. Tarjan, R.E., Depth first search and linear graph algorithms, SIAM Jour. Comp., Vol. 1, No. 2, pp. 146–160 (1972).
9. Tarjan, R.E., Data Structures and Network Algorithms, Regional Conference Series in Applied math. 44, 1983.

Computing Disjoint Paths with Length Constraints

Spyros Tragoudas[1] and Yaakov L. Varol[2]

[1] Computer Science Dept., Southern Illinois University, Carbondale IL 62901, USA
[2] Computer Science Dept., University of Nevada, Reno NV 89557-0148, USA

Abstract. We show that the problem of computing a pair of disjoint paths between nodes s and t of an undirected graph, each having at most K, $K \in Z^+$, edges is NP-complete. A heuristic for its optimization version is given whose performance is within a constant factor from the optimal. It can be generalized to compute any constant number of disjoint paths. We also generalize an algorithm in [1] to compute the maximum number of edge disjoint paths of the shortest possible length between s and t. We show that it is NP–complete to decide whether there exist at least K, $K \in Z^+$, disjoint paths that may have at most $S + 1$ edges, where S is the minimum number of edges on any path between s and t. In addition, we examine a generalized version of the problem where disjoint paths are routed either between a node pair $(s1, t1)$ or a node pair $(s2, t2)$. We show that it is NP–hard to find the maximum number of disjoint paths that either connect pair $(s1, t1)$ the shortest way or $(s2, t2)$ the shortest way.

1 Introduction

We consider the problem of computing a set P of edge or node disjoint paths in an undirected graph $G = (V, E)$ of n nodes and m edges, between nodes $s, t \in V$, so that every path $p \in P$ has bounded length. The length $l(p)$ of a path p is defined as the sum of the edges in the path. If the edges are assigned lengths $w(e)$, then $l(p) = \sum_{e \in p} w(e)$. We are interested in minimizing the quantity $\max_{p \in P} l(p)$. We assume here that all lengths are positive integers. Edge or node disjoint paths are useful in communication networks because they ensure reliability of transmission between a source and a sink [4, 5]. In addition, bounding the length of the longest path ensures that the noise interference is under control [1]. Our results are presented on edge disjoint paths but can be modified to apply to the problem of finding node disjoint paths.

The special case where $|P| = 2$ is of particular interest. In [1] it was shown that the problem of finding a pair of bounded length edge disjoint paths is NP-complete in a more general graph instance where the edges have been assigned lengths which are *non polynomial to the input size*. This is not a very realistic case, however, since in practice the weights on the edges are polynomial functions of the input size.

We note that the problem of computing a pair of paths is also of interest to similar problems in networking. For example, the authors of [5] have studied the simpler problem of finding a pair of edge (or node) disjoint paths so that the sum of their lengths is minimum, and they present a fast polynomial time algorithm. Earlier, a less efficient polynomial time algorithm had been presented in [4] for the more generalized version where the goal is to compute k edge (or node) disjoint paths so that the sum of their lengths is minimum. Observe, however, that the problem studied in [4, 5] does not provide any guarantee on the noise interference and the algorithms in [4, 5] cannot ensure any performance guarantee with respect to our objective function.

In this paper (Section 2) we show that it is NP–complete to decide whether a graph G contains a pair of edge disjoint paths between s and t such that neither has more than K edges, $K \in Z^+$. This is an improvement over a weaker result in [1] for a more general problem. Moreover, we show that the algorithm in [4] serves as a heuristic for the optimization version of this problem. It finds $k \geq 2$ edge disjoint paths p_i, $1 \leq i \leq k$, such that no path has total length more than $k \cdot OPT - k + 1$, where OPT is the optimal cost. If the algorithm in [4] fails to return k paths, then the input instance does not allow for k edge disjoint paths.

In Section 3, we study another problem variation. We are interested in finding the maximum possible number of edge disjoint paths whose length is less than a given bound, preferably close to the minimum possible. We present an algorithm, which generalizes an idea in [1], that finds all disjoint paths of length S, where S is the minimum length on any path between s and t. Our algorithm handles instances with arbitrary lengths on the edges whereas the algorithm in [1] has polynomial time complexity only if the lengths on the edges of G are polynomial. We also show that it is NP–complete to determine whether at least K, $K \in Z^+$, edge disjoint paths exist so that no path has more than $S + 1$ edges.

Section 4 studies a generalized version of the problem. Here the goal is to compute edge disjoint paths connecting either node $s1$ with $t1$ or node $s2$ with $t2$ so that the maximum path length is minimized. It is NP–complete to determine whether there exists a path connecting s_1 with t_1 and another edge disjoint path connecting s_2 with t_2, such that both paths have less than L edges, $L \in Z^+$. This problem is a generalization of the problem in Section 2. We are also interested about the variation where we want to determine if there exist at least K, $K \in Z^+$, edge disjoint paths that either connect node $s1$ with $t1$ in the least expensive way or node $s2$ with $t2$ in the least expensive way. We show that this problem is also NP–complete. An even more generalized variation involves $k > 2$ pairs (s_i, t_i) of source/target nodes among which we want to find edge disjoint paths. Section 5 concludes.

2 Finding two or more edge disjoint paths with bounded maximum path length

We show that the following decision problem, defined on a graph $G = (V, E)$ with unit weights on its edges, is NP-complete.

Length Bounded Pair of Edge disjoint paths (LBPE)

Instance: Graph $G = (V, E)$, specified nodes s and t, positive integer $L < |V|$.

Question: Does G contain a pair of edge disjoint paths from s to t neither involving more than L edges?

We reduce from the following NP-complete problem [2].

Maximum 2-Satisfiability (MAX-2SAT)

Instance: Set U of variables, collection C of clauses over U such that each clause $c \in C$ has $|c| = 2$, positive integer $K \leq |C|$.

Question: Is there a truth assignment for U that simultaneously satisfies at least K of the clauses in C.

Theorem 1. *The LBPE problem is NP-complete even for graphs with maximum node degree 3.*

Proof. The problem is trivially in NP. Next, we show that the MAX-2SAT problem polynomially reduces to the *LBPE* problem. For the sake of simplicity, we show that the MAX-2SAT problem reduces to a modified version of the *LBPE* problem on $G = (V, E)$ where the edges $(u, v) \in E$, $u, v \in V$, have lengths which are polynomial functions of $|U|$ and $|C|$. Observe here that in the original formulation of *LBPE* problem each edge has unit weight. In order to be compatible with the latter, every edge (u, v) with length $l(u, v) > 1$ in the constructed instance below must be substituted with a path $(u, u_1, u_2, ..., u_{|l(u,v)-1|}, v)$ consisting of $|l(u, v)|$ unit-weighted edges, and such that each node u_i, $1 \leq i \leq |l(u, v) - 1|$, has degree 2. (For the sake of simplicity, in the description of the construction below we sometimes assign polynomial weights on some edges but using this type of local transformation we always result to a graph instance with unit weights on all the edges.)

Consider an arbitrary labeling of the variables and the clauses in the MAX-2SAT instance, i.e., $U = \{u_1, u_2, ..., u_m\}$, and $C = \{c_1, c_2, ..., c_n\}$, where $m = |U|$ and $n = |C|$. Given an instance of the MAX-2SAT problem we construct an instance of the LBPE problem as described below:

G contains the special nodes s and t, and has $|C|$ *clause components* G_{c_i}, one for each clause c_i, as well as $|U|$ *variable components* G_{u_j}, one for each variable u_j.

A clause component G_c, $c \in C$, consists of 6 nodes. Two of the nodes are labeled c_{in} and c_{out}. The other four nodes in G_c are related to the two literals l_1 and l_2 in the clause c and are called $c\_l_k\_in$ and $c\_l_k\_out$, $k = 1, 2$. Thus, if x and y are two variables and the clause c is $(x' + y)$ then the clause component G_c would have the nodes c_{in}, c_{out}, $c\_x'\_in$, $c\_x'\_out$, c_y_in, and c_y_out. With regard to the edges, G_c will have 6 *inner-clause* edges: $(c\_in, c\_l_k\_in)$ $(c\_out, c\_l_k\_out)$, of length 1, and $(c\_l_k\_in, c\_l_k\_out)$ of length 4, for $k = 1, 2$.

Furthermore, the clause components are connected with *intra-clause* edges: (c_i_out, c_{i+1}_in), $(c_n\_out, t))$ of length 1, and (s, c_1_in) of length L_1, to be specified later. Finally, there are edges of length 1 connecting the 4 nodes of the form c_l_in and c_l_out to corresponding nodes in the variable components depending on the literal assignment in c. Thus, the clause components account for $6 \cdot |C|$ nodes and $6 \cdot |C|$ inner-clause edges, $4 \cdot |C|$ edges to connect to variable components, and $|C| + 1$ intra-clause edges. See Fig. 1 for the structure of G_c.

For each variable $u \in \{u_1, u_2, ..., u_m\}$, we construct a variable component G_u which has the nodes u_in and u_out, as well as, $u\_c_j\_in$, $u\_c_j\_out$, $u'\_c_j\_in$, $u'\_c_j\_out$, for $1 \leq j \leq |C|$. In other words, each variable component has $2 + 4 \cdot |C|$ nodes, and together all variable components introduce $2 \cdot |U| + 4 \cdot |C| \cdot |U|$, or $2 \cdot m + 4 \cdot n \cdot m$ nodes.

Let L_2 be a length to be specified later. The *inner–variable* edges of the variable component G_u are:

- $(u\_c_j\_in, u\_c_j\_out)$ and $(u'\_c_j\_in, u'\_c_j\_out)$, for $j = 1, 2, ..., n$, and each of length 1.
- $(u\_c_j\_out, u\_c_{j+1}\_in)$ and $(u'\_c_j\_out, u'\_c_{j+1}\_in)$, for $j = 1, 2, ..., n-1$, and each of length L_2.
- $(u\_in, u\_c_1\_in)$, $(u\_in, u'\_c_1\_in)$, $(u\_c_n\_out, u\_out)$, and $(u'\_c_n\_out, u\_out)$, each of length L_2.

Thus the total number of inner–variable edges in all the variable components is $2 \cdot |C| \cdot |U| + 2 \cdot (|C| - 1) \cdot |U| + 4 \cdot |U| = 2 \cdot n \cdot m + 2 \cdot (n-1) \cdot m + 4 \cdot m$. The following *intra–variable* edges connect the various variable components to each other and to s and t: (s, u_1_in), (u_i_out, u_{i+1}_in) for $1 \leq i \leq m - 1$, and (u_m_out, t). These $(|U| + 1) = (m + 1)$ edges each have unit length. See Fig. 2 for the construction of variable component G_u.

Finally, as mentioned earlier there are $4 \cdot |C| = 4 \cdot n$ edges of length 1 connecting clause components and variable components. If a clause c has a literal l over a variable u (l is u or l is u') then the clause component G_c is connected to the variable component G_u via the edges (c_l_in, l_c_in) and (c_l_out, l_c_out). Since there are n clauses each with two literals, this amounts to $2 \cdot 2 \cdot n = 4 \cdot n$ edges.

Thus, the total number of nodes in G for the *LBPE* instance being implemented is $2 + 6 \cdot |C| + 2 \cdot |U| + 4 \cdot |C| \cdot |U| = 2 + 6 \cdot n + 2 \cdot m + 4 \cdot n \cdot m$. Furthermore, the total number of edges in G is $11 \cdot |C| + 4 \cdot |C| \cdot |U| + 3 \cdot |U| + 2 = 11 \cdot n + 4 \cdot n \cdot m + 3 \cdot m + 2$. Figure 3 illustrates a constructed G from a MAX–2SAT instance, where $U = \{u, v, w\}$ and $C = (u + v) \cdot (u' + v) \cdot (v' + w)$ and $K = 2$.

The weights L_1 and L_2, and the value L for the constructed instance of the *LBPE* decision problem on G will be set in such a way that the MAX–2SAT instance is satisfied if and only if the constructed *LBPE* instance on G is satisfied.

Clearly, there can only be two edge disjoint paths starting at node s. The path that starts with (s, c_1_in) will be called the *clause path*, and the other, starting with (s, u_1_in), will be referred to as the *variable path*.

The clause path, whose first edge has weight L_1, *will be forced not to include any* inner–variable *edges that have length* L_2. Recall that at least K of the clauses in C have to be satisfied. Thus, a clause path must have at least K sub–paths of the form $\{(c_i\_in, c_i\_l\_in), (c_i\_l\_in, l\_c_i\_in), (l\_c_i\_in, l\_c_i\_out), (l\_c_i\_out, c_i\_l\_out), (c_i\_l\_out, c_i\_out), (c_i\_out, c_{i+1}\_in)\}$, where l is a literal in clause c_i. (If $i = |C|$, then $c_{i+1}\_in = t$.) Each such subpath is called an *enforcer of clause c_i* and has length 6. As we show later, they guarantee that at least K clauses are satisfied. The value of L_2 will enforce the clause path to go through every clause component G_c, and to have at least K clause enforcers. Equivalently, a clause path may have at most $|C| - K$ subpaths made up of inner and intra–clause edges. Such a path, which we call a *non–enforcer of* c_i, has the form $\{(c_i\_in, c_i\_l\_in),$ $(c_i\_l\_in, c_i\_l\_out), (c_i\_l\_out, c_i\_out), (c_i\_out, c_{i+1}\_in)\}$, where l is a literal of clause c_i, and its length is 7. Such a path cannot necessarily guarantee that clause c_i is satisfied.

We enforce the above as follows. First we set L, the length upper bound of the *LBPE* instance to be $L = L_1 + 6 \cdot K + 7 \cdot (|C| - K) = L_1 + 7 \cdot n - K$, and $L_2 = 7 \cdot |C| = 7 \cdot n$. Clearly, the clause path cannot include an edge with weight L_2 since its length would be greater than L. (Its length would be at least $L_1 + 7 \cdot |C| + L_2 > L$.) Similarly, its length would exceed L if it has less than K clause enforcers.

Having routed the clause path, the routing of the variable path is restricted. It cannot contain any clause component nodes c_in or c_out since such nodes are in the clause path and have degree 3. If the variable path contains an edge of the form (l_c_in, c_l_in), it must then follow edge (c_l_in, c_l_out) and come back to the variable component using (c_l_out, l_c_out). However, as we show later, the length assignment on L_1 would prevent the variable path to contain such subpaths. Thus, a variable path starting with (s, u_1_in), must go through every variable component, and in each contain either all the nodes on the right or the left of G_u involving only one of the literals associated with u.

In particular, if the variable path contains the nodes of G_u that correspond to nodes $u'\_c_j\_in, u'\_c_j\_out, 1 \leq j \leq |C|$, then we assign variable u to be true (and u' to be false), otherwise it is assigned to be false (and u' to be true). Clearly, the length of the variable path is easily shown to be precisely $(((|C| + 1) \cdot L_2 + |C|) \cdot |U| + 1 + |U| = ((n + 1) \cdot 7 \cdot n + n) \cdot m + 1 + m$. We would like the latter quantity to be no more than L. In fact, we insist that it be equal to L, which quickly sets the value of L_1 since $((n + 1) \cdot 7 \cdot n + n) \cdot m + 1 + m = L_1 + 7 \cdot n - K$, resulting in $L_1 = 1 + m - 7 \cdot n + K + 8 \cdot n \cdot m + 7 \cdot n^2 \cdot m$.

The rest of the proof follows directly: If the *LBPE* instance on G is satisfied then the MAX-2SAT instance is satisfied. Simply observe that the clause path must have at least K enforcers of clause c_i. Each such subpath is satisfied by the assignment on the variable that corresponds to literal l in the clause c_i. See also Fig. 4. It is also easy to see that if the MAX-2SAT instance is satisfied then one can select a clause path and a variable path as outlined above in order to satisfy the *LBPE* instance on G.

In the remaining, we briefly show that the algorithm in [4] (or for the special case of $k = 2$ the algorithm in [5]) serves as an efficient heuristic for the problem of computing k edge disjoint paths so that the maximum path length is no more than k times the path length in an optimal solution. We assume that all lengths are positive integers. For this, simply observe that the algorithm in [4] returns k edge disjoint paths for which the sum of their lengths is minimum and equal to SUM_OPT. Clearly, an optimal solution to our problem has cost OPT such that $OPT \geq SUM\_OPT/k$. Since all edge lengths are positive, every path has length at least one and thus the most costly path in the solution returned by the algorithm in [4] has length at most $SUM\_OPT - k + 1 \leq OPT \cdot k - k + 1$. If we allow for nonnegative integer weights on the edges, the same heuristic returns a bound of $OPT \cdot k$.

3 Many disjoint paths of short length

Another interesting problem variation is to obtain as many short length edge disjoint paths as possible. We assume again that the lengths on the edges are positive integers. In Section 3.1 we consider the problem of finding the maximum number of paths of the shortest possible length between the source s and the destination t. Let S be the number of edges of a shortest path in G. We initially present a polynomial time algorithm that finds the maximum number of paths of length S. In Section 3.2 we examine the more generalized problem where the paths are allowed to have either length S or $S+1$. We show that it is an NP–complete problem to determine whether there exist at least K, $K \in Z^+$, edge disjoint paths whose length is either the shortest or one more unit than the shortest, even if all lengths on the edges are uniform.

3.1 Many disjoint paths of shortest length

The algorithm that computes all edge disjoint paths of shortest length uses a Dijkstra–like computation on G. Although we assume here that all edge weights are positive integers, the algorithm of this section can be modified to handle instances where the weights are nonnegative integers. We note that in [1] an algorithm for the same problem was presented which has pseudo–polynomial worst case time complexity.

Our approach consists of two phases. Phase 1 consists of a systematic traversal of G, starting from node s, the source. We maintain a directed graph $\bar{G} = (\bar{V}, \bar{E})$ consisting of the nodes visited by our expansion algorithm, and the directed edges that reflect the direction of the traversal. Every node in $\bar{G}$ is assigned a nonnegative weight $l()$ which controls the expansion procedure. $\bar{G}$ initially contains only s, the source, which is assigned a weight 0, i.e., $l(s) = 0$. All edges of the undirected graph G are set as unmarked. In the following description, we use (u, v) to represent an undirected edge of G, and $< u, v >$ to represent a directed edge of $\bar{G}$ from node u to node v.

Phase 1 is a basic loop that every time inserts one directed edge on $\bar{G}$. (An iteration of the loop may not necessarily add a new node on $\bar{G}$.) The basic loop consults the weight $l(v)$ of each node $v \in \bar{V}$ as well as the weight $w(u, v)$ of each unmarked edge $(u, v) \in E$ that is incident to a node already in $\bar{G}$. We select edge $< u, v >$ for which the sum $l(u) + w(u, v)$ is minimum, and we mark the respective edge $(u, v) \in E$. If node $v \in \bar{V}$ then $l(v)$ is already equal to $l(u) + w(u, v)$ and we simply insert edge $< u, v >$ in $\bar{E}$. If $v \notin \bar{V}$, then we insert $v \in \bar{V}$, edge $< u, v > \in \bar{E}$, and set $l(v) = l(u) + w(u, v)$. The process terminates when an edge (u, t) is selected, where t is the target node, such that $l(u) + w(u, t) > l(t)$.

Figure 5 illustrates the construction of $\bar{G}$. Every directed path in $\bar{G}$ from s to any other node $v \in \bar{V}$ is a shortest path from s to v. (The proof is similar to that for Dijkstra's shortest path algorithm and is omitted.) Note that when the weights $w()$ are positive, $\bar{G}$ is acyclic.

Phase 2 applies a maximum flow algorithm on $\bar{G}$, where the capacity of each edge is set to be 1. The flow paths returned by Phase 2 form our solution. The length of each such path is S, and they form a maximum set of edge disjoint paths. We conclude:

Theorem 2. *We can compute in polynomial time the maximum number of edge disjoint paths of shortest length between nodes s and t of a graph G.*

3.2 Many non–necessarily shortest length disjoint paths

This section shows that the problem of computing the maximum number of disjoint paths is NP–hard if we allow for either paths of shortest length S or of length $S+1$. More precisely, we show the following theorem.

Theorem 3. *The problem of deciding whether a graph G has K or more, $K \in Z^+$, edge disjoint paths between nodes s and t, each having at most $S + 1$ edges is NP–complete.*

Proof. The problem is clearly in NP. The reduction is similar to that in Theorem 3.2 of [1] for a different decision problem. We reduce from a restricted version of the 3–SAT instance where the number of occurrences of variable u_i in the instance is equal to that of $\bar{u}_i$. Let m, n, r_i be the number of variables, clauses, and appearances of variable u_i (or $\bar{u}_i$) in the 3–SAT instance, respectively. Let $r = \sum_i r_i$. This variation of 3–SAT was shown to be NP–complete in [1]. Given such an instance of 3–SAT we construct a graph G with the property that the 3–SAT instance is satisfied if and only if G has $n+r$ edge disjoint paths of length at most $S + 1$.

Graph G contains subgraphs G_i, $1 \leq i \leq m$, each associated with variable u_i, that have only s and t in common. Node $u_{i,k}$ (resp. $\bar{u}_{i,k}$) of G_i corresponds to the kth occurrence of u_i (resp. $\bar{u}_i$) in the 3–SAT instance. Graph G_i consists of $r_i + 1$ components: the α_G_i component, and r_i components $\beta_k\_G_i$, $1 \leq k \leq r_i$. Any two $\beta_k\_G_i$ components share node s. The $\beta_k\_G_i$ and α_G_i components share nodes s, $u_{i,k}$, and $\bar{u}_{i,k}$. Figures 6(a) and 6(b) show the α_G_i and a $\beta_k\_G_i$ component, respectively. Besides the latter components, G contains nodes $c_1, ..., c_n$. It has edges (c_i, t), $1 \leq i \leq n$, and also $(U_{i,k}, c_j)$ (resp. $(\bar{U}_{i,k}, c_j)$) if the kth occurrence of u_i (resp. $\bar{u}_i$) is in the jth clause. Figure 7 shows G for the 3–SAT instance $(\bar{u}_1 + \bar{u}_2 + u_3) \cdot (u_1 + \bar{u}_2 + \bar{u}_3) \cdot (u_1 + u_2 + \bar{u}_3) \cdot (\bar{u}_1 + u_2 + u_3)$.

Assume that G has $n + r$ edge disjoint paths each having at most $S + 1 = 6$ edges. Clearly, every path that contains edges of a $\beta_k\_G_i$ component must also contain some node c_i, otherwise its length is more than $S + 1$. Since the degree of t is exactly $n + r$, each edge incident to t must participate in a path. These observations imply that r paths must contain nodes $w_{i,k}$ and $y_{i,k}$ of the α_G_i

components, and that each α_G_i component must contribute exactly r_i paths. The structure of α_G_i guarantees that for the latter to happen either all paths must contain only nodes $u_{i,k}$, $1 \leq k \leq r_i$, or only nodes $\bar{u}_{i,k}$. Each of the remaining n paths contains a node c_i, $1 \leq i \leq n$, and a node $u_{i,k}$ (or $\bar{u}_{i,k}$) that is not participating in one of the first r paths. If $u_{i,k}$ (resp. $\bar{u}_{i,k}$) is on the same path with c_j then the jth clause is satisfied by assigning u_i to true (resp. false). A similar argument can be used to show that if the 3–SAT instance is satisfiable then G has $n + r$ edge disjoint paths each having at most $S + 1$ edges.

4   More than one source and destination

This examines a generalized problem formulation where we want to find disjoint paths that either connect a pair of nodes s_1 and t_1 or a pair of nodes s_2, t_2. We show that it is an NP–complete problem to find at least K, $K \in Z^+$, shortest length edge disjoint paths connecting either s_1 and t_1 or s_2 and t_2. In particular we have:

Theorem 4. *It is NP–complete to decide whether a graph G has K or more, $K \in Z^+$, edge disjoint paths connecting either node $s1$ with $t1$ in the shortest possible way or node $s2$ with $t2$ in the shortest possible way.*

Proof. The proof of this theorem is along the lines of Theorem 3. In particular, we reduce from the same 3–SAT variation as in Theorem 4 and that the following modifications in the construction of Theorem 3 suffice to show Theorem 4.

We relabel s as $s1$ and t as $t2$. In addition, we collapse all the $a_{i,k}$ nodes in the $\beta_k\_G_i$ components to a single node $s2$, thus introducing parallel edges from $s1$ to $s2$. Similarly, we collapse all the $z_{i,f}$, $1 \leq f \leq r_i$, nodes in the α_G_i components to a single node $t1$. Again, parallel edges are introduced between $t1$ and $t2$. It is easy to see that in the constructed instance the length of a shortest path between $s1$ and $t1$ or $s2$ and $t2$ is 5. However, using arguments as in Theorem 3 it can be easily shown that we have $n + r$ disjoint paths either between $s1$ and $t1$ or between $s2$ and $t2$, each having 5 edges if and only if the 3–SAT instance is satisfiable. The above construction can be also modified to avoid parallel edges.

Another variation of this generalized problem formulation would be (following the direction of Section 2) to find a pair of disjoint paths, the first connecting nodes s_1 and t_1 and the second connecting nodes s_2 and t_2 such that the maximum of the two path lengths is minimized. It can be easily shown, by restriction form the LBPE problem in Section 2, that it is NP–complete to determine whether there exist two such paths so that the maximum of the two lengths is less than an integer K. In [6] a heuristic is proposed for the optimization version of this problem formulation.

Finally, a generalization is to consider $k > 2$ pairs of nodes (s_i, t_i), $1 \leq i \leq k$, and to compute $K \geq k$ disjoint paths connecting the above pairs of nodes. This is a very difficult problem, however. Note that if we insist that the paths are node disjoint it is NP–complete even to determine whether k paths exist

[2]. However, when $k = 2$ a polynomial time has been reported in [2, 3] that determines whether a graph has two node disjoint paths.

5 Conclusions

We have shown that it is NP–complete to determine whether a graph has two edge disjoint paths connecting nodes s and t so that each has at most K edges, $K \in Z^+$. However, an efficient heuristic for its optimization version is given. Although we can find, in polynomial time, the maximum number of edge disjoint paths of shortest length between s and t, it is NP–complete to determine if G has at least K, $K \in Z^+$, edge disjoint paths connecting s and t whose length does not exceed the shortest by more than one unit. We also consider a generalization where the paths are connecting either nodes s_1 and $t1$ or s_2 and t_2. For this generalized problem, we also showed that it is NP–complete to determine whether there exist K or more, $K \in Z^+$, edge disjoint paths connecting $(s1, t1)$ or $(s2, t2)$ with minimum cost. All the results can be modified to apply to the similar problem of finding node disjoint paths. They can also be modified to apply to directed graphs.

References

1. Itai A., Perl Y., Shiloach Y.: The Complexity of Finding Maximum Disjoint Paths with Length Constraints. Networks **12** (1982) 277–286
2. Garey M.R., Johnson D.S.: Computers and Intractability. A Guide to the Theory of NP-Completeness. W.H. Freeman and Company, New York, NY, 1979
3. Shiloach Y.: The Two Paths Problem is Polynomially Solvable. Report STAN-CS-78-654. Computer Science Department, Stanford University, Stanford, CA
4. Suurballe, J.W.: Disjoint Paths in a Network. Networks **4** (1974) 125–145
5. Suurballe, J.W., Tarjan, R.E.: A Quick Method for Finding Shortest Pairs of Disjoint Paths. Networks **14** (1984) 325–336
6. Tragoudas, S., Varol Y.L.: On the Computation of Disjoint Paths with Length Constraints. Technical Report 96-6. Computer Science Department, Southern Illinois Univerity, Carbondale, IL 62901

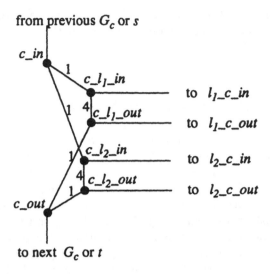

Fig. 1. Component G_c
(The numbers on the edges indicate lenghts)

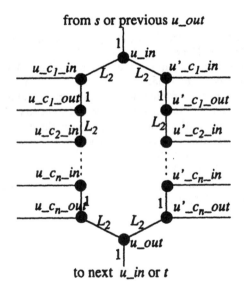

Fig. 2. Component G_u
(The numbers on the edges indicate lenghts)

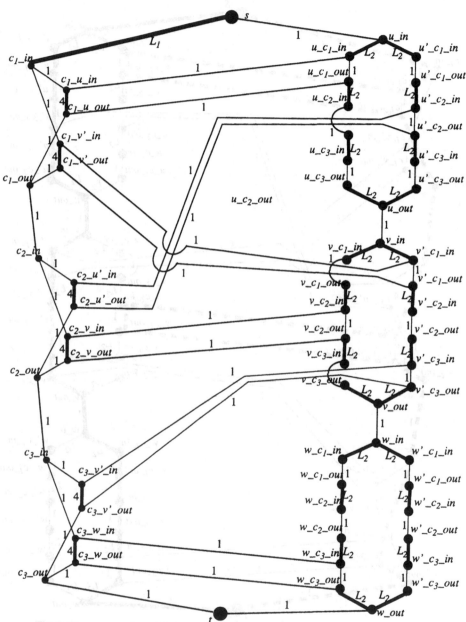

Fig. 3. The construction of G for the MAX-2SAT instance $(u+v')\,(u'+v)\,(v'+w)$ with $K=2$. Thick edges have lengths greater than 1. The numbers on the edges represent lengths. $L_1 = 21$, $L_2 = 246$, and $L = 265$.

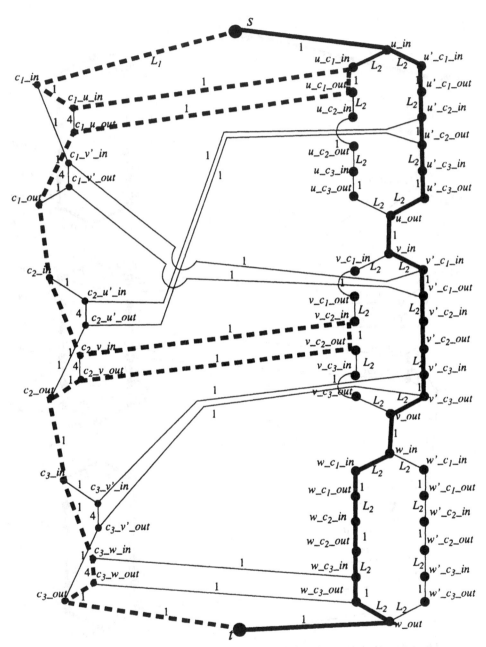

Fig. 4. Two disjoint paths (▬ ▬ ▬ ▬ , ▬▬▬▬) in the *LBPE* instance that have length at most *L* and result in satisfying the first two clauses in the MAX-2SAT instance with $K = 2$.

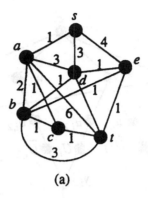

(a)

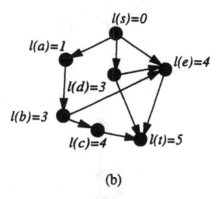

(b)

Fig. 5. The graphs of the algorithm in Section 3.1
(a) Graph $\underline{G}=(\underline{V},\underline{E})$
(b) Graph $\overline{G}=(V,E)$

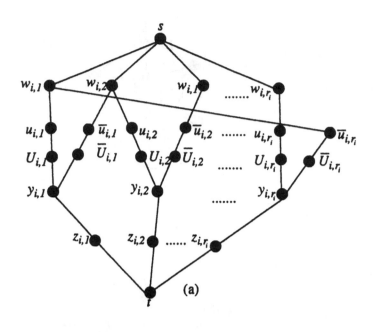

(a)

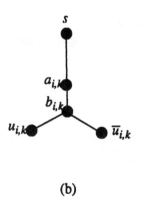

(b)

Fig. 6. On the construction for the theorem of Section 3.2.

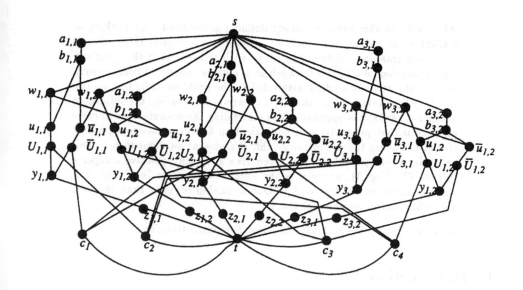

Fig. 7. The complete graph G for the 3--SAT instance in the proof of Theorem 3

Generalized Edge-Rankings of Trees
(Extended Abstract)

Xiao Zhou[1], Md. Abul Kashem[2] and Takao Nishizeki[2]

[1] Education Center for Information Processing
[2] Graduate School of Information Sciences
Tohoku University, Sendai 980-77, Japan.
Email: zhou@ecip.tohoku.ac.jp, kashem@nishizeki.ecei.tohoku.ac.jp
nishi@ecei.tohoku.ac.jp

Abstract. In this paper we newly define a generalized edge-ranking of a graph G as follows: for a positive integer c, a c-edge-ranking of G is a labeling (ranking) of the edges of G with integers such that, for any label i, deletion of all edges with labels $> i$ leaves connected components, each having at most c edges with label i. The problem of finding an optimal c-edge-ranking of G, that is, a c-edge-ranking using the minimum number of ranks, has applications in scheduling the manufacture of complex multi-part products; it is equivalent to finding a c-edge-separator tree of G having the minimum height. We present an algorithm to find an optimal c-edge-ranking of a given tree T for any positive integer c in time $O(n^2 \log \Delta)$, where n is the number of vertices in T and Δ is the maximum vertex-degree of T. Our algorithm is faster than the best algorithm known for the case $c = 1$.

Key words: Algorithm, Edge-ranking, Separator tree, Tree.

1 Introduction

An *edge-ranking* of a graph G is a labeling of edges of G with positive integers such that every path between two edges with the same label i contains an edge with label $j > i$ [8, 5]. Clearly an edge-labeling is an edge-ranking if and only if, for any label i, deletion of all edges with labels $> i$ leaves connected components, each having at most one edge with label i. The *edge-ranking problem* is to find an edge-ranking of a given graph G using the minimum number of ranks (labels). The problem seems to be NP-complete in general [3]. However, polynomial-time algorithms have been reported for trees. Iyer *et al.* [8] have given an $O(n \log n)$ time approximation algorithm for finding an edge-ranking of trees T using at most twice the minimum number of ranks, where n is the number of vertices in T. P. de la Torre *et al.* [3] have given an exact algorithm to solve the edge-ranking problem for trees in time $O(n^3 \log n)$ by means of a two-layered greedy method. Although Deogun and Peng [5] have claimed an exact $O(n^3)$ time algorithm, several flaws in [5] are pointed out in [4].

In this paper we newly define a generalization of an ordinary edge-ranking. For a positive integer c, a *c-edge-ranking* (or a *c-ranking* for short) of a graph G

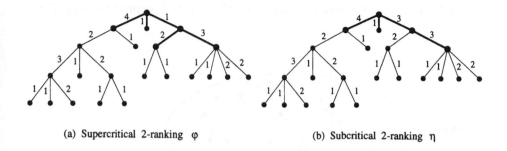

(a) Supercritical 2-ranking φ (b) Subcritical 2-ranking η

Fig. 1. Two optimal 2-rankings of a tree T.

is a labeling of the edges of G with integers such that, for any label i, deletion of all edges with labels $> i$ leaves connected components, each having at most c edges with label i. Clearly an ordinary edge-ranking is a 1-edge-ranking. The integer label of an edge is called the *rank* of the edge. A c-ranking of G using the minimum number of ranks is called an *optimal c-ranking* of G. The *c-ranking problem* is to find an optimal c-ranking of a given graph G. Fig. 1 depicts two optimal 2-rankings φ and η of a tree T using four ranks, where the ranks are drawn next to the edges and the 2-rankings φ and η are different only for the label of the rightmost edge incident to the root. Connected components obtained from T by deleting all edges with labels $> i$ for the 2-ranking φ of Fig. 1(a) are drawn in ovals in Fig. 2.

The problem of finding an optimal c-ranking of a graph G has applications in scheduling the parallel assembly of a complex multi-part product from its components, where the vertices of G correspond to the parts and the edges correspond to assembly operations [5, 7, 8]. Let us consider a robot with $c + 1$ hands which can connect at most $c + 1$ connected components at a time. If we have as many robots as we need, then the problem of minimizing the number of steps required for the parallel assembly of a product using the robots is equivalent to finding an optimal c-ranking of the graph G. Fig. 2 shows that one can assemble in parallel a product of Fig. 1(a) in four steps using seven robots of three hands. Note that, among ten connected components in Step 1, there are seven connected components each of which is not an isolated vertex. In each step robots simultaneously connect at most three connected components of the previous step.

The c-edge-ranking problem for a graph G is also equivalent to finding a c-edge-separator tree of G having the minimum height. Consider the process of starting with a connected graph G and partitioning it recursively by deleting at most c edges from each of the remaining connected components until the graph has no edge. The tree representing the recursive decomposition is called a *c-edge-separator tree* of G. Thus a c-edge-separator tree corresponds to a parallel computation scheme based on the process above, and an optimal c-ranking of G provides a parallel computation scheme having the minimum computation time

[9]. Fig. 2 illustrates a 2-edge-separator tree of the tree T depicted in Fig. 1(a).

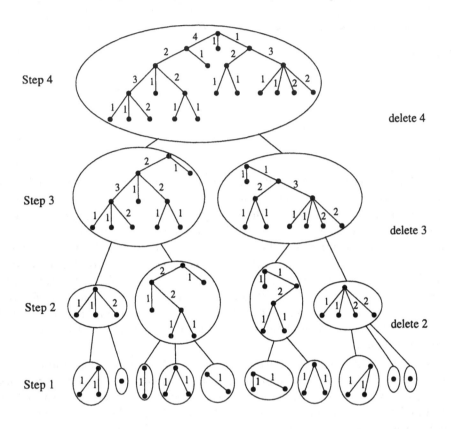

Fig. 2. A 2-edge-separator tree of the tree T in Fig. 1(a).

The vertex-ranking problem [6] and the c-vertex-ranking problem [12] for a graph G are defined similarly. A *c-vertex-ranking* of a graph G is a labeling of the vertices of G with positive integers such that, for any label i, deletion of all vertices with labels $> i$ leaves connected components, each having at most c vertices with label i. The vertex-ranking problem, that is, 1-vertex-ranking problem, is NP-complete in general [2, 10]. However, Iyer *et al.* presented an $O(n \log n)$ time algorithm to solve the vertex-ranking problem for trees [6]. Then Schäffer obtained a linear-time algorithm by refining their algorithm and its analysis [11]. On the other hand, Zhou *et al.* gave a linear-time algorithm to solve the c-vertex-ranking problem for trees [12]. Very recently Bodlaender *et al.* have given a polynomial-time algorithm to solve the vertex-ranking problem for graphs with bounded treewidth [2].

In this paper we give an algorithm to solve the c-edge-ranking problem on trees. The algorithm does not use a complicated greedy method, and finds an optimal c-ranking of a given tree T by means of "bottom-up tree computation" on T as the previous ranking algorithms [3, 6]: an optimal c-ranking of a subtree rooted at an internal vertex of T is updated from those of the subtrees rooted at its children. It is the first polynomial-time algorithm for the generalized edge-ranking problem, and takes time $O(n^2 \log \Delta)$ for any positive integer c, where Δ is the maximum vertex-degree of T. It should be noted that c is not always bounded. Thus our algorithm is faster than the best algorithm of complexity $O(n^3 \log n)$ known for the ordinary edge-ranking problem [3]. Our algorithm uses several new techniques as well as ones employed in [3, 6] for the ordinary ranking problems, and uses data-structures which can be constructed from lists of "visible edges" by radix sorting. The early conference versions [13, 14] of the paper claimed the bound $O(n \log n)$ for the ordinary edge-ranking problem on trees, but there were flaws in the algorithms. This paper corrects the flaws and extends the algorithms for the c-ranking problem.

2 Preliminaries

In this section we define some terms and present easy observations.

Let $G = (V, E)$ denote a graph with vertex set V and edge set E. We often denote by $V(G)$ and $E(G)$ the vertex set and the edge set of G, respectively. An edge joining vertices u and v is denoted by (u, v). We denote by n the number of vertices in a tree $T = (V, E)$ and by Δ the maximum vertex-degree of T. One may assume that $n \geq 3$ and hence $\Delta \geq 2$. T is a "free tree," but we regard T as a "rooted tree" for convenience sake: an arbitrary vertex of T is designated as the *root* of T. The maximal subtree of T rooted at a vertex $w \in V$ is denoted by $T(w)$. Let $e = (v, w)$ be an edge in T such that w is a child of v. Then the tree obtained from $T(w)$ by adding e is denoted by $T(e)$. We denote by $T - T(e)$ the tree obtained from T by deleting all edges and all vertices of $T(e)$ except v. We will use notions as: root, internal vertex, child and leaf in their usual meaning.

Let φ be an edge-labeling of a tree T with positive integers. The label (rank) of an edge $e \in E$ is represented by $\varphi(e)$. An edge e of T is *visible from vertex w under φ* if e is contained in $T(w)$ and every edge in the path from w to e has a rank $\leq \varphi(e)$. An edge visible from the root of T is often called a *visible edge* for short. In Fig. 1 all visible edges are drawn in thick lines. For a subtree T' of T, we denote by $\varphi|T'$ a restriction of φ to $E(T')$: let $\varphi' = \varphi|T'$, then $\varphi'(e) = \varphi(e)$ for $e \in E(T')$.

One can easily observe that the following lemma holds as in the case of the ordinary edge-ranking [3].

Lemma 1. *An edge-labeling φ of a tree T is a c-ranking of T if and only if*

   (a) *$\varphi|T(w)$ is a c-ranking of $T(w)$ for every child w of the root of T; and*
   (b) *no more than c edges of the same rank are visible from the root under φ.* □

This lemma immediately implies:

Corollary 2. *An edge-labeling φ of a tree T is a c-ranking if and only if no more than c edges of the same rank are visible from each vertex of T under φ.* $\square$

Since it is not so simple to find an optimal c-ranking, we focus on specific types of optimal c-rankings. Before formally defining them, we need to define some terms.

One may assume without loss of generality that a c-ranking φ of tree T uses consecutive positive integers starting from 1 as the ranks. Thus the largest rank is equal to the number of ranks used by φ. Iyer *et al.* introduced an idea of a "critical list" to solve the ordinary vertex-ranking problem [6]. We define the list $L(\varphi)$ of a *c-ranking* φ of tree T to be a list containing the ranks of all edges visible from the root, that is,

$$L(\varphi) = \{\varphi(e) \mid e \in E \text{ is visible from the root}\}.$$

The ranks in the list $L(\varphi)$ are sorted in non-increasing order. The list $L(\varphi)$ may contain same ranks with repetition $\leq c$. For an integer l we denote by $count(L(\varphi), l)$ the number of l's contained in $L(\varphi)$, that is, the number of visible edges with rank l. By Lemma 1 $count(L(\varphi), l) \leq c$ for each rank l. The 2-ranking φ in Fig. 1(a) has the list $L(\varphi) = \{4, 3, 2, 1, 1\}$, and hence $count(L(\varphi), 4) = count(L(\varphi), 3) = count(L(\varphi), 2) = 1$ and $count(L(\varphi), 1) = 2$. On the other hand, the 2-ranking η in Fig. 1(b) has the list $L(\eta) = \{4, 3, 3, 1\}$.

For a list L and an integer i, we define a sublist $[i \leq L]$ of L as follows:

$$[i \leq L] = \{l \in L \mid i \leq l\}.$$

Similarly we define sublists $[i < L]$, $[L \leq i]$ and $[L < i]$ of L for the integer i. For lists L and L' we use $L \subseteq L'$ and $L \cup L'$ in their usual meaning in which we regard L, L' and $L \cup L'$ as multi-sets.

We define the *lexicographical order* $\prec$ on sequences (lists) of positive integers as follows: let $A = \{a_1, a_2, \cdots, a_p\}$ and $B = \{b_1, b_2, \cdots, b_q\}$ be two sequences (lists) of positive integers, then $A \prec B$ if there exists an integer i such that

(a) $a_j = b_j$ for all $1 \leq j < i$, and
(b) either $a_i < b_i$ or $p < i \leq q$.

We write $A \preceq B$ if either $A = B$ or $A \prec B$.

We are now ready to define notions of optimality. A *critical c-ranking* φ of tree T is defined to be a c-ranking with the lexicographically smallest list $L(\varphi)$. Every critical c-ranking φ is optimal, because all edges of the largest rank are visible and hence the topmost rank in $L(\varphi)$ is equal to the number of ranks used by φ. The optimal c-ranking φ depicted in Fig. 1(a) is indeed critical, but the optimal c-ranking η in Fig. 1(b) is not critical since $L(\varphi) \prec L(\eta)$. P. de la Torre *et al.* used the concept of "supercritical ranking" to solve the ordinary edge-ranking problem [3]. We adapt the same notion for c-rankings. We define a c-ranking φ of tree T to be *supercritical* if the restriction $\varphi|T(v)$ is critical for every vertex v of T. Thus a supercritical c-ranking is critical and hence optimal. The c-ranking φ depicted in Fig. 1(a) is indeed supercritical. A c-ranking φ of tree T is *subcritical* if $\varphi|T(v)$ is critical for every vertex v of T except the root. Thus a c-ranking φ of T is subcritical if and only if $\varphi|T(v)$ is supercritical for every child v of the root of T. Furthermore a c-ranking φ of T is supercritical if and only if φ is subcritical and critical. Although a supercritical c-ranking is subcritical, a subcritical c-ranking is not always critical and is not always optimal. For example, the optimal c-ranking η in Fig. 1(b) is subcritical, but is not critical.

The definitions above immediately imply that every tree has a critical c-ranking, but do not immediately imply that every tree has a supercritical c-ranking. However, we will show in Section 3 that every tree does in fact have a supercritical c-ranking. Since any supercritical c-ranking is optimal, it suffices to give an algorithm for finding a supercritical c-ranking of T as we do in Section 3.

3   Optimal c-ranking

The main result of this paper is the following theorem.

Theorem 3. *For any positive integer c an optimal c-ranking of a tree T can be found in time $O(n^2 \log_2 \Delta)$, where n is the number of vertices in T and Δ is the maximum vertex-degree of T.*

In the remaining of this section we give an algorithm to find a supercritical c-ranking of a tree T in time $O(n^2 \log \Delta)$. Our algorithm uses the technique of "bottom-up tree computation." That is, it repeats the following operation for each internal vertex v of a tree T from leaves to the root: constructs a supercritical c-ranking φ of subtree $T(v)$ from those $\varphi_1, \varphi_2, \cdots, \varphi_d$ of the subtrees $T(w_1), T(w_2), \cdots, T(w_d)$ rooted at the children $w_1, w_2, \cdots, w_d$ of v. For such a strategy to succeed, it is favorable that a supercritical c-ranking φ of $T(v)$ could be extended from $\varphi_1, \varphi_2, \cdots, \varphi_d$, that is, φ could be obtained from $\varphi_1, \varphi_2, \cdots, \varphi_d$ by appropriately labeling the edges $e_i = (v, w_i)$, $1 \leq i \leq d$, without changing the labeling of the subtrees. We indeed prove that this actually holds as in the case of ordinary edge-rankings [3], and give an algorithm to find a supercritical c-ranking φ of $T(v)$ from φ_i, $1 \leq i \leq d$.

For notational convenience we may assume that v is the root of tree T and hence $T(v) = T$. Let $E(v) = \{e_i = (v, w_i) \mid 1 \leq i \leq d\}$, and let $L_i = L(\varphi_i)$. We first have the following lemma.

Lemma 4. *Every tree T has a supercritical c-ranking φ. Furthermore for any supercritical c-rankings φ_i of $T(w_i)$, $1 \leq i \leq d$, T has a supercritical c-ranking φ which is an extension of φ_i, that is, $\varphi|T(w_i) = \varphi_i$ for every i, $1 \leq i \leq d$.*   □

Lemma 4 implies the existence of a supercritical c-ranking φ, but the proof (omitted in this extended abstract) does not yield a polynomial-time algorithm to decide the d ranks of the edges $e_i \in E(v)$.

In the remaining of this section we present a sequence of lemmas and corollaries which help us to decide the d ranks in time $O(dn \log(d + 1))$. Repeating the same operation for all internal vertices v of T from bottom to top, we can find a supercritical c-ranking of T in time

$$O\left(\sum_{v \in V} d(v)n \log(d(v)+1)\right) = O(n^2 \log \Delta),$$

where $d(v)$ is the number of children of v in T. Note that $\sum_{v \in V} d(v) = n-1$.

We decide the d ranks of the edges $e_i, 1 \le i \le d$, in non-increasing order as in the case of the ordinary edge-ranking [3]. For a c-ranking φ of tree T, let m_φ be the maximum rank of edges in $E(v)$, that is, $m_\varphi = \max\{\varphi(e_i) \mid e_i \in E(v)\}$. Let n_φ be the number of edges in $E(v)$ labeled by m_φ. We then have the following lemma.

Lemma 5. *Let φ and η be two subcritical c-rankings of tree T. If $(m_\varphi, n_\varphi) \prec (m_\eta, n_\eta)$, then $L(\varphi) \prec L(\eta)$.* □

The following corollary is an immediate consequence of Lemma 5.

Corollary 6. *The following (a) and (b) hold:*

(a) *if φ is a supercritical c-ranking of tree T and η is a subcritical c-ranking of T, then $(m_\varphi, n_\varphi) \preceq (m_\eta, n_\eta)$; and*
(b) *every supercritical c-ranking φ of T has the same pair (m_φ, n_φ).* □

We thus denote by $\beta_{\sup}$ the same value m_φ for all supercritical c-rankings φ of T, and call $\beta_{\sup}$ *the super rank of tree T*. Then Corollary 6 immediately implies the following.

Corollary 7. *The super rank $\beta_{\sup}$ of tree T is equal to the minimum integer β for which T has a subcritical c-ranking φ of $m_\varphi = \beta$.* □

We will later give an algorithm SUPER-RANK to find the super rank $\beta_{\sup}$ of tree T using Corollary 7.

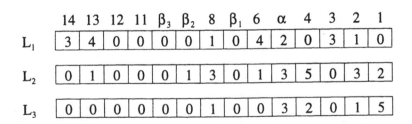

	14	13	12	11	β_3	β_2	8	β_1	6	α	4	3	2	1
L_1	3	4	0	0	0	0	1	0	4	2	0	3	1	0
L_2	0	1	0	0	0	1	3	0	1	3	5	0	3	2
L_3	0	0	0	0	0	0	1	0	0	3	2	0	1	5

Fig. 3. Illustration of three lists for the case $c = 5$.

Let β be a positive integer. Let $j, 1 \le j \le d$, be an index such that sublist $[L_j < \beta]$ is the lexicographically largest among all $[L_i < \beta], 1 \le i \le d$, that is, if e_j was labeled by β then the sublist of ranks in L_j hidden by β would be lexicographically largest. Note that the larger $[L_j < \beta]$ is, the smaller is the list of the c-ranking extended from $\varphi_1, \varphi_2, \cdots, \varphi_d$ by labeling e_j with β. The index j depends on β. For example, Fig. 3 illustrates a case in which $d = 3$ and $c = 5$, and if $\beta = \beta_1 = 7$ then $j = 1$, and if $\beta = \beta_2 = 9$ then $j = 2$. In Fig. 3 a number in the array corresponding to list L_i represents $count(L_i, \gamma)$ for its column number γ. Then the following lemma holds.

Lemma 8. *Let j be the index defined for β as above, and assume that tree T has a subcritical c-ranking η of $m_\eta = \beta$. Then T has a subcritical c-ranking φ such that $\varphi(e_j) = m_\varphi = \beta$ and $L(\varphi) \preceq L(\eta)$.* □

The following corollary is an immediate consequence of Lemma 8.

Corollary 9. *Let j be the index defined for $\beta = \beta_{\sup}$, that is, let $[L_j < \beta_{\sup}]$ be the lexicographically largest one among all sublists $[L_i < \beta_{\sup}]$, $1 \le i \le d$. Then tree T has a supercritical c-ranking φ such that $\varphi(e_j) = \beta_{\sup}$.* □

Thus, once $\beta_{\sup}$ is decided, one can easily decide an edge e_j to be labeled by $\beta_{\sup}$. Furthermore we have the following lemma.

Lemma 10. *Let j be the index defined for $\beta = \beta_{\sup}$, and let $T' = T - T(e_j)$. Then any supercritical c-ranking φ' of T' can be extended to a supercritical c-ranking φ of T as follows:*

$$\varphi(e) = \begin{cases} \beta_{\sup} & \text{if } e = e_j; \\ \varphi_j(e) & \text{if } e \in E(T(w_j)); \text{ and} \\ \varphi'(e) & \text{if } e \in E(T'). \end{cases}$$

□

Using Corollary 9 and Lemma 10 one can easily verify that the following algorithm UPDATE correctly decides the ranks of $e_1, e_2, \cdots, e_d$ if algorithm SUPER-RANK, given later, could correctly find the super rank of trees. An algorithm similar to UPDATE is used to find an ordinary edge-ranking in [3].

Procedure UPDATE(v);
  begin
1 $T' := T(v)$;
2 **for** $d := d(v)$ **downto** 1 **do**   { decide the $d(v)$ ranks in non-increasing
                       order, where $d(v)$ is the number of children of v in T }
    begin
3   let $w_{i_1}, w_{i_2}, \cdots, w_{i_d}$ be the children of v in T';
4   find the super rank $\beta_{\sup}$ of tree T' by the algorithm SUPER-RANK;
                             { SUPER-RANK will be given later }
5   find an index i_j, $1 \le j \le d$, such that $[L_{i_j} < \beta_{\sup}]$ is the
      lexicographically largest one among the d lists $[L_{i_k} < \beta_{\sup}]$, $1 \le k \le d$;
6   label edge $e_{i_j} = (v, w_{i_j})$ with $\beta_{\sup}$;   { cf. Corollary 9 }
7    $T' := T' - T'(e_{i_j})$;   { cf. Lemma 10 }

end
end;

Clearly line 1 can be done in time $O(1)$. Lines 3–7 are executed $d = d(v)$ times. One execution of lines 3 and 5–7 can be done in time $O(n)$. Therefore if one execution of line 4, i.e., SUPER-RANK, takes time $O(n \log(d+1))$, then the algorithm UPDATE runs in time $O(dn \log(d+1))$ as we claim. Thus it suffices to give the algorithm SUPER-RANK for finding β_{sup} of a tree in time $O(n \log(d+1))$.

By Corollary 7, in order to find β_{sup} we need to check the existence of a subcritical c-ranking φ with $m_\varphi = \beta$ for some integers β. A necessary and sufficient condition for the existence will be given in Lemma 12 below, which is a key lemma in the paper to make the algorithm faster and simpler than the best algorithm known for the ordinary edge-ranking. For a c-ranking φ of T, let $c_\varphi = count(L(\varphi), m_\varphi)$, then $c_\varphi \leq c$. Before presenting Lemma 12 we prove the following lemma.

Lemma 11. *Assume that φ is a subcritical c-ranking of tree T and that β is an integer such that $\beta \geq m_\varphi$ and $count(\cup_{i=1}^d L_i, \beta) \leq c-1$. Then T has a subcritical c-ranking φ^* such that*

(i) $m_{\varphi^*} = \beta$; *and*
(ii) *if $\beta > m_\varphi$ then $c_{\varphi^*} = 1 + count(\cup_{i=1}^d L_i, \beta)$.*

Proof. Let $\varphi(e_k) = m_\varphi$ for some k, $1 \leq k \leq d$. We then modify φ to φ^* by labeling e_k with β; all the other labels remain the same. Clearly φ^* is a subcritical c-ranking of T and $m_{\varphi^*} = \beta$. If $\beta > m_\varphi$, then $n_{\varphi^*} = 1$ and $c_{\varphi^*} = 1 + count(\cup_{i=1}^d L_i, \beta)$. □

The necessary and sufficient condition in Lemma 12 is recursive. That is, the existence of a subcritical c-ranking φ of a tree T is reduced to the existence of a subcritical c-ranking φ' of a smaller tree T' satisfying an additional condition: $c_{\varphi'} \leq c''$ for some c'', $1 \leq c'' \leq c$. We are now ready to present Lemma 12.

Lemma 12. *Let $1 \leq c' \leq c$, and let j be the index defined for β as before. Tree T has a subcritical c-ranking φ such that $\varphi(e_j) = m_\varphi = \beta$ and $c_\varphi \leq c'$ if and only if the following three conditions hold:*

(a) $count(\cup_{i=1}^d L_i, \beta) \leq c' - 1$;
(b) $count(\cup_{i=1}^d L_i, \gamma) \leq c$ *for all ranks $\gamma \in [\beta < L_j]$; and*
(c) *if $d \geq 2$, then tree $T' = T - T(e_j)$ has a subcritical c-ranking φ' such that*
  (i) *if $count(\cup_{i=1}^d L_i, \beta) \leq c' - 2$, then $m_{\varphi'} = \beta$ and $c_{\varphi'} \leq c''$ where $c'' = c' - n_j - 1$ and $n_j = count(L_j, \beta)$; and*
  (ii) *if $count(\cup_{i=1}^d L_i, \beta) = c' - 1$, then there is an integer β', $1 \leq \beta' \leq \beta - 1$, such that $count(\cup_{\substack{i=1 \\ i \neq j}}^d L_i, \beta') \leq c - 1$, and furthermore $m_{\varphi'} = \beta'$ for the largest β' among all such integers.*

Proof. $\Longrightarrow$: Let φ be a subcritical c-ranking of T such that $\varphi(e_j) = m_\varphi = \beta$ and $c_\varphi \leq c'$. Since one or more edges in $E(v)$ including e_j are labeled with β, we have $1 + count(\cup_{i=1}^d L_i, \beta) \leq c_\varphi \leq c'$, and hence (a) holds true. Since any integer $\gamma > \beta$ satisfies $c \geq count(L(\varphi), \gamma) = count(\cup_{i=1}^d L_i, \gamma)$, (b) holds true. Thus it suffices to prove (c) and one may assume that $d \geq 2$. Let $\varphi' = \varphi|T'$, then φ' is a subcritical c-ranking of T' and $m_{\varphi'} \leq m_\varphi = \beta$. Since $count(\cup_{i=1}^d L_i, \beta) \leq c' - 1$, there are the following two cases.

Case (i): $count(\cup_{i=1}^d L_i, \beta) \leq c' - 2$.

If $m_{\varphi'} = \beta$, then $c_{\varphi'} + n_j + 1 = c_\varphi \leq c'$ and hence $c_{\varphi'} \leq c' - n_j - 1 = c''$ as required. Thus one may assume that $\beta > m_{\varphi'}$. Clearly $count(\cup_{\substack{i=1 \\ i \neq j}}^d L_i, \beta) \leq c - 1$.

Applying Lemma 11 to T', one can know that T' has a subcritical c-ranking φ^* such that $m_{\varphi^*} = \beta$ and $c_{\varphi^*} \leq c''$, because

$$c_{\varphi^*} = 1 + count(\cup_{\substack{i=1 \\ i \neq j}}^d L_i, \beta) = 1 + count(\cup_{i=1}^d L_i, \beta) - n_j \leq 1 + c' - 2 - n_j = c''.$$

Case (ii): $count(\cup_{i=1}^d L_i, \beta) = c' - 1$.

In this case, no edge in $E(v) - \{e_j\}$ is labeled with β by φ', and hence $1 \leq m_{\varphi'} \leq \beta - 1$. Since $1 + count(\cup_{\substack{i=1 \\ i \neq j}}^d L_i, m_{\varphi'}) \leq c_{\varphi'} \leq c$, we have $count(\cup_{\substack{i=1 \\ i \neq j}}^d L_i, m_{\varphi'}) \leq c - 1$. Thus we have $\beta' \geq m_{\varphi'}$ since β' is the largest integer such that $1 \leq \beta' \leq \beta - 1$ and $count(\cup_{\substack{i=1 \\ i \neq j}}^d L_i, \beta') \leq c - 1$. Therefore, applying Lemma 11 to T', one can know that T' has a subcritical c-ranking φ^* with $m_{\varphi^*} = \beta'$.

$\Longleftarrow$: Suppose that (a), (b) and (c) hold. Extend the subcritical c-ranking φ' of T' to an edge-labeling φ of T as follows:

$$\varphi(e) = \begin{cases} \beta & \text{if } e = e_j; \\ \varphi_j(e) & \text{if } e \in E(T(w_j)); \text{ and} \\ \varphi'(e) & \text{otherwise.} \end{cases}$$

Then $\varphi(e_j) = m_\varphi = \beta$ and

$$L(\varphi) = L(\varphi') \cup \{\beta\} \cup [\beta \leq L_j]. \tag{1}$$

We now claim that every $\gamma \geq 1$ satisfies

$$count(L(\varphi), \gamma) \leq c. \tag{2}$$

If either $1 \leq \gamma < \beta$ or $\gamma > \beta$ and $\gamma \notin L_j$, then by (1) and (c) we have $count(L(\varphi), \gamma) = count(L(\varphi'), \gamma) \leq c$. On the other hand, if $\gamma \in [\beta < L_j]$, then by (1), (b) and $m_{\varphi'} \leq \beta$ we have $count(L(\varphi), \gamma) = count(\cup_{i=1}^d L_i, \gamma) \leq c$. Thus it suffices to prove $count(L(\varphi), \beta) \leq c$. If either $d = 1$ or $d \geq 2$ and $m_{\varphi'} < \beta$, then by (a) we have

$$c_\varphi = count(L(\varphi), \beta) = count(\cup_{i=1}^d L_i, \beta) + 1 \leq c' \leq c.$$

If $d \geq 2$ and $m_{\varphi'} = \beta$, then by (c) φ' should satisfy (i) and hence $c_{\varphi'} \leq c''$ and consequently

$$c_\varphi = count(L(\varphi), \beta) = c_{\varphi'} + n_j + 1 \leq c'' + n_j + 1 = c' \leq c.$$

Thus we have verified (2) and $c_\varphi \leq c'$. By (2) and Lemma 1 φ is a c-ranking of T, and is subcritical. $\qquad\square$

Let $B = \{\gamma \mid 1 \leq \gamma \leq n-1$ and $count(\cup_{i=1}^{d} L_i, \gamma) \leq c-1\}$. Then T necessarily has a subcritical c-ranking φ such that $m_\varphi = \beta$ for some integer $\beta \in B$; for example, label $e_1, e_2, \cdots, e_d$ by the smallest d positive integers larger than all labels used for $\varphi_1, \varphi_2, \cdots, \varphi_d$, then the resulting subcritical c-ranking φ of T satisfies $m_\varphi \in B$. Note that the labeling of the $n-1$ edges of T with $1, 2, \cdots, n-1$ is a trivial c-ranking of T. Thus B is a set of integers eligible for β_{sup}. From Lemma 12 one can easily derive the following recursive algorithm CHECK to decide whether T has a subcritical c-ranking φ of $m_\varphi = \beta$ and $c_\varphi \leq c'$ for a given integer $\beta \in B$ by checking the conditions (a), (b) and (c) in Lemma 12.

```
    Procedure CHECK(T, β, c′);
    begin
 8  if count(∪ᵈᵢ₌₁ Lᵢ, β) > c′ − 1
        then return false              { the condition (a) does not hold}
        else begin                     {the condition (a) holds}
 9      let w₁, w₂, ···, w_d be the children of v in T;         {d ≥ 1}
10      let j, 1 ≤ j ≤ d, be an index for which [L_j < β] is the lexicographically
            largest among all [L_k < β], 1 ≤ k ≤ d;
11      if count(∪ᵈᵢ₌₁ Lᵢ, γ) > c for a rank γ ∈ [β < L_j]
        then return false              {the condition (b) does not hold}
        else                           {the condition (b) holds}
12      if the root v of T has exactly one child    {obviously (c) holds}
            then return true
            else begin                             {d ≥ 2}
13          T′ := T − T(e_j);
14          if count(∪ᵈᵢ₌₁ Lᵢ, β) ≤ c′ − 2 then begin
15          n_j := count(L_j, β);
16          c″ := c′ − n_j − 1;
            CHECK(T′, β, c″)          {recursively check the condition (c)}
            end else                  {count(∪ᵈᵢ₌₁ Lᵢ, β) = c′ − 1}
17          if there is no integer β′, 1 ≤ β′ ≤ β−1, s.t. count(∪ᵈᵢ₌₁ Lᵢ, β′) ≤ c−1
                                                              i≠j
            then return false         {the condition (c) does not hold}
            else begin
18          let β′, 1 ≤ β′ ≤ β − 1, be the largest one among all integers
                such that count(∪ᵈᵢ₌₁ Lᵢ, β′) ≤ c − 1;
                                     i≠j
            CHECK(T′, β′, c)          {recursively check the condition (c)}
            end
        end
    end
    end;
```

By Corollary 7 and Lemma 12, $\beta_{\sup}$ is the smallest integer β satisfying the conditions (a), (b) and (c) for $c' = c$. Therefore we have the following algorithm SUPER-RANK to find $\beta_{\sup}$ of tree T.

Procedure SUPER-RANK(T);
begin
19 $B := \{\gamma \mid 1 \leq \gamma \leq n - 1$ and $count(\cup_{i=1}^d L_i, \gamma) \leq c - 1\}$;
20 choose the smallest integer $\beta \in B$ satisfying the conditions (a), (b) and (c)
     by executing CHECK(T, β, c) for all $\beta \in B$;
21 $\beta_{\sup} := \beta$
end;

We will prove the following lemma later.

Lemma 13. *The algorithm CHECK takes $O(n)$ time.*

By Lemma 13 the straightforward implementation of SUPER-RANK takes $O(n^2)$ time since $|B| \leq n - 1$. P. de la Torre *et al.* used a binary search to decide $\beta_{\sup}$ [3]. We also use a binary search, but the range of our search is bounded by d. Thus we can improve the time-complexity $O(n^2)$ to $O(n \log(d+1))$ as follows.

If $count(\cup_{i=1}^d L_i, \gamma) \leq c$ for all ranks $\gamma \in \cup_{i=1}^d L_i$, let $\alpha = 0$. Otherwise, let α be the largest rank γ such that $count(\cup_{i=1}^d L_i, \gamma) \geq c + 1$. Then $\beta_{\sup} \geq \alpha + 1$. Let $\beta_1 < \beta_2 < \cdots < \beta_d$ be the smallest d integers such that $\beta_i \geq \alpha + 1$ and $count(\cup_{i=1}^d L_i, \beta_i) \leq c - 1$. For the example in Fig. 3 $\alpha = 5$, $\beta_1 = 7$, $\beta_2 = 9$ and $\beta_3 = 10$. Then the following Lemma 14 holds.

Lemma 14. *Tree T has a subcritical c-ranking φ such that $m_\varphi \leq \beta_d$, and T has no subcritical c-ranking η such that $m_\eta < \beta_1$.*

Proof. Let

$$\varphi(e) = \begin{cases} \beta_i & \text{if } e = e_i, 1 \leq i \leq d; \text{ and} \\ \varphi_i(e) & \text{if } e \in E(T(w_i)), 1 \leq i \leq d. \end{cases}$$

Then the definition of $\beta_1, \beta_2, \cdots, \beta_d$ implies that there are no more than c edges of the same rank visible from v under φ. Therefore, by Lemma 1, φ is a c-ranking of T. Clearly φ is subcritical and $m_\varphi = \beta_d$.

Let η be any subcritical c-ranking of T. Since η is a c-ranking, $m_\eta > \alpha$ and m_η satisfies $count(\cup_{i=1}^d L_i, m_\eta) \leq c - 1$. Therefore, by the definition of β_1, we have $\beta_1 \leq m_\eta$. □

Corollary 7, Lemmas 11, 12 and 14 immediately imply the following lemma.

Lemma 15. *The following (a) and (b) hold:*

(a) *$\beta_{\sup} \in \{\beta_1, \beta_2, \cdots, \beta_d\}$; and*
(b) *let i be an integer such that $1 \leq i \leq d$, then $\beta_{\sup} \in \{\beta_1, \beta_2, \cdots, \beta_i\}$ if and only if T has a subcritical c-ranking η of $m_\eta = \beta_i$.* □

Replace line 19 of Procedure SUPER-RANK with the following:

$$B := \{\beta_1, \beta_2, \cdots, \beta_d\}.$$

Then, using the binary search technique, one can find the smallest integer β at line 20 by calling CHECK at most $\lceil \log(d+1) \rceil$ times. Clearly line 19 can be done in time $O(n)$, and line 21 can be done in time $O(1)$. Therefore the algorithm SUPER-RANK can be done in time $O(n \log(d+1))$ if Lemma 13 is correct.

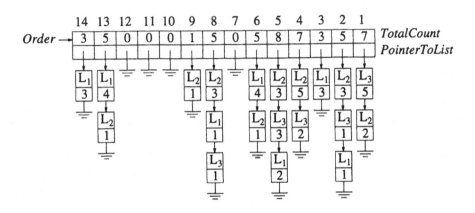

Fig. 4. Illustration of data-structure *Order* for the example of Fig. 3.

In the remaining of this section we prove Lemma 13. We use an array *Order* to support the operations at lines 10, 17 and 18. The array *Order* consists of records, each of which contains two items of data: *TotalCount* and *PointerToList*. The array *Order* is of length m, where m is the largest integer in $\cup_{i=1}^{d} L_i$. For each integer l, $1 \leq l \leq m$, the item *TotalCount* of record *Order*[l] is the sum of all $count(L_i, l)$, $1 \leq i \leq d$. The item *PointerToList* of record *Order*[l] points to a list $\mathcal{L}_l$ of records containing two fields: one of the fields stores the names of lists L_i such that $count(L_i, l) \geq 1$ and the other stores $count(L_i, l)$, $1 \leq i \leq d$. Furthermore we assume that list $\mathcal{L}_l$ links the names of the lists L_i in lexicographically non-increasing order of $[L_i \leq l]$. This is one of the key ideas behind the speed-up of our algorithm. We also use d lists $Same[i]$, $1 \leq i \leq d$; $Same[i]$ links all the same names L_i's appearing in the lists $\mathcal{L}_l$, $1 \leq l \leq m$. The data-structure *Order* is illustrated in Fig. 4 for the example of Fig. 3. Lists $Same[i]$ are not drawn in Fig. 4 for simplicity. One can construct the data-structure *Order* by radix-sorting a set of d lists, $\{(l, count(L_i, l)) \mid m \geq l \geq 1 \text{ and } count(L_i, l) \geq 1\}$, $1 \leq i \leq d$, in lexicographically non-increasing order. Since T has $n-1$ edges, the d lists contain at most $n-1$ elements in total. Furthermore $0 \leq count(L_i, l) \leq c$. Therefore the radix sorting can be done in time $O(c \cdot m + n)$ [1]. Clearly an optimal c-ranking of tree T uses at most

$\lceil (n-1)/c \rceil$ ranks, and hence $m \leq \lceil (n-1)/c \rceil$. Thus $Order$ can be constructed in time $O(c \cdot m + n) = O(n)$.

We are now ready to prove Lemma 13.

Proof of Lemma 13

Clearly lines 8, 9 and 12 can be executed total in time $O(n)$ during one execution of CHECK(T, β, c').

We now verify that one execution of lines 11 and 13–16 can be done in time $O(|E(T(e_j))|)$. Scanning $Order[l]$ for all ranks $l \in [\beta < L_j]$, one can execute line 11 in time $O(|E(T(e_j))|)$ since L_j contains at most $|E(T(e_j))|$ ranks. Traversing list $Same[j]$, one can easily update the data-structure $Order$ for a new tree T' at line 13 in time $O(|E(T(e_j))|)$. Lines 14–16 can be easily done in $O(1)$ time. Thus we have verified that one execution of lines 11 and 13–16 can be done in time $O(|E(T(e_j))|)$. Since

$$O\left(\sum_{1 \leq j \leq d} |E(T(e_j))| \right) = O(|E(T)|) = O(n),$$

lines 11 and 13–16 can be executed total in time $O(n)$.

We next claim that lines 10, 17 and 18 can be done total in time $O(n)$. Scan array $Order[l]$ from $l = \beta - 1$ in decreasing order of l until $l = l_k$ such that $TotalCount$ of $Order[l_k]$ is ≥ 1. Then the top element of list $\mathcal{L}_{l_k}$ is L_j. Thus the index j in line 10 can be easily found. One can search β' in lines 17 and 18 by scanning array $Order[l]$ from $l = \beta - 1$ in decreasing order of l until $l = \beta'$ such that $TotalCount$ of $Order[\beta']$ is $\leq c - 1$. (Note that the data-structure $Order$ in lines 17 and 18 has been updated for the new tree T' at line 13.) If there is such an integer β', then one can similarly find the index j' for β' by scanning $Order[l]$ from $l = \min\{\beta' - 1, l_k\}$. Thus one can execute lines 10, 17 and 18 total in time $O(n)$ since we scan $Order$ at most twice.

Thus we have shown that the algorithm CHECK can be done total in $O(n)$ time. $\square$

4 Conclusion

We newly define a generalized edge-ranking of a graph, called a c-ranking, and give an algorithm for finding an optimal c-ranking of a given tree T in time $O(n^2 \log \Delta)$ for any positive integer c, where n is the number of vertices in T and Δ is the maximum vertex-degree of T. This is the first polynomial-time algorithm for the generalized edge-ranking problem. Our algorithm is faster than the best algorithm of complexity $O(n^3 \log n)$ known for the ordinary edge-ranking problem [3], and is simple in a sense that it does not use a two-layered greedy method.

We may replace the positive integer c by a function $f : \mathbf{N} \to \mathbf{N}$ to define a more generalized edge-ranking of a graph as follows: an f-ranking of a graph G is a labeling of the edges of G with integers such that, for any label i, deletion of all edges with labels $> i$ leaves connected components, each having at most

$f(i)$ edges with label i. By some trivial modifications of our algorithm for the c-ranking of a tree, we can find an optimal f-ranking of a given tree in the same time complexity of $O(n^2 \log \Delta)$.

Our work raises several interesting open problems:

1. Is there a faster sequential algorithm for trees ?
2. What is the complexity of the c-ranking problem for general graphs ?
3. What is the complexity of parallel algorithms for c-rankings of trees or general graphs ?

Acknowledgement. We would like to thank Professor A. Nakayama for his careful reading and valuable comments on an early version of the paper.

References

1. A. V. Aho, J. E. Hopcroft, and J. D. Ullman, *The Design and Analysis of Computer Algorithms*, Addison-Wesley, Reading, MA, 1974.
2. H. Bodlaender, J. S. Deogun, K. Jansen, T. Kloks, D. Kratsch, H. Müller, and Zs. Tuza, Rankings of graphs, *Proc. of the International Workshop on Graph-Theoretic Concepts in Computer Science, Lecture Notes in Computer Science, Springer Verlag*, **903** (1994), pp. 292–304.
3. P. de la Torre, R. Greenlaw, and A. A. Schäffer, Optimal edge ranking of trees in polynomial time, *Algorithmica*, **13** (1995), pp. 592–618.
4. P. de la Torre, R. Greenlaw, and A. A. Schäffer, A note on Deogun and Peng's edge ranking algorithm, *Technical Report 93-13, Dept. of Computer Science, Univ. of New Hampshire, Durham, New Hampshire 03824, USA*, 1993.
5. J. S. Deogun and Y. Peng, Edge ranking of trees, *Congressus Numerantium*, **79** (1990), pp. 19–28.
6. A. V. Iyer, H. D. Ratliff, and G. Vijayan, Optimal node ranking of trees, *Information Processing Letters*, **28** (1988), pp. 225–229.
7. A. V. Iyer, H. D. Ratliff, and G. Vijayan, Parallel assembly of modular products – an analysis, *Technical Report PDRC, Technical Report 88-06, Georgia Institute of Technology*, 1988.
8. A. V. Iyer, H. D. Ratliff, and G. Vijayan, On an edge-ranking problem of trees and graphs, *Discrete Applied Mathematics*, **30** (1991), pp. 43–52.
9. N. Megiddo, Applying parallel computation algorithms in the design of serial algorithms, *Journal of the ACM*, **30** (1983), pp. 852–865.
10. A. Pothen, The complexity of optimal elimination trees, *Technical Report CS-88-13, Pennsylvania State University, USA*, 1988.
11. A. A. Schäffer, Optimal node ranking of trees in linear time, *Information Processing Letters*, **33** (1989/90), pp. 91–96.
12. X. Zhou, H. Nagai, and T. Nishizeki, Generalized vertex-rankings of trees, *Information Processing Letters*, **56** (1995), pp. 321–328.
13. X. Zhou and T. Nishizeki, An efficient algorithm for edge-ranking trees, *Proc. of the 2nd. European Symp. on Algorithms, Lecture Notes in Computer Science, Springer-Verlag*, **855** (1994), pp. 118–129.
14. X. Zhou and T. Nishizeki, Finding optimal edge-rankings of trees, *Proc. of the 6th Annual ACM-SIAM Symp. on Discrete Algorithms*, 1995, pp. 122–131.

List of Participants

Paola Alimonti
alimon@dis.uniroma1.it

Giorgio Ausiello
ausiello@dis.uniroma1.it

Luitpold Babel
babel@statistik.tu-muenchen.de

Michel Bauderon
bauderon@labri.u-bordeaux.fr

Sergej L. Bezrukov
sb@uni-paderborn.de

Adrienne Broadwater
axb0650@ucs.usl.edu

Krzysztof Bryś
brys@alpha.im.pw.edu.pl

Chi-Cang Chen
ccchen@cse.ttit.edu.tw

Serafino Cicerone
cicerone@iinf02.ing.univaq.it

Pierluigi Crescenzi
piluc@dsi.uniroma1.it

Fabrizio d'Amore
damore@dis.uniroma1.it

Daniele Giorgio Degiorgi
degiorgi@inf.ethz.ch

Hristo Djidjev
hristo@cs.rice.edu

Silvana Di Vincenzo
divince@dis.uniroma1.it

Kemal Efe
efe@cacs.usl.edu

Paolo G. Franciosa
pgf@dis.uniroma1.it

Daniele Frigioni
frigioni@univaq.it

Roberto Giaccio
giaccio@dis.uniroma1.it

Jens Gustedt
gustedt@math.TU-Berlin.DE

Torben Hagerup
torben@mpi-sb.mpg.de

Dagmar Handke
Dagmar.Handke@uni-konstanz.de

Stephan Hartmann
hartmann@math.tu-berlin.de

Michel Hurfin
hurfin@irisa.fr

Yoshihide Igarashi
igarashi@comp.cs.gunma-u.ac.jp

Helene Jacquet
jacquet@labri.u-bordeaux.fr

Wolfram Kahl
kahl@informatik.unibw-muenchen.de

Damon Kaller
kaller@cs.sfu.ca

Abul Kashem
kashem@nishizeki.ecei.tohoku.ac.jp

Lefteris Kirousis
kirousis@cti.gr

Bettina Klintz
klinz@opt.math.tu-graz.ac.at

Evangelos Kranakis
kranakis@scs.carleton.ca

Sven-Oliver Krumke
krumke@informatik.uni-wuerzburg.de

Ludek Kučera
ludek@kam.ms.mff.cuni.cz

Ewa Malesińska
malesin@math.tu-berlin.de

Alberto Marchetti-Spaccamela
alberto@dis.uniroma1.it

Ernst Mayr
mayr@informatik.tu-muenchen.de

Rolf Möhring
moehring@math.tu-berlin.de

Haiko Muller
hm@minet.uni-jena.de

Manfred Nagl
nagl@i3.informatik.rwth-aachen.de

Umberto Nanni
nanni@dis.uniroma1.it

Falk Nicolai
nicolai@marvin.informatik.uni-duisburg.de

Hartmut Noltemeier
noltemei@informatik.uni-wuerzburg.de

Francesco Parisi-Presicce
parisi@dsi.uniroma1.it

Giulio Pasqualone
pasqualo@dis.uniroma1.it

Andrzej Pelc
pelc@uqah.uquebec.ca

David Peleg
peleg@wisdom.weizmann.ac.il

Arunabha Sen
arunabha.sen@asu.edu

Swamy Sitarama
sitarama@cps.msu.edu

Ondrej Sýkora
sykorao@savba.sk

Thomas Szymczak
szymczak@informatik.uni-rostock.de

Tadao Takaoka
takaoka@cis.ibaraki.ac.jp

Gottfried Tinhofer
gottin@statistik.tu-muenchen.de

Yaakov Varol
varol@dws024.cs.unr.edu

Dorothea Wagner
Dorothea.Wagner@uni-konstanz.de

Peter Widmayer
widmayer@inf.ethz.ch

Andrew Yao
yao@cs.princeton.edu

Frances Yao
yao@parc.xerox.com

List of Authors

List of WG Proceedings

WG '75   U. Pape (Ed.): *Graphen-Sprachen und Algorithmen auf Graphen.* 1. Fachtagung Graphentheoret. Konzepte der Informatik, Hanser, Munich, 1976, 236 pages, ISBN 3-446-12215-X.

WG '76   H. Noltemeier (Ed.): *Graphen, Algorithmen, Datenstrukturen.* Proc. of WG '76, Graphtheoretic Concepts in Computer Science, Hanser, Munich, 1977, 336 pages, ISBN 3-446-12330-4.

WG '77   J. Mühlbacher (Ed.): *Datenstructuren, Graphen, Algorithmen.* Proc. of WG '77, Hanser, Munich, 1978, 368 pages, ISBN 3-446-12526-3.

WG '78   M. Nagl and H.-J. Schneider (Eds.): *Graphs, Data Structures, Algorithms.* Proc. of WG '78, Hanser, Munich, 1979, 320 pages, ISBN 3-446-12748-3.

WG '79   U. Pape (Ed.): *Discrete Structures and Algorithms.* Proc. of WG '79, Hanser, Munich, 1980, 270 pages, ISBN 3-446-13135-3.

WG '80   H. Noltemeier (Ed.): *Graphtheoretic Concepts in Computer Science.* Proc. of WG '80, Lecture Notes in Computer Science 100, Springer-Verlag, Berlin, 1981, 403 pages, ISBN 0-387-10291-4.

WG '81   J. Mühlbacher (Ed.): *Proc. of the 7th Conf. Graphtheoretic Concepts in Computer Science (WG '81).* Hanser, Munich, 1982, 355 pages, ISBN 3-446-13538-3.

WG '82   H.-J. Schneider and H. Göttler (Eds.): *Proc. of the 8th Conf. Graphtheoretic Concepts in Computer Science (WG '82).* Hanser, Munich, 1983, 280 pages, ISBN 3-446-13778-5.

WG '83   M. Nagl and J. Perl (Eds.): *Proc. WG '83, Workshop on Graphtheoretic Concepts in Computer Science.* Trauner, Linz, 1984, 397 pages, ISBN 3-853-20311-6.

WG '84   U. Pape (Ed.): *Proc. WG '84, Workshop on Graphtheoretic Concepts in Computer Science.* Trauner, Linz, 1985, 381 pages, ISBN 3-853-20334-5.

WG '85   H. Noltemeier (Ed.): *Graphtheoretic Concepts in Computer Science.* Proc. WG '85, Trauner, Linz, 1986, 443 pages, ISBN 3-853-20357-4.

WG '86   G. Tinhofer and G. Schmidt (Eds.): *Graph-Theoretic Concepts in Computer Science.* Proc. WG '86, Lecture Notes in Computer Science 246, Springer-Verlag, Berlin, 1987, 305 pages, ISBN 0-387-17218-1.

WG '87   H. Göttler and H.-J. Schneider (Eds.): *Graph-Theoretic Concepts in Computer Science*. Proc. WG '87, Lecture Notes in Computer Science 314, Springer-Verlag, Berlin, 1988, 254 pages, ISBN 0-387-19422-3.

WG '88   J. van Leeuwen (Ed.): *Graph-Theoretic Concepts in Computer Science*. Proc. WG '88, Lecture Notes in Computer Science 344, Springer-Verlag, Berlin, 1989, 457 pages, ISBN 0-387-50728-0.

WG '89   M. Nagl (Ed.): *Graph-Theoretic Concepts in Computer Science*. Proc. WG '89, Lecture Notes in Computer Science 411, Springer-Verlag, Berlin, 1990, 374 pages, ISBN 0-387-52292-1.

WG '90   M. Möhring (Ed.): *Graph-Theoretic Concepts in Computer Science*. Proc. WG '90, Lecture Notes in Computer Science 484, Springer-Verlag, Berlin, 1991, 360 pages, ISBN 0-387-53832-1.

WG '91   G. Schmidt and R. Berghammer (Eds.): *Graph-Theoretic Concepts in Computer Science*. Proc. WG '91, Lecture Notes in Computer Science 570, Springer-Verlag, Berlin, 1992, 253 pages, ISBN 0-387-55121-2.

WG '92   E. W. Mayr (Ed.): *Graph-Theoretic Concepts in Computer Science*. Proc. WG '92, Lecture Notes in Computer Science 657, Springer-Verlag, Berlin, 1993, 350 pages, ISBN 0-387-56402-0.

WG '93   J. van Leeuwen (Ed.): *Graph-Theoretic Concepts in Computer Science*. Proc. WG '93, Lecture Notes in Computer Science 790, Springer-Verlag, Berlin, 1994, 431 pages, ISBN 0-387-57889-4.

WG '94   E. W. Mayr, G. Schmidt and G. Tinhofer (Eds.): *Graph-Theoretic Concepts in Computer Science*. Proc. WG '94, Lecture Notes in Computer Science 903, Springer-Verlag, Berlin, 1995, 414 pages, ISBN 3-540-59071-4.

WG '95   M. Nagl (Ed.): *Graph-Theoretic Concepts in Computer Science*. Proc. WG '95, Lecture Notes in Computer Science 1017, Springer-Verlag, Berlin, 1995, 406 pages, ISBN 3-540-60618-1.

WG '96   F. d'Amore, P. G. Franciosa and A. Marchetti-Spaccamela (Eds.): *Graph-Theoretic Concepts in Computer Science*. Proc. WG '96, Lecture Notes in Computer Science, Springer-Verlag, Berlin, this volume.

Lecture Notes in Computer Science

For information about Vols. 1–1122

please contact your bookseller or Springer-Verlag